physica status solidi (b)

basic research

Volumes 145 to 150

Subject and Author Index 1988

Volumes 105 to 110

physica status solidi (a)

applied research

AKADEMIE-VERLAG BERLIN

ISSN 0370-1972/0031-8965 phys. stat. sol. (b)/(a), Berlin **R21** (1988) 1—510

International Classification System for Physics*)

60. Condensed matter: structure, mechanical and thermal properties
    61. Structure of liquids and solids; crystallography
    (see also 68.20. Solid surface structure, 71. Electron states)
    62. Mechanical and acoustical properties of condensed matter
    (see also 61.70. Defects in crystals, 68.30. Surfaces and interfaces)
    63. Lattice dynamics and crystal statistics
    (see also 65. Thermal properties, 66.70. Thermal conduction, 68.30. Dynamics of surface and interface vibrations, 78.30. Infrared and Raman spectra)
    64. Equations of state, phase equilibria, and phase transitions
    65. Thermal properties of condensed matter
    (see also 63. Lattice dynamics; for thermodynamic properties of quantum fluids, see 67.40; for thermal properties of solid helium, see 67.80)
    66. Transport properties of condensed matter (nonelectronic)
    67. Quantum fluids and solids; liquid and solid helium
    68. Surfaces and interfaces; thin films and whiskers
    (for impact phenomena, see 79; for crystal growth, see 61.50)

70. Condensed matter: electronic structure; electrical, magnetic, and optical properties
    71. Electronic states
    (see also 63. Lattice dynamics, 73. Electronic structure and electrical properties of surfaces, interfaces, and thin films)
    72. Electronic transport in condensed matter
    (for surfaces, interfaces, and thin films, see 73)
    73. Electronic and electrical properties of surfaces, interfaces, and thin films
    74. Superconductivity
    75. Magnetic properties and materials
    76. Magnetic resonances and relaxation in condensed matter; Mössbauer effect
    77. Dielectric properties and materials
    (for conductivity phenomena, see 72.20 and 72.80)
    78. Optical properties and condensed-matter spectroscopy and other interactions of matter with particles and radiation
    (for phonon spectra, see 63)
    79. Electron and ion emission by liquids and solids; impact phenomena

85. Devices**)

---

*) Excerpt; reproduced with permission of International Council for Scientific and Technical Information (ICSTI).

**) Outside the ICSTI Classification for Physics.

(The Substance Classification is given on cover three)

physica status solidi (b)
basic research

physica status solidi (a)
applied research

SUBJECT AND AUTHOR INDEX

Volumes 145 to 150, 1988

Volumes 105 to 110, 1988

ISSN 0370-1972/0031-8965

Edited by
E. BORCHARDT

AKADEMIE-VERLAG · BERLIN

Subscriptions and orders for this Subject and Author Index
or single copies of physica status solidi (b) and physica status solidi (a) should be directed
in the GDR:

to the Postzeitungsvertrieb or
to the Akademie-Verlag Berlin, Leipziger Str. 3—4, DDR-1086 Berlin;

in the other socialist countries:

to a book-shop for foreign language literature or
to the competent news-distributing agency;

in the FRG and BERLIN (WEST):

to a book-shop or
to the wholesale distributing agency Kunst und Wissen, Erich Bieber, OHG, Wilhelmstr. 4—6, P. O. Box 102844, D-7000 Stuttgart 1;

in the other Western European countries:

to Kunst und Wissen, Erich Bieber GmbH, General Wille-Str. 4, CH-8002 Zürich;

in USA and CANADA:

to VCH Publishers, Inc., 303 NW 12th Avenue, Deerfield Beach, FL 33442—1788, USA;

in other countries:

to the international book- and journal-selling trade,
to Buchexport, Volkseigener Außenhandelsbetrieb der Deutschen Demokratischen Republik, Postfach 160, DDR-7010 Leipzig, or
to the Akademie-Verlag Berlin, Leipziger Str. 3—4, DDR-1086 Berlin.

Schriftleiter und verantwortlich für den Inhalt der Zeitschriften:
Professor Dr. E. Gutsche, Leipziger Str. 3—4, Postfach 1233, DDR-1086 Berlin.

Herausgeber und verantwortlich für den Inhalt des Registers:
Dipl.-Phys. E. Borchardt, Leipziger Str. 3—4, Postfach 1233, DDR-1086 Berlin.

Verlag:
Akademie-Verlag Berlin, Leipziger Str. 3—4, DDR-1086 Berlin; Fernruf: 2236221 und 2236229; Telex-Nr.: 114420;
Bank: Staatsbank der DDR, Berlin, Kto.-Nr.: 6836-26-20712.

Veröffentlicht unter der Lizenznummer 1310 und 1620 des Presseamtes beim Vorsitzenden des Ministerrates der Deutschen Demokratischen Republik.

Gesamtherstellung:
VEB Druckerei „Thomas Müntzer", DDR-5820 Bad Langensalza.

Bestellnummer des Registers: 1068/85 R 21.

Printed in the German Democratic Republic.
AN (EDV) 20831/20735

Preface

The twentyfirst index volume contains volumes 145 to 150 of "physica status solidi (b) — basic research" as well as volumes 105 to 110 of "physica status solidi (a) — applied research", i.e. all volumes having appeared in 1988 containing 1397 papers.

Beginning from January 1986 the International Classification System for Physics (Condensed Matter, items 60 and 70) of the International Council for Scientific and Technical Information, Abstracting Board completed by item 85 and a Substance Classification (both outside the ICSTI Classification for Physics) will be used for quoting papers in physica status solidi. Nevertheless, before taking over these subject classifications into the index volume they were once more carefully revised and in some cases improved and completed in order to get a most uniform and clear classification for all papers. The average number of quotations per paper in this index volume is about 2.7.

Again I have to express my thanks to the readers for their interest, and I will always readily take up any critical remarks and proposals for improvements. I am especially obliged to Prof. Dr. E. Gutsche, Dr. H.-J. Hänsch, Dr. H. Lange, Dr. S. Oberländer, and Dipl. Phys. K. Müller for giving me advice and unselfish help. Furthermore I have to thank Mrs. E. Schwan for her exact and reliable secretary's work at the manuscript.

Berlin, June 1989

Eva Borchardt

International Classification System for Physics*)

60. Condensed matter: structure, mechanical and thermal properties
61. Structure of liquids and solids; crystallography
    (see also 68.20. Solid surface structures, 71. Electron states)
    61.10. X-ray determination of structures
        (for specific determinations, see 61.55 to 61.80)
    61.12. Neutron determination of structures
        (for specific determinations, see 61.55 to 61.80)
    61.14. Electron determination of structures
        (for specific determinations, see 61.55 to 61.80)
    61.16. Other determination of structures
        (for specific determination, see 61.55 to 61.80)
    61.20. Classical, semiclassical, and quantum theories of liquid structure
        (for electronic states, see 71; for liquid helium, see 67)
    61.25. Studies of specific liquid structures
    61.30. Liquid crystals
    61.40. Amorphous and polymeric materials
    61.50. Crystalline state (including molecular motions in solids)
        (for magnetic structure and spin systems, see 75.25)
    61.55. Specific structures of elements and alloys
    61.60. Specific structures of inorganic compounds
    61.65. Specific structures of organic compounds
    61.70. Defects in crystals
        (see also 61.80. Radiation damage, 62. Mechanical and acoustical properties,
        71.55. Impurities and defect levels, 76.30. EPR of color centers and other defects,
        78.50. Impurity and defect absorption in solids)
    61.80. Radiation damage and other irradition effects
        (for techniques of structure determination, see 61.10 to 61.16; for electron and ion impact phenomena, see 79.20)
    61.90. Other topics in structure of liquids and solids

62. Mechanical and acoustical properties of condensed matter
    (see also 61.70. Defects in crystals, 68.30. Surfaces and interfaces)
    62.10. Mechanical properties of liquids
        (for viscosity of liquids, see also 66.20)
    62.20. Mechanical properties of solids (related to microscopic structure)
    62.30. Mechanical and elastic waves
    62.40. Anelasticity, internal friction, and mechanical resonances
        For thermomechanical effects, see 65.70
        For magnetomechanical effects, see 75.80
        For piezoelectric effects, see 77.60
        For elasto-optical effects, see 78.20
    62.50. High-pressure and shock-wave effects in solids
    62.60. Acoustical properties of liquids
        For lattice dynamics, phonons, see 63
        For second sound in quantum fluids, see 67.40
    62.65. Acoustical properties of solids
        For magnetoacoustic effects in metals, see 75.80
        For acoustoelectric effects, see 72.50
        For acousto-optical effects, see 78.20
    62.80. Ultrasonic relaxation
        (see also 74.30. Ultrasonic attenuation in superconductors)
    62.90. Other topics in mechanical and acoustical properties of condensed matter

63. Lattice dynamics and crystal statistics
(see also 65. Thermal properties, 66.70. Thermal conduction, 68.30. Dynamics of surface and interface vibrations, 78.30. Infrared and Raman spectra)
63.10. General theory
63.20. Phonons and vibrations in crystal lattices
63.50. Vibrational states in disordered systems
63.70. Statistical mechanics of lattice vibrations
(see also 65. Thermal properties of condensed matter, and 66.70. Thermal conduction)
63.75. Statistical mechanics of displacive phase-transitions
For order–disorder and statistical mechanics of model systems, see 64.60
For crystallographic aspects of polymorphic and order–disorder transformations, see 61.50
63.90. Other topics in lattice dynamics and crystal statistics

64. Equations of state, phase equilibria, and phase transitions
64.10. General theory of equations of state and phase equilibria
64.30. Equations of state of specific substances
(see also 65.70. Thermal expansion)
64.60. General studies of phase transitions
(for magnetic, superconducting, and quantum fluid critical phenomena, see 75, 74, and 67, respectively)
64.70. Phase equilibria, phase transitions, and critical points of specific substances
64.75. Solubility, segregation, and mixing
64.80. Other phase properties of systems
64.90. Other topics in equations of state, phase equilibria, and phase transitions

65. Thermal properties of condensed matter
(see also 63. Lattice dynamics; for thermodynamic properties of quantum fluids, see 67.40; for thermal properties of solid helium, see 67.80)
65.20. Heat capacities of liquids
65.40. Heat capacities of solids
(for specific heat of superconductors, see 74.30; for specific heat of magnetic systems, see 75.40)
65.50. Thermodynamic properties and entropy
65.70. Thermal expansion and thermomechanical effects
(see also 64.30. Equations of state)
For thermal conduction in nonmetallic liquids, see 66.60; for nonmetallic solids, see 66.70
For electronic thermal conduction, see 72.10 and 72.20
For thermal conductivity of superconductors, see 74.30
For pyroelectric and electrocaloric effects, see 77.70
65.90. Other topics in thermal properties of condensed matter

66. Transport properties of condensed matter (nonelectronic)
66.10. Diffusion and ionic conduction in liquids
66.20. Diffusive momentum transport
(for viscosity of liquids, see also 62.10)
66.30. Diffusion in solids
66.60. Thermal conduction in nonmetallic liquids
(for thermal conduction in liquid metals, see 72.15)
66.70. Nonelectronic thermal conduction and heat-pulse propagation in nonmetallic solids
(for thermal conduction in solid metals, see 72.15; for statistical mechanics of lattice vibrations, see 63.70)
66.90. Other topics in nonelectronic transport properties

67. Quantum fluids and solids; liquid and solid helium
    - 67.20. Quantum effects on the structure and dynamics of nondegenerate fluids
    - 67.40. Boson degeneracy and superfluidity of helium-4
    - 67.50. Fermi fluids; liquid helium-3
    - 67.60. Mixed systems; liquid helium-3, -4 mixtures
    - 67.70. Films (including physical adsorption)
    - 67.80. Solid helium and related quantum crystals
    - 67.90. Other topics in quantum fluids
      (e.g., neutron-star matter)

68. Surfaces and interfaces; thin films and whiskers
    (for impact phenomena, see 79; for crystals growth, see 61.50)
    - 68.10. Fluid surfaces and fluid–fluid interfaces
    - 68.15. Liquid thin films
    - 68.20. Solid surface structures
    - 68.25. Mechanical and acoustical properties of solid surfaces and interfaces
    - 68.30. Dynamics of solid surfaces and interface vibrations
    - 68.40. Surface energy of solids; thermodynamic properties
    - 68.45. Solid–fluid interface processes
    - 68.48. Solid–solid interfaces (including bicrystals)
      (for grain boundaries, see 61.70)
    - 68.55. Thin film growth, structure, and epitaxy
    - 68.60. Physical properties of thin films: nonelectronic
    - 68.70. Whiskers and dendrites: growth, structure, and nonelectronic properties
    - 68.90. Other topics in the structure and nonelectronic properties of surfaces and thin films

70. Condensed matter: electronic structure; electrical, magnetic, and optical properties

71. Electronic states
    (see also 63. Lattice dynamics, 73. Electronic structure and electrical properties of surfaces, interfaces, and thin films)
    - 71.10. General theories and computational techniques
    - 71.20. Electronic density of states determinations
      (including energy states of liquid semiconductors)
      (see also 65.40. Electronic heat capacity)
    - 71.25. Nonlocalized single-particle electronic states
    - 71.30. Metal–insulator transitions
    - 71.35. Excitons and related phenomena (including electron–hole drops)
    - 71.36. Polaritons
      (including photon–phonon and photon–magnon interactions)
    - 71.38. Polarons and electron–phonon interactions
      (see also 63.20, phonon–electron interactions in lattices)
    - 71.45. Collective effects
    - 71.50. Localized single-particle electronic states (excluding impurities)
    - 71.55. Impurity and defect levels
    - 71.60. Positron states
      (see also 78.70. Positron annihilation)
    - 71.70. Level splitting and interactions
      (see also 75.10. — in magnetic phenomena, 75.30. Exchange and superexchange interactions, 73.20. Electronic surface states)
    - 71.90. Other topics in electron states

72. Electronic transport in condensed matter
(for surfaces, interfaces, and thin films, see 73)
 72.10. Theory of electronic transport; scattering mechanism
 72.15. Electronic conduction in metals and alloys
 72.20. Conductivity phenomena in semiconductors and insulators
 (for nonelectronic conduction, see 66.70)
 72.30. High-frequency effects; plasma effects
 72.40. Photoconduction and photovoltaic effects; photodielectric effects
 72.50. Acoustoelectric effects
 72.55. Magnetoacoustic effects
 72.60. Mixed conductivity and conductivity transitions
 72.70. Noise processes and phenoma
 72.80. Conductivity of specific semiconductors and insulators
 72.90. Other topics in electronic transport in condensed matter

73. Electronic structure and electrical properties of surfaces, interfaces, and thin films
 73.20. Electronic surface states
 (for emission and impact phenomena, see 79)
 73.25. Surface conductivity
 73.30. Surface double layers, Schottky barriers, and work functions
 73.40. Interfaces
 73.60. Electronic properties of thin films
 73.90. Other topics in electrical properties of surfaces, interfaces, and thin films

74. Superconductivity
 74.10. Occurrence, critical temperature
 74.20. Theory
 74.30. General properties
 74.40. Fluctuations and critical effects
 74.50. Tunneling phenomena, Josephson effect, and proximity effects
 74.55. Type-I superconductivity
 74.60. Type-II superconductivity
 74.70. Superconducting materials
 74.90. Other topics in superconductivity

75. Magnetic properties and materials
 75.10. General theory and models of magnetic ordering
 (see also 71.25. Nonlocalized single-particle electronic states, 71.70. Level splitting and interactions)
 75.20. Diamagnetism and paramagnetism
 75.25. Spin arrangements in magnetically ordered materials
 (neutron studies, etc.)
 75.30. Magnetically ordered materials, other intrinsic properties
 (for critical point effects, see 75.40)
 75.40. Critical-point effects, specific heats, short-range order
 (see also 65.40. Heat capacities)
 75.50. Studies of specific magnetic materials
 75.60. Domain effects, magnetization curves, and hysteresis
 75.70. Magnetic films and plates
 75.80. Magnetomechanical and magnetoelectric effects, magnetostriction
 For galvanomagnetic effects, see 72.15 and 72.20
 For magneto-optical effects, see 78.20
 75.90. Other topics in magnetic properties and materials

76. Magnetic resonances and relaxation in condensed matter; Mössbauer effect
    76.20. General theory of resonances and relaxation
    76.30. Electron paramagnetic resonance and relaxation
    76.40. Diamagnetic and cyclotron resonances
    76.50. Ferromagnetic, antiferromagnetic, and ferrimagnetic resonances; spin-wave resonance (see also 75.30. Spin waves)
    76.60. Nuclear magnetic resonance and relaxation
    76.70. Magnetic double resonances and cross effects
    76.80. Mössbauer effect; other γ-ray spectroscopy
    76.90. Other topics in magnetic resonances and relaxation

77. Dielectric properties and materials
    (for conductivity phenomena, see 72.20 and 72.80)
    77.20. Permittivity
    77.30. Polarization and depolarization effects
    77.40. Dielectric loss and relaxation
    77.50. Dielectric breakdown and space-charge effects
    77.55. Dielectric thin films
    77.60. Piezoelectricity and electrostriction
    (for piezo-optical effects, see 78.20)
    77.70. Pyroelectric and electrocaloric effects
    77.80. Ferroelectricity and antiferroelectricity
    77.85. Electrical resonances
    77.90. Other topics in dielectric properties and materials

78. Optical properties and condensed-matter spectroscopy and other interactions of matter with particles and radiation
    (for phonon spectra, see 63)
    78.20. Optical properties and materials
    78.30. Infrared and Raman spectra and scattering
    78.35. Brillouin and Rayleigh scattering
    78.40. Visible and ultraviolet spectra
    78.45. Stimulated emission
    78.50. Impurity and defect absorption in solids
    78.55. Photoluminescence
    78.60. Other luminescence and radiative recombination
    For photoconduction and photovoltaic effects, see 72.40
    78.65. Optical properties of thin films
    78.70. Other interactions of matter with particles and radiation
    78.90. Other topics in optical properties of condensed matter and other interactions of matter with particles and radiation

79. Electron and ion emission by liquids and solids; impact phenomena
    79.20. Impact phenomena, including electron spectra and sputtering
    79.40. Thermoionic emission
    79.60. Photoemission and photoelectron spectra
    79.70. Field emission and field ionization
    79.75. Exoelectron emission
    79.80. Resonance tunneling
    79.90. Other topics in emission and impact phenomena in condensed matter

85. Devices**)

**) Outside the ICSTI Classification for Physics.

Substance Classification (according to physica status solidi)

S1. Transition metals and their alloys and compounds
(for oxides, see S10)
- S1.1. Fe
- S1.2. Other 3d metals
- S1.3. Noble metals
- S1.4. Lanthanides
- S1.5. Actinides
- S1.61. Carbides, silicides, germanides
- S1.62. Nitrides, phosphides, arsenides
- S1.63. Sulfides, selenides, tellurides

S2. Alkali metals and their alloys and compounds
(for alkali halides, see S9.11)

S3. Alkaline earth metals and their alloys and compounds
(for alkaline earth halides, see S9.12)

S4. Other metals and their alloys

S5. Nonmetallic elements
- S5.11. Silicon
- S5.12. Germanium
- S5.2. Solidified Gases

S6. Group IV compounds
(for transition metal compounds, see S1.61)

S7. Group V compounds
(for transition metal compounds, see S1.62)
- S7.1. III–V compounds
- S7.11. Phosphides
- S7.12. Arsenides
- S7.13. Antimonides
- S7.15. Mixed crystals
- S7.16. Higher compounds

S8. Chalcogenides
(for transition metal compounds, see S1.63)
- S8.1. II–VI compounds
- S8.11. Sulfides
- S8.12. Selenides
- S8.13. Tellurides
- S8.15. Mixed crystals
- S8.16. Higher compounds

S9. Halides
- S9.11. Alkali halides
- S9.12. Alkaline earth halides
- S9.15. Mixed crystals
- S9.16. Higher compounds

S10. Oxides
- S10.1. Simple oxides
- S10.15. Mixed crystals

S11. Other higher compounds
- S11.1. Salts of oxy-acids
- S11.2. Spinels, garnets, ferrites

S12. Organic compounds
- S12.1 Quasi-one-dimensional systems

For comments on the use of the classification scheme see p. 10.

Some Comments on the Use of the Subject and Author Index

The "International Classification System for Physics" (the condensed matter sections 60 and 70) completed by a "Substance Classification" (according to physica status solidi) is used. On cover two the second-level headings of the condensed matter sections and on pages 4 to 8 the third-level headings are reproduced. As a help for using the subject index on pages 11 and 12 in addition also a series of important fourth-level headings is given. The substance classification is published on cover three and on page 9.

All papers will be quoted under each relevant section. There is made a distinction between one or two principal entries (refering to the main subject matter of the paper) and secondary entries (refering to subsidiary subject matter and substance, respectively). In the principal entries the title of the paper is given in its original language and in English. In addition, every principal entry is marked by * in front of the title of the paper. In the secondary entries the title is only given in English.

On the line following the title the subject classification is given (in parantheses) in the same manner as in the volumes. Each entry gives the volume, the number of the issue, the initial and final page numbers, and the year in which the paper appeared. A paper of physica status solidi (b) or physica status solidi (a) is marked by (b) or (a) in front of the line, respectively. The letter "K" before the page numbers indicates a short note published in the offset part of the journal.

Review articles appear, at the head of the list of entries under a given subject and are marked. These articles are listed in the chronological order of their publication, irrespective of being a paper of physica status solidi (b) or physica status solidi (a). The subject classification for a review article consisting of several parts is based on the article as a whole rather than on the individual parts.

The entries for original papers, short notes, and errata are made together in a single list for each subject and follow the entries for review articles. These are again ordered according to data of publication, irrespective of being a paper of physica status solidi (b) or physica status solidi (a). Original papers are sometimes published in several parts under the same principal title (in general with special subtitles). The various parts of papers of this type are treated separately and each part is classified according to its particular contents.

In quoting errata the following rule is obeyed. If the erratum refers to a paper quoted in the preceding index, it is listed in the chronological order of its publication otherwise the erratum is quoted together with the paper to which it refers.

The author index gives the names of the authors in alphabetical order. Each name is followed by the bibliographic details of the paper. A paper of physica status solidi (b) or physica status solidi (a) is again marked by (b) or (a) in front of the number of the volume, respectively. In the last column the subject classification of the paper is presented in the same manner as in the volumes. When several papers have been published by the same author they are arranged according to the dates of publication, irrespective of being a paper of physica status solidi (b) or physica status solidi (a). To find a paper of your interest please choose first the special entry of the subject index and then use the volume number and page.

In using the subject index attention should be given to the following details:

61.10 to 61.16 Theory and experimental techniques of determination of structures
— for specific determinations see 61.55 to 61.80
61.16 Other determination of structures
— electron microscopy and field-ion microscopy determinations
— EPR and NMR determinations
61.25 Studies of specific liquid structures
— macromolecular and polymer solutions
— molten salts
— liquid metals
61.40 Amorphous and polymeric materials
— glasses, polymers, elastomers, plastics
61.50 Crystalline state (including molecular motions in solids)
— physics of crystal growth
— crystal symmetry: models, space groups, crystalline systems and classes
— crystal morphology and orientation
— crystallographic aspects of polymorphic and order–disorder transformations
61.70 Defects in crystals
— grain and twin boundaries
— stacking faults and other planar or extended defects
— doping and implantation of impurities
— impurity concentration, distribution, and gradients
61.80 Radiation damage and other irradiation effects
- channeling, blocking, energy loss of particles
— X-rays, γ-rays, electrons and positrons, neutrons, atoms and molecules, ions
— for ion implantation see 61.70
63.20 Phonons and vibrations in crystal lattices
— phonon–phonon, phonon–electron, and phonon–defect interactions
64.60 General studies of phase transitions
— order–disorder and statistical mechanics of model systems
— equilibrium properties near single critical points, critical exponents
— dynamic critical phenomena
— multicritical points
— metastable phases
64.80 Other phase properties of systems
— stoichiometry and homogeneity
— microstructure
66.30 Diffusion in solids
— theory of diffusion and ionic conduction in solids
— self-diffusion and ionic conduction in non-metals
— diffusion, migration, and displacement of impurities and other defects
71.25 Nonlocalized single-particle electronic states
— techniques of band-structure calculation
— measurements of Fermi surface parameters (including dHvA, magnetoacoustic, positron annihilation, and cyclotron resonance studies, etc.)
— effective mass and *g*-factors
— band structure of crystalline metals, semiconductors, and insulators
71.45 Collective effects
— exchange, correlation, dielectric and magnetic functions, plasmons
— Fermi-Thomas models
— calculations of total electronic binding energy
71.70 Level splitting and interactions
— crystal and ligand fields
— spin–orbit coupling, Zeeman, Stark, and strain splitting
— exchange interactions
— nuclear states and interactions

72.10 Theory of electronic transport; scattering mechanism
— general formulation of transport theory
— scattering by phonons, magnons, and other nonlocalized excitations
— scattering by point defects, dislocations, surfaces, and other imperfections (including Kondo effect)
72.15 Electronic conduction in metals and alloys
— galvanomagnetic and other magnetotransport effects
— thermomagnetic and thermoelectric effects
— collective modes
— scattering mechanism and Kondo effect
72.20 Conductivity phenomena in semiconductors and insulators
— scattering mechanism
— piezoresistance
— galvanomagnetic and other magnetotransport effects
— thermomagnetic and thermoelectric effects
73.40 Interfaces
— contact resistance, contact potential, rectification
— tunneling: general
— p–n junctions
— MIS, MIM, MSM, SIS, and SMS structures
75.30 Magnetically ordered materials, other intrinsic properties
— saturation moments and magnetic susceptibility
— spin waves
— exchange and superexchange interactions
— anisotropy
— magnetic impurity interactions
— magnetic phase boundaries
— magnetocaloric effect
75.50 Studies of specific magnetic materials
— ferromagnetism
— antiferromagnetics, ferrimagnetics
— amorphous magnetic materials
— magnetic liquids
75.60 Domain effects, magnetization curves, and hysteresis
— high coercitivity materials
— fine-particle systems
— magnetic aftereffects
— magnetic annealing and temperature-hysteresis effects
78.20 Optical properties and materials
— general theory
— optical constants and parameters
— optical rotatory power
— birefringence
— piezo-, elasto-, acousto-, electro-, magneto-, and thermo-optical effects
78.60 Other luminescence and radiative recombination
— electroluminescence
— cathodo- and iono-luminescence
— thermoluminescence
— sono- and tribo-luminescence
— chemiluminescence
78.70 Other interactions of matter with particles and radiation
— positron annihilation
— X-ray scattering, absorption, and emission
— microwave and radiofrequency (including resonances)
79.20 Impact phenomena, including electron spectra and sputtering
— electron impact: Auger and secondary emission
— atom, molecule, and ion impact
— atomic and molecular beam interactions

Subject Index

60. Condensed matter: structure, mechanical and thermal properties

61. Structure of liquids and solids; crystallography
(see also 68.20. Solid surface structures, 71. Electron states)

61.10. X-ray determination of structures
(for specific determinations, see 61.55 to 61.80)

A. P. DESHPANDE, V. B. SAPRE, and C. MANDE
*Study of the Environment of Copper in Some Ternary Chalcogenides by EXAFS Spectroscopy
(61.10; S1.2; S8.16)
(b) Vol. **145**, No. 1, p. 77—88 (1988).

H.-R. HÖCHE, J. NIEBER, M. CLAUSNITZER, and G. MATERLIK
*Modification of Specularly Reflected X-Ray Intensity by Grazing Incidence Coplanar Bragg-Case Diffraction
(61.10; S5.12)
(a) Vol. **105**, No. 1, p. 53—60 (1988).

J. HÄRTWIG, V. HOLÝ, R. KITTNER, J. KUBĚNA, and V. LERCHE
*Investigation of Quartz Crystals with Non-Homogeneous Distributions of Impurity Atoms by X-Ray Methods
(61.10 and 61.70; S10.1)
(a) Vol. **105**, No. 1, p. 61—75 (1988).

T. BAUMBACH, H.-G. BRÜHL, U. PIETSCH, and H. TERAUCHI
*Characterization of $Ga_{1-x}Al_xAs/GaAs$ Superlattices and Thin Single Layers by X-Ray Diffraction
(61.10 and 68.55; S7.12; S7.15)
(a) Vol. **105**, No. 1, p. 197—205 (1988).

K. SUMIYAMA and Y. NAKAMURA
Nonequilibrium $Fe_{1-x}Co_x$ Alloys Produced by Vapor Quenching
(75.70; 61.10; 76.80; S1.1)
(a) Vol. **105**, No. 1, p. 291—295 (1988).

V. HOLÝ and J. HÄRTWIG
*The Role of Diffuse Scattering on Microdefects in the Precise Lattice Parameter Measurement
(61.10 and 61.50; 61.70; S5.11)
(b) Vol. **145**, No. 2, p. 363—372 (1988).

T. Itoh, Y. Iwabuchi, and M. Kataoka
*Study on the Size and Shape of CuCl Microcrystals Embedded in Alkali-Chloride Matrices and Their Correlation with Exciton Confinement
(61.10 and 71.35; 78.20; S9)
(b) Vol. **145**, No. 2, p. 567—577 (1988).

A. P. Bezirganyan and P. A. Bezirganyan
*Solution of the Two-Dimensional Stationary Schrödinger Equation with Cosine-Like Coefficient (in View of X-Ray Diffraction)
(61.10 and 78.70; 68.20)
(a) Vol. **105**, No. 2, p. 345—355 (1988).

M. Polcarová, K. Godwod, J. Bąk-Misiuk, S. Kadečkova, and J. Brádler
Lattice Parameters of Fe–Si Alloy Single Crystals
(61.55; 61.10; S1.1; S1.61)
(a) Vol. **106**, No. 1, p. 17—23 (1988).

I. R. Entin
*Dynamical and Kinematical X-Ray Diffraction in Crystals Strongly Disturbed by Ultrasonic Vibrations
(61.10)
(a) Vol. **106**, No. 1, p. 25—30 (1988).

V. G. Kohn
*X-Ray Standing Waves under the Conditions of Multiple Diffraction
(61.10)
(a) Vol. **106**, No. 1, p. 31—39 (1988).

E. Z. Arshakyan, A. O. Aboyan, and P. A. Bezirganyan
*X-Ray Interferometric Investigations of Structural Distortions in Semiconductor Crystals Caused by Constant Electric and Magnetic Fields
(61.10)
(a) Vol. **106**, No. 1, p. 41—47 (1988).

I. R. Entin and I. A. Smirnova
*The Effect of X-Ray Beam Intensity Oscillations as a Function of the Amplitude of Ultrasound Excited in a Crystal
(61.10; 62.60; S5.11)
(a) Vol. **106**, No. 2. p. 339—350 (1988).

N. Ganev and I. Kraus
*On the Application of X-Ray Tensometry to States of Stress Due to Tangential Machining of Metal Surfaces
(61.10 and 62.20; 68.25)
(a) Vol. **106**, No. 2, p. 467—472 (1988).

E. Förster, P. Glas, K. Goetz, St. Joksch, P. V. Nickles, M. Schnürer, and I. Will
*About the Sensitivity of in Situ Diffraction Measurements with X-Radiation of a Laser-Produced Plasma
(61.10; 65.70; 68.20; S5.11)
(a) Vol. **107**, No. 1, p. 85—93 (1988).

I. A. Smirnova and I. R. Entin
*Extinction Length Variation in a Distored Crystal
(61.10)
(b) Vol. **147**, No. 2, p. 459—469 (1988).

V. Fritzsche
Approximations for the Scattering of Excited Electrons in Solids
(79.60; 61.10; 68.20)
(b) Vol. **147**, No. 2, p. 485—494 (1988).

U. Pietsch
*The Advantage of Internal Force Calculations in the X-Ray Crystal Structure Analysis
(61.10 and 62.20; S5.12; S7.12)
(b) Vol. **147**, No. 2, p. 517—527 (1988).

A. Tomás, M. Guymont, M. P. Pardo, M. Guittard, and J. Flahaut
X-Ray Diffraction and Electron Microscopy Studies of α- and β-Ga_2S_3
(61.60; 61.10; 61.16; S8)
(a) Vol. **107**, No. 2, p. 775—784 (1988).

H. Rhan and U. Pietsch
Investigation of Nanometer Layer Heterostructures by X-Ray Grazing Incidence Diffraction
(68.55; 61.10; S7.15)
(a) Vol. **107**, No. 2, p. K93—K98 (1988).

P. Rennert and Nguyen Van Hung
*Curved Wave and Multiple Scattering Corrections in Calculated EXAFS Spectra
(61.10 and 78.70; 71.10)
(b) Vol. **148**, No. 1, p. 49—61 (1988).

H. Schaeben
Entropy Optimization in Texture Goniometry. I. Methodology
(61.50; 61.10; 61.12)
(b) Vol. **148**, No. 1, p. 63—72 (1988).

V. V. Kochelab, V. B. Molodkin, S. I. Olikhovskii, and M. E. Osinovskii
*The Thickness Dependence of the X-Ray Diffuse Scattering Intensity for Crystals with Microdefects at Laue-Case Diffraction
(61.10; 61.70; S5.12)
(a) Vol. **108**, No 1, p. 67—79 (1988).

G. B. Demishev, S. S. Kabalkina, and T. N. Kolobyanina
*X-Ray Studies of Thallium Chalogenides TlS and TlSe up to 37 GPa
(61.10; 64.70; S8)
(a) Vol. **108**, No. 1, p. 89—95 (1988).

G. Wallner, E. Burkel, H. Metzger, and J. Peisl
*Grazing Incidence X-Ray Diffraction from Si with an Implantation Induced Amorphous Surface Layer
(61.10; 68.20; S5.11)
(a) Vol. **108**, No. 1, p. 129—133 (1988).

H. R. Höche and J. Nieber
*Transformation of the Angular-Dependent Form of X-Ray Diffraction into the Photon-Energy-Dependent Form for Grazing Incidence
(61.10)
(a) Vol. **108**, No. 2, p. 489—493 (1988).

V. V. Aristov, U. Winter, A. Yu. Nikulin, S. V. Redkin, A. A. Snigirev, P. Zaumseil, and V. A. Yunkin
*Interference Thickness Oscillations of an X-Ray Wave on Periodically Profiled Silicon
(61.10 and 68.20; S5.11)
(a) Vol. **108**, No. 2, p. 651—655 (1988).

R. G. Gabrielyan and H. A. Aslanian
*On the Theory of X-Ray Diffraction by Surface Acoustic Waves
(61.10 and 62.30)
(a) Vol. **108**, No. 2, p. K85—K88 (1988).

J. Härtwig and V. Lerche
*Anisotropic Deformation of a Crystal Plate and Its Analysis with X-Ray Diffraction Methods
(61.10; 61.70; 62.20; S10.1)
(a) Vol. **109**, No. 1, p. 79—91 (1988).

A. P. Bezirganyan
*X-Ray Reflection from and Transmission through a Plane-Parallel Dielectric Plate with Cosine-Like Polarizability (Symmetrical Laue Case)
(61.10 and 78.70; 68.20)
(a) Vol. **109**, No. 1, p. 101—110 (1988).

G. T. Baumbach, H. Rhan, and U. Pietsch
*Detection of the Active Layer of $A^{III}B^{V}$ Semiconductor Quantum-Well Structures by High Resolution X-Ray Diffractometry
(61.10 and 68.55; S7.12; S7.15)
(a) Vol. **109**, No. 1, p. K7—K10 (1988).

I. V. Polikarpov and V. V. Skadorov
*X-Ray Bragg Dynamical Diffraction in Crystals Weakly Deformed by Ultrasonic Waves
(61.10)
(b) Vol. **149**, No. 2, p. 435—440 (1988).

J. Gastaldi, C. Jourdan, G. Grange, and C. L. Bauer
"In Situ" Determination of Grain Boundary Migration Rates by Synchrotron White Beam X-Ray Topography
(61.70; 61.10; S4)
(a) Vol. **109**, No. 2, p. 403—411 (1988).

G. B. Mitra and Lily Chakravartti
*Variation of the Polarisation Ratio with Different Diffraction Parameters
(61.10)
(a) Vol. **109**, No. 2, p. K89—K92 (1988).

V. A. Bushuev, A. Yu. Kazimirov, and M. V. Kovalchuk
*Coherent Compton Effect under Conditions of X-Ray Dynamical Laue Diffraction
(61.10 and 78.70; S5.11)
(b) Vol. **150**, No. 1, p. 9—18 (1988).

R. V. Vedrinskii, L. A. Bugaev, and I. G. Levin
Single and Multiple Electron Scattering Description in X-Ray Absorption Spectrum Theory and the Problem of Solid Structure Investigation
(78.70; 61.10; 71.10; S9.11; S10.1; S11.1)
(b) Vol. **150**, No. 1, p. 307—314 (1988).

S. Matthies, K. Helming, T. Steinkopff, and K. Kunze
*Standard Distributions for the Case of Fibre Textures
(61.10; 61.12; 61.14)
(b) Vol. **150**, No. 1, p. K1—K5 (1988).

O. Pacherová and R. Bubáková
*A Many-Beam Diffraction Effect in the CoKα_1 Spectral Lineshape
(61.10; S5.12)
(a) Vol. **110**, No. 1, p. 35—42 (1988).

J. M. Holender, J. Sołtys, G. Wendrock, and P. Czurratis
Non-Isothermal Studies of the Eutectoidal Decomposition of the Alloy Al–59 at% Zn
(64.70; 61.10; 61.55; S1.2; S4)
(a) Vol. **110**, No. 1, p. 123—132 (1988).

S. S. Jiang, Y. Qiu, and G. S. Green
*The Simulation of Modulated I_3 Fringes in an X-Ray Section Topograph of a Stacking Fault
(61.10; 61.70; S5)
(a) Vol. **110**, No. 2, p. 323—330 (1988).

P. H. Bezirganyan, V. G. Aslanyan, and S. E. Bezirganyan
*Interpretation of Interference Patterns Obtained from X-Ray Laue Interferometers in Primary Plane Waves
(61.10)
(a) Vol. **110**, No. 2, p. 359—373 (1988).

V. A. Ganshin, Yu. N. Korkishko, and T. V. Morozova
Properties of Proton Exchanged Optical Waveguiding Layers in $LiNbO_3$ and $LiTaO_3$
(61.70 and 78.20; 61.10; S11.1)
(a) Vol. **110**, No. 2, p. 397—402 (1988).

61.12. Neutron determination of structures
(for specific determinations, see 61.55 to 61.80)

H. Schaeben
Entropy Optimization in Texture Goniometry. I. Methodology
(61.50; 61.10; 61.12)
(b) Vol. **148**, No. 1, p. 63—72 (1988).

W. Kullmann, H. Rauh, R. Geick, U. Steigenberger, and W. F. Kuhs
*Thermal Parameters and Probability Density Functions of the Narrow Gap Semiconductor Bi_2Te_3 under Hydrostatic Pressure
(61.12 and 61.60; 61.50; 65.90; S8)
(b) Vol. **148**, No. 2, p. 445—456 (1988).

S. Ligenza and M. Konwicki
Cluster Structure of the Li–Ti–Zn Ferrite System Studied by Neutron Scattering
(75.50; 61.12; S11.2)
(a) Vol. **109**, No. 1, p. 319—327 (1988).

S. Matthies, K. Helming, T. Steinkopff, and K. Kunze
Standard Distributions for the Case of Fibre Textures
(61.10; 61.12; 61.14)
(b) Vol. **150**, No. 1, p. K1—K5 (1988).

61.14. Electron determination of structures
(for specific determinations, see 61.55 to 61.80)

H. S. Kim and S. S. Sheinin
*On the Relationship between Diffracted Beam Intensity and Bloch Function Orthogonality in the Dynamical Theory of Electron Diffraction
(61.14)
(a) Vol. **105**, No. 1, p. 45—52 (1988).

G. Kästner
*Approximations to the Systematic Many-Beam Case of High-Energy Electron Diffraction. I. Transition from Two-Beam to Many-Beam Interaction
(61.14)
(b) Vol. **145**, No. 2, p. 373—390 (1988).

G. Kästner
*Approximations to the Systematic Many-Beam Case of High-Energy Electron Diffraction. II. Improved Three-Beam and Four-Beam Solutions and Extinction Data
(61.14)
(b) Vol. **146**, No. 1, p. 63—74 (1988).

K. Z. Botros, S. M. Saleh, A. H. G. El-Dhaher, and S. Hajaig
*On the Visibility of Edge Fringes in Strong Beam Images of Intrinsic and Extrinsic Stacking Faults in F.C.C. Structures
(61.14 and 61.70; S1.2)
(a) Vol. **106**, No. 1, p. 49—55 (1988).

J. Q. You and T. B. Hu
*A Global Cut-and-Project Method to Construct Generalized Fibonacci Lattices and Quasilattices
(61.14; 61.90)
(b) Vol. **147**, No. 2, p. 471—484 (1988).

W. Cao, H. Q. Ye, and K. H. Kuo
A New Octagonal Quasicrystal and Related Crystalline Phases in Rapidly Solidified Mn_4Si
(61.55; 61.14; S1.61)
(a) Vol. **107**, No. 2, p. 511—519 (1988).

D. Colaitis
*Modulation par interfaces dans l'évolution quasi-continue de systèmes à deux sortes de couches
(Modulation by Interfaces in the Quasi-Continuous Evolution of Two-Layer Systems)
(61.14)
(a) Vol. **107**, No. 2, p. 531—550 (1988).

H. Hashimoto and A. Kumao
*Electron Microscope Image Contrast Formed by Electrons from Elastic–Inelastic and Inelastic–Elastic Scattering Processes
(61.14)
(a) Vol. **107**, No. 2, p. 611—618 (1988).

A. Gómez and V. M. Castaño
*Unified Approach to the High-Energy Approximation in Transmission Electron Microscopy
(61.14)
(a) Vol. **107**, No. 2, p. 845—850 (1988).

D. Mercurio, B. Frit, G. Harburn, B. H. Parry, R. P. Williams, and R. J. D. Tilley
*Diffraction Patterns from Ordered Intergrowth Phases in the Bi_2O_3–TeO_2 System
(61.14 and 64.80; S10.15)
(a) Vol. **108**, No. 1, p. 111—122 (1988).

M. Verwerft, G. van Tendeloo, J. van Landuyt, W. Coene, and S. Amelinckx
*Trial Model for the Tilting Scheme in $AgNbO_3$ Derived by Electron Diffraction and Imaging
(61.14; S11.1)
(a) Vol. **109**, No. 1, p. 67—78 (1988).

K. Watanabe, Y. Kikuchi, K. Hiratsuka, and H. Yamaguchi
*A New Approach for *n*-Beam Lattice Image Calculation
(61.14)
(a) Vol. **109**, No. 1, p. 119—126 (1988).

H. Blumtritt, G. N. Panin, E. B. Yakimov, and J. Heydenreich
*SEM–EBIC Investigations of Grain Boundaries in Cadmium Telluride
(61.14; 61.70; S8.13)
(a) Vol. **109**, No. 1, p. K3—K6 (1988).

S. Matthies, K. Helming, T. Steinkopff, and K. Kunze
Standard Distributions for the Case of Fibre Textures
(61.10; 61.12; 61.14)
(b) Vol. **150**, No. 1, p. K1—K5 (1988).

R. DORBEZ, B. YANGUI, M. BEN SALEM, M. H. PISCHEDDA, et G. NIHOUL
*Effets tridimensionnels observés dans des images électroniques de l'oxyde de néodyme hexagonal (Three-Dimensional Effects in the Electronic Images of Hexagonal Neodymium Oxide)
(61.14; S10.1)
(a) Vol. **110**, No. 1, p. 43—54 (1988).

D. KASHCHIEV and YU. O. KANTER
*Oscillations of Specular Beam Intensity in Reflection Diffraction from the Surface of a Growing Epitaxial Film. A Theoretical Study
(61.14 and 68.55; S1.2)
(a) Vol. **110**, No. 1, p. 61—76 (1988).

61.16 Other determination of structures

(for specific determinations, see 61.55 to 61.80)

H. W. ZANDBERGEN, R. GRONSKY, and G. THOMAS
Surface Decomposition of Superconducting $YBa_2Cu_3O_7$
(68.20; 61.16; S10)
(a) Vol. **105**, No. 1, p. 207—218 (1988).

C. BOULESTEIX, M. BEN SALEM, B. YANGUI, Z. KANG, and L. EYRING
Thickness of Interfaces between Twins, Glide Domains, and Grain Boundaries in Oxides from HREM Studies (Sm_2O_3, Nd_2O_3, Pr_7O_{12}, $Tb_{12}O_{22}$, and $YBa_2Cu_3O_{7-x}$)
(68.48; 61.16; S10)
(a) Vol. **107**, No. 2, p. 469—480 (1988).

A. BOURRET, J. L. ROUVIÈRE, and J. SPENDELER
Direct Observation of Atomic Columns in Semiconductors by HREM at 400 kV
(61.55 and 61.60; 61.16; S5.11; S5.12; S7.1; S8.1)
(a) Vol. **107**, No. 2, p. 481—501 (1988).

L. A. BURSILL and XUDONG FAN
Surface Structure of the Ceramic Superconductor $YBa_2Cu_3O_7$
(68.20; 61.16; 74.70; S10)
(a) Vol. **107**, No. 2, p. 503—510 (1988).

W. COENE, A. F. DE JONG, D. VAN DYCK, G. VAN TENDELOO, and J. VAN LANDUYT
*Digital Image Processing for High Resolution Electron Microscopy
(61.16; 61.70)
(a) Vol. **107**, No. 2, p. 521—530 (1988).

P. DELAVIGNETTE, TH. KARAKOSTAS, G. NOUET, and F. W. SCHAPINK
Grain Boundary Orientation Determination in Transmission Electron Microscopy
(61.70; 61.16)
(a) Vol. **107**, No. 2, p. 551—558 (1988).

M. DE GRAEF, L. DELAEY, and D. BRODDIN
High Resolution Electron Microscopic Study of the X-Phase in Cu–Al and Cu–Al–Zn Alloys
(61.55; 61.16; S1.2; S4)
(a) Vol. **107**, No. 2, p. 597—609 (1988).

A. LOISEAU and C. VANNUFFEL
$TiAl_2$: A Reentrant Phase in the Ti–Al System
(64.70; 61.16; S1.2; S4)
(a) Vol. **107**, No. 2, p. 655—671 (1988).

M. H. LORETTO
*Analytical Electron Microscopy of Advanced Materials
(61.16)
(a) Vol. **107**, No. 2, p. 673—680 (1988).

PING LU and DAVID J. SMITH
The Observation of In Situ and Ex Situ Oxidation Processes for ZnTe Surfaces by High-Resolution Electron Microscopy
(68.20; 61.16; S8.13)
(a) Vol. **107**, No. 2, p. 681—691 (1988).

R. PORTIER, M. GUYMONT, and D. GRATIAS
*Dark Field Equivalency Conditions for a Tilted Incident Beam
(61.16)
(a) Vol. **107**, No. 2, p. 739—744 (1988).

E. RUEDL, G. VALDRÈ, P. DELAVIGNETTE, and U. VALDRÈ
Structural and Analytical Electron Microscopy of Dislocation Loops and Precipitates Formed during Ageing of an α-Bombarded Mn–Cr Austenitic Steel
(61.55 and 61.70; 61.16; S1.2)
(a) Vol. **107**, No. 2, p. 745—758 (1988).

A. D. STALIOS, P. DELAVIGNETTE, and R. DE BATIST
*Transmission Electron Microscopy Observation of Crystal Defects in a Natural Diopside Mineral
(61.16 and 61j70; S11.2)
(a) Vol. **107**, No. 2, p. 759—767 (1988).

A. TOMAS, M. GUYMONT, M. P. PARDO, M. GUITTARD, and J. FLAHAUT
X-Ray Diffraction and Electron Microscopy Studies of α- and β-Ga_2S_3
(61.60; 61.10; 61.16; S8)
(a) Vol. **107**, No. 2, p. 775—784 (1988).

P. WERNER and J. HEYDENREICH
Experimental Study of Structure Modification of $2M_1$-Muscovite by High-Resolution Electron Microscopy
(61.60; 61.16; S11)
(a) Vol. **107**, No. 2, p. 807—816 (1988).

H. W. ZANDBERGEN and G. THOMAS
*HREM Investigations of Defects in Sintered $YBa_2Cu_3O_{7-\delta}$
(61.16 and 61.70; 74.70; S10.15)
(a) Vol. **107**, No. 2, p. 825—843 (1988).

C. SARRAZIN, R. J. GABORIAUD, and J. P. RIVIÈRE
TEM Investigations of Ni–Cu Thin Film Coatings Obtained by Multilayer Technique, Coevaporation, and Ion Beam Assited Deposition
(68.55; 61.16; S1.2)
(a) Vol. **107**, No. 2, p. 867—871 (1988).

Y. KIKUCHI, K. WATANABE, S. NAITOH, K. HIRATSUKA, and H. YAMAGUCHI
HREM Image of a Vacancy in Copper
(61.70; 61.16; S1.2)
(a) Vol. **108**, No. 2, p. 509—517 (1988).

O. S. RAJORA and A. E. CURZON
A Study of the Diffusion of Indium into TaS_2 and $TaSe_2$ in a Scanning Electron Microscope
(66.30; 61.16; S1.63)
(a) Vol. **108**, No. 2, p. 529—536 (1988).

A. ARMIGLIATO, A. BOURRET, S. FRABBONI, and A. PARISINI
*Contrast of Small SiX Particles in Silicon by Computed HREM Images
(61.16; S5.11)
(a) Vol. **109**, No. 1, p. 53—60 (1988).

YE TAO, MEI-SHEN HU, and DUAN FENG
Direct Observation of α–β Phase Transition in $KNbW_2O_9$ by Transmission Electron Microscopy
(64.70; 61.16; S11)
(a) Vol. **109**, No. 2, p. 435—444 (1988).

L. NISTOR, A. DE VEIRMAN, and J. VAN LANDUYT
*Cross Section Transmission Electron Microscopy Studies of Zr Samples Laser Irradiated in Air
(61.16; S1)
(a) Vol. **110**, No. 1, p. K1—K3 (1988).

T. SHIOZAWA and T. KOBAYASHI
High Resolution Electron Microscopic Study of Silver Sulfide Microcrystals Formed on Silver Bromide Emulsion Grains
(61.60; 61.16; S1.63; S9)
(a) Vol. **110**, No. 2, p. 375—382 (1988).

E. G. DONI and G. L. BLERIS
*Study of Special Triple Junctions and Faceted Boundaries by Means of the CSL Model
(61.16 and 61.70)
(a) Vol. **110**, No. 2, p. 383—395 (1988).

61.20. Classical, semiclassical, and quantum theories of liquid structure
(for electronic states, see 71; for liquid helium, see 67)

J. Maćkowiak
A Model of Crystallization in a One-Dimensional Classical Liquid
(64.60; 61.20)
(b) Vol. **145**, No. 2, p. 419—424 (1988).

C. Regnaut
*Linearised Version of the Optimised Random Phase Approximation. Application to the Structure of Dense Fluids and Metallic Potentials
(61.20; 61.25)
(b) Vol. **146**, No. 2, p. 413—425 (1988).

61.25. Studies of specific liquid structures

Xian-Wu Zou, Zhun-Zhi Jin, and De-Cheng Tian
*Structure Factors of Liquid Zn, Cd, and Hg
(61.25; S1; S1.2)
(b) Vol. **145**, No. 1, p. 73—76 (1988).

C. Regnaut
Linearised Version of the Optimised Random Phase Approximation. Application to the Structure of Dense Fluids and Metallic Potentials
(61.20; 61.25)
(b) Vol. **146**, No. 2, p. 413—425 (1988).

Zhun-Zhi Jin, Xian-Wu Zou, and Yong-Qiao Yao
*Dependence of Structure Factors upon Temperature for Liquid Metals, Zn and Hg
(61.25; S1; S1.2)
(b) Vol. **147**, No. 1, p. 45—49 (1988).

R. V. Gopala Rao and Ratna Das
*Structure Factor Computation of RbCl and RbBr
(61.25; S9.11)
(a) Vol. **110**, No. 2, p. 347—357 (1988).

61.30. Liquid crystals

Yuan-Chao Chu, T. Tsang, E. Rahimzadeh, and L. Yin
*Solid Phase Structures of Cyano-Biphenyl Liquid Crystals
(61.30; S12)
(a) Vol. **105**, No. 1, p. K1—K6 (1988).

61.40. Amorphous and polymeric materials

E. KOKMEIJER, E. HUIZER, B. J. THIJSSE, and A. VAN DEN BEUKEL
Resistivity Changes during Structural Relaxation of Amorphous $Fe_{40}Ni_{40}B_{20}$
(72.15; 61.40; S1.1)
(a) Vol. **105**, No. 1, p. 235—242 (1988).

G. BRAUER, W. HERTLE, and A. G. BALOGH
*γ-Irradiation of Polyethylene Studied by Positron Annihilation
(61.40 and 78.70; 61.80; S12)
(a) Vol. **105**, No. 1, p. K7—K12 (1988).

P. ŠVEC and P. DUHAJ
*Some Comments on the Determination of Kinetic Crystallization Parameters of Metallic Glasses
(61.40; 61.50)
(a) Vol. **105**, No. 2, p. 319—324 (1988).

MING MAO and YIZHEN HE
*Two-Stage Structural Relaxation of the Metallic Glass $Fe_{46}Ni_{31}V_1Si_8B_{14}$
(61.40; 65.40; S1.1; S1.2)
(a) Vol. **105**, No. 2, p. 325—333 (1988).

J. P. ALLEMAND, F. FOUQUET, J. PEREZ and B. DE GUILLEBON
Evolution of the Mechanical and Micromechanical Properties as a Function of Quench Rate in a Metallic Fe–Cr–C–P–Si Glass
(62.20; 61.40; 75.80; S1.1)
(a) Vol. **105**, No. 2, p. 427—439 (1988).

G. PASTOR, P. TEJEDOR, I. JIMÉNEZ, E. DOMÍNGUEZ, M. TORRES, and J. V. GARCÍA-RAMOS
*Low Pressure Chemical Vapour Deposition Amorphous Silicon Behaviour under Annealing
(61.40 and 68.55; 78.30; S5.11)
(a) Vol. **106**, No. 1, p. 11—16 (1988).

A. I. POPOV, I. A. DOMORYAD, and N. I. MICHALEV
*Structural Modification of the Arsenic Chalcogenide Glasses under γ-Radiation
(61.40; 62.20; S8)
(a) Vol. **106**, No. 2, p. 333—337 (1988).

L. N. ALEKSANDROV
*Mechanical Stress in the Crystalization Process of Amorphous Semiconductors by Pulse Action
(61.40 and 65.70; S5.11)
(a) Vol. **106**, No. 2, p. K135—K138 (1988).

H. HE
*Influence of Rings on Electronic Structure in Topologically Disordered System
(61.40 and 71.20)
(b) Vol. **147**, No. 1, p. 185—190 (1988).

S. RAM
*Magnetic and Magnetic Resonance Studies on Oxygen-Deficient $SrFeO_{3-y}$ Perovskites
(61.40; 75.60; 76.30; S11.2)
(a) Vol. **107**, No. 1, p. 55—62 (1988).

L. TICHÝ and P. NAGELS
*Non-Isothermal Crystallization of Bi_2S_3 from Glassy $(GeS_2)_{0.5}(Bi_2S_3)_{0.5}$
(61.40; 64.60; S8)
(a) Vol. **107**, No. 2, p. 769—774 (1988).

S. KOKAI, I. KIKUCHI, O. UEMURA, and T. SATOW
*Short-Range Order in Amorphous $GeSe_2$
(61.40; S8)
(a) Vol. **108**, No. 2, p. K89—K92 (1988).

MING MAO, YUANSHENG WANG, and YIZHEN HE
*Crystallization of Metallic Glass $Fe_{46}Ni_{31}VSi_8B_{14}$
(61.40 and 64.70; S1.1; S1.2)
(a) Vol. **109**, No. 1, p. 39—51 (1988).

H. WIECZOREK and W. FUCHS
Transient Photocurrents in a-Si:H Diodes: Effects of Deep Trapping
(72.40 and 73.40; 61.40; 71.55; S5.11)
(a) Vol. **109**, No. 1, p. 245—253 (1988).

ZONGQUAN LI, HUI SHEN, LANPING YUE, and YIZHEN HE
*A Special Metastable Phase of Metallic Glass $Pd_{77.5}Ni_6Si_{16.5}$
(61.40; S1.61)
(a) Vol. **109**, No. 1, p. K1—K2 (1988).

R. ROUSINA and G. K. SHIVAKUMAR
Effect of Heat Treatment on the Optical Absorption of Amorphous Indium Selenide Films
(78.65; 61.40; S8)
(a) Vol. **110**, No. 1, p. K57—K60 (1988).

GUOAN WU, YUANDA DONG, KEQIN XIAO, YIZHEN HE, LONGSHU WANG, and JIAQING HE
*An Investigation of Fast Neutron Irradiation Effects on the Structure of $Cu_{50}Ti_{50}$ Metallic Glass
(61.40; 64.70; S1.2)
(a) Vol. **110**, No. 2, p. 317—321 (1988).

61.50. Crystalline state (including molecular motions in solids)
(for magnetic structure and spin systems, see 75.25)

J. Griesche, R. Enderlein, and D. Schikora
*Orientation and Strain in Heteroepitaxial Growth **(Review Article)**
(61.50 and 68.48; 68.20; S5.11; S7.12; S8.13)
(a) Vol. **109**, No. 1, p. 11—38 (1988).

V. I. Dimitrov
*Monte Carlo Determination of Short-Range Parameters for Multicomponent Solid Solutions
(61.50; 64.75)
(b) Vol. **145**, No. 1, p. 65—72 (1988).

S. Chandra and H. C. Verma
*Charge Shift Model Calculation of the Cohesive Energy of Magnesium
(61.50; S3)
(b) Vol. **145**, No. 1, p. K1—K3 (1988).

M. M. Aleksandrova, V. D. Blank, A. E. Golobokov, Yu. S. Konyaev, A. Yu. Zerr, and E. I. Estrin
*Amorphisation of Gallium Antimonide under the Conditions of Shear Deformation under Pressure
(61.50 and 62.20; S7.13)
(a) Vol. **105**, No. 1, p. K29—K32 (1988).

V. Holý and J. Härtwig
*The Role of Diffuse Scattering on Microdefects in the Precise Lattice Parameter Measurement
(61.10 and 61.50; 61.70; S5.11)
(b) Vol. **145**, No. 2, p. 363—372 (1988).

R. A. Evarestov and V. P. Smirnov
*Double-Valued Band Representations of Monoclinic and Orthorhombic Space Groups
(61.50; 71.50)
(b) Vol. **145**, No. 2, p. 391—399 (1988).

S. L. Chaplot
*Potential Barriers for Molecular Rotations in Solids
(61.50; S11.1)
(b) Vol. **145**, No. 2, p. K89—K93 (1988).

P. Švec and P. Duhaj
Some Comments on Determination of Kinetic Crystallization Parameters of Metallic Glasses
(61.40; 61.50)
(a) Vol. **105**, No. 2, p. 319—324 (1988).

D. M. Vanderwalker
*The Formation of Hydrides in Titanium
(61.50; 61.70; S1.2)
(a) Vol. **105**, No. 2, p. K77—K80 (1988).

B. R. SINGH, H. S. KOTHARI, O. P. DAGA, J. K. SINGH, G. S. T. RAO, J. D. JAIN, and W. S. KHOKLE
*X-Ray Diffraction and EDS Analysis of $Y_1Ba_2Cu_3O_{9-y}$ Superconducting Compound
(61.50 and 74.70; S10.15)
(a) Vol. **106**, No. 1, p. K53—K56 (1988).

M. DE GRAEF and L. DELAEY
*A New Empirical Two-Body Interaction Potential Fitted to Second- and Third-Order Elastic Moduli
(61.50; 62.20)
(b) Vol. **146**, No. 2, p. 427—437 (1988).

XIAOQING PAN and DUAN FENG
*Direct Observation of Nucleation of Discommensurations in Barium Sodium Niobate
(61.50; S11.1)
(a) Vol. **106**, No. 2, p. K117—K121 (1988).

O. E. YANOVITSKII and E. S. KRYACHKO
*Model for Orientational Defects in Quasi-One-Dimensional Ice Crystals
(61.50 and 61.70)
(b) Vol. **147**, No. 1, p. 69—81 (1988).

S. D. CHATURVEDI, R. C. DIXIT, and A. A. S. SANGACHIN
*Analysis of Crystal Binding and Thermoelastic Properties of Cuprous Halides, Silver Halides, and Thallous Halides
(61.50 and 63.90; 62.20; 65.70; S9)
(b) Vol. **147**, No. 1, p. 115—121 (1988).

J. HORÁK, Z. STARÝ, AND J. KLIKORKA
*Relations between Structure, Bonding, and Nature of Point Defects in Layered Crystals of Tetradymite Structure
(61.50 and 61.70; 72.80; S8; S8.15)
(b) Vol. **147**, No. 2, p. 501—510 (1988).

H. SCHAEBEN
*Entropy Optimization in Texture Goniometry. I. Methodology
(61.50; 61.10; 61.12)
(b) Vol. **148**, No. 1, p. 63—72 (1988).

QING GU and XI-MAO BAO
*Effect of the Structure of Laser Crystallized Silicon-on-Insulator on Impurity Diffusion
(61.50; 66.30; 72.80; S5.11)
(a) Vol. **108**, No. 1, p. 81—87 (1988).

M. HEYTING, B. J. THIJSSE, and A. VAN DEN BEUKEL
*Change of Electrical Resistance during Crystallization of Icosahedral $Al_{86}Mn_{14}$
(61.50 and 72.15; S4)
(a) Vol. **108**, No. 1, p. 97—101 (1988).

N. M. PLAKIDA and W. SALEJDA
The Improper Ferroelastic Phase Transition in Superionic $Rb_3H(SeO_4)_2$ Crystals
(64.70; 61.50; S11)
(b) Vol. **148**, No. 2, p. 473—481 (1988).

S. Yu. Korostelev, S. G. Psakhe, and V. E. Panin
*A Molecular-Dynamical Study of the Changes in the Atomic Structure of a Material Induced by a Shock Wave with a Distorted Front
(61.50 and 62.50; S1.1)
(b) Vol. **148**, No. 2, p. 483—488 (1988).

Halicioglu, H. O. Pamuk, and S. Erkoc
Interatomic Potentials with Multi-Body Interactions
(63.10; 61.50; 68.40; S5.11)
(b) Vol. **149**, No. 1, p. 81—92 (1988).

E. Anastassakis and E. Liarokapis
*A Macroscopic Application of the Wigner-Eckart Theorem
(61.50 and 62.20)
(b) Vol. **149**, No. 1, p. K1—K5 (1988).

P. Mohazzabi and F. Behroozi
*Vibrational Energy as a Function of Interatomic Distance in Rare-Gas Solids. A Universal Relationship
(61.50 and 65.40; S5.2)
(b) Vol. **149**, No. 2, p. 495—501 (1988).

K. Kumar, I. S. Arora, and J. Shanker
*On the Narayan-Ramaseshan Interionic Potentials
(61.50; S9.11)
(b) Vol. **150**, No. 1, p. K7—K10 (1988).

F. J. Bremer, M. Beyss, and H. Wenzl
*The Order–Discorder Transition of the Intermetallic Phase Ni_3Al
(61.50; S1.2)
(a) Vol. **110**, No. 1, p. 77—82 (1988).

V. Yu. Balandin, A. V. Dvurechenskii, and L. N. Aleksandrov
Intermediate Crystallization of Amorphous Layers of Silicon at Plused Laser Annealing
(64.60; 61.50; S5.11)
(a) Vol. **110**, No. 1, p. 133—139 (1988).

61.55. Specific structures of elements and alloys

T. Bhattacharya, A. Manna, and A. K. Ghosh
Construction of Electronic Charge Density in Solids. Calculation of Atomic Scattering Factors of Elements and Energy of Formation of Alloys
(71.10; 61.55; S1)
(b) Vol. **145**, No. 1, p. 181—186 (1988).

A. Extremera
The Magnetic Behaviour of $NaHg_4$ Peritectic
(75.30; 61.55; S1.3)
(a) Vol. **105**, No. 1, p. 281—284 (1988).

M. Polcarová, K. Godwod, J. Bąk-Misiuk, S. Kadečková, and J. Brádler
*Lattice Parameters of Fe–Si Alloy Single Crystals
(61.55; 61.10; S1.1; S1.61)
(a) Vol. **106**, No. 1, p. 17—23 (1988).

A. V. Andreeva, A. V. Bazhenov, N. A. Bulenkov, and A. A. Firsova
*Structural and Photoelectrical Properties of a Third-Order Twin Boundary ($\Sigma = 27$) in Silicon
(61.55; 61.70; 72.20; S5.11)
(a) Vol. **106**, No. 2, p. 351—362 (1988).

P. Czurratis, R. Kroggel, V. Syneček, M. Simerska, and H. Löffler
Dislocation Loops in Al–Zn and Al–Zn–Mg Alloys
(61.70; 61.55; S1.2; S4)
(a) Vol. **106**, No. 2, p. 407—418 (1988).

R. E. Hummel, W. M. Goho, and V. Gerold
*The Influence of Percipitation on the Optical Properties and Electronic Structure of Copper–Cobalt Alloys
(61.55 and 78.20; S1.2)
(a) Vol. **106**, No. 2, p. 641—649 (1988).

Jinghua Tian, Jingkui Liang, Dayu Yang, Hanjie Fan, and Fanghua Li
*The Study of the One-Dimensional Composition-Incommensurate Structure of $Pr_{1+\varepsilon}Fe_4B_4$
(61.55; S1.1; S1.4)
(a) Vol. **107**, No. 1, p. 45—53 (1988).

An Hu, Shu-yi Zhang, Xiang-yang Yuan, Qing Shen, Zhong-nan Lu, and Duan Feng
Lattice Expansion and Phonon Softening in Nb/Cu Superlattices
(63.20 and 68.30; 61.55; S1.2)
(a) Vol. **107**, No. 1, p. 153—159 (1988).

P. Grima, M. Quintero, G. S. Pérez, R. Tovar, and J. C. Woolley
*Phase Diagram and Lattice Parameter Values for the $Hg_{2x}(AgIn)_{1-x}Te_2$ Alloys
(61.55 and 64.70; S8.15)
(a) Vol. **107**, No. 1, p. 165—169 (1988).

A. Bourret, J. L. Rouvière, and J. Spendeler
*Direct Observation of Atomic Columns in Semiconductors by HREM at 400 kV
(61.55 and 61.60; 61.16; S5.11; S5.12; S7.1; S8.1)
(a) Vol. **107**, No. 2, p. 481—501 (1988).

W. Cao, H. Q. Ye, and K. H. Kuo
*A New Octagonal Quasicrystal and Related Crystalline Phases in Rapidly Solidified Mn_4Si
(61.55; 61.14; S1.61)
(a) Vol. **107**, No. 2, p. 511—519 (1988).

M. de Graef, L. Delaey, and D. Broddin
*High Resolution Electron Microscopic Study of the X-Phase in Cu–Al and Cu–Al–Zn Alloys
(61.55; 61.16; S1.2; S4)
(a) Vol. **107**, No. 2, p. 597—609 (1988).

Y. HAYASHI
*Ordering of Ni_3Mn and Diffusivity of Hydrogen in the Alloy
(61.55 and 66.30; S1.2)
(a) Vol. **107**, No. 2, p. 619—624 (1988).

J. MENON and C. SURYANARAYANA
*The Al–Co Decagonal Phase
(61.55; S1.2; S4)
(a) Vol. **107**, No. 2, p. 693—708 (1988).

M. NAKAGAWA and K. YASUDA
*Phase Changes and Structural Features of Cu_3Au II in a Gold–Copper–Silver Ternary Alloy during Isothermal Ageing
(61.55 and 64.70; S1.3)
(a) Vol. **107**, No. 2, p. 709—717 (1988).

E. RUEDL, G. VALDRÈ, P. DELAVIGNETTE, and U. VALDRÈ
*Structural and Analytical Electron Microscopy of Dislocation Loops and Percipitates Formed during Ageing of an α-Bombarded Mn–Cr Austenitic Steel
(61.55 and 61.70; 61.16; S1.2)
(a) Vol. **107**, No. 2, p. 745—758 (1988).

CHU THIEN TRUONG, G. WENDROCK, and H. LÖFFLER
*On the Existence Field of Guinier-Preston Zones in the Al–Zn System
(61.55; 64.70; S4)
(a) Vol. **108**, No. 1, p. 103—110 (1988).

P. CZURRATIS, E. KORNGIEBEL, G. JACOBI, and H. LÖFFLER
The Effect of Additions of Mn and Cr and the Applied Solidification Technique on the Structure of an Al-2.2 at%Zn-1.4 at% Mg Alloy
(61.70; 61.55; 62.20; S4)
(a) Vol. **108**, No. 1, p. 171—182 (1988).

B. SCHÖNFELD, L. REINHARD, G. KOSTORZ, and W. BÜHRER
*Short-Range Order and Atomic Displacements in Ni–20 at% Cr Single Crystals
(61.55 and 64.80; S1.2)
(b) Vol. **148**, No. 2, p. 457—471 (1988).

H.-G. FRITSCHE, D. BONCHEV, and O. MEKENYAN
*Are Small Clusters of Inert-Gas Atoms Polyhedra of Minimum Surfaces?
(61.55; S5.2)
(b) Vol. **148**, No. 2, p. K101—K104 (1988).

R. MANAILA, V. FLORESCU, A. JIANU, and A. BADESCU
*Icosahedral Phases in the Al–Fe–Cr Alloy System
(61.55; S1.1; S1.2)
(a) Vol. **109**, No. 1, p. 61—66 (1988).

L. N. KOSTADINOV and D. D. DOBREV
*On the Structure of the C_8 Cubic Modification of Carbon
(61.55; S5)
(a) Vol. **109**, No. 2, p. K85—K88 (1988).

J. M. HOLENDER, J. SOŁTYS, G. WENDROCK, and P. CZURRATIS
Non-Isothermal Studies of the Eutectoidal Decomposition of the Alloy Al–59 at% Zn
(64.70; 61.10; 61.55; S1.2; S4)
(a) Vol. **110**, No. 1, p. 123—132 (1988).

A. KOWALCZYK and A. WRZECIONO
Magnetic and Crystallographic Properties of $R_2Fe_{14-x}Cr_xB$ Compounds (R = Y, Nd, and Gd)
(75.30; 61.55; S1.1; S1.2)
(a) Vol. **110**, No. 1, p. 241—245 (1988).

U. D. KULKARNI, S. MURALIDHAR, and S. BANERJEE
*Computer Simulation of the Early Stages of Ordering in Ni–Mo Alloys
(61.55; S1.2)
(a) Vol. **110**, No. 2, p. 331—345 (1988).

61.60. Specific structures of inorganic compounds

P. P. LOTTICI, G. ANTONIOLI, and C. RAZZETTI
*EXAFS in $Zn_xCd_{1-x}Ga_2S_4$ Defect Chalcopyrite Solid Solution
(61.60 and 78.70; S8.16)
(b) Vol. **145**, No. 2, p. 401—407 (1988).

A. E. CURZON and O. S. RAJORA
*A Study of NbS_xSe_{2-x} Using Electron Channeling Patterns
(61.60; 61.80; S8.15)
(a) Vol. **105**, No. 2, p. 335—343 (1988).

K. S. ALEKSANDROV, S. V. MELNIKOVA, I. N. FLEROV, A. D. VASILEV, A. I. KRUGLIK, and I. T. KOKOV
Phase Transitions in Cs_2CdI_4 Single Crystals
(64.70; 61.60; 65.40; 77.20; 78.20; S9.16)
(a) Vol. **105**, No. 2, p. 441—446 (1988).

Y. SRINIVASAVA RAO and C. S. SUNANDANA
Characterization of Diamagnetically Substituted (Na^+, Te^{4+}, Cl^-) Lithium Ferrite
(75.50; 61.60; S11.2)
(a) Vol. **106**, No. 1, p. 217—225 (1988).

Y. H. LIANG, J. CHEN, L. C. WANG, Y. YIAN, X. Q. PAN, M. WEI, G. Q. WANG, X. JING, S. Y. DING, D. FENG, and Q. R. ZHANG
*A HREM Study on the Defects in Y–Ba–Cu–O Superconductors
(61.60 and 74.70; S10.15)
(a) Vol. **107**, No. 1, p. 63—72 (1988).

B. DOMENGÈS, M. HERVIEU, C. MICHEL, A. MAIGNAN, and B. RAVEAU
*$La_3Ba_3Cu_6O_{14+x}$: A Single Phase or a Mixture?
(61.60; 64.80; S10.15)
(a) Vol. **107**, No. 1, p. 73—84 (1988).

M. Roy, R. N. P. Chowdhury, and H. N. Acharya
*Structural and Thermal Properties of $(Sm_{1-x}Nd_x)_2(MoO_4)_3$
(61.60 and 65.90; S10)
(a) Vol. **107**, No. 1, p. K1—K5 (1988).

M. Bredol and V. Leute
*The Crystal Structure of Ga_2Se_2Te
(61.60; S8.16)
(a) Vol. **107**, No. 1, p. K7—K10 (1988).

M. Koyano, S. Ōhara, H. Negishi, M. Sasaki, M. Inoue, M. Nomura, and H. Fujiwara
*Effect of Pressure on Electrical Resistivity and Crystal Structure of Quasi-Two-Dimensional Mo_4O_{11}
(61.60 and 72.20; S10)
(b) Vol. **147**, No. 2, p. 559—565 (1988).

A. Bourret, J. L. Rouvière, and J. Spendeler
*Direct Observation of Atomic Columns in Semiconductors by HREM at 400 kV
(61.55 and 61.60; 61.16; S5.11; S5.12; S7.1; S8.1)
(a) Vol. **107**, No. 2, p. 481—501 (1988).

A. Tomas, M. Guymont, M. P. Pardo, M. Guittard, and J. Flahaut
*X-Ray Diffraction and Electron Microscopy Studies of α- and β-Ga_2S_3
(61.60; 61.10; 61.16; S8)
(a) Vol. **107**, No. 2, p. 775—784 (1988).

P. Werner and J. Heydenreich
*Experimental Study of Structure Modification of $2M_1$-Muscovite by High-Resolution Electron Microscopy
(61.60; 61.16; S11)
(a) Vol. **107**, No. 2, p. 807—816 (1988).

G. A. Wiegers, R. J. Haange, and F. van Bolhuis
*The Crystal Structure of Stage-2 4H–$Ag_{0.22}NbS_2$
(61.60; S8)
(a) Vol. **107**, No. 2, p. 817—824 (1988).

J. Mahy, G. A. Wiegers, F. van Bolhuis, A. Diedering, and R. J. Haange
*X-Ray and Electron Diffraction Study of Intercalates Ag_xTaS_2
(61.60; S8)
(a) Vol. **107**, No. 2, p. 873—887 (1988).

V. P. Dymont, G. I. Makovetskii, and V. M. Ryzhkovskii
*Neutron Diffraction Determination of Mn and Zn Structure Positions in MnZnSb
(61.60; S7)
(a) Vol. **107**, No. 2, p. K89—K91 (1988).

R. Flagmeyer, V. Gottschalch, and H. Frey
*Internal Distortion in GaInAsP Quaternary Alloys Studied by Ion Channeling
(61.60; S7.16)
(a) Vol. **108**, No. 1, p. 123—128 (1988).

W. Kullmann, H. Rauh, R. Geick, U. Steigenberger, and W. F. Kuhs
*Thermal Parameters and Probability Density Functions of the Narrow Gap Semiconductor Bi_2Te_3 under Hydrostatic Pressure
(61.12 and 61.60; 62.50; 65.90; S8)
(b) Vol. **148**, No. 2, p. 445—456 (1988).

Z. Starý, L. Beneš, and J. Horák
*Interaction of Anti-Site Defects with Ge Impurity Atoms in Sb_2Te_3 Single Crystals
(61.60; 61.70; S8)
(a) Vol. **109**, No. 1, p. 93—99 (1988).

M. Levalois et G. Allais
*Etude structurale, par diffraction de R-X, des liaisons dans les semiconducteurs ternaires $ZnSiAs_2$, $ZnGeAs_2$ et $ZnSnAs_2$
(X-Ray Diffraction Study of Bonding in the Ternary Semiconductors $ZnSiAs_2$, $ZnGeAs_2$, and $ZnSnAs_2$)
(61.60; S7.16)
(a) Vol. **109**, No. 1, p. 111—118 (1988).

E. A. Zhilinskaya, V. N. Lazukin, E. A. Bychkov, and S. B. Rosenkov
*Correlation between the Structures of Ge–S Glasses and that of Impurity Mn^{2+} Complexes
(61.60 and 71.55; 76.30; S8.16)
(b) Vol. **149**, No. 2, p. 427—433 (1988).

J. L. Sánchez Llamazares, S. García García, S. Díaz Castañón, and N. Suarez Almodovar
On the Effect of Al Addition on the Phase Composition and Magnetic Properties of Nd–Fe–B Permanent Magnet Alloys
(75.30; 61.60; S1.1)
(a) Vol. **109**, No. 2, p. 569—573 (1988).

J. K. Liang, S. S. Xie, Y. L. Zhang, G. C. Che, G. H. Rao, J. Q. Huang, and X. R. Cheng
*Crystal Structure and Superconductivity of the $Tl_{2+x}Ba_2Ca_{2-x}Cu_3O_{10}$ Compound
(61.60 and 74.70; S11)
(a) Vol. **110**, No. 1, p. 55—60 (1988).

V. Bhaskara Sastry, K. Satyanarayana Murthy, N. T. Mamedov, and Prabhavathy
*Powder Data on $TlInSe_2$, $TlInTe_2$, and $TlGaTe_2$
(61.60; S8.16)
(a) Vol. **110**, No. 1, p. K5—K7 (1988).

T. Shiozawa and T. Kobayashi
*High Resolution Electron Microscopic Study of Silver Sulfide Microcrystals Formed on Silver Bromide Emulsion Grains
(61.60; 61.16; S1.63; S9)
(a) Vol. **110**, No. 2, p. 375—382 (1988).

61.65. Specific structures of organic compounds

J. Wąsicki, A. V. Belushkin, V. G. Khomenko, I. Natkaniec, Yu. V. Telezhenko, and T. Wasiutyński
Low-Temperature Structure and Lattice Dynamics of 4,4′-Difluorobiphenyl
(63.20; 61.65; S12)
(b) Vol. **145**, No. 2, p. 445—453 (1988).

61.70. Defects in crystals

(see also 61.80. Radiation damage. 62. Mechanical and acoustical properties, 71.55. Impurities and defect levels, 76.30. EPR of color centers and other defects, 78.50. Impurity and defect absorption in soils)

V. S. Malyshevskii
Interaction of Channeled Electrons with Lattice Excitations **(Review Article)**
(61.80; 61.70; 63.20; 78.70)
(b) Vol. **147**, No. 1, p. 11—44 (1988).

Y. Madhavan, K. Ramachandran, and T. M. Haridasan
Mean Square Displacements of Neighbours and Phonon Resonance Due to a Vacancy in HgTe and CdTe
(63.20; 61.70; S8.13)
(b) Vol. **145**, No. 1, p. 139—144 (1988).

J. Härtwig, V. Holý, R. Kittner, J. Kuběna, and V. Lerche
*Investigation of Quartz Crystals with Non-Homogeneous Distributions of Impurity Atoms by X-Ray Methods
(61.10 and 61.70; S10.1)
(a) Vol. **105**, No. 1, p. 61—75 (1988).

A. Rizzo, C. de Blasi, M. Catalano, and P. Cavaliere
*Dislocations in $A^{III}B^{VI}$ Single Crystals
(61.70; S8)
(a) Vol. **105**, No. 1, p. 101—112 (1988).

J. Jabłoński, J. Wojciechowski, and K. Kucharski
*The Nucleation Kinetics of Platelike Oxide Precipitates in Cz-Silicon
(61.70; S5.11)
(a) Vol. **105**, No. 1, p. 113—121 (1988).

C. A. C. Sequeira and J. M. B. Fernandez Diniz
Conduction Processes in Oxide Solid Electrolytes
(66.30 and 72.20; 61.70; S10)
(a) Vol. **105**, No. 1, p. 123—130 (1988).

L. M. Clarebrough and C. T. Forwood
*Burgers Vectors of Secondary Grain Boundary Dislocations in a $\Sigma3$ Boundary in α-Iron
(61.70; S1.1)
(a) Vol. **105**, No. 1, p. 131—138 (1988).

M. Lazaridou and C. Varotsos
*Defect Formation and Migration Entropies in Alkaline Earth Fluorides
(61.70; S9.12)
(a) Vol. **105**, No. 1, p. K13—K15 (1988).

F. C. Frank
*On the Burgers Circuit
(61.70)
(a) Vol. **105**, No. 1, p. K21—K23 (1988).

M. J. Marcinkowski
*Burgers Circuit Perspectives
(61.70)
(a) Vol. **105**, No. 1, p. K25—K27 (1988).

P. P. Pal-Val, V. D. Natsik, and S. Kadečková
Low-Temperature Dislocation Relaxations in High-Purity Iron Single Crystals Deformed at 295 and 78 K
(62.40; 61.70; S1.1)
(a) Vol. **105**, No. 1, p. K37—K40 (1988).

V. Holý and J. Härtwig
The Role of Diffuse Scattering on Microdefects in the Precise Lattice Parameter Measurement
(61.10 and 61.50; 61.70; S5.11)
(b) Vol. **145**, No. 2, p. 363—372 (1988).

J. Kreissl and W. Gehlhoff
Electron Paramagnetic Resonance of the Mn_4^0 Cluster in Silicon
(71.55 and 76.30; 61.70; S5.11)
(b) Vol. **145**, No. 2, p. 609—617 (1988).

J. Tatarkiewicz and A. M. Witowski
*Hydrogen Vibrations in ZnTe
(61.70 and 63.20; 78.30; S8.13)
(b) Vol. **145**, No. 2, p. K95—K98 (1988).

Wen-Chen Zheng
*Determination of the Local Thermal Expansion Coefficient in the Vicinity of Cr^{3+} Ions Associated with Vacancies in MgO
(61.70 and 65.70; 71.70; S10.1)
(b) Vol. **145**, No. 2, p. K99—K102 (1988).

S. V. Nistor, A. Bouwen, and D. Schoemaker
*ESR of Pb^{3+} Centers in Alkali Chlorides
(61.70 and 76.30; S9.11)
(b) Vol. **145**, No. 2, p. K135—K139 (1988).

I. Ya. Dekhtyar, E. G. Madatova, A. Čížek, and M. Šob
*Relation between the Strength Characteristics of Iron and Nickel and the Electronic Structure of Defects by Positron Annihilation Spectroscopy
(61.70; 78.70; S1.1; S1.2)
(a) Vol. **105**, No. 2, p. 357—363 (1988).

C. T. Forwood and L. M. Clarebrough
*Rigid Body Displacements at a Faceted Σ3 Boundary in α-Iron
(61.70; S1.1)
(a) Vol. **105**, No. 2, p. 365—375 (1988).

D. Köhler, G. Koschek, and E. Kubalek
Vacancy Diffusion Profiles at the Grain Boundaries of Electronic Ceramics
(66.30; 61.70; 78.60; S10.1)
(a) Vol. **105**, No. 2, p. 377—385 (1988).

C. Vijayan and Y. V. G. S. Murti
*Ionic Thermocurrents Due to Z-Centres in KCl:Gd^{2+}
(61.70; 77.30; 77.40; 78.50; S9.11)
(a) Vol. **105**, No. 2, p. 397—402 (1988).

M. H. Xu, S. Patu, and Z. G. Wang
The Improvement in Fatigue Life of Pure Polycrystalline Nickel by Nitrogen Ion Implantation
(62.20; 61.70; S1.2)
(a) Vol. **105**, No. 2, p. 419—425 (1988).

G. Irmer and W. Siegel
*Carrier Concentration Profiles across GaP Grain Boundaries Observed by Raman Measurements
(61.70 and 72.20; 78.30; S7.11)
(a) Vol. **105**, No. 2, p. 549—553 (1988).

F. R. Kessler, U. Barkow, R. Nies, and N. Unzner
Optical Profile Analysis of Diffused and Ion-Implanted Silicon
(78.20; 61.70; 66.30; 78.50; S5.11)
(a) Vol. **105**, No. 2, p. 627—635 (1988).

D. M. Vanderwalker
The Formation of Hydrides in Titanium
(61.50; 61.70; S1.2)
(a) Vol. **105**, No. 2, p. K77—K80 (1988).

E. G. Bagduev, E. V. Korovkin, G. I. Peresada, E. G. Ponyatovskii, and M. Sh. Shikhsaidov
*Dark Instability of F-Centres in Pressure-Treated KCl and KBr Crystals
(61.70; 78.60; S9.11)
(b) Vol. **146**, No. 1, p. 75—80 (1988).

T. C. T. Ting
*Line Forces and Dislocations in Anisotropic Elastic Composite Wedges and Spaces
(61.70; 62.20)
(b) Vol. **146**, No. 1, p. 81—90 (1988).

L. E. Nagli and S. V. Dyachenko
Influence of v_c^--Vacancy on Luminescence of Pb^+ Centres in Alkali Halides
(78.55; 61.70; S9.11)
(b) Vol. **146**, No. 1, p. 295—301 (1988).

J. Prakash and F. Fischer
TSD in Pressed Pellets of NaCl:HCO_2^-
(77.30; 61.70; S9.11)
(b) Vol. **146**, No. 1, p. K71—K73 (1988).

K. Z. Botros, S. M. Saleh, A. H. G. El-Dhaher, and S. Hajaig
*On the Visibility of Edge Fringes in Strong Beam Images of Intrinsic and Extrinsic Stacking Faults in F.C.C. Structures
(61.14 and 61.70; S1.2)
(a) Vol. **106**, No. 1, p. 49—55 (1988).

R. Sommerfeldt, L. Holtzmann, E. Krätzig, and B. C. Grabmaier
*Influence of Mg Doping and Composition on the Light-Induced Charge Transport in $LiNbO_3$
(61.70; 72.40; 78.20; S11.1)
(a) Vol. **106**, No. 1, p. 89—98 (1988).

G.-H. Wang, M.-K. Teng, D.-X. Shen, C.-Y. Yi, Y.-Y. Zhou, Y.-Y. Lu, H.-W. Wang, Y.-Z. Zhu, and L. Dou
*Positron Annihilation Lifetime and Doppler Broadening Studies of Electron-Irradiated Polypropylene
(61.70 and 78.70; S12)
(a) Vol. **106**, No. 1, p. K1—K6 (1988).

S. V. Bulyarskii and V. P. Oleinikov
*Thermodynamics of Defect Formation and Defect Interaction in Compound Semiconductors
(61.70)
(b) Vol. **146**, No. 2, p. 439—447 (1988).

L. N. Kristoforov and A. V. Zolotaryuk
*Dynamics of Ionic and Bonding Defects in Quasi-One-Dimensional Hydrogen-Bonded Chains
(61.70 and 63.90; 66.90; 71.45; S12.1)
(b) Vol. **146**, No. 2, p. 487—501 (1988).

S. B. S. Sastry and G. Muralidharan
Thermoluminescence and Fluorescence Studies on $RbBr:Gd^{2+}$
(78.60; 61.70; S9.11)
(b) Vol. **146**, No. 2, p. 727—731 (1988).

A. V. Andreeva, A. V. Bazhenov, N. A. Bulenkov, and A. A. Firsova
Structural and Photoelectrical Properties of a Third-Order Twin Boundary ($\Sigma = 27$) in Silicon
(61.55; 61.70; 72.20; S5.11)
(a) Vol. **106**, No. 2, p. 351—362 (1988).

V. G. Gorshkov, Yu. K. Danileiko, V. V. Osiko, A. V. Sidorin, N. V. Veselovskaya, Yu. V. Dankovskii, and B. L. Shklyar
*Mechanical Microstresses in Dislocation-Free Floating-Zone Silicon Monocrystals
(61.70; 62.20; S5.11)
(a) Vol. **106**, No. 2, p. 363—369 (1988).

U. R. Kevorkyan
*The Simulation of Inactivation Reactions of Point Defect Interactions with Impurities and Impurity-Defect Complexes
(61.70; S1.1)
(a) Vol. **106**, No. 2, p. 379—386 (1988).

H. M. Eissa
*The Threefold Dislocation Nodes in the Basal Plane of H. C. P. Single Crystals
(61.70)
(a) Vol. **106**, No. 2, p. 387—397 (1988).

YU. S. NECHAEV and K. H. G. EGZIABHER
On Gas Permeability, Diffusivity, and Characteristics of Near-Dislocation Segregation Regions in Metals
(66.30; 61.70; 64.75; S1.1; S1.2)
(a) Vol. **106**, No. 2, p. 399—406 (1988).

P. CZURRATIS, R. KROGGEL, V. SYNEČEK, M. SIMERSKA, and H. LÖFFLER
*Dislocation Loops in Al–Zn and Al–Zn–Mg Alloys
(61.70; 61.55; S1.2; S4)
(a) Vol. **106**, No. 2, p. 407—418 (1988).

G. DLUBEK, A. DLUBEK, R. KRAUSE, O. BRÜMMER, K. FRIEDLAND, and R. RENTZSCH
*Point Defects in GaAs Studied by Correlated Positron Lifetime, Optical, and Electrical Measurements. I. Native Point Defects and Their Complexes in As-Grown GaAs
(61.70; 72.80; 78.55; 78.70; S7.12)
(a) Vol. **106**, No. 2, p. 419—432 (1988).

F. W. SCHAPINK and F. D. TICHELAAR
*Grain Boundary Structures and Associated Dislocations in Cubic Ordered Alloys
(61.70)
(a) Vol. **106**, No. 2, p. 433—440 (1988).

J. D. STEPHENSON
*Quasi-Realtime Observation of Partial Dislocation Growth in $(0\bar{1}.0)$ Synthetic Quartz near the α–β Phase Transition Temperature (T_c) Using Electric Potential Switching
(61.70; 64.60; S10.1)
(a) Vol. **106**, No. 2, p. 441—449 (1988).

J. BĄK-MISIUK, H. G. BRÜHL, W. PASZKOWICZ, and U. PIETSCH
Temperature Dependence of the Lattice Constant in Doped and Nonstoichiometry GaAs, $GaAs_{1-x}P_x$, and GaP
(65.40; 61.70; S7.11; S7.12)
(a) Vol. **106**, No. 2, p. 451—457 (1988).

W. SIENICKI
Investigation of Tungsten Disulphide and Diselenide Intercalated by Thallium
(72.80; 61.70; S1.63)
(a) Vol. **106**, No. 2, p. K157—K159 (1988).

DI-HUA DING, ZICHENG DENG, and SHUHONG LI
*The Elastic Fields of Disclinations in a Planar Phase Boundary
(61.70)
(b) Vol. **147**, No. 1, p. 51—67 (1988).

O. E. YANOVITSKII and E. S. KRYACHKO
*Model for Orientational Defects in Quasi-One-Dimensional Ice Crystals
(61.50 and 61.70)
(b) Vol. **147**, No. 1, p. 69—81 (1988).

K. EFTAXIAS, C. VAROTSOS, and V. HADJICONTIS
*Cation Vacancy Migration Entropy in Alkali Halides
(61.70; S9.11)
(b) Vol. **147**, No. 1, p. 83—88 (1988).

D. T. Britton and P. C. Rice-Evans
*Fast Fourier Transform Deconvolution of Positron Annihilation Doppler-Broadened Lineshapes
(61.70 and 78.70; S1.2; S5)
(b) Vol. **147**, No. 1, p. 89—96 (1988).

A. V. Egranov and A. I. Nepomnyachikh
*Excitonic Processes and Thermoluminescence in NaCl Crystals Doped with Mg or Mn
(61.70 and 71.35; 78.60; S9.11)
(b) Vol. **147**, No. 1, p. 267—272 (1988).

M. Kumeda, M. Kiwaki, and T. Shimizu
*Temperature Dependence of the E′ Center Creation in Silica Glasses
(61.70; 76.30; S10.1)
(b) Vol. **147**, No. 1, p. K1—K4 (1988).

D. M. Zayachuk, P. M. Starik, and V. A. Shenderovskii
*The Effect of Nonstoichiometric Tellurium on the Properties of Lead and Tin Tellurides
(61.70; 72.20; 75.20; S8.13)
(a) Vol. **107**, No. 1, p. 95—99 (1988).

A. Henry, K. Saminadayar, J. L. Pautrat, and N. Magnea
*The Kinetics of the Formation of the Thermal Donors in Silicon. Effect of Various Parameters
(61.70; 72.80; 78.30; S5.11)
(a) Vol. **107**, No. 1, p. 101—110 (1988).

S. Ohtaki and I. Hashimoto
*Numerical Evaluation for Stage II_A Recovery in Cold-Worked Aluminium
(61.70; 62.20; 72.15; S4)
(a) Vol. **107**, No. 1, p. 123—130 (1988).

V. E. Borisenko, D. I. Zarovskii, and V. V. Tokarev
*Influence of Argon Implantation on the Formation of Platinum Silicides
(61.70 and 68.55; S1.61)
(a) Vol. **107**, No. 1, p. K33—K35 (1988).

J. Horak, Z. Starý, and J. Klikorka
*Relations between Structure, Bonding, and Nature of Point Defects in Layered Crystals of Tetradymite Structure
(61.50 and 61.70; 72.80; S8; S8.15)
(b) Vol. **147**, No. 2, p. 501—510 (1988).

C. Ruiz-Mejía
About the F Center in Cesium Halides
(71.55; 61.70; S9.11)
(b) Vol. **147**, No. 2, p. 593—600 (1988).

A. B. Vassilikou-Dova and G. Lehmann
*EPR of V^{4+} in Apophyllite
(61.70 and 76.30; S11)
(b) Vol. **147**, No. 2, p. 691—697 (1988).

H. BENDER and J. VANHELLEMONT
*Rod-Like Defects in Silicon: Coesite or Hexagonal Silicon?
(61.70; S5.11)
(a) Vol. **107**, No. 2, p. 455—467 (1988).

W. COENE, A. F. DE JONG, D. VAN DYCK, G. VAN TENDELOO, and J. VAN LANDUYT
Digital Image Processing for High Resolution Electron Microscopy
(61.16; 61.70)
(a) Vol. **107**, No. 2, p. 521—530 (1988).

P. DELAVIGNETTE, TH. KARAKOSTAS, G. NOUET, and F. W. SCHAPINK
*Grain Boundary Orientation Determination in Transmission Electron Microscopy
(61.70; 61.16)
(a) Vol. **107**, No. 2, p. 551—558 (1988).

N. FRANGIS and C. MANOLIKAS
*Composition Faults in $Zn_mIn_2S_{3+m}$ Crystals
(61.70; S8.16)
(a) Vol. **107**, No. 2, p. 589—595 (1988).

A. LEFEBVRE and G. VANDERSCHAEVE
*The Origin of Microtwinning in Room Temperature Indented Undoped and n-Doped GaAs Crystals
(61.70 and 62.20; S7.12)
(a) Vol. **107**, No. 2, p. 647—653 (1988).

E. RUEDL, G. VALDRÈ, P. DELAVIGNETTE, and U. VALDRÈ
*Structural and Analytical Electron Microscopy of Dislocation Loops and Precipitates Formed during Ageing of an α-Bombarded Mn–Cr Austenitic Steel
(61.55 and 61.70; 61.16; S1.2)
(a) Vol. **107**, No. 2, p. 745—758 (1988).

A. D. STALIOS, P. DELAVIGNETTE, and R. DE BATIST
*Transmission Electron Microscopy Observation of Crystal Defects in a Natural Diopside Mineral
(61.16 and 61.70; S11.2)
(a) Vol. **107**, No. 2, p. 759—767 (1988).

G. TORRES-VILLASENOR and G. VAN TENDELOO
*Guinier-Preston Zones in the α-Phase of Superplastic Zn–Al Alloy
(61.70; 62.20; S1.2; S4)
(a) Vol. **107**, No. 2, p. 785—790 (1988).

H. R. WENK, S. J. SHAFFER, and G. VAN TENDELOO
*Planar Defects in Low Temperature Quartz
(61.70 and 64.80; S10)
(a) Vol. **107**, No. 2, p. 799—805 (1988).

H. W. ZANDBERGEN and G. THOMAS
*HREM Investigations of Defects in Sintered $YBa_2Cu_3O_{7-\delta}$
(61.16 and 61.70; 74.70; S10.15)
(a) Vol. **107**, No. 2, p. 825—843 (1988).

M. Reiche, J. Reichel, and W. Nitzsche
*Correlations between Thermal Donor Formation, Rod-Like Defect Formation, and Oxygen Reduction during Low-Temperature Annealing of CZ-Grown Silicon
(61.70; S5.11)
(a) Vol. **107**, No. 2, p. 851—865 (1988).

V. V. Kochelab, V. B. Molodkin, S. I. Olikhovskii, and M. E. Osinovskii
The Thickness Dependence of the X-Ray Diffuse Scattering Intensity for Crystals with Microdefects at Laue-Case Diffraction
(61.10; 61.70; S5.12)
(a) Vol. **108**, No. 1, p. 67—79 (1988).

M. Hashimoto, H. Saka, T. Imura, S. Jitsukawa, and K. Suzuki
*Effect of Alloying on the Void Swelling in Cu-Base and Ag-Base Alloys
(61.70; 64.75; S1.2; S1.3)
(a) Vol. **108**, No. 1, p. 163—170 (1988).

P. Czurratis, E. Korngiebel, G. Jacobi, and H. Löffler
*The Effect of Additions of Mn and Cr and the Applied Solidification Technique on the Structure of an Al–2.2at%Zn–1.4at%Mg Alloy
(61.70; 61.55; 62.20; S4)
(a) Vol. **108**, No. 1, p. 171—182 (1988).

V. V. Litvinov, G. V. Palchik, and V. I. Urenev
Extrinsic Self-Trapping and Negative *U* of Oxygen-Related Thermal Donors in Germanium
(71.55; 61.70; S5.12)
(a) Vol. **108**, No. 1, p. 311—321 (1988).

V. P. Goltsev and S. I. Zhukova
*The Loss of Strength of Al-Alloy after Sb^+ Implantation
(61.70 and 62.20; S4)
(a) Vol. **108**, No. 1, p. K1—K4 (1988).

N. Q. Khan, M. Fried, G. Battistig, Z. Laczik, T. Lohner, E. Jároli, V. Schiller, and J. Gyulai
*Ellipsometric and Ion Backscattering Measurements of the Properties of Silicon-On-Insulator Structure Formed by Thermally Activated Redistribution of High-Dose Ion Implanted Nitrogen
(61.70 and 68.48; 78.65; S5.11)
(a) Vol. **108**, No. 1, p. K35—K40 (1988).

J. Piqueras, F. Domínguez-Adame, and B. Mendez
*SEM-CL of Reaction Bonded SiC
(61.70 and 78.60; S1.61)
(a) Vol. **108**, No. 1, p. K81—K84 (1988).

T. S. Shi, Y. G. Zhao, P. X. Cai, H. F. Gu, and L. M. Xie
*IR Absorption Peaks and Oxygen Vacancies in $YBa_2Cu_3O_x$
(61.70 and 78.30; 78.50; S10)
(b) Vol. **148**, No. 2, p. 715—721 (1988).

V. V. Frolov and B. N. Mukashev
*On the Origin of the Hydrogen-Related Complex Corresponding to the 2223 cm^{-1} (at 10 K) Infrared Band in c-Si:H
(61.70; 78.30; S5.11)
(b) Vol. **148**, No. 2, p. K105—K107 (1988).

O. Eibl, P. Pongratz, P. Skalicky, and H. Schmelz
*Dislocations in $BaTiO_3$ Ceramics
(61.70; S11.1)
(a) Vol. **108**, No. 2, p. 495—502 (1988).

E. Susi, A. Poggi, G. Lulli, C. Monteverdi, and M. Merli
*Lifetime and Crystal Order in Annealed CZ Silicon
(61.70; 72.20; S5.11)
(a) Vol. **108**, No. 2, p. 503—508 (1988).

Y. Kikuchi, K. Watanabe, S. Naitoh, K. Hiratsuka, and H. Yamaguchi
*HREM Image of a Vacancy in Copper
(61.70; 61.16; S1.2)
(a) Vol. **108**, No. 2, p. 509—517 (1988).

Ch. Schnittler and G. Teichmann
*A Kinetic Investigation of Cluster Formation in Semiconductors at Elevated Temperatures
(61.70; 64.75; S5.11)
(a) Vol. **108**, No. 2, p. 519—528 (1988).

P. J. Ratcliffe and R. A. Collins
*Thermal Oxidation of Ion Implanted Copper
(61.70; 73.40; S1.2)
(a) Vol. **108**, No. 2, p. 537—543 (1988).

X. L. Yuan and S. W. S. McKeever
*Impurity Clustering Effects in Magnesium-Doped Lithium Fluoride
(61.70; 78.50; 78.60; S9.11)
(a) Vol. **108**, No. 2, p. 545—551 (1988).

U. Jendrich and P. Haasen
*Internal Friction Due to Kink Motion and Kink Pair Formation in Intrinsic and n-Doped Germanium
(61.70 and 62.40; S5.12)
(a) Vol. **108**, No. 2, p. 553—568 (1988).

S. E. Urreta de Pereyra, H. Bertorello, and A. A. Ghilarducci de Salva
Precipitation and Grain Boundary Internal Friction Peaks in Al–Mg–Si
(62.40; 61.70; S4
(a) Vol. **108**, No. 2, p. 577—586 (1988).

V. B. Shikin and N. I. Shikina
*Charged Dislocations in Semiconductors
(61.70 and 71.55; S5.11; S5.12)
(a) Vol. **108**, No. 2, p. 669—681 (1988).

J. M. Neto and T. Abritta
Luminescence and Site Occupation of Cr^{3+} in Sodium–β″-Alumina
(78.55; 61.70; S10)
(a) Vol. **108**, No. 2, p. 795—802 (1988).

B. NARASHIMHA, K. V. RAO, and R. N. P. CHAUDHURY
*Optical Absorption of Quenched and γ-Ray Irradiated BaF_2 Single Crystals Later Subjected to High Electric Fields or Laser Excitation
(61.70 and 78.50; S9.12)
(a) Vol. **108**, No. 2, p. K97—K100 (1988).

P. S. GLADKOV and K. B. OZANYAN
Photoconductivity in Bulk and Epitaxial GaAs:V
(71.55 and 72.40; 61.70; S7.12)
(a) Vol. **108**, No. 2, p. K125—K130 (1988).

H. KRAUSE, H.-P. BÄR, and H. SCHULZE
*Correlations between Current–Voltage Characteristics and Visible Defect Channels in Thin SiO_2 Films after Field Stress
(61.70 and 72.20; S10.1)
(a) Vol. **108**, No. 2, p. K131—K133 (1988).

H.-P. BÄR, H. KRAUSE, and H. SCHULZE
*On the Nature of Structural Defects in SiO_2 Films Visible with the Aid of the Liquid Crystal Technique
(61.70 and 72.20; S10.1)
(a) Vol. **108**, No. 2, p. K135—K137 (1988).

N. KRISTIANPOLLER, B. TRIEMAN, R. CHEN, and Y. KIRSH
*The Application of Thermally Stimulated Processes to the Study of Defects in Perovskite Type Fluorides
(61.70; 72.20; 78.60; S9.16)
(b) Vol. **149**, No. 1, p. 45—54 (1988).

S. MURAMATSU
Electron-Lattice Interaction Parameters Determined from Absorption and MCD Spectra of the F Centers in CsCl and CsBr
(71.55 and 78.50; 61.70; S9.11; S9.12)
(b) Vol. **149**, No. 1, p. K19—K23 (1988).

K. INABE, S. NAKAMURA, and N. TAKEUCHI
*Optical Bleaching of BaFCl Crystals X-Irradiated at Room Temperature
(61.70 and 78.50; S9.15)
(b) Vol. **149**, No. 1, p. K67—K71 (1988).

J. HÄRTWIG and V. LERCHE
Anisotropic Deformation of a Crystal Plate and Its Analysis with X-Ray Diffraction Methods
(61.10; 61.70; 62.20; S10.1)
(a) Vol. **109**, No. 1, p. 79—91 (1988).

Z. STARÝ, L. BENEŠ, and J. HORÁK
Interaction of Anti-Site Defects with Ge Impurity Atoms in Sb_2Te_3 Single Crystals
(61.60; 61.70; S8)
(a) Vol. **109**, No. 1, p. 93—99 (1988).

HONG ZHANG, XIAO-GAN WANG, and YAO-XIAN FU
*Study of the Movement of Oxygen Vacancies in the Orthorhombic Phase of $YBa_2Cu_3O_{7-x}$ by Positron Doppler Broadening Spectroscopy
(61.70; 74.70; 78.70; S10.15)
(a) Vol. **109**, No. 1, p. 135—139 (1988).

J. OLFE and H. NEUHÄUSER
*Dislocation Groups, Multipoles, and Friction Stresses in α-CuZn Alloys
(61.70; 62.20; S1.2)
(a) Vol. **109**, No. 1, p. 149—160 (1988).

H. BLUMTRITT, G. N. PANIN, E. B. YAKIMOV, and J. HEYDENREICH
SEM-EBIC Investigations of Grain Boundaries in Cadmium Telluride
(61.14; 61.70; S8.13)
(a) Vol. **109**, No. 1, p. K3—K6 (1988).

A. E. LIGACHEV, O. S. KUZMIN, G. V. POTEMKIN, and V. P. YANOVSKII
*Structure and Properties of Implanted Steels
(61.70 and 62.20; S1.1)
(a) Vol. **109**, No. 1, p. K11—K14 (1988).

B. DECAMPS and M. CONDAT
*$a/2$ ⟨110⟩ Dislocations at the γ/γ' Interface in a Ni-Based Superalloy
(61.70; S1.2)
(a) Vol. **109**, No. 1, p. K15—K18 (1988).

E. RADZHABOV
*Reaction of OH^- and F Centers in Sodium Halides
(61.70 and 71.55; 78.50; S9.11)
(b) Vol. **149**, No. 2, p. 441—446 (1988).

W. A. OATES and K. MIKA
*Application of the Local Environment Model to the Thermodynamics of Large Hard Core Interstitial Solutes in Binary Alloys
(61.70 and 65.50)
(b) Vol. **149**, No. 2, p. 457—464 (1988).

M. WAUTELET, E. D. GEHAIN, and P. QUENON
*Modelling the Role of Electronically Excited States in Laser-Assisted Phenomena
(61.70; 79.20; S5.11)
(b) Vol. **149**, No. 2, p. 465—475 (1988).

T. SOMA, N. ATARASHI, and H.-MATSUO KAGAYA
*Stable Configuration of an Interstitial Atom with Relaxation Effect in Si and Ge
(61.70; S5.11; S5.12)
(b) Vol. **149**, No. 2, p. 477—482 (1988).

P. C. HÄGELE, S. WOLF, and C. SCHMID
*Mass Defects in Polyethylene and the Role of Mass Weighting in Phonon Green's Function Theory
(61.70 and 63.20; S12)
(b) Vol. **149**, No. 2, p. 521—532 (1988).

S. NARA
A Note on the Icosahedral Morphology of Ultrafine Metal Particles
(68.55; 61.70; 73.20; S1.3)
(b) Vol. **149**, No. 2, p. 555—564 (1988).

A. Hamaídia and A. Hachimi
Colour Centres in Alkali Metal Doped CaF_2
(78.20 and 78.50; 61.70; 78.55; S9.12)
(b) Vol. **149**, No. 2, p. 711—724 (1988).

L. Bosi and M. Nimis
*An Analysis of the Decay Properties of the F and M Centers in LiF Crystals
(61.70; 78.50; S9.11)
(b) Vol. **149**, No. 2, p. K89—K94 (1988).

J. Tatarkiewicz and M. Stutzmann
*Hydrogen Vibrations in Semiconductors and Insulators
(61.70 and 63.20)
(b) Vol. **149**, No. 2, p. K95—K99 (1988).

K. Inabe, S. Nakamura, and N. Takeuchi
*Optical Bleaching of F-Centers in BaFCl Crystals after X-Irradiation at Room Temperature
(61.70; 78.50; S9.15)
(b) Vol. **149**, No. 2, p. K101—K104 (1988).

E. M. Gololobov, A. S. Gurevich, E. L. Mager, Z. V. Mezhevich, N. A. Prytkova, Zh. M. Tomilo, D. M. Turtsevich, and N. M. Shimanskaya
*Orthorhombic Lattice Distortion of $YBa_2Cu_3O_{7-y}$ and the Superconducting Transition Temperature
(61.70 and 74.70; S10.15)
(b) Vol. **149**, No. 2, p. K145—K148 (1988).

S. Fujita, K. Maeda, and S. Hyodo
*Radiation Enhanced Dislocation Glide Effect on Indentation-Induced Fracture of GaAs Single Crystals
(61.70 and 62.20; S7.12)
(a) Vol. **109**, No. 2, p. 383—393 (1988).

V. E. Borisenko and S. G. Yudin
*Implanted Impurity Behaviour in Silicon Subjected to Transient Heating
(61.70 and 66.30; S5.11)
(a) Vol. **109**, No. 2, p. 395—402 (1988).

J. Gastaldi, C. Jourdan, G. Grange, and C. L. Bauer
*"In Situ" Determination of Grain Boundary Migration Rates by Synchrotron White Beam X-Ray Topography
(61.70; 61.10; S4)
(a) Vol. **109**, No. 2, p. 403—411 (1988).

V. D. Rodríguez and C. Ortiz
*Thermally Stimulated Relaxation in NaF X-Irradiated at Room Temperature
(61.70 and 78.60; S9.11)
(a) Vol. **109**, No. 2, p. 413—419 (1988).

J. R. Gavarri, R. Mokrani, G. Vacquier, and C. Boulesteix
*Relations between Anisotropic Defects, Structural Evolution and van der Waals Bonding in 2H-$NbSe_2$
(61.70 and 65.70; 62.20; S8)
(a) Vol. **109**, No. 2, p. 445—454 (1988).

J. Krynicki and H. Rzewuski
*Deep Electronic Levels in Carbon-Implanted Silicon
(61.70 and 71.55; S5.11)
(a) Vol. **109**, No. 2, p. 531—536 (1988).

H. Scibior, I. Brylowska, P. Mazurek, and M. Subotowicz
Influence of Ion Implantation on the Luminescence of GaAs
(78.55 and 78.60; 61.70; S7.12)
(a) Vol. **109**, No. 2, p. 597—601 (1988).

N. T. Bagraev and I. S. Polovtsev
*Reactions for Iron Centers in Silicon under Optical Pumping
(61.70 and 71.55; 72.40; 76.30; 76.60; S5.11)
(b) Vol. **150**, No. 1, p. 19—24 (1988).

A. de Andrés
Applications of a Point Polarizable Ion Model to Suzuki Phase Precipitates in Alkali Halides
(63.20; 61.70; 78.30; S9.11)
(b) Vol. **150**, No. 1, p. 31—37 (1988).

K. Löbe and R. Boyn
*Sm^{3+} in ZnS on Lattice Sites and Wurtzite-Type Interstitial Sites?
(61.70 and 71.55; 71.70; 78.55; S8.11)
(b) Vol. **150**, No. 1, p. 191—200 (1988).

S. B. S. Sastry and G. Muralidharan
Thermoluminescence and Other Optical Studies on γ-Irradiated RbCl:Gd^{2+}
(78.60; 61.70; S9.11)
(b) Vol. **150**, No. 1, p. 315—322 (1988).

A. Hamaïdia
Colour Centres in Alkali Metal Doped SrF_2
(78.20 and 78.50; 61.70; 76.30; S9.12)
(b) Vol. **150**, No. 1, p. 323—336 (1988).

A. P. Filippov
*The Influence of Impurities on the Dislocation Splitting in Semiconductors
(61.70)
(a) Vol. **110**, No. 1, p. 83—88 (1988).

B. I. Smirnov and I. A. Smirnov
*Defect Structure and Physico-Mechanical Properties of SmS Single Crystals
(61.70; 62.20; 72.20; S8)
(a) Vol. **110**, No. 1, p. 89—96 (1988).

N. I. Tarbaev, J. Schreiber, and G. A. Shepelskii
*Physical Properties of $A^{II}B^{VI}$ Semiconductor Crystals after Plastic Deformation at Low Temperature
(61.70 and 71.55; 72.40; 78.50; 78.55; S8.1)
(a) Vol. **110**, No. 1, p. 97—106 (1988).

K. Meinel, M. Klaua, and H. Bethge
*On Twin and Stacking Fault Formation during the Epitaxial Film Growth of F.C.C. Materials on (111) Substrates
(61.70 and 68.55; S1.3)
(a) Vol. **110**, No. 1, p. 189—196 (1988).

S. S. Jiang, Y. Qiu and G. S. Green
The Simulation of Modulated I_3 Fringes in an X-Ray Section Topograph of a Stacking Fault
(61.10; 61.70; S5)
(a) Vol. **110**, No. 2, p. 323—330 (1988).

E. G. Doni and G. L. Bleris
*Study of Special Triple Junctions and Faceted Boundaries by Means of the CSL Model
(61.16 and 61.70)
(a) Vol. **110**, No. 2, p. 383—395 (1988).

V. A. Ganshin, Yu. N. Korkishko, and T. V. Morozova
*Properties of Proton Exchanged Optical Waveguiding Layers in $LiNbO_3$ and $LiTaO_3$
(61.70 and 78.20; 61.10; S11.1)
(a) Vol. **110**, No. 2, p. 397—402 (1988).

A. Serra and N. de Diego
*Characterization of Defects in Deformed Titanium
(61.70; 78.70; S1.2)
(a) Vol. **110**, No. 2, p. 409—414 (1988).

R. B. McLellan and M. L. Wasz
*Carbon–Vacancy Interactions in B.C.C. Iron
(61.70; S1.1)
(a) Vol. **110**, No. 2, p. 421—427 (1988).

J. Pellegrino and J. M. Galligan
The Photoplastic Effect in II–VI Compounds. Mercury Cadmium Telluride
(72.20; 61.70; S8.15)
(a) Vol. **110**, No. 2, p. 565—573 (1988).

A. Stesmans and G. de Vos
MgO Powder Containing Low Concentrations of Isotopically Pure $^{52}Cr^{3+}$ Ions. Its Applicaion as ESR Marker
(76.30; 61.70; S10.1)
(a) Vol. **110**, No. 2, p. 615—629 (1988).

H. Ohyama and K. Nemoto
*Recovery Mechanism of Lattice Defects Formed in the Collector Region for Electron-Irradiated npn Si Transistors
(61.70 and 73.40; 85; S5.11)
(a) Vol. **110**, No. 2, p. 677—686 (1988).

O. V. Aleksandrov, V. V. Kozlovskii, V. V. Popov, and B. E. Samorukov
*Diffusion of Impurities from Implanted Silicon Layers by Rapid Thermal Annealing
(61.70 and 66.30; S5.11)
(a) Vol. **110**, No. 2, p. K61—K65 (1988).

61.80. Radiation damage and other irradiation effects

(for techniques of structure determination, see 61.10 to 61.16; for electron and ion impact phenomena, see 79.20)

V. S. MALYSHEVSKII
*Interaction of Channeled Electrons with Lattice Excitations **(Review Article)**
(61.80; 61.70; 63.20; 78.70)
(b) Vol. **147**, No. 1, p. 11—44 (1988).

V. M. KOROL
* Sodium-Ion Implantation into Silicon **(Review Article)**
(61.80; 66.30; 72.80; 73.40; S5.11)
(a) Vol. **110**, No. 1, p. 9—34 (1988).

S. SHINDO and H. MINOWA
*Theory of Higher-Order Corrections to the Electronic Stopping Power of Solids
(61.80)
(b) Vol. **145**, No. 1, p. 89—101 (1988).

N. M. KABACHNIK, V. N. KONDRATEV, and O. V. CHUMANOVA
*The Impact-Parameter Dependence of Electron Stopping in Proton Collisions with Gas and Solid Atoms
(61.80; S4; S5.11)
(b) Vol. **145**, No. 1, p. 103—116 (1988).

R. URBAN, P. EHRHARDT, W. SCHILLING, H. R. SCHOBER, and H. LAUTER
*Erratum to: Investigation of the Resonant Vibration Modes of Radiation Induced Interstitial Atoms in Aluminium by Diffuse Inelastic Neutron Scattering
(61.80 and 63.20; S4)
(b) Vol. **145**, No. 1, p. 351 (1988).

S. V. DYACHENKO, L. E. NAGLI, and F. V. PIROGOV
*Radiation Defect Production Processes in Activated KCl Crystals
(61.80; 78.70; S9.11)
(a) Vol. **105**, No. 1, p. 87—95 (1988).

L. A. KAZAKEVICH, V. I. KUZNETSOV, and P. F. LUGAKOV
*Clusters of Radiation Defects in Silicon with Dislocations
(61.80; S5.11)
(a) Vol. **105**, No. 1, p. 97—100 (1988).

I. J. JABR, J. D. MEYER, K. A. AL-SALEH, and N. S. SALEH
*Ion-Beam Mixing of Te/Au. Metastable Phase Formation
(61.80 and 64.75; 73.60; S1.3; S5)
(a) Vol. **105**, No. 1, p. 177—181 (1988).

G. BRAUER, W. HERTLE, and A. G. BALOGH
γ-Irradiation of Polyethylene Studied by Positron Annihilation
(61.40 and 78.70; 61.80; S12)
(a) Vol. **105**, No. 1, p. K7—K12 (1988).

V. N. Brudnyi and V. V. Peschev
*Electron Traps in n-GaAs Irradiated with High Electron Beam Fluxes at High Temperatures
(61.80 and 71.55; S7.12)
(a) Vol. **105**, No. 1, p. K57—K60 (1988).

Tran Thoai Duy Bao
Plasmon Excitation in Metal Surface Layers by Fast Electrons
(71.45 and 73.20; 61.80; S4)
(b) Vol. **145**, No. 2, p. 509—515 (1988).

A. E. Curzon and O. S. Rajora
A Study of NbS_xSe_{2-x} Using Electron Channeling Patterns
(61.60; 61.80; S8.15)
(a) Vol. **105**, No. 2, p. 335—343 (1988).

E. Sobeslavsky, H. U. Jäger, U. Kreissig, W. Skorupa, and K. Wollschläger
*Nitrogen Profile Modification in High Dose Implantation Synthesis of Silicon Nitride
(61.80; S1.62; S5.11)
(a) Vol. **105**, No. 2, p. 387—396 (1988).

D. Schalch, A. Scharmann, and R. Wolfrat
*Hydrogen-Ion-Implanted Silicon Nitride and Silicon Oxynitride Films
(61.80 and 78.65; S6)
(a) Vol. **105**, No. 2, p. K81—K86 (1988).

V. N. Brudnyi and I. V. Kamenskaya
*The Electrical Properties and Fermi Level Pinning in Proton-Irradiated GaSb
(61.80 and 72.80; S7.13)
(a) Vol. **105**, No. 2, p. K141—K144 (1988).

M. Dębowska, J. Ch. Abbé, and G. Duplâtre
*Positron Annihilation Studies in Zeolites
(61.80 and 78.70; S11)
(b) Vol. **146**, No. 1, p. 91—96 (1988).

R. Klabes, A. Thomas, G. Kluge, W. Beyer, R. Grötzschel, and P. Süptitz
*Ion-Beam Induced Silver Doping in the $Ag_2Se/GeSe_2$-Resist System
(61.80 and 68.55; 66.30; S8)
(a) Vol. **106**, No. 1, p. 57—65 (1988).

S. Dhanuskodi and N. Hariharan
*Radiation Damage of Bisglycine Cadmium Chloride Single Crystals. An EPR Study
(61.80; 76.30; S12)
(a) Vol. **106**, No. 1, p. 67—72 (1988).

C. Ascheron, H. A. Klose, W. Frentrup, and M. Griepentrog
*Gettering of Copper in Proton- and Helium-Bombarded Buried Regions of Gallium Phosphide
(61.80; S7.11)
(a) Vol. **106**, No. 1, p. 73—79 (1988).

G. Dlubek, C. Ascheron, R. Krause, H. Erhard, and D. Klimm
*Positron Study of Vacancy Defects in Proton and Neutron Irradiated GaP, InP, and Si
(61.80; 78.70; S5.11; S7.11)
(a) Vol. **106**, No. 1, p. 81—88 (1988).

Yu. Dmitriev, V. Koshkin, and U. Ulmanis
*Defocusing of Atomic Knock-Ons in Crystals with Low Local Symmetry of Structural Elements
(61.80; 72.80; S8)
(a) Vol. **106**, No. 1, p. K7—K11 (1988).

E. M. Shpilveski and D. A. Gorbachevskii
*Argon Ion Effect on Solid Phase Reaction in Copper–Tin, Copper–Antimony Films
(61.80 and 68.48; 66.30; S1.3; S4)
(a) Vol. **106**, No. 1, p. K23—K26 (1988).

O. B. Alekseevskii, S. A. Vorobiev, and N. S. Kostyukov
*Radiation-Induced Conductivity of Dielectric Solids Irradiated by Intensive Fluxes of Fast Neutrons
(61.80 and 72.20)
(b) Vol. **146**, No. 2, p. K109—K113 (1988).

K. Havancsak, G. Szenes, V. A. Skuratov, Yun Dun Man, W. Malinowski, and V. A. Shchegolev
*233 MeV Ne Ion Irradiation Effects on the Mechanical Properties of Copper
(61.80; 62.20; S1.2)
(a) Vol. **106**, No. 2, p. 371—378 (1988).

I. V. Marchishin, V. N. Ovsyuk, and S. B. Sevastianov
Investigation of Radiation Defects in Hyperabrupt p—n Junctions
(71.55 and 73.40; 61.80)
(a) Vol. **106**, No. 2, p. 557—568 (1988).

A. Unkroth, W. Andrä, and G. Götz
Composition Dependence of Magnetization of Ion Implanted Cobalt Layers
(75.70; 61.80; S1.2; S1.61; S1.62)
(a) Vol. **106**, No. 2, p. 589—594 (1988).

J. Czerbniak, M. Moneta, B. Pawłowski, and M. Piwowarski
*The Influence of Thermal Vibrations on RBS Proton Spectra Obtained from Subsurface Regions of Si Crystals
(61.80 and 68.30; S5.11)
(b) Vol. **147**, No. 1, p. 123—126 (1988).

G. Dlubek, A. Dlubek, R. Krause, and O. Brümmer
*Point Defects in GaAs Studied by Correlated Positron Lifetime, Optical, and Electrical Measurements. II. Point Defects in GaAs Irradiated with Fast Neutrons
(61.80; 78.50; 78.55; 78.70; S7.12)
(a) Vol. **107**, No. 1, p. 111—121 (1988).

B. E. Mdivanyan and M. Sh. Shikhsaidov
Photostimulated Enhancement of Dislocation Glide in Gallium Arsenide Crystals
(62.20; 61.80; S7.12)
(a) Vol. **107**, No. 1, p. 131—140 (1988).

A. J. Abu El-Haija, K. A. Al-Saleh, N. A. Halim, J. M. Khalifeh, and N. S. Saleh
*Ion Beam Induced Mixing of Cu/Si System Using Electrical Resistivity and RBS Measurements
(61.80 and 73.60; S1.2; S5.11)
(a) Vol. **107**, No. 1, p. 253—260 (1988).

G. M. Gusakov, A. A. Komarnitskii, and A. S. Em
*Dynamics of Optical Parameters of Silicon and Germanium Single Crytals during Two-Pulsed Nanosecond Laser Irradiation
(61.80 and 68.55; 64.70; 78.20; S5.11; S5.12)
(a) Vol. **107**, No. 1, p. 261—271 (1988).

H. Ohyama and K. Nemoto
*Effect of Lattice Defects in the Collector Region of npn Si Transistors on the Degradation of h_{FE}
(61.80 and 73.40; 85; S5.11)
(a) Vol. **107**, No. 1, p. 429—439 (1988).

E. Jároli, B. Pécz, R. Veresegyházy, F. Pászti, T. Lohner, M. Fried, I. Mojzes, and J. Gyulai
*Effect of Ion Beam Mixing on the Evolution of Arsenic from the Au–GaAs System
(61.80 and 68.48; S1.3; S7.12)
(a) Vol. **107**, No. 1, p. K15—K17 (1988).

Guoan Cheng, Hengde Li, and Baixin Liu
*Enhanced Adhesion in Metal/Oxide Pairs by Ion Irradiation
(61.80 and 68.48; S1.1; S1.2; S10.1)
(a) Vol. **107**, No. 1, p. K29—K32 (1988).

Yu. M. Annenkov, V. F. Stolyarenko, and T. S. Frangulian
*Influence of Li Ions on the Accumulation of Radiation Defects in KBr under Pulsed Electron Irradiation
(61.80; S9.11)
(b) Vol. **147**, No. 2, p. K109—K113 (1988).

D. B. Tran Thoai and E. Zeitler
*Inelastic Scattering of Fast Electrons by Small Metal Particles
(61.80 and 73.20; 71.45; S4)
(a) Vol. **107**, No. 2, p. 791—797 (1988).

E. Zlateva and I. Gutzow
*Radiation-Induced Volume Changes in Thermoluminescent-Active Materials
(61.80 and 78.60; S9.12)
(a) Vol. **107**, No. 2, p. K99—K103 (1988).

P. I. Gaiduk, F. F. Komarov, N. L. Prokhorenko, and V. S. Solovev
*Structure of Silicon after B^+ and P^+ Implantation and Rapid Thermal Annealing
(61.80; S5.11)
(a) Vol. **107**, No. 2, p. K105—K107 (1988).

V. I. Kuznetsov, P. F. Lugakov, and V. V. Lukyanitsa
*Influence of Irradiation Temperature on Radiation Defect Accumulation in Silicon
(61.80; 71.55; S5.11)
(a) Vol. **108**, No. 1, p. 143—148 (1988).

A. SUBBARAYAN, C. BALASUBRAMANIAN, and SA. K. NARAYANDASS
*Laser-Induced Damage Study of Vacuum Deposited Perylene Thin Films
(61.80 and 79.20; S12)
(a) Vol. **108**, No. 1, p. K31—K34 (1988).

I. ŠKORVÁNEK, B. IDZIKOWSKI, A. ZENTKO, and E. MOSINIEWICZ-SZABLEWSKA
Influence of Neutron Irradiation on the Magnetic Properties of FeNiCrMoSiB Amorphous Alloys
(75.50; 61.80; S1.1; S1.2)
(a) Vol. **108**, No. 2, p. 747—751 (1988).

D. LAPRAZ, P. IACCONI, Y. SAYADI, P. KELLER, J. BARTHE, and G. PORTAL
Some Thermoluminescence Properties of an α-Al_2O_3 Sample. Sensitization Effects
(78.60; 61.80; S10)
(a) Vol. **108**, No. 2, p. 783—794 (1988).

W. WESCH, E. WENDLER, N. P. KEKELIDSE, and V. V. TARBA
*Investigation of the Amorphization Process in GaAs and GaP by Means of Optical Spectroscopy
(61.80; 78.50; S7.11; S7.12)
(a) Vol. **108**, No. 2, p. K93—K96 (1988).

S. OHZAWA, H. NITTA, and Y. H. OHTSUKI
*Background Spectrum Enhancement in Channeling Radiation
(61.80)
(b) Vol. **149**. No. 1, p. 41—43 (1988).

V. N. BAIER, V. M. KATKOV, and V. M. STRAKHOVENKO
*Incoherent Radiation and Pair Creation in Crystals
(61.80 and 78.70)
(b) Vol. **149**, No. 1, p. 403—415 (1988).

V. A. KHRAMTSOV, V. N. LOMASOV, YA. YA. PILKEVICH, M. P. VLASENKO, and L. S. VLASENKO
*Accumulation and Transformation of Radiation Defects in Silicon under Different Doses and Intensities of Electron Irradiation
(61.80; 76.30; S5.11)
(a) Vol. **109**, No. 1, p. 127—134 (1988).

G. A. KACHURIN, I. E. TYSHCHENKO, E. WIESER, and CH. WEISE
*Redistribution of Boron in Silicon by High-Temperature Irradiation with Heavy Ions
(61.80 and 66.30; S5.11)
(a) Vol. **109**, No. 1, p. 141—148 (1988).

I. A. BACHILO, F. F. KOMAROV, A. P. NOVIKOV, and S. A. PETROV
*Structural Analysis of Silicon Doped with High Doses of C^+ Ions
(61.80 and 68.55; S1.61; S5.11)
(a) Vol. **109**, No. 1, p. 231—238 (1988).

C. A. LONDOS and J. GRAMMATIKAKIS
*Notes on the Carbon-Associated Deep Level Complex in Irradiated Silicon
(61.80; 71.55; S5.11)
(a) Vol. **109**, No. 2, p. 421—426 (1988).

B. A. Budkevich, I. A. Ges, S. P. Zhvavyi, G. D. Ivlev, and V. A. Pilipovich
*Modification of Electrochromic WO_3 Films by Pulsed Ultraviolet Laser Irradiation
(61.80 and 73.60; 68.60; S10.1)
(a) Vol. **110**, No. 1, p. 155—164 (1988).

H. Ohyama and K. Nemoto
*Process of Defect Formation in the Collector Region of an Electron-Irradiated npn Si Transistor
(61.80 and 73.40; 85; S5.11)
(a) Vol. **110**, No. 1, p. 301—308 (1988).

A. Galeckas, V. Netiksis, M. Petrauskas, and J. Vaitkus
Reflectivity and Dynamic Gratings in Implanted Si Induced by Picosecond Laser Pulses
(78.20; 61.80; S5.11)
(b) Vol. **150**, No. 2, p. 743—748 (1988).

P. F. Lugakov and T. A. Lukashevich
*Peculiarities of Radiation Defect Accumulation under High-Flux γ-Ray and Electron Irradiation of Silicon
(61.80; 72.80; S5.11)
(a) Vol. **110**, No. 2, p. 403—408 (1988).

D. V. Morgan, H. Thomas, W. T. Anderson, P. Thompson, A. Christou, and D. J. Diskett
*High Temperature Metallisation for GaAs Device Processing
(61.80 and 73.40; S1.61; S4; S7.12)
(a) Vol. **110**, No. 2, p. 531—536 (1988).

61.90. Other topics in structure of liquids and solids

J. Q. You and T. B. Hu
A Global Cut-and-Project Method to Construct Generalized Fibonacci Lattices and Quasilattices
(61.14; 61.90)
(b) Vol. **147**, No. 2, p. 471—484 (1988).

62. Mechanical and acoustical properties of condensed matter
(see also 61.70. Defects in crystals, 68.30. Surfaces and interfaces)

62.20. Mechanical properties of solids
(related to microscopic structure)

T. Ohgaku and N. Takeuchi
*The Relation of the Blaha Effect with Internal Friction for Alkali Halide Crystals
(62.20; 62.40; S9.11)
(a) Vol. **105**, No. 1, p. 153—159 (1988).

M. M. Aleksandrova, V. D. Blank, A. E. Golobokov, Yu. S. Konyaev, A. Yu. Zerr, and E. I. Estrin
*Amorphisation of Gallium Antimonide under the Conditions of Shear Deformation under Pressure
(61.50 and 62.20; S7.13)
(a) Vol. **105**, No. 1, p. K29—K32 (1988).

S. C. Goyal and A. K. Sharma
*Fourth-Order Elastic Constants of Fluorite Structure Solids
(62.20; S9.12)
(b) Vol. **145**, No. 2, p. 435—443 (1988).

M. H. Xu, S. Patu, and Z. G. Wang
*The Improvement in Fatigue Life of Pure Polycrystalline Nickel by Nitrogen Ion Implantation
(62.20; 61.70; S1.2)
(a) Vol. **105**, No. 2, p. 419—425 (1988).

J. P. Allemand, F. Fouquet, J. Perez et B. de Guillebon
*Evolution des propriétés mécaniques et micromécaniques d'un verre métallique Fe–Cr–C–P–Si avec la vitesse d'élaboration
(Evolution of the Mechanical and Micromechanical Properties as a Function of Quench Rate in a Metallic Fe–Cr–C–P–Si Glass)
(62.20; 61.40; 75.80; S1.1)
(a) Vol. **105**, No. 2, p. 427—439 (1988).

A. V. Golubkov, V. M. Egorov, T. S. Orlova, V. M. Sergeeva, B. I. Smirnov, and I. A. Smirnov
*The Semiconductor Metal Phase Transition under Uniaxial Compression $Sm_{1-x}Gd_xS$ Single Crystals
(62.20 and 64.70; S8)
(a) Vol. **105**, No. 2, p. K93—K96 (1988).

S. P. Nikanorov, Yu. A. Burenkov, A. B. Lebedev, A. V. Golubkov, T. B. Zhukova, and I. A. Smirnov
*Temperature Dependence of Elastic Properties of $Sm_{1-x}Gd_xS$ Single Crystals in the Vicinity of the "Black-Gold" Phase Transition
(62.20; S8)
(a) Vol. **105**, No. 2, p. K103—K106 (1988).

H.-J. Föckel and Th. Goldberg
*Role of Eigenstresses in the Propagation of a Persistent Slip Band Nucleus near a Free Surface
(62.20)
(a) Vol. **105**, No. 2, p. K111—K114 (1988).

K. Rajanna and S. Mohan
*Response of Thin Film Strain Gauge to Tensile and Compressive Strains by Cantilever Technique
(62.20 and 73.60; S1.2)
(a) Vol. **105**, No. 2, p. K181—K184 (1988).

T. C. T. Ting
Line Forces and Dislocations in Anisotropic Elastic Composite Wedges and Spaces
(61.70; 62.20)
(b) Vol. **146**, No. 1, p. 81—90 (1988).

S. K. S. Chauhan, R. Agrawal, P. C. Pathak, B. R. K. Gupta, and R. P. Goyal
Thermal Effects on Thermophysical Properties of Alkali Cyanide Crystals
(65.50; 62.20; 65.70; S11)
(b) Vol. **146**, No. 1, p. 97—103 (1988).

M. Kumari and N. Dass
The Melting Laws
(64.70; 62.20; S2; S5.2; S9)
(b) Vol. **146**, No. 1, p. 105—110 (1988).

R. K. Singh and N. V. K. Prabhakar
*Phase Transition and Anharmonic Properties of Semimagnetic Semiconductors
(62.20 and 64.30; S8.15)
(b) Vol. **146**, No. 1, p. 111—116 (1988).

T. R. Middya, D. Roy, and A. N. Basu
*Multiple-Scattering-Theory Calculation of Elastic and Dielectric Properties of Alkali-Halide Solid Solutions
(62.20 and 77.20; S9.11)
(b) Vol. **146**, No. 1, p. 117—124 (1988).

T. Ohgaku and N. Takeuchi
*Temperature Dependence of the Effective Stress for KCl Single Crystals
(62.20; S9.11)
(a) Vol. **106**, No. 1, p. K19—K21 (1988).

M. de Graef and L. Delaey
A New Empirical Two-Body Interaction Potential Fitted to Second- and Third-Order Elastic Moduli
(61.50; 62.20)
(b) Vol. **146**, No. 2, p. 427—437 (1988).

R. P. Goyal, O. P. Gupta, and B. R. K. Gupta
Theory of Compression in Some Fluorite Structure Crystals
(64.30; 62.20; 65.50; S9.12; S11)
(b) Vol. **146**, No. 2, p. 475—480 (1988).

A. I. Popov, I. A. Domoryad, and N. I. Michalev
Structural Modifcation of the Arsenic Chalcogenide Glasses under γ-Radiation
(61.40; 62.20; S8)
(a) Vol. **106**, No. 2, p. 333—337 (1988).

V. G. Gorshkov, Yu. K. Danileiko, V. V. Osiko, A. V. Sidorin, N. V. Veselovskaya, Yu. V. Dankovskii, and B. L. Shklyar
Mechanical Microstresses in Dislocation-Free Floating-Zone Silicon Monocrystals
(61.70; 62.20; S5.11)
(a) Vol. **106**, No. 2, p. 363—369 (1988).

K. Havancsak, G. Szenes, V. A. Skuratov, Yun Dun Man, W. Malinowski, and V. A. Shchegolev
233 MeV Ne Ion Irradiation Effects on the Mechanical Properties of Copper
(61.80; 62.20; S1.2)
(a) Vol. **106**, No. 2, p. 371—378 (1988).

N. Ganev and I. Kraus
*On the Application of X-Ray Tensometry to States of Stress Due to Tangential Machining of Metal Surfaces
(61.10 and 62.20; 68.25)
(a) Vol. **106**, No. 2, p. 467—472 (1988).

S. D. Chaturvedi, R. C. Dixit, and A. A. S. Sangachin
Analysis of Crystal Binding and Thermoelastic Properties of Cuprous Halides, Silver Halides, and Thallous Halides
(61.50 and 63.90; 62.20; 65.70; S9)
(b) Vol. **147**, No. 1, p. 115—121 (1988).

B. I. Craig and P. V. Smith
Determination of the Bulk Properties of Silicon by the Periodic LUC-MINDO Method
(71.10; 62.20; S5.11)
(b) Vol. **147**, No. 1, p. K31—K35 (1988).

S. Ohtaki and I. Hashimoto
Numerical Evaluation for Stage II_A Recovery in Cold-Worked Aluminium
(61.70; 62.20; 72.15; S4)
(a) Vol. **107**, No. 1, p. 123—130 (1988).

B. E. Mdivanyan and M. Sh. Shikhsaidov
*Photostimulated Enhancement of Dislocation Glide in Gallium Arsenide Crystals
(62.20; 61.80; S7.12)
(a) Vol. **107**, No. 1, p. 131—140 (1988).

Yu. V. Pisarevskii, I. M. Silvestrova, R. Voszka, Á. Péter, I. Földvári, and J. Janszky
*Elastic and Acoustic Properties of $ZnWO_4$ Single Crystal
(62.20 and 62.65; S11.1)
(a) Vol. **107**, No. 1, p. 161—164 (1988).

M. Z. Butt and I. M. Ghauri
*Effect of Short-Range Order on the Temperature Dependence of Plastic Flow in α-Brasses
(62.20; S1.2)
(a) Vol. **107**, No. 1, p. 187—195 (1988).

P. Fernández, J. Llopis, and J. Piqueras
*Annealing Effects on Mechanically Damaged ZnO Ceramics
(62.20; 78.60; S10.1)
(a) Vol. **107**, No. 1, p. 197—203 (1988).

P. A. Maksimiyuk, V. A. Gley, M. Hamerský, P. Lukáč, and Z. Trojanová
*Temperature Dependence of Young's Modulus of α-Zirconium Polycrystals
(62.20; S1)
(a) Vol. **107**, No. 1, p. K11—K13 (1988).

R. Flagmeyer, K. Lenkeit, T. Baumbach, Yu. O. Kanter, and A. A. Fedorov
Characterization of $In_xGa_{1-x}As/GaAs$ Strained Layer Superlattices by Ion Backscattering-Channeling and X-Ray Diffraction
(68.48; 62.20; S7.12; S7.15)
(a) Vol. **107**, No. 1, p. K19—K24 (1988).

U. Pietsch
*The Advantage of Internal Force Calculations in the X-Ray Crystal Structure Analysis
(61.10 and 62.20; S5.12; S7.12)
(b) Vol. **147**, No. 2, p. 517—527 (1988).

D. Hans, A. A. S. Sangachin, and M. Kumar
*Elastic, Dielectric, and Photoelastic Constants of Alkali Halide Crystals
(62.20; 63.90; 77.20; S9.11)
(b) Vol. **147**, No. 2, p. 529—536 (1988).

A. Lefebvre and G. Vanderschaeve
*The Origin of Microtwinning in Room Temperature Indented Undoped and n-Doped GaAs Crystals
(61.70 and 62.20; S7.12)
(a) Vol. **107**, No. 2, p. 647—653 (1988).

G. Torres-Villasenor and G. van Tendeloo
Guinier-Preston Zones in the α-Phase of Superplastic Zn–Al Alloy
(61.70; 62.20; S1.2; S4)
(a) Vol. **107**, No. 2, p. 785—790 (1988).

A. Sachs
*Relations between Third Order Isothermal and Adiabatic Elastic Constants for an Isotropic Material
(62.20; S4)
(b) Vol. **148**, No. 1, p. 101—105 (1988).

R. Ramji Rao and A. Padmaja
*Fourth-Order Elastic Constants of Sodium
(62.20; S2)
(b) Vol. **148**, No. 1, p. K1—K3 (1988).

P. Czurratis, E. Korngiebel, G. Jacobi, and H. Löffler
The Effect of Additions of Mn and Cr and the Applied Solidification Technique on the Structure of an Al–2.2 at % Zn–1.4 at % Mg Alloy
(61.70; 61.55; 62.20; S4)
(a) Vol. **108**, No. 1, p. 171—182 (1988).

E. Ya. Baiguzin, A. I. Melker, and A. I. Mikhailin
*Atomic Mechanisms of Fracture Nucleation and Fracture Development in Two-Dimensional Crystals in Thermodynamic Equilibrium. I. One-Phase Systems
(62.20)
(a) Vol. **108**, No. 1, p. 205—218 (1988).

D. C. Jiles
Variation of the Magnetic Properties of AISI 4140 Steels with Plastic Strain
(75.60; 62.20; S1.1)
(a) Vol. **108**, No. 1, p. 417—429 (1988).

V. P. Goltsev and S. I. Zhukova
*The Loss of Strength of Al-Alloy after Sb^+ Implantation
(61.70 and 62.20; S4)
(a) Vol. **108**, No. 1, p. K1—K4 (1988).

C. Prieto, R. Ramirez, J. A. Gonzalo, and W. Windsch
*Brillouin Scattering Determination of the Elastic Constants of Tris-Sarcosine Calcium Chloride at Room Temperature
(62.20; S12)
(a) Vol. **108**, No. 1, p. K9—K13 (1988).

M. Cinal and S. Olszewski
Quantum-Statistical Approach to Elastic Constants in Metals Influenced by Magnetization
(75.80; 62.20; S1; S1.1; S1.2)
(b) Vol. **148**, No. 2, p. 699—707 (1988).

P. V. Reddy
*High-Temperature Elastic Behaviour of Mn–Mg Mixed Ferrites
(62.20; 75.50; S11.2)
(a) Vol. **108**, No. 2, p. 607—611 (1988).

N. Singh, N. S. Banger, and S. P. Singh
*Elastic Constants and Phonon Spectra of Alkali Metals. Dynamical Treatment with Temperature-Dependent Two-Body Interaction
(62.20; S2)
(b) Vol. **149**, No. 1, p. 63—68 (1988).

H. Günther
*On Lorentz Symmetries in Solids
(62.20)
(b) Vol. **149**, No. 1, p. 101—109 (1988).

R. K. Singh, D. C. Gupta, and S. P. Sanyal
*Anharmonic Properties of IV–VI Compound Semiconductors
(62.20; S8)
(b) Vol. **149**, No. 1, p. 121—125 (1988).

E. Anastassakis and E. Liarokapis
*A Macroscopic Application of the Wigner-Eckart Theorem
(61.50 and 62.20)
(b) Vol. **149**, No. 1, p. K1—K5 (1988).

J. Härtwig and V. Lerche
Anisotropic Deformation of a Crystal Plate and Its Analysis with X-Ray Diffraction Methods
(61.10; 61.70; 62.20; S10.1)
(a) Vol. **109**, No. 1, p. 79—91 (1988).

J. Olfe and H. Neuhäuser
Dislocation Groups, Multipoles, and Friction Stresses in α-CuZn Alloys
(61.70; 62.20; S1.2)
(a) Vol. **109**, No. 1, p. 149—160 (1988).

V. M. Koleshko and I. V. Kiryushin
*Deformation Mechanism Maps and Gettering Diagrams for Single-Crystal Silicon
(62.20; S5.11)
(a) Vol. **109**, No. 1, p. 161—169 (1988).

P. Francois, A. Lefebvre, and G. Vanderschaeve
*Low Temperature Plasticity of Brittle Materials. A New Device for Compressive Testing under Confining Pressure
(62.20; S7.12)
(a) Vol. **109**, No. 1, p. 187—192 (1988).

G. Gosmanová and I. Kraus
*X-Ray Analysis of Residual Stress State with Variable Components of Stress Tensor
(62.20; 68.20)
(a) Vol. **109**, No. 1, p. 199—204 (1988).

A. E. Ligachev, O. S. Kuzmin, G. V. Potemkin, and V. P. Yanovskii
*Structure and Properties of Implanted Steels
(61.70 and 62.20; S1.1)
(a) Vol. **109**, No. 1, p. K11—K14 (1988).

J. M. Galligan and P. Haasen
*Two Trapping Centers in the Photoplastic Effect in CdS
(62.20; S8.11)
(a) Vol. **109**, No. 1, p. K19—K22 (1988).

S. Fujita, K. Maeda, and S. Hyodo
*Radiation Enhanced Dislocation Glide Effect on Indentation-Induced Fracture of GaAs Single Crystals
(61.70 and 62.20; S7.12)
(a) Vol. **109**, No. 2, p. 383—393 (1988).

J. R. Gavarri, R. Mokrani, G. Vacquier, and C. Boulesteix
Relations between Anisotropic Defects, Structural Evolution and van der Waals Bonding in 2H-$NbSe_2$
(61.70 and 65.70; 62.20; S8)
(a) Vol. **109**, No. 2, p. 445—454 (1988).

R. Kühnert, B. Michel, and K. Henniger
*Second Order Moiré Analysis
(62.20)
(a) Vol. **109**, No. 2, p. K93—K95 (1988).

D. Hans
*Third-Order Elastic Constants for Alkaline Earth Oxides and Fluorides
(62.20; S9.12; S10.1)
(b) Vol. **150**, No. 1, p. 47—51 (1988).

B. R. K. Gupta
*Acoustic Mode Vibrational Anharmonicity in the Intermediate Valent Compound Samarium Sulfide
(62.20; S8)
(b) Vol. **150**, No. 1, p. 59—63 (1988).

B. I. Smirnov and I. A. Smirnov
Defect Structure and Physico-Mechanical Properties of SmS Single Crystals
(61.70; 62.20; 72.20; S8)
(a) Vol. **110**, No. 1, p. 89–96 (1988).

A. Z. Mohamed, M. M. Mostafa, M. S. Sakr, and A. A. El-Daly
*The Role of Grain Size on the Work-Hardening Characteristics of Cu–Al Solid Solutions
(62.20; S1.2)
(a) Vol. **110**, No. 1, p. K13—K17 (1988).

H.-A. Kuhn and H. G. Sockel
*Comparison between Experimental Determination and Calculation of Elastic Properties of Nickel-Base Superalloys between 25 and 1200 °C
(62.20; S1.2)
(a) Vol. **110**, No. 2, p. 449—458 (1988).

62.30. Mechanical and elastic waves

M. BOBETH and G. DIENER
*Effective Scattering and Damping of Elastic Waves in Multi-Component Mixtures
(62.30)
(b) Vol. **145**, No. 1, p. 123—138 (1988).

P. M. OSSI
*Elastic Anomalies and Martensite Nucleation in β-Phase Alloys
(62.30 and 64.70; S1.2; S1.3)
(a) Vol. **108**, No. 2, p. 587—598 (1988).

R. G. GABRIELYAN and H. A. ASLANIAN
*On the Theory of X-Ray Diffraction by Surface Acoustic Waves
(61.10 and 62.30)
(a) Vol. **108**, No. 2, p. K85—K88 (1988).

A. S. BAKAI and N. P. LAZAREV
Diffusion of Implanted Hydrogen in a Silicon Film under Stress of a Standing Sound Wave
(66.30; 62.30; S5.11)
(a) Vol. **108**, No. 2, p. K101—K103 (1988).

M. CZAJKOWSKI, M. DROZDOWSKI, and M. KOZIELSKI
*Propagation of Plane Elastic Waves in $LiCsSO_4$ Single Crystals
(62.30 and 63.20; S11.1)
(a) Vol. **110**, No. 2, p. 437—442 (1988).

62.40. Anelasticity, internal friction, and mechanical resonances

For thermomechanical effects, see 65.70
For magnetomechanical effects, see 75.80
For piezoelectric effects, see 77.60
For elasto-optical effects, see 78.20

T. OHGAKU and N. TAKEUCHI
The Relation of the Blaha Effect with Internal Friction for Alkali Halide Crystals
(62.20; 62.40; S9.11)
(a) Vol. **105**, No. 1, p. 153—159 (1988).

P. P. PAL-VAL, V. D. NATSIK, and S. KADEČKOVÁ
*Low-Temperature Dislocation Relaxations in High-Purity Iron Single Crystals Deformed at 295 and 78 K
(62.40; 61.70; S1.1)
(a) Vol. **105**, No. 1, p. K37—K40 (1988).

LITIAN WANG and TINGSUI GE (T. S. KÊ)
*Nonlinear Internal Friction in Mn–Cu Martensite
(62.40; 64.70; S1.2)
(a) Vol. **105**, No. 2, p. 447—453 (1988).

J. W. Moroń and M. Kłosek
*Remarks on the Nature of the High Internal Friction Peak near T_g in Glassy Alloys
(62.40; S1.1; S1.2)
(a) Vol. **106**, No. 1, p. K13—K17 (1988).

I. T. Suprun
*Influence of Pipe Diffusion of Defects on the Amplitude Dependence of Hysteresis Internal Friction
(62.40; 66.30)
(a) Vol. **107**, No. 1, p. 141—152 (1988).

M. A. Kenawy, G. Saad, G. Graiss, and A. Fawzy
*The Effect of Excitation-Induced Fatigue on the Recovery of Internal Friction of Al–1.0 wt% Mn–0.28 wt% Fe Alloy
(62.40; S4)
(a) Vol. **107**, No. 1, p. 171—176 (1988).

B. L. Cheng and T. S. Kê
*Bamboo Boundary Internal-Friction Peak in 99.9999% Aluminium and the Effect of Cold Work on the Peak
(62.40; S4)
(a) Vol. **107**, No. 1, p. 177—185 (1988).

M. Kobiyama, S. Takamura, and K. Nakata
*Internal Friction in Single-Crystal Al Alloys after Electron Irradiation at Low Temperature
(62.40; S4)
(a) Vol. **108**, No. 1, p. 219—224 (1988).

U. Jendrich and P. Haasen
*Internal Friction Due to Kink Motion and Kink Pair Formation in Intrinsic and n-Doped Germanium
(61.70 and 62.40; S5.12)
(a) Vol. **108**, No. 2, p. 553—568 (1988).

T. S. Kê, L. D. Zhang, B. L. Cheng, and A. W. Zhu
*On the Origin of the Macrocrystalline Internal-Friction Peak (the Bamboo Boundary Peak)
(62.40; S4)
(a) Vol. **108**, No. 2, p. 569—575 (1988).

S. E. Urreta de Pereyra, H. Bertorello, and A. A. Ghilarducci de Salva
*Precipitation and Grain Boundary Internal Friction Peaks in Al–Mg–Si
(62.40; 61.70; S4)
(a) Vol. **108**, No. 2, p. 577—586 (1988).

Jin-Song Zhu, R. Schaller, and W. Benoit
*Internal Friction Transitory Effects Associated with Martensitic Transformation in Ni–Ti Alloys
(62.40 and 64.70; S1.2)
(a) Vol. **108**, No. 2, p. 613—618 (1988).

G. Graiss, G. Saad, and A. Fawzy
*Effect of Loading on the Recovery of Internal Friction of an Al–1.0 wt% Mn–0.28 wt% Fe Alloy
(62.40; S4)
(a) Vol. **110**, No. 2, p. 481—487 (1988).

62.50. High-pressure and shock-wave effects in solids

T. Soma, K. Kamada, and H.-Matsuo Kagaya
The Pressure Effect on the Phase Diagrams of the Ge–Sn and Si–Sn Systems
(64.75; 62.50; S6)
(b) Vol. **147**, No. 1, p. 109—114 (1988).

K. Bielecki
A High Pressure EPR Study of Cr^{3+} Doped Deuterated Potassium Aluminium Alum
(76.30; 62.50; S11.1)
(b) Vol. **148**, No. 1, p. K53—K56 (1988).

W. Kullmann, H. Rauh, R. Geick, U. Steigenberger, and W. F. Kuhs
Thermal Parameters and Probability Density Functions of the Narrow Gap Semiconductor Bi_2Te_3 under Hydrostatic Pressure
(61.12 and 61.60; 62.50; 65.90; S8)
(b) Vol. **148**, No. 2, p. 445—456 (1988).

S. Yu. Korostelev, S. G. Psakhe, and V. E. Panin
*A Molecular-Dynamical Study of the Changes in the Atomic Structure of a Material Induced by a Shock Wave with a Distorted Front
(61.50 and 62.50; S1.1)
(b) Vol. **148**, No. 2, p. 483—488 (1988).

M. Maćkowiak and P. Koziol
Effect of Pressure on the Symmetric Hydrogen Bond in $(CCl_3COO)_2HK$
(76.60; 62.50; S11)
(a) Vol. **108**, No. 2, p. 739—745 (1988).

A. A. Opalenko and A. A. Kornilova
Anisotropy of the Mössbauer Effect in β-Sn under Pressure and the Lattice Dynamics
(63.20 and 76.80; 62.50; S4)
(b) Vol. **150**, No. 1, p. 25—29 (1988).

A. K. Garg
High-Pressure Raman Spectroscopic Study of the Ice Ih → Ice IX Phase Transition
(64.70; 62.50; 78.30; S10)
(a) Vol. **110**, No. 2, p. 467—480 (1988).

62.60. Acoustical properties of liquids

For lattice dynamics, phonons, see 63
For second sound in quantum fluids, see 67.40

I. R. ENTIN and I. A. SMIRNOVA
The Effect of X-Ray Beam Intensity Oscillations as a Function of the Amplitude of Ultrasound Excited in a Crystal
(61.10; 62.60; S5.11)
(a) Vol. **106**, No. 2, p. 339—350 (1988).

62.65. Acoustical properties solids

For magnetoacoustic effects in metals, see 75.80
For acoustoelectric effects, see 72.50
For acousto-optical effects, see 78.20

YU. V. PISAREVSKII, I. M. SILVESTROVA, R. VOSZKA, Á. PÉTER, I. FÖLDVÁRI, and J. JANSZKY
*Elastic and Acoustic Properties of $ZnWO_4$ Single Crystal
(62.20 and 62.65; S11.1)
(a) Vol. **107**, No. 1, p. 161—164 (1988).

R. MIZERIS, J. GRIGAS, V. SAMULIONIS, V. SKRITSKI, A. I. BARANOV, and L. A. SHUVALOV
Microwave and Ultrasonic Investigations of Superionic Phase Transitions in $CsDSO_4$ and $CsDSeO_4$
(64.70; 62.65; 77.20; S11.1)
(a) Vol. **110**, No. 2, p. 429—436 (1988).

62.80. Ultrasonic relaxation

(see also 74.30. Ultrasonic attenuation in superconductors)

V. L. AKSENOV, S. FLACH, and N. M. PLAKIDA
*A Theoretical Study of Ultrasonic Anomalies in $La_{2-x}Sr_xCuO_4$
(62.80 and 63.75; S10)
(b) Vol. **149**, No. 1, p. K7—K12 (1988).

63. Lattice dynamics and crystal statistics

(see also 65. Thermal properties, 66.70. Thermal conduction, 68.30. Dynamics of surface and interface fibrations, 78.30. Infrared and Raman spectra)

63.10. General theory

J. M. WESSELINOWA
On the Theory of the Broad Spin Response in Uranium Compounds
(75.10; 63.10; S1.5)
(b) Vol. **145**, No. 1, p. K65—K69 (1988).

F. KH. ABDULLAEV, A. A. ABDUMALIKOV, and E. N. TSOI
*Soliton and Breather Dynamics in a Nematic Liquid Crystal under the Action of Constant and Time-Varying Magnetic Fields
(63.10; 71.45)
(b) Vol. **146**, No. 2, p. 457—466 (1988).

T. HALICIOGLU, H. O. PAMUK, and S. ERKOC
*Interatomic Potentials with Multi-Body Interactions
(63.10; 61.50; 68.40; S5.11)
(b) Vol. **149**, No. 1, p. 81—92 (1988).

P. KLEINERT and H. SCHLEGEL
*Perturbation Expansion of a Simple Disordered System
(63.10)
(b) Vol. **149**, No. 2, p. K111—K115 (1988).

H. BÖTTGER and M. THEUERKAUF
*Phonon Localization in Discordered Systems. Application of the Potential-Well Analogy Method
(63.10 and 63.50)
(b) Vol. **150**, No. 1, p. 73—83 (1988).

63.20. Phonons and vibrations in crystal lattices

V. S. MALYSHEVSKII
Interaction of Channeled Electrons with Lattice Excitations (**Review Article**)
(61.80; 61.70; 63.20; 78.70)
(b) Vol. **147**, No. 1, p. 11—44 (1988).

I. V. BODNAR, G. F. SMIRNOVA, T. V. SMIRNOVA, YU. A. ALESHCHENKO, and L. K. VODOPYANOV
*Lattice Vibrations of $CuGaSe_2$, $CuInSe_2$ Ternary Compounds and $CuGa_xIn_{1-x}Se_2$ Solid Solutions
(63.20; 78.30; S8.16)
(b) Vol. **145**, No. 1, p. 117—122 (1988).

Y. Madhavan, K. Ramachandran, and T. M. Haridasan
*Mean Square Displacements of Neighbours and Phonon Resonance Due to a Vacancy in HgTe and CdTe
(63.20; 61.70; S8.13)
(b) Vol. **145**, No. 1, p. 139—144 (1988).

R. Urban, P. Ehrhardt, W. Schilling, H. R. Schober, and H. Lauter
*Erratum to: Investigation of the Resonant Vibration Modes of Radiation Induced Interstitial Atoms in Aluminum by Diffuse Inelastic Neutron Scattering
(61.80 and 63.20; S4)
(b) Vol. **145**, No. 1, p. 351 (1988).

P. W. Mauriz and E. L. Albuquerque
*Calculation of the Debye-Weller Factor for Mössbauer Surface Atoms: Anisotropic Effects
(63.20; 76.80)
(b) Vol. **145**, No. 1, p. K5—K8 (1988).

G. Irmer, M. Herms, J. Monecke, and H. Bock
*Raman Scattering from LO-Phonon-Plasmon Modes in $GaAs_{1-x}P_x$ Crystals
(63.20 and 78.30; 71.45; S7.15)
(b) Vol. **145**, No. 1, p. K79—K82 (1988).

J. Wąsicki, A. V. Belushkin, V. G. Khomenko, I. Natkaniec, Yu. V. Telezhenko, and T. Wasiutyński
*Low-Temperature Structure and Lattice Dynamics of 4,4'-Difluorobiphenyl
(63.20; 61.65; S12)
(b) Vol. **145**, No. 2, p. 445—453 (1988).

J. Tatarkiewicz and A. M. Witowski
*Hydrogen Vibrations in ZnTe
(61.70 and 63.20; 78.30; S8.13)
(b) Vol. **145**, No. 2, p. K95—K98 (1988).

T. D. Ibragimov and N. T. Mamedov
*Two-Phonon Absorption Spectra and Phase Transition in $TlGaSe_2$
(63.20 and 64.70; 78.30; S8.16)
(b) Vol. **145**, No. 2, p. K103—K105 (1988).

B. N. Onwuagba
*Phonon Spectrum in Metastable F.C.C. Lanthanum
(63.20; S1.4)
(b) Vol. **146**, No. 1, p. 131—136 (1988).

V. L. Aksenov, S. Flach, and N. M. Plakida
*A Microscopical Model of Structural Phase Transitions in La_2CuO_4
(63.20 and 63.75; S10)
(b) Vol. **146**, No. 1, p. K1—K5 (1988).

P. ČAPKOVÁ, M. MERISALO, M. LAITINEN, V. VALVODA, and L. DOBIÁŠOVÁ
*Thermal Vibrations and Thermal Expansion of $ZrC_{0.98}$ Studied by X-Ray Diffraction
(63.20 and 65.70; S1.61)
(a) Vol. **106**, No. 1, p. 107—114 (1988).

E. L. ALBUQUERQUE, P. FULCO, and D. R. TILLEY
*Lattice Dynamics for Superlattices
(63.20)
(b) Vol. **146**, No. 2, p. 449—456 (1988).

Z. KLESZCZEWSKI and J. BODZENTA
*Phonon–Phonon Interaction in Gadolinium–Gallium Garnet Crystals
(63.20; S11.2)
(b) Vol. **146**, No. 2, p. 467—474 (1988).

V. B. GUPTA and H. C. GUPTA
*Phonon Dipersion Curves of the Superconducting A15 Compounds Nb_3Ge and $Nb_3(Al–Ge)$
(63.20; S1.2)
(b) Vol. **146**, No. 2, p. 481—486 (1988).

P. P. LOTTICI
*A MREI Model for the Zincblende-Like Phonons in the Mixed Defect Chalcopyrite $ZnGa_2(S_xSe_{1-x})_4$
(63.20; 78.30; S8.16)
(b) Vol. **146**, No. 2, p. 503—510 (1988).

J. PETZELT, S. KAMBA, S. PAČESOVÁ, E. POLLERT, J. ŠRÁMEK, O. SMRČKOVÁ, and D. SÝKOROVÁ
*Infrared Reflectivity of High-T_c $YBa_2Cu_3O_{7-x}$ and Related Ceramics
(63.20 and 78.30; 74.70; S10.15)
(b) Vol. **146**, No. 2, p. 743—755 (1988).

A. ANEDDA, G. BONGIOVANNI, and E. FORTIN
Raman Scattering from Vibrational Modes in Layered $Cd_xHg_{1-x}I_2$ Solid Solution
(78.30; 63.20; S9.15)
(b) Vol. **146**, No. 2, p. 757—765 (1988).

E. JAHNE, O. GOEDE, and V. WEINHOLD
*Far Infrared Reflectivity of Tetrahedrally and Octahedrally Coordinated MnS
(63.20 and 78.30; S8.11)
(b) Vol. **146**, No. 2, p. K157—K160 (1988).

B. H. BAIRAMOV, V. V. TOPOROV, N. V. AGRINSKAYA, E. A. SAMEDOV, G. IRMER, and J. MONECKE
*Raman Scattering by Coupled LO-Phonon-Plasmon Modes in CdTe
(63.20 and 78.30; S8.13)
(b) Vol. **146**, No. 2, p. K161—K165 (1988).

V. Dvořák
*Erratum to: On the Structural Phase Transition in La_2CuO_4
(63.20 and 64.70; S10)
(b) Vol. **146**, No. 2, p. K167 (1988).

M. D. Fontana, A. Ridah, and G. E. Kugel
Frequency Dependence of the Soft Mode in $KNbO_3$
(77.80 and 78.30; 63.20; S11.1)
(b) Vol. **147**, No. 1, p. 441—447 (1988).

R. Ramji Rao and A. Rajput
*Phonon Dispersion Relations in Silver and Copper
(63.20; S1.2; S1.3)
(b) Vol. **147**, No. 1, p. K11—K14 (1988).

An Hu, Shu-yi Zhang, Xiang-yang Yuan, Qing Shen, Zhong-nan Lu, and Duan Feng
*Lattice Expansion and Phonon Softening in Nb/Cu Superlattices
(63.20 and 68.30; 61.55; S1.2)
(a) Vol. **107**, No. 1, p. 153—159 (1988).

Ningsheng Luo, Wenlan Xu, and Shuechu Shen
*Application of the Embedded Atom Method to Phonon Dispersions and Local Surface Vibrations in Copper
(63.20; 68.30; S1.2)
(b) Vol. **147**, No. 2, p. 511—516 (1988).

J. Giordano et J. L. Sauvajol
*Etude de l'évolution avec la température des modes actifs en diffusion Raman du ZnF_2
(Study of the Temperature Dependence of the Raman Active Modes of ZnF_2)
(63.20 and 78.30; S9)
(b) Vol. **147**, No. 2, p. 537—545 (1988).

I. Ya. Fugol, A. M. Ratner, L. V. Khashchina, and I. M. Pritula
Infrared Absorption and Anharmonicity of Crystalline Nitrogen
(78.30; 63.20; S5.2)
(b) Vol. **147**, No. 2, p. 757—766 (1988).

W. Lu, H. J. Ye, Z. Y. Yu, S. Y. Zhang, Y. Fu, W. L. Xu, S. C. Shen, and W. Giriat
Local and Quasi-Local Modes of Fe and Zn in CdTe
(78.30; 63.20; S8.15)
(b) Vol. **147**, No. 2, p. 767—778 (1988).

W. S. Tse, C. S. Fang, K. F. Pai, and W. Lih
A Raman Investigation of the Polymorphism of Potassium Nitrite
(78.30; 63.20; S11)
(b) Vol. **147**, No. 2, p. 791—795 (1988).

A. Overberg and F. Fischer
Effect of Magnetic Field on $^3A_{1u} \rightarrow {}^1A_{1g}$ Luminescence of Pb^{2+} in CaF_2 and SrF_2 and of Au^- in NaCl
(78.55; 63.20; S9.11; S9.12)
(b) Vol. **147**, No. 2, p. 811—821 (1988).

M. H. Kuok and S. H. Tang
A Raman Study of $KCdBr_3$ Single Crystals
(78.30; 63.20; S9.16)
(b) Vol. **147**, No. 2, p. K195—K199 (1988).

H. Poelman, J. Vennik, and G. Dalmai
*Temperature Effects on the V_2O_5 Surface Phonon Spectrum
(63.20 and 68.30; S10.1)
(a) Vol. **107**, No. 2, p. 731—737 (1988).

R. Enderlein, F. Bechstedt, and G. Gerecke
*The Complete Set of Polar Optic Phonon Modes of Superlattices. Confined Bulk and Interface Modes
(63.20 and 68.30; 73.40)
(b) Vol. **148**, No. 1, p. 173—183 (1988).

G. I. Bersuker
*Dynamics in the System of Dipolar Impurity Centers in Crystals
(63.20 and 71.55; S11)
(b) Vol. **148**, No. 1, p. 243—249 (1988).

N. Singh, N. S. Banger, and S. P. Singh
*Phonon Frequencies of B.C.C. Li: An Application of the Local Pseudopotential
(63.20; S2)
(b) Vol. **148**, No. 1, p. K5—K9 (1988).

T. D. Ibragimov and N. T. Mamedov
*Two-Phonon Absorption in $TlInTe_2$ and $TlGaTe_2$
(63.20 and 78.30; S8.16)
(b) Vol. **148**, No. 1, p. K89—K92 (1988).

K. R. Allekhverdiev, N. D. Akhmedzade, M. M. Tagiev, M. M. Shirinov, and S. Häseler
*Long-Wavelength IR-Active Phonons in the $TlInS_2$–$TlInSe_2$ System
(63.20 and 78.30; S8.16)
(b) Vol. **148**, No. 1, p. K93—K96 (1988).

J. Otto
*Stress-Induced Dichroism of the Vibrational Lines and Phonon Sidebands of SH^- Centres in CsBr
(63.20; 78.30; S9.11)
(b) Vol. **148**, No. 2, p. 489—499 (1988).

ZIJING LIN and KELIN WANG
*A Simple Scheme for Studies of Surface Phonons of Covalent Semiconductors. I. Calculation of Force Constants
(63.20 and 68.30; S5)
(b) Vol. **148**, No. 2, p. 525—532 (1988).

D. PRAKASH and J. C. UPADHYAYA
*Three-Body Effect on the Lattice Dynamics of Pt and Co
(63.20; S1.2)
(b) Vol. **148**, No. 2, p. K109—K112 (1988).

I. A. GOICHUK, V. V. KUKHTIN, and E. G. PETROV
*Orientational Vibrations of Two-Dimensional Dipole Lattice on Support
(63.20; 75.10)
(b) Vol. **149**, No. 1, p. 55—61 (1988).

S. S. KUSHWAHA, M. P. DIKSHIT, and S. K. SRIVASTAVA
*A Model for Lattice Dynamics of Lanthanides
(63.20; S1.4)
(b) Vol. **149**, No. 1, p. 69—79 (1988).

E. JÄGER, M. MOSSA, and K. HEHL
*Difference Equations in the Theory of Biphonons
(63.20; 68.30; 71.35)
(b) Vol. **149**, No. 1, p. 111—120 (1988).

V. R. BELOSLUDOV, M. YU. LAVRENTIEV, and S. A. SYSKIN
*Dynamical Properties of the Molecular Crystals with Electrostatic Interaction Taken into Account. Low Pressure Ice Phases (I_h and I_c)
(63.20; S10)
(b) Vol. **149**, No. 1, p. 133—142 (1988).

ZIJING LIN and KELIN WANG
*A Simple Scheme of Studies of Surface Phonons of Covalent Semiconductors. II. Surface Phonons at Ideal and Relaxed Si(110) Surface
(63.20 and 68.30; S5.11)
(b) Vol. **149**, No. 1, p. 143—148 (1988).

R. BALTRAMIEJŪNAS and A. ŽUKAUSKAS
*Manifestation of Hot Phonons in the Luminescence Spectra of Highly Photoexcited $A^{II}B^{VI}$ Crystals
(63.20 and 78.55; 71.45; S8.12)
(b) Vol. **149**, No. 1, p. 337—346 (1988).

S. K. MISHRA and T. N. SINGH
*Lattice Dynamics of B.C.C. Transition Metals
(63.20; S1; S1.1)
(b) Vol. **149**, No. 2, p. 503—510 (1988).

P. C. Hägele, S. Wolf, and C. Schmid
*Mass Defects in Polyethylene and the Role of Mass Weighting in Phonon Green's Function Theory
(61.70 and 63.20; S12)
(b) Vol. **149**, No. 2, p. 521—532 (1988).

J. Tatarkiewicz and M. Stutzmann
*Hydrogen Vibrations in Semiconductors and Insulators
(61.70 and 63.20)
(b) Vol. **149**, No. 2, p. K95—K99 (1988).

K. Kumazaki and K. Imai
*Far-Infrared Reflection and Raman Scattering Spectra in γ-InSe
(63.20 and 78.30; S8)
(b) Vol. **149**, No. 2, p. K183—K186 (1988).

A. A. Opalenko and A. A. Kornilova
*Anisotropy of the Mössbauer Effect in β-Sn under Pressure and the Lattice Dynamics
(63.20 and 76.80; 62.50; S4)
(b) Vol. **150**, No. 1, p. 25—29 (1988).

A. de Andrés
*Applications of a Point Polarizable Ion Model to Suzuki Phase Precipitates in Alkali Halides
(63.20; 61.70; 78.30; S9.11)
(b) Vol. **150**, No. 1, p. 31—37 (1988).

J. Wąsicki, A. M. Gurevich, V. N. Eropkin, J. Mayer, I. Natkaniec, Yu. V. Telezhenko, and V. G. Khomenko
*Low-Temperature Heat Capacity and Phonon Spectra of 4,4'-Difluorobiphenyl
(63.20 and 65.40; S12)
(b) Vol. **150**, No. 1, p. 39—45 (1988).

S. Zekeng, B. Prevot, and C. Schwab
*Raman Determination of the Faust-Henry Coefficient of GaAs in the 1.9 to 2.7 eV Range at Ordinary and Low Temperatures
(63.20 and 78.30; S7.12)
(b) Vol. **150**, No. 1, p. 65—72 (1988).

R. Enderlein
*Macroscopic Dynamic Theory of Polar Interface Modes of Superlattices
(63.20 and 68.30)
(b) Vol. **150**, No. 1, p. 85—101 (1988).

M. Czajkowski, M. Drozdowski, and M. Kozielski
*Propagation of Plane Elastic Waves in $LiCsSO_4$ Single Crystals
(62.30 and 63.20; S11.1)
(a) Vol. **110**, No. 2, p. 437—442 (1988).

63.50. Vibrational states in disordered systems

I. I. Reshina
Raman Study of Disorder in Laser-Annealed III–V Semiconductors
(78.30; 63.50; S7.11; S7.12)
(a) Vol. **106**, No. 1, p. 261—270 (1988).

P. P. Lottici
*Phonon and Fraction Dynamics in Amorphous Arsenic
(63.50; S5)
(b) Vol. **146**, No. 2, p. K81—K85 (1988).

B. K. Ghosh and Bal K. Agrawal
*Vibrational Structure of Hydrogenated Amorphous Silicon Carbide Alloys
(63.50; S6)
(b) Vol. **147**, No. 1, p. 97—102 (1988).

I. D. Mikhailov and A. P. Loghinov
*Density of Vibrational States of Deuterized Polyethylene with Markovian Ordering
(63.50; S12)
(b) Vol. **147**, No. 1, p. K15—K19 (1988).

C. Wiecko
*Perturbative Approach to Dilute Off-Diagonal Defects
(63.50)
(b) Vol. **148**, No. 1, p. 73—80 (1988).

M. Rani, A. Pratap, and N. S. Saxena
*Collective Excitations in Liquid Cesium Using a Theory of Phonons in Amorphous Solids
(63.50; S2)
(b) Vol. **149**, No. 1, p. 93—99 (1988).

H. Böttger and M. Theuerkauf
*Phonon Localization in Disordered Systems. Application of the Potential-Well Analogy Method
(63.10 and 63.50)
(b) Vol. **150**, No. 1, p. 73—83 (1988).

Z. Wang
*The Density of Vibrational States and Fracton Dimensionality for a Periodic Fractal Lattice
(63.50)
(b) Vol. **150**, No. 1, p. K15—K17 (1988).

63.70. Statistical mechanics of lattice vibrations

(see also 65. Thermal properties of condensed matter, and 66.70. Thermal conduction)

D. B. Sirdeshmukh and K. G. Subhadra
*The Pressure Dependence of Debye Temperatures of Alkali Halides
(63.70; S9.11)
(b) Vol. **150**, No. 1, p. K11—K14 (1988).

63.75. Statistical mechanics of displacive phase-transitions

For order–disorder and statistical mechanics of model systems, see 64.60
For crystallographic aspects of polymorphic and order–disorder transformations, see 61.50

N. Kristoffel and P. Konsin
*Vibronic Theory of Structural Phase Transitions and Displacive Ferroelectrics (**Review Article**)
(63.75 and 77.80; 64.70; S8; S11)
(b) Vol. **149**, No. 1, p. 11—40 (1988).

V. L. Aksenov, S. Flach, and N. M. Plakida
*A Microscopical Model of Structural Phase Transitions in La_2CuO_4
(63.20 and 63.75; S10)
(b) Vol. **146**, No. 1, p. K1—K5 (1988).

V. L. Aksenov, S. Flach, and N. M. Plakida
*A Theoretical Study of Ultrasonic Anomalies in $La_{2-x}Sr_xCuO_4$
(62.80 and 63.75; S10)
(b) Vol. **149**, No. 1, p. K7—K12 (1988).

63.90. Other topics in lattice dynamics and crystal statistics

M. Kumar, A. K. Pachauri, D. S. Chaturvedi, and A. K. Sharma
Equation of State and Pressure Dependence of the Grüneisen Parameter for NaCl Crystal
(64.30; 63.90; S9.11)
(b) Vol. **146**, No. 1, p. 125—130 (1988).

L. N. Kristoforov and A. V. Zolotaryuk
*Dynamics of Ionic and Bonding Defects in Quasi-One-Dimensional Hydrogen-Bonded Chains
(61.70 and 63.90; 66.90; 71.45; S12.1)
(b) Vol. **146**, No. 2, p. 487—501 (1988).

S. D. Chaturvedi, R. C. Dixit, and A. A. S. Sangachin
*Analysis of Crystal Binding and Thermoelastic Properties of Cuprous Halides, Silver Halides, and Thallous Halides
(61.50 and 63.90; 62.20; 65.70; S9)
(b) Vol. **147**, No. 1, p. 115—121 (1988).

T. Takeuchi
*Note on MD Calculation of $RbAg_4I_5$
(63.90; S9.16)
(b) Vol. **147**, No. 1, p. K9—K10 (1988).

D. Hans, A. A. S. Sangachin, and M. Kumar
Elastic, Dielectric, and Photoelastic Constants of Alkali Halide Crystals
(62.20; 63.90; 77.20; S9.11)
(b) Vol. **147**, No. 2, p. 529—536 (1988).

64. Equations of state, phase equilibria, and phase transitions

64.10. General theory of equations of state and phase equilibria

W. Vogelsberger, H.-G. Fritsche, and E. Müller
Size Dependence of Surface Thermodynmic Parameters of Microclusters and the Location of the Gibbs' "Surface of Tension"
(68.40; 64.10; S5.2)
(b) Vol. **148**, No. 1, p. 155—164 (1988).

Li-Rong Chen and Tzu-Tung Yen
*On the Equation of Solid State at High Pressure
(64.10)
(a) Vol. **108**, No. 1, p. K15—K18 (1988).

Ariel Fernández
*Center Manifold and Phase-Ordering Dynamics for the Onset of Nonequilibrium Organizations
(64.10)
(b) Vol. **149**, No. 1, p. 127—132 (1988).

64.30. Equations of state specific substances

(see also 65.70. Thermal expansion)

R. K. Singh and N. V. K. Prabhakar
*Phase Transition and Anharmonic Properties of Semimagnetic Semiconductors
(62.20 and 64.30; S8.15)
(b) Vol. **146**, No. 1, p. 111—116 (1988).

M. Kumar, A. K. Pachauri, S. D. Chaturvedi, and A. K. Sharma
*Equation of State and Pressure Dependence of the Grüneisen Parameter for NaCl Crystal
(64.30; 63.90; S9.11)
(b) Vol. **146**, No. 1, p. 125—130 (1988).

R. P. Goyal, O. P. Gupta, and B. R. K. Gupta
*Theory of Compression in Some Fluorite Structure Crystals
(64.30; 62.20; 65.20; S9.12; S11)
(b) Vol. **146**, No. 2, p. 475—480 (1988).

Wen-Chen Zheng
*On the Equatiо́ns of State of $ZnSiF_6 \cdot 6\,H_2O:Ni^{2+}$ Crystal Determined from the Pressure Dependence of EPR Spectra
(64.30; 71.70; 76.30; S11)
(b) Vol. **149**, No. 2, p. K105—K109 (1988).

64.60. General studies of phase transitions

(for magnetic, superconducting, and quantum fluid critical phenomena, 75, 74, and 67, respectively)

J. Maćkowiak
*A Model of Crystallization in a One-Dimensional Classical Liquid
(64.60; 61.20)
(b) Vol. **145**, No. 2, p. 419—424 (1988).

M. V. Chernitskii, A. E. Nikiforov, and A. I. Krotkii
*Microscopic Mechanisms of Paramagnetic Probe–Lattice Interaction for the System $RbCdF_3:Cr^{3+}$ near the Structural Phase Transition
(64.60 and 76.30; S9.16)
(b) Vol. **145**, No. 2, p. 425—434 (1988).

B. Pilawa, G. Hess, H. G. Kahle, and A. Kasten
Linear Birefringence Investigations of the Jahn-Teller Distorted Phase in the Mixed Crystal Systems $(Tb_x, Tm_{1-x})VO_4$ and $(Tb_x, Yb_{1-x})VO_4$
(78.20; 64.60; S11.1)
(b) Vol. **145**, No. 2, p. 729—739 (1988).

J. D. Stephenson
Quasi-Realtime Observation of Partial Dislocation Growth in (01.1) Synthetic Quartz near the α–β Phase Transition Temperature (T_c) Using Electric Potential Switching
(61.70; 64.60; S10.1)
(a) Vol. **106**, No. 2, p. 441—449 (1988).

A. I. Gusev, A. A. Rempel, and V. N. Lipatnikov
*Magnetic Susceptibility and Atomic Ordering in Tantalum Carbide
(64.60; 75.30; S1.61)
(a) Vol. **106**, No. 2, p. 459—466 (1988).

G. Dolino, F. Mogeon, and P. Bastie
*Birefringence and γ-Ray Study of the Irreversible Behaviour of the Incommensurate Phase of Quartz
(64.60; 78.20; S10.1)
(a) Vol. **107**, No. 2, p. 559—570 (1988).

S. C. Farmer and A. H. Heuer
Surface Phase Transformation in Monoclinic ZrO_2
(68.20; 64.60; S10)
(a) Vol. **107**, No. 2, p. 571—578 (1988).

L. Tichý and P. Nagels
Non-Isothermal Crystallization of Bi_2S_3 from Glassy $(GeS_2)_{0.5}(Bi_2S_3)_{0.5}$
(61.40; 64.60; S8)
(a) Vol. **107**, No. 2, p. 769—774 (1988).

V. D. Sajfert, Lj. D. Mašković, R. P. Djajić-Jovanović, and U. F. Kozmidis-Luburić
*Phase Transition in 1D KDP-Type Ferroelectrics
(64.60; 77.80; S11.1)
(b) Vol. **148**, No. 1, p. 81—90 (1988).

A. E. Aliev, A. Sh. Akramov, L. N. Fershtat, and P. K. Khabibullaev
*Mechanism of a Superion Phase Transition in α-$LiIO_3$
(64.60; 77.40; S11.1)
(a) Vol. **108**, No. 1, p. 189—196 (1988).

M. Funke and S. Trimper
Low Temperature Behaviour of the Random Field Ising Model
(75.40; 64.60)
(b) Vol. **148**, No. 2, p. K151—K156 (1988).

V. Yu. Balandin, A. V. Dvurechenskii, and L. N. Aleksandrov
*Intermediate Crystallization of Amorphous Layers of Silicon at Pulsed Laser Annealing
(64.60; 61.50; S5.11)
(a) Vol. **110**, No. 1, p. 133—139 (1988).

Li-Rong Chen
*Quasiregular Model and Phase Equilibrium Diagram of a Binary System
(64.60)
(a) Vol. **110**, No. 1, p. K19—K24 (1988).

64.70. Phase equilibria, phase transitions, and critical points of specific substances

N. Kristoffel and P. Konsin
Vibronic Theory of Structural Phase Transitions and Displacive Ferroelectrics (**Review Article**)
(63.75 and 77.80; 64.70; S8; S11)
(b) Vol. **149**, No. 1, p. 11—40 (1988).

E. D. Yakushkin and V. N. Anisimova
*Phase Transition in Monoclinic KD_2PO_4 Crystal
(64.70; 77.20; S11.1)
(a) Vol. **105**, No. 1, p. 139—143 (1988).

A. A. Sukhovskii, I. P. Aleksandrova, Yu. N. Moskvich, and O. V. Rozanov
*Phase Diagram of the Monoclinic and Orthorhombic Ammonium Hydrogen Selenate
(64.70; 76.60; S11.1)
(a) Vol. **105**, No. 1, p. 145—152 (1988).

H. PYKACZ, Z. CZAPLA, and J. MRÓZ
*Dilatometric Study of the Phase Transitions in NH_4HSeO_4
(64.70; 65.70; S11)
(a) Vol. **105**, No. 1, p. K33—K35 (1988).

T. D. IBRAGIMOV and N. T. MAMEDOV
*Two-Phonon Absorption Spectra and Phase Transition in $TlGaSe_2$
(63.20 and 64.70; 78.30; S8.16)
(b) Vol. **145**, No. 2, p. K103—K105 (1988).

K. S. ALEKSANDROV, S. V. MELNIKOVA, I. N. FLEROV, A. D. VASILEV, A. I. KRUGLIK, and I. T. KOKOV
*Phase Transitions in Cs_2CdI_4 Single Crystals
(64.70; 61.60; 65.40; 77.20; 78.20; S9.16)
(a) Vol. **105**, No. 2, p. 441—446 (1988).

LITIAN WANG and TINGSUI GE (T. S. KÊ)
Nonlinear Internal Friction in Mn–Cu Martensite
(62.40; 64.70; S1.2)
(a) Vol. **105**, No. 2, p. 447—453 (1988).

A. V. GOLUBKOV, V. M. EGOROV, T. S. ORLOVA, V. M. SERGEEVA, B. I. SMIRNOV, and I. A. SMIRNOV
*The Semiconductor–Metal Phase Transition under Uniaxial Compression $Sm_{1-x}Gd_xS$ Single Crystals
(62.20 and 64.70; S8)
(a) Vol. **105**, No. 2, p. K93—K96 (1988).

I. A. AKSENOV, L. A. MAKOVETSKAYA, G. P. POPELNYUK, and I. P. SHILOVICH
*Phase Diagram of $CuAl_xIn_{1-x}S_2$ Solid Solutions
(64.70; S8.16)
(a) Vol. **105**, No. 2, p. K97—K102 (1988).

B. PURA, J. PRZEDMOJSKI, K. WENTOWSKA, R. DĄBROWSKI, and K. CZUPRYŃSKI
*X-Ray Critical Scattering Study of Nematic–Smectic A Phase Transition in 8DBT/120CB Mixture
(64.70; S12)
(a) Vol. **105**, No. 2, p. K107—K110 (1988).

M. KUMARI and N. DASS
*The Melting Laws
(64.70; 62.20; S2; S5.2; S9)
(b) Vol. **146**, No. 1, p. 105—110 (1988).

V. DVOŘÁK
*Erratum to: On the Structural Phase Transition in La_2CuO_4
(63.20 and 64.70; S10)
(b) Vol. **146**, No. 2, p. K167 (1988).

G. D. IVLEV, V. L. MALEVICH, and V. V. ZHIDKOV
*Melting Temperature of Amorphized Silicon Heated by Nanosecond Laser Radiation
(64.70; S5.11)
(a) Vol. **106**, No. 2, p. K123—K127 (1988).

V. M. TALANOV
*Specific Points on Spinel Phase Diagrams
(64.70; S11.2)
(a) Vol. **106**, No. 2, p. K129—K133 (1988).

H. PFEIFFER
*On Oxygen Ordering in the High Temperature Superconductor Y–Ba–Cu–O
(64.70 and 74.70; S10.15)
(a) Vol. **106**, No. 2, p. K161—K164 (1988).

R. ALLUB and S. M. MUJIBUR RAHMAN
*On the Phase Stability of Rb under External Pressure
(64.70; S2)
(b) Vol. **147**, No. 1, p. K5—K8 (1988).

P. GRIMA, M. QUINTERO, G. S. PÉREZ, R. TOVAR, and J. C. WOOLLEY
*Phase Diagram and Lattice Parameter Values for the $Hg_{2x}(AgIn)_{1-x}Te_2$ Alloys
(61.55 and 64.70; S8.15)
(a) Vol. **107**, No. 1, p. 165—169 (1988).

M. QUINTERO, P. GRIMA, R. TOVAR, G. S. PÉREZ, and J. C. WOOLLEY
*Phase Relations and the Effects of Ordering in $(AgIn)_{1-z}Mn_{2z}Te_2$ and $(CuIn)_{1-z}Mn_{2z}Te_2$ Alloys
(64.70; S8.15)
(a) Vol. **107**, No. 1, p. 205—211 (1988).

G. M. GUSAKOV, A. A. KOMARNITSKII, and A. S. EM
Dynamics of Optical Parameters of Silicon and Germanium Single Crystals during Two-Pulsed Nanosecond Laser Irradiation
(61.80 and 68.55; 64.70; 78.20; S5.11; S5.12)
(a) Vol. **107**, No. 1, p. 261—271 (1988).

A. LOISEAU and C. VANNUFFEL
*$TiAl_2$: A Reentrant Phase in the Ti–Al System
(64.70; 61.16; S1.2; S4)
(a) Vol. **107**, No. 2, p. 655—671 (1988).

M. NAKAGAWA and K. YASUDA
*Phase Changes and Structural Features of Cu_3Au II in a Gold–Copper–Silver Ternary Alloy during Isothermal Ageing
(61.55 and 64.70; S1.3)
(a) Vol. **107**, No. 2, p. 709—717 (1988).

G. B. DEMISHEV, S. S. KABALKINA, and T. N. KOLOBYANINA
X-Ray Studies of Thallium Chalcogenides TlS and TlSe up to 37 GPa
(61.10; 64.70; S8)
(a) Vol. **108**, No. 1, p. 89—95 (1988).

CHU THIEN TRUONG, G. WENDROCK, and H. LÖFFLER
On the Existence Field of Guinier-Preston Zones in the Al–Zn System
(61.55; 64.70; S4)
(a) Vol. **108**, No. 1, p. 103—110 (1988).

R. A. Yakshibaev, N. N. Mukhamadeeva, and R. F. Almukhametov
*Phase Transformations and Ionic Transport in $Cu_{2-\delta}Te$ Superionic Conductor
(64.70 and 66.30; S8)
(a) Vol. **108**, No. 1, p. 135—141 (1988).

D. Baltrūnas, B. Vengalis, S. Motiejūnas, and Shiktorov
*Effect of a Deviation from Stoichiometry on the Mössbauer Parameters of $Sn_{0.8}Ge_{0.2}Te$ during Phase Transition
(64.70; 76.80; S8.15)
(a) Vol. **108**, No. 1, p. 197—204 (1988).

N. M. Plakida and W. Salejda
*The Improper Ferroelastic Phase Transition in Superionic $Rb_3H(SeO_4)_2$ Crystals
(64.70; 61.50; S11)
(b) Vol. **148**, No. 2, p. 473—481 (1988).

P. M. Ossi
*Elastic Anomalies and Martensite Nucleation in β-Phase Alloys
(62.30 and 64.70; S1.2; S1.3)
(a) Vol. **108**, No. 2, p. 587—598 (1988).

Jin-Song Zhu, R. Schaller, and W. Benoit
*Internal Friction Transitory Effects Associated with Martensitic Transformation in Ni–Ti Alloys
(62.40 and 64.70; S1.2)
(a) Vol. **108**, No. 2, p. 613—618 (1988).

V. Parizek and T. A. Chuong
Influence of In on Transport Properties and Phase Transition in $Pb_{1-x}Sn_xTe$ Thin Films
(73.60; 64.70; S8.16)
(a) Vol. **108**, No. 2, p. 643—650 (1988).

Ming Mao, Yuansheng Wang, and Yizhen He
*Crystallization of Metallic Glass $Fe_{46}Ni_{31}VSi_8B_{14}$
(61.40 and 64.70; S1.1; S1.2)
(a) Vol. **109**, No. 1, p. 39—51 (1988).

Jun Chen and Duan Feng
*TEM Study of Phases and Domains in $NaNbO_3$ at Room Temperature
(64.70; 77.80; S10)
(a) Vol. **109**, No. 1, p. 171—185 (1988).

J. Minge and T. Krajewski
*EPR Study of SO_4^- Radicals in $K_4LiH_3(SO_4)_4$ Crystals
(64.70; 76.30; S11.1)
(a) Vol. **109**, No. 1, p. 193—198 (1988).

L. N. Aleksandrov, V. Yu. Balandin, A. V. Dvurechenskii, and O. A. Kulyasova
*Melting of Multilayered Structures under Pulse Heating (Computational Experiment)
(64.70; S5.11; S10.1)
(a) Vol. **109**, No. 1, p. K27—K29 (1988).

B. S. Evseev
*Non-Equilibrium Gradient-Zone Crystallization in Semiconductors
(64.70; 66.70; 68.45)
(a) Vol. **109**, No. 1, p. K31—K34 (1988).

A. V. Zaretskii, V. F. Petrenko, and V. A. Chesnakov
The Protonic Conductivity of Heavily KOH-Doped Ice
(66.90 and 77.20; 64.70; S10)
(a) Vol. **109**, No. 2, p. 373—381 (1988).

Jun Chen and Duan Feng
*In Situ TEM Studies of Para–Ferro Phase Transitions in $NaNbO_3$
(64.70; 77.80; S11.1)
(a) Vol. **109**, No. 2, p. 427—434 (1988).

Ye Tao, Mei-shen Hu, and Duan Feng
*Direct Observation of α–β Phase Transition in $KNbW_2O_9$ by Transmission Electron Microscopy
(64.70; 61.16; S11)
(a) Vol. **109**, No. 2, p. 435—444 (1988).

J. M. Holender, J. Sołtys, G. Wendrock, and P. Czurratis
*Non-Isothermal Studies of the Eutectoidal Decomposition of the Alloy Al–59 at% Zn
(64.70; 61.10; 61.55; S1.2; S4)
(a) Vol. **110**, No. 1, p. 123—132 (1988).

J.-W. Schünemann, K. Bärner, H. Fjellvag, A. Kjekshus, J. Eigemann, E. Gmelin, and U. Sondermann
*Classification of Martensitic-Like Transformations in $Cr_xMn_{1-x}As_{1-y}P_y$ Crystals
(64.70; 75.40; S7.15)
(a) Vol. **110**, No. 1, p. 141—154 (1988).

K. Polański, I. Zasada, K. Dolecki, and E. Misiak
*The Structural Phase Transition in Cobalt Monocrystals Investigated by Means of SEM
(64.70 and 75.60; S1.2)
(a) Vol. **110**, No. 1, p. 231—240 (1988).

Guoan Wu, Yuanda Dong, Keqin Xiao, Yizhen He, Longshu Wang, and Jiaqing He
An Investigation of Fast Neutron Irradiation Effects on the Structure of $Cu_{50}Ti_{50}$ Metallic Glass
(61.40; 64.70; S1.2)
(a) Vol. **110**, No. 2, p. 317—321 (1988).

R. Mizeris, J. Grigas, V. Samulionis, V. Skritski, A. I. Baranov, and L. A. Shuvalov
*Microwave and Ultrasonic Investigations of Superionic Phase Transitions in $CsDSO_4$ and $CsDSeO_4$
(64.70; 62.65; 77.20; S11.1)
(a) Vol. **110**, No. 2, p. 429—436 (1988).

H. Hempel, H. Maack, and G. Sorge
*Ferroelastic Phase Transition in $LiRb_4H(SO_4)_3 \cdot H_2SO_4$. Landau-Theory and Experimental Results
(64.70 and 77.80; S11)
(a) Vol. **110**, No. 2, p. 459—465 (1988).

A. K. Garg
*High-Pressure Raman Spectroscopic Study of the Ice Ih → Ice IX Phase Transition
(64.70; 62.50; 78.30; S10)
(a) Vol. **110**, No. 2, p. 467—480 (1988).

Xingzhao Ding and Yizhen He
*A Study of Structural Relaxation in Glassy $Pd_{77.5}Cu_{6.0}Si_{16.5}$ by Microcalorimetric Measurements
(64.70 and 65.50; S1.3)
(a) Vol. **110**, No. 2, p. K67—K71 (1988).

R. Poprawski
*Comments on the Results of Dilatometric Studies of NH_4HSeO_4 Crystals
(64.70 and 77.80; S11.1)
(a) Vol. **110**, No. 2, p. K73—K76 (1988).

Huimin Shao, Huaqin Wang, Yuanfu Hsia, Rongchuan Liu, and Xin Jin
*Tetragonal-to-Orthorhombic Phase Transition of Fe-Substituted $YBa_2Cu_3O_7$
(64.70 and 74.70; S10.15)
(a) Vol. **110**, No. 2, p. K89—K91 (1988).

64.75. Solubility, segregation, and mixing

V. I. Dimitrov
Monte Carlo Determination of Short-Range Parameters for Multicomponent Solid Solutions
(61.50; 64.75)
(b) Vol. **145**, No. 1, p. 65—72 (1988).

I. J. Jabr, J. D. Meyer, K. A. Al-Saleh, and N. S. Saleh
*Ion-Beam Mixing of Te/Au. Metastable Phase Formation
(61.80 and 64.75; 73.60; S1.3; S5)
(a) Vol. **105**, No. 1, p. 177—181 (1988).

K. Meinel, M. Klaua, and H. Bethge
Segregation and Sputter Effects on Perfectly Smooth (111) and (100) Surfaces of Au–Ag Alloys Studied by AES
(68.40 and 68.55; 64.75; 79.20; S1.3)
(a) Vol. **106**, No. 1, p. 133—144 (1988).

Yu. S. Nechaev and K. H. G. Egziabher
On Gas Permeability, Diffusivity, and Characteristics of Near-Dislocation Segregation Regions in Metals
(66.30; 61.70; 64.75; S1.1; S1.2)
(a) Vol. **106**, No. 2, p. 399—406 (1988).

T. Soma, K. Kamada, and H.-Matsuo Kagaya
*The Pressure Effect on the Phase Diagrams of the Ge–Sn and Si–Sn Systems
(64.75; 62.50; S6)
(b) Vol. **147**, No. 1, p. 109—114 (1988).

L. N. Aleksandrov, S. V. Lozovskii, and S. Y. Knyazev
Silicon Zone Sublimation Regrowth
(68.55; 64.75; 71.55; S5.11)
(a) Vol. **107**, No. 1, p. 213—223 (1988).

M. Hashimoto, H. Saka, T. Imura, S. Jitsukawa, and K. Suzuki
Effect of Alloying on the Void Swelling in Cu-Base and Ag-Base Alloys
(61.70; 64.75; S1.2; S1.3)
(a) Vol. **108**, No. 1, p. 163—170 (1988).

J. Szablewski and B. Kuźnicka
*Activation Energy for Precipitation in Unformed and Deformed Cu-Cr Alloys
(64.75 and 66.30; S1.2)
(a) Vol. **108**, No. 1, p. K5—K8 (1988).

I. Manna and S. K. Pabi
*A Resistometric Study of Migrating Boundary Characteristics in Pb–Sn Alloys
(64.75 and 68.48; S4)
(a) Vol. **108**, No. 1, p. K19—K23 (1988).

Ch. Schnittler and G. Teichmann
A Kinetic Investigation of Cluster Formation in Semiconductors at Elevated Temperatures
(61.70; 64.75; S5.11)
(a) Vol. **108**, No. 2, p. 519—528 (1988).

M. Quintero, P. Grima, J. E. Avon, G. Lamarche, and J. C. Woolley
*Phase Diagram, Optical Energy Gap, and Magnetic Susceptibility of $(CuIn)_{1-z}Mn_{2z}Se_2$ Alloys
(64.75; 75.30; S8.16)
(a) Vol. **108**, No. 2, p. 599—606 (1988).

M. Sacchi, M. Antonini, and M. Prudenziati
*Ru-Metal Segregation in Borosilicate Glasses
(64.75; S10.1)
(a) Vol. **109**, No. 1, p. K23—K26 (1988).

T. Kanadani and A. Sakakibara
*Short-Range Clustering in the Al–3 mass% Mg Alloy
(64.75; S4)
(a) Vol. **110**, No. 1, p. K9—K11 (1988).

L. J. Huang, E. Ma, and B. X. Liu
*The Band Structure and the Anomalous Alloying Behavior of Noble Metal Based Systems
(64.75; 71.25; S1.3)
(a) Vol. **110**, No. 2, p. 443—448 (1988).

64.80. Other phase properties of systems

H. J. Fischbeck
A New Electronic Pinning Mechanism for Sliding Charge Density Waves
(71.20; 64.80; 71.38; 71.45)
(b) Vol. **146**, No. 2, p. 555—572 (1988).

B. Domengès, M. Hervieu, C. Michel, A. Maignan, and B. Raveau
$La_3Ba_3Cu_6O_{14+x}$: A Single Phase or a Mixture?
(61.60; 64.80; S10.15)
(a) Vol. **107**, No. 1, p. 73—84 (1988).

H. R. Wenk, S. J. Shaffer, and G. van Tendeloo
*Planar Defects in Low Temperature Quartz
(61.70 and 64.80; S10)
(a) Vol. **107**, No. 2, p. 799—805 (1988).

D. Mercurio, B. Frit, G. Harburn, B. H. Parry, R. P. Williams, and R. J. D. Tilley
*Diffraction Patterns from Ordered Intergrowth Phases in the Bi_2O_3—TeO_2 System
(61.14 and 64.80; S10.15)
(a) Vol. **108**, No. 1, p. 111—122 (1988).

B. Schönfeld, L. Reinhard, G. Kostorz, and W. Bührer
*Short-Range Order and Atomic Displacements in Ni–20 at% Cr Single Crystals
(61.55 and 64.80; S1.2)
(b) Vol. **148**, No. 2, p. 457—471 (1988).

65. Thermal properties of condensed matter

(see also 63. Lattice dynamics; for thermodynamic properties of quantum fluids, see 67.40; for thermal properties of solid helium, see 67.80)

65.40. Heat capacities of solids

(for specific heat of superconductors, see 74.30; for specific heat of magnetic systems, see 75.40)

K. K. Mamedov, A. Yu. Yangirov, A. G. Guseinov, and A. M. Abdullaev
*Heat Capacity and Phase Transitions in Highly Anisotropic $A^{III}B^{VI}$-Type Semiconductors and Their Analogs at Low Temperatures **(Review Article)**
(65.40; S8)
(a) Vol. **106**, No. 2, p. 315—331 (1988).

Ming Mao and Yizhen He
Two-Stage Structural Relaxation of the Metallic Glass $Fe_{46}Ni_{31}V_1Si_8B_{14}$ 325
(61.40; 65.40; S1.1; S1.2)
(a) Vol. **105**, No. 2, p. 325—333 (1988).

R. Tsumura, D. Ríos Jara, M. Chavez, L. Rodríguez, T. Akachi, and R. Escudero
*Specific Heat Measurements of the β_1 and β_1' Phases in a Cu–Zn–Al Alloy
(65.40; S1.2)
(a) Vol. **105**, No. 2, p. 411—418 (1988).

K. S. Aleksandrov, S. V. Melnikova, I. N. Flerov, A. D. Vasilev, A. I. Kruglik, and I. T. Kokov
Phase Transitions in Cs_2CdI_4 Single Crystals
(64.70; 61.60; 65.40; 77.20; 78.20; S9.16)
(a) Vol. **105**, No. 2, p. 441—446 (1988).

J. Bąk-Misiuk, H. G. Brühl, W. Paszkowicz, and U. Pietsch
*Temperature Dependence of the Lattice Constant in Doped and Nonstoichiometric GaAs, $GaAs_{1-x}P_x$, and GaP
(65.40; 61.70; S7.11; S7.12)
(a) Vol. **106**, No. 2, p. 451—457 (1988).

E. Heiner
Nonequilibrium Thermalized Stage of Noble Metals
(71.10; 65.40; 71.38; S1.3)
(b) Vol. **148**, No. 2, p. 599—609 (1988).

P. Mohazzabi and F. Behroozi
*Vibrational Energy as a Function of Interatomic Distance in Rare-Gas Solids. A Universal Relationship
(61.50 and 65.40; S5.2)
(b) Vol. **149**, No. 2, p. 495—501 (1988).

J. Wąsicki, A. M. Gurevich, V. N. Eropkin, J. Mayer, I. Natkaniec, Yu. V. Telezhenko, and V. G. Khomenko
*Low-Temperature Heat Capacity and Phonom Spectra of 4,4′-Difluorobiphenyl
(63.20 and 65.40; S12)
(b) Vol. **150**, No. 1, p. 39—45 (1988).

65.50. Thermodynamic properties and entropy

S. K. S. Chauhan, R. Agrawal, P. C. Pathak, B. R. K. Gupta, and R. P. Goyal
*Thermal Effects on Thermophysical Properties of Alkali Cyanide Crystals
(65.50; 62.20;, 65.70; S11)
(b) Vol. **146**, No. 1, p. 97—103 (1988).

R. P. Goyal, O. P. Gupta, and B. R. K. Gupta
Theory of Compression in Some Fluorite Structure Crystals
(64.30; 62.20; 65.50; S9.12; S11)
(b) Vol. **146**, No. 2, p. 475—480 (1988).

J. Mimkes
Thermal Expansion, Heat of Fusion, and Electron Density of Monovalent Metals
(65.70; 65.50; S1.2; S1.3; S2)
(b) Vol. **148**, No. 1, p. 91—100 (1988).

W. A. Oates and K. Mika
*Application of the Local Environment Model to the Thermodynamics of Large Hard Core Interstitial Solutes in Binary Alloys
(61.70 and 65.50)
(b) Vol. **149**, No. 2, p. 457—464 (1988).

Nguyen Tang and Vu Van Hung
*Investigation of the Thermodynamic Properties of Anharmonic Crystals by the Momentum Method. I. General Results for Face-Centred Cubic Crystals
(65.50)
(b) Vol. **149**, No. 2, p. 511—519 (1988).

R. V. Gopala Rao and A. Satpathy
*On the Evaluation of Entropy and Thermodynamic Properties of 1:1 Type of Molten Salts through the Generalized Debye Inverse Screening Length
(65.50; S9.11)
(b) Vol. **150**, No. 1, p. 53—58 (1988).

Xingzhao Ding and Yizhen He
*A Study of Structural Relaxation in Glassy $Pd_{77.5}Cu_{6.0}Si_{16.5}$ by Microcalorimetric Measurements
(64.70 and 65.50; S1.3)
(a) Vol. **110**, No. 2, p. K67—K71 (1988).

65.70. Thermal expansion and thermomechanical effects

(see also 64.30 Equations of state)
For thermal conduction in nonmetallic liquids, see 66.60; for nonmetallic solids, see 66.70
For electronic thermal conduction, see 72.10 and 72.20
For thermal conductivity of superconductors, see 74.30
For pyroelectric and electrocaloric effects, see 77.70

H. Pykacz, Z. Czapla, and J. Mróz
Dilatometric Study of the Phase Transitions in NH_4HSeO_4
(64.70; 65.70; S11)
(a) Vol. **105**, No. 1, p. K33—K35 (1988).

Wen-Chen Zheng
*Determination of the Local Thermal Expansion Coefficient in the Vicinity of Cr^{3+} Ions Associated with Vacancies in MgO
(61.70 and 65.70; 71.70; S10.1)
(b) Vol. **145**, No. 2, p. K99—K102 (1988).

S. K. S. Chauhan, R. Agrawal, P. C. Pathak, B. R. K. Gupta, and R. P. Goyal
Thermal Effects on Thermophysical Properties of Alkali Cyanide Crystals
(65.50; 62.20; 65.70; S11)
(b) Vol. **146**, No. 1, p. 97—103 (1988).

S. U. Jen and C. J. Weng
*Thermal Expansion of $(FeM)_{80}B_{14}Si_6$ Metallic Glasses (M = V, Cr, Mn)
(65.70; 75.80; S1.61)
(a) Vol. **106**, No. 1, p. 99—105 (1988).

P. Čapková, M. Merisalo, M. Laitinen, V. Valvoda, and L. Dobiášová
*Thermal Vibrations and Thermal Expansion of $ZrC_{0.98}$ Studied by X-Ray Diffraction
(63.20 and 65.70; S1.61)
(a) Vol. **106**, No. 1, p. 107—114 (1988).

L. N. Aleksandrov
*Mechanical Stress in the Crystallization Process of Amorphous Semiconductors by Pulse Action
(61.40 and 65.70; S5.11)
(a) Vol. **106**, No. 2, p. K135—K138 (1988).

S. D. Chaturvedi, R. C. Dixit, and A. A. S. Sangachin
Analysis of Crystal Binding and Thermoelastic Properties of Cuprous Halides, Silver Halides, and Thallous Halides
(61.50 and 63.90; 62.20; 65.70; S9)
(b) Vol. **147**, No. 1, p. 115—121 (1988).

E. Förster, P. Glas, K. Goetz, St. Joksch, P. V. Nickles, M. Schnürer, and I. Will
About the Sensitivity of in Situ Diffraction Measurements with X-Radiation of a Laser-Produced Plasma
(61.10; 65.70; 68.20; S5.11)
(a) Vol. **107**, No. 1, p. 85—93 (1988).

J. Mimkes
*Thermal Expansion, Heat of Fusion, and Electron Density of Monovalent Metals
(65.70; 65.50; S1.2; S1.3; S2)
(b) Vol. **148**, No. 1, p. 91—100 (1988).

P. Deus, U. Voland, and H. Neumann
*Low-Temperature Thermal Expansion of $ZnSiAs_2$
(65.70; S7.16)
(a) Vol. **108**, No. 1, p. 225—231 (1988).

J. R. Gavarri, R. Mokrani, G. Vacquier, and C. Boulesteix
*Relations between Anisotropic Defects, Structural Evolution and van der Waals Bonding in 2H-$NbSe_2$
(61.70 and 65.70; 62.20; S8)
(a) Vol. **109**, No. 2, p. 445—454 (1988).

A. Tempel, M. Mäder, A. Zehe, and R. Grötzschel
Thermal Mismatch Biased Rhombohedral Structure of Strained Epitaxial CaF_2 Films on Si (111)
(68.55; 65.70; S9.12)
(a) Vol. **109**, No. 2, p. 493—501 (1988).

S. B. Feodosov, I. A. Gospodarev, and E. S. Syrkin
*Anomalies of the Linear Expansion Coefficient in Highly Anisotropic Crystals at Low Temperatures
(65.70)
(b) Vol. **150**, No. 1, p. K19—K23 (1988).

E. A. Eivazov, A. F. Safarov, and A. O. Ismailov
*Thermal Properties of $Co_{1-x}Cu_xCr_2S_4$ in the Magnetic Phase Transition Region
(65.70; 75.50; S8)
(a) Vol. **110**, No. 1, p. K45—K49 (1988).

65.90. Other topics in thermal properties of condensed matter

M. Roy, R. N. P. Chowdhury, and H. N. Acharya
*Structural and Thermal Properties of $(Sm_{1-x}Nd_x)_2(MoO_4)_3$
(61.60 and 65.90; S10)
(a) Vol. **107**, No. 1, p. K1—K5 (1988).

W. Kullmann, H. Rauh, R. Geick, U. Steigenberger, and W. F. Kuhs
Thermal Parameters and Probability Density Functions of the Narrow Gap Semiconductor Bi_2Te_3 under Hydrostatic Pressure
(61.12 and 61.60; 62.50; 65.90; S8)
(b) Vol. **148**, No. 2, p. 445—456 (1988).

66. Transport properties of condensed matter (nonelectronic)

66.30. Diffusion in solids

V. M. Korol
Sodium-Ion Implantation into Silicon **(Review Article)**
(61.80, 66.30; 72.80; 73.40; S5.11)
(a) Vol. **110**, No. 1, p. 9—34 (1988).

J. G. Fleming and D. A. Stevenson
*Interdiffusion in Mercury Cadmium Telluride Evaluated from Vapor Phase Growth Kinetics
(66.30; 68.55; S8.15)
(a) Vol. **105**, No. 1, p. 77—85 (1988).

C. A. C. Sequeira and J. M. B. Fernandez Diniz
*Conduction Processes in Oxide Solid Electrolytes
(66.30 and 72.20; 61.70; S10)
(a) Vol. **105**, No. 1, p. 123—130 (1988).

T. Bremer, W. Heiland, A. Klekamp, and E. Krätzig
*SIMS Investigations of Iron Profiles in $LiNbO_3$:Ti Waveguides
(66.30 and 79.20; 78.20; S11.1)
(a) Vol. **105**, No. 1, p. K17—K20 (1988).

H. Bakker and J. P. A. Westerveld
*Expression for the Diffusion Coefficient in Intermetallic Compounds with Cubic Structures
(66.30)
(b) Vol. **145**, No. 2, p. 409—417 (1988).

D. Köhler, G. Koschek, and E. Kubalek
*Vacancy Diffusion Profiles at the Grain Boundaries of Electronic Ceramics
(66.30; 61.70; 78.60; S10.1)
(a) Vol. **105**, No. 2, p. 377—385 (1988).

N. Kapoor and A. N. Nigam
*Influence of Hydrogen Diffusion on Thermal Conductivity of Nickel
(66.30 and 72.15; S1.2)
(a) Vol. **105**, No. 2, p. 403—409 (1988).

F. R. Kessler, U. Barkow, R. Nies, and N. Unzner
Optical Profile Analysis of Diffused and Ion-Implanted Silicon
(78.20; 61.70; 66.30; 78.50; S5.11)
(a) Vol. **105**, No. 2, p. 627—635 (1988).

V. Hadjicontis, C. A. Londos, and K. Eftaxias
*Correlation of Diffusivities of Various Elements in Silicon
(66.30; S5.11)
(a) Vol. **105**, No. 2, p. K87—K92 (1988).

R. KLABES, A. THOMAS, G. KLUGE, W. BEYER, R. GRÖTZSCHEL, and P. SÜPTITZ
Ion-Beam Induced Silver Doping in the $Ag_2Se/GeSe_2$-Resist System
(61.80 and 68.55; 66.30; S8)
(a) Vol. **106**, No. 1, p. 57—65 (1988).

E. M. SHPILEVSKII and D. A. GORBACHEVSKII
Argon Ion Effect on Solid Phase Reaction in Copper–Tin, Copper–Antimony Films
(61.80 and 68.48; 66.30; S1.3; S4)
(a) Vol. **106**, No. 1, p. K23—K26 (1988).

YU. S. NECHAEV and K. H. G. EGZIABHER
*On Gas Permeability, Diffusivity, and Characteristics of Near-Dislocation Segregation Regions in Metals
(66.30; 61.70; 64.75; S1.1; S1.2)
(a) Vol. **106**, No. 2, p. 399—406 (1988).

K. MEINEL, M. KLAUA, CH. AMMER, and H. BETHGE
*AES Investigations on Low-Temperature Interdiffusion in Thin Au Films on Ag(111) Substrates
(66.30 and 68.48; 68.60; S1.3)
(a) Vol. **106**, No. 2, p. 493—507 (1988).

I. T. SUPRUN
Influence of Pipe Diffusion of Defects on the Amplitude Dependence of Hysteresis Internal Friction
(62.40; 66.30)
(a) Vol. **107**, No. 1, p. 141—152 (1988).

K. KONAN, G. GALIBERT, and J. CALAS
*Silver Photodissolution and Optical Properties of $GeSe_3$ Vitreous Glass Thin Films
(66.30 and 78.65; 73.60; S8)
(a) Vol. **107**, No. 1, p. 273—279 (1988).

P. SEKAR and T. M. HARIDASAN
*Diffusion Coefficient of F^- Ions in Perfect β-PbF_2 Crystals
(66.30; S9)
(b) Vol. **147**, No. 2, p. 495—500 (1988).

F. H. M. SPIT, P. I. LOEFF, and H. BAKKER
*On the Use of the ^{95}Zr Radioisotope for Diffusion Measurements
(66.30; S1)
(b) Vol. **147**, No. 2, p. K105—K107 (1988).

Y. HAYASHI
*Ordering of Ni_3Mn and Diffusivity of Hydrogen in the Alloy
(61.55 and 66.30; S1.2)
(a) Vol. **107**, No. 2, p. 619—624 (1988).

C. VAROTSOS, K. EFTAXIAS, and V. HADJICONTIS
*Correlation of the Diffusion Coefficients of Various Elements Diffusing in Ferromagnetic and Paramagnetic α-Fe
(66.30; S1.1)
(a) Vol. **107**, No. 2, p. K109—K114 (1988).

R. E. Hummel
*On the Current Density Dependence of Electromigration in Thin Films
(66.30 and 85; S4)
(a) Vol. **107**, No. 2, p. K175—K179 (1988).

Qing Gu and Xi-Mao Bao
Effect of the Structure of Laser Crystallized Silicon-on-Insulator on Impurity Diffusion
(61.50; 66.30; 72.80; S5.11)
(a) Vol. **108**, No. 1, p. 81—87 (1988).

R. A. Yakshibaev, N. N. Mukhamadeeva, and R. F. Almukhametov
*Phase Transformations and Ionic Transport in $Cu_{2-\delta}Te$ Superionic Conductor
(64.70 and 66.30; S8)
(a) Vol. **108**, No. 1, p. 135—141 (1988).

H. Kahnt, F. Schirrmeister, and A. Feltz
*Model of Photostimulated Diffusion of Ag in Amorphous Chalcogenides
(66.30; 72.20; S8)
(a) Vol. **108**, No. 1, p. 149—153 (1988).

C. A. C. Sequeira and J. M. B. Fernandes Diniz
*Transport Equations in Common Solid State Electrochemical Situations
(66.30 and 72.10; S10)
(a) Vol. **108**, No. 1, p. 155—162 (1988).

G. Jeandel, A. Warin, and G. Morlot
*Dielectric Constant and ITC Measurements of $BaF_2:Dy^{3+}$
(66.30 and 77.50; S9.12)
(a) Vol. **108**, No. 1, p. 183—187 (1988).

J. Szablewski and B. Kuźnicka
*Activation Energy for Precipitation in Undeformed and Deformed Cu–Cr Alloys
(64.75 and 66.30; S1.2)
(a) Vol. **108**, No. 1, p. K5—K8 (1988).

O. S. Rajora and A. E. Curzon
*A Study of the Diffusion of Indium into TaS_2 and $TaSe_2$ in a Scanning Electron Microscope
(66.30; 61.16; S1.63)
(a) Vol. **108**, No. 2, p. 529—536 (1988).

A. S. Bakai and N. P. Lazarev
*Diffusion of Implanted Hydrogen in a Silicon Film under Stress of a Standing Sound Wave
(66.30; 62.30; S5.11)
(a) Vol. **108**, No. 2, p. K101—K103 (1988).

G. A. KACHURIN, I. E. TYSHCHENKO, E. WIESER, and CH. WEISE
*Redistribution of Boron in Silicon by High-Temperature Irradiation with Heavy Ions
(61.80 and 66.30; S5.11)
(a) Vol. **109**, No. 1, p. 141—148 (1988).

P. ENDERS and U. SCHADE
*Atomic Diffusion in Tetrahedral Semiconductors. Approximative Treatment of Diffusion Theories
(66.30; S7.1)
(b) Vol. **149**, No. 2, p. 447—456 (1988).

G. NEUMANN, C. TUIJN, G. DE VRIES, and H. BAKKER
*Diffusion Mechanisms in B.C.C. Metals. Calculation of the Correlation Factor for Mixed Nearest and Next-Nearest Neighbour Vacancy Jumps
(66.30; S1)
(b) Vol. **149**, No. 2, p. 483—493 (1988).

V. E. BORISENKO and S. G. YUDIN
*Implanted Impurity Behaviour in Silicon Subjected to Transient Heating
(61.70 and 66.30; S5.11)
(a) Vol. **109**, No. 2, p. 395—402 (1988).

K. MEINEL, M. KLAUA, and CH. AMMER
*Determination of Concentration-Dependent Diffusion Coefficients from Thin Film/Substrate Interdiffusion
(66.30 and 68.48)
(a) Vol. **109**, No. 2, p. 525—529 (1988).

K. MEINEL and M. KLAUA
*On Pore Formation in Thin Film/Substrate Interdiffusion at Low Temperatures
(66.30 and 68.48; 68.60; S1.3)
(a) Vol. **109**, No. 2, p. K97—K100 (1988).

A. SCHRÖDER and M. FELLER-KNIEPMEIER
*The Ferromagnetic Diffusion Anomaly of the Diffusion System Iron-Antimony
(66.30; 75.50; S1.1)
(a) Vol. **110**, No. 1, p. 107—121 (1988).

X. M. XIE, T. G. CHEN, and J. HUANG
*Diffusivity of Oxygen in the Orthorhombic $YBa_2Cu_3O_y$ Phase
(66.30; S10)
(a) Vol. **110**, No. 2, p. 415–419 (1988).

O. V. ALEKSANDROV, V. V. KOZLOVSKII, V. V. POPOV, and B. E. SAMORUKOV
*Diffusion of Impurities from Implanted Silicon Layers by Rapid Thermal Annealing
(61.70 and 66.30; S5.11)
(a) Vol. **110**, No. 2, p. K61—K65 (1988).

66.70. Nonelectronic thermal conduction and heat-pulse propagation in nonmetallic solids
(for thermal conduction in solid metals, see 72.15; for statistical mechanics of lattice vibrations, see 63.70)

P. Svoboda and P. Vašek
*Low Temperature Thermal Conductivity of Metallic Glasses $Fe_xNi_{80-x}B_{20}$
(66.70; 72.15; S1.1)
(a) Vol. **106**, No. 1, p. 115—121 (1988).

G. Pompe and E. Hegenbarth
*Thermal Conductivity of Amorphous Si at Low Temperatures
(66.70; S5.11)
(b) Vol. **147**, No. 1, p. 103—108 (1988).

B. S. Evseev
Non-Equilibrium Gradient-Zone Crystallization in Semiconductors
(64.70; 66.70; 68.45)
(a) Vol. **109**, No. 1, p. K31—K34 (1988).

V. V. Gorbachev, V. M. Durasov, R. B. Żezin, E. V. Ivakin, A. S. Rubanov, and N. A. Tatyanina
*Light-Induced Thermal Gratings in Natural Diamond
(66.70 and 78.35; S6)
(b) Vol. **150**, No. 2, p. 901—905 (1988).

66.90. Other topics in nonelectronic transport properties

L. N. Kristoforov and A. V. Zolotaryuk
Dynamics of Ionic Bonding Defects in Quasi-One-Dimensional Hydrogen-Bonded Chains
(61.70 and 63.90; 66.90; 71.45; S12.1)
(b) Vol. **146**, No. 2, p. 487—501 (1988).

A. V. Zaretskii, V. F. Petrenko, and V. A. Chesnakov
*The Protonic Conductivity of Heavily KOH-Doped Ice
(66.90 and 77.20; 64.70; S10)
(a) Vol. **109**, No. 2, p. 373—381 (1988).

67. Quantum fluids and solids, liquid and solid helium

67.40. Boson degeneracy und superfluidity of helium-4

R. Bahadur and C. M. Kachhava
*Green Function Formalism and Superfluid Properties of ^{4}He
(67.40; S5.2)
(b) Vol. **147**, No. 2, p. 583—587 (1988).

67.90. Other topics in quantum fluids
(e.g., neutron-star matter)

M. Mertig, V. G. Luppov, and Yu. K. Pilipenko
Nuclear Polarization of Atomic Hydrogen
(76.60; 67.90)
(b) Vol. **149**, No. 2, p. 697—709 (1988).

68. Surfaces and interfaces, thin films and whiskers
(for impact phenomena, see 79; for crystal growth, see 61.50)

68.10. Fluid surfaces and fluid–fluid interfaces

I. T. Yakubov, A. G. Khrapak, V. V. Pogosov, and S. A. Trigger
*Thermodynamical Characteristics of Liquid Metal Droplets
(68.10; 73.30; S2)
(b) Vol. **145**, No. 2, p. 455—466 (1988).

68.20. Solid surface structures

J. Griesche, R. Enderlein, and D. Schikora
Orientation and Strain in Heteroepitaxial Growth (**Review Article**)
(61.50 and 68.48; 68.20; S5.11; S7.12; S8.13)
(a) Vol. **109**, No. 1, p. 11—38 (1988).

H. W. Zandbergen, R. Gronsky, and G. Thomas
*Surface Decomposition of Superconducting $YBa_2Cu_3O_7$
(68.20; 61.16; S10)
(a) Vol. **105**, No. 1, p. 207—218 (1988).

A. P. Bezirganyan and P. A. Bezirganyan
Solution of the Two-Dimensional Stationary Schrödinger Equation with Cosine-Like Coefficient (in View of X-Ray Diffraction)
(61.10 and 78.70; 68.20)
(a) Vol. **105**, No. 2, p. 345—355 (1988).

E. Förster, P. Glas, K. Goetz, St. Joksch, P. V. Nickles, M. Schnürer, and I. Will
About the Sensitivity of in Situ Diffraction Measurements with X-Radiation of a Laser-Produced Plasma
(61.10; 65.70; 68.20; S5.11)
(a) Vol. **107**, No. 1, p. 85—93 (1988).

V. Fritsche
Approximations for the Scattering of Excited Electrons in Solids
(79.60; 61.10; 68.20)
(b) Vol. **147**, No. 2, p. 485—494 (1988).

L. A. Bursill and Xu Dong Fan
*Surface Structure of the Ceramic Superconductor $YBa_2Cu_3O_7$
(68.20; 61.16; 74.70; S10)
(a) Vol. **107**, No. 2, p. 503—510 (1988).

S. C. Farmer and A. H. Heuer
*Surface Phase Transformation in Monoclinic ZrO_2
(68.20; 64.60; S10)
(a) Vol. **107**, No. 2, p. 571—578 (1988).

R. Hoogewijs, P. de Volder, L. Fiermans, and J. Vennik
*Room Temperature Oxidation in Air of Polycrystalline CuZn
(68.20; 79.60; S1.2; S10)
(a) Vol. **107**, No. 2, p. 625—631 (1988).

Ping Lu and David J. Smith
*The Observation of In Situ and Ex Situ Oxidation Processes for ZnTe Surfaces by High-Resolution Electron Microscopy
(68.20; 61.16; S8.13)
(a) Vol. **107**, No. 2, p. 681—691 (1988).

H.-J. Ou and J. M. Cowley
*The Surface Reaction of Pd/MgO Studied by Scanning Reflection Electron Microscopy
(68.20; 68.48; S1; S10)
(a) Vol. **107**, No. 2, p. 719—729 (1988).

G. Wallner, E. Burkel, H. Metzger, and J. Peisl
Grazing Incidence X-Ray Diffraction from Si with an Implantation Induced Amorphous Surface Layer
(61.10; 68.20; S5.11)
(a) Vol. **108**, No. 1, p. 129—133 (1988).

R. Z. Bakhtizin and V. G. Valeev
Microroughness Field Emission Electron Energy Distribution and Electron–Phonon Interaction Spectroscopy near the Metal Surface
(79.70; 68.20; S1; S4)
(a) Vol. **108**, No. 1, p. 251—255 (1988).

V. V. Aristov, U. Winter, A. Yu. Nikulin, S. V. Redkin, A. A. Snigirev, P. Zaumseil, and V. A. Yunkin
*Interference Thickness Oscillations of an X-Ray Wave on Periodically Profiled Silicon
(61.10 and 68.20; S5.11)
(a) Vol. **108**, No. 2, p. 651—655 (1988).

B. Wolf and A. Zehe
*A SIMS Analysis of Hydrogen Involved Reactions on InSb Surfaces
(68.20; 68.30; S7.13)
(b) Vol. **149**, No. 1, p. 149—156 (1988).

A. P. Bezirganyan
X-Ray Reflection from and Transmission through a Plane-Parallel Dielectric Plate with Cosine-Like Polarizability (Symmetrical Laue Case)
(61.10 and 78.70; 68.20)
(a) Vol. **109**, No. 1, p. 101—110 (1988).

G. Gosmanová and I. Kraus
X-Ray Analysis of Residual Stress State with Variable Components of Stress Tensor
(62.20; 68.20)
(a) Vol. **109**, No. 1, p. 199—204 (1988).

R. Z. Bakhtizin and Yu. M. Yumaguzin
*High-Energy Tails in the Total Energy Distribution of Field Emitted Electrons from Tungsten
(68.20 and 79.70; S1)
(b) Vol. **150**, No. 1, p. 103—108 (1988).

P. Mrozek, M. Menyhárd, A. Jabłoński, and G. Tyuliev
*Surface Composition of the Ordered Fe–Co Alloys
(68.20; 79.20; S1.1; S1.2)
(a) Vol. **110**, No. 2, p. 495—507 (1988).

68.25. Mechanical and acoustical properties of solid surfaces and interfaces

P. Ziesche and R. Kaschner
*Rigorous Theorems on Surface Stress and Relaxation Forces at Surfaces and Edges
(68.25 and 68.40; 71.10)
(b) Vol. **145**, No. 1, p. K9—K13 (1988).

N. Ganev and I. Kraus
On the Application of X-Ray Tensometry to States of Stress Due to Tangential Machining of Metal Surfaces
(61.10 and 62.20; 68.25)
(a) Vol. **106**, No. 2, p. 467—472 (1988).

68.30. Dynamics of solid surfaces and interface vibrations

K. S. Srivastava, A. Sinha, R. Srivastava, and A. Tandon
*Interaction between Two Surface Excitations in Degenerate Polar Semiconductors
(68.30 and 71.45; S7.1)
(b) Vol. **146**, No. 1, p. 141—147 (1988).

J. Czerbniak, M. Moneta, B. Pawłowski, and M. Piwowarski
*The Influence of Thermal Vibrations on RBS Proton Spectra Obtained from Subsurface Regions of Si Crystals
(61.80 and 68.30; S5.11)
(b) Vol. **147**, No. 1, p. 123—126 (1988).

K. Yonashiro, T. Tomoyose, M. Yamashiro, and M. Kobayashi
Surface Modes of a Type II Semiconductor Superlattice
(71.36 and 71.45; 68.30; 73.40)
(b) Vol. **147**, No. 1, p. 141—148 (1988).

An Hu, Shu-yi Zhang, Xiang-yang Yuan, Qing Shen, Zhong-nan Lu, and Duan Feng
*Lattice Expansion and Phonon Softening in Nb/Cu Superlattices
(63.20 and 68.30; 61.55; S1.2)
(a) Vol. **107**, No. 1, p. 153—159 (1988).

NINGSHENG LUO, WENLAN XU, and SHUECHU SHEN
Application of the Embedded Atom Method to Phonon Dispersions and Local Surface Vibrations in Copper
(63.20; 68.30; S1.2)
(b) Vol. **147**, No. 2, p. 511—516 (1988).

H. POELMAN, J. VENNIK, and G. DALMAI
*Temperature Effects on the V_2O_5 Surface Phonon Spectrum
(63.20 and 68.30; S10.1)
(a) Vol. **107**, No. 2, p. 731—737 (1988).

J. S. PAN and H. B. PAN
*Size-Quantum Effect of the Energy of a Charge Carrier in a Semiconductor Crystallite
(68.30 and 77.50; 71.38)
(b) Vol. **148**, No. 1, p. 129—141 (1988).

R. ENDERLEIN, F. BECHSTEDT, and G. GERECKE
*The Complete Set of Polar Optic Phonon Modes of Superlattices. Confined Bulk and Interface Modes
(63.20 and 68.30; 73.40)
(b) Vol. **148**, No. 1, p. 173—183 (1988).

ZIJING LIN and KELIN WANG
*A Simple Scheme for Studies of Surface Phonons of Covalent Semiconductors. I. Calculation of Force Constants
(63.20 and 68.30; S5)
(b) Vol. **148**, No. 2, p. 525—532 (1988).

E. JÄGER, M. MOSSA, and K. HEHL
Difference Equations in the Theory of Biphonons
(63.20; 68.30; 71.35)
(b) Vol. **149**, No. 1, p. 111—120 (1988).

ZIJING LIN and KELIN WANG
*A Simple Scheme for Studies of Surface Phonons of Covalent Semiconductors. II. Surface Phonons at Ideal and Relaxed Si(110) Surface
(63.20 and 68.30; S5.11)
(b) Vol. **149**, No. 1, p. 143—148 (1988).

B. WOLF and A. ZEHE
A SIMS Analysis of Hydrogen Involved Reactions on InSb Surfaces
(68.20; 68.30; S7.13)
(b) Vol. **149**, No. 1, p. 149—156 (1988).

R. ENDERLEIN
*Macroscopic Dynamic Theory of Polar Interface Modes of Superlattices
(63.20 and 68.30)
(b) Vol. **150**, No. 1, p. 85—101 (1988).

68.40. Surface energy of solids; thermodynamic properties

P. ZIESCHE and R. KASCHNER
*Rigorous Theorems in Surface Stress and Relaxation Forces at Surfaces and Edges
(68.25 and 68.40; 71.10)
(b) Vol. **145**, No. 1, p. K9—K13 (1988).

K. MEINEL, M. KLAUA, and H. BETHGE
*Segregation and Sputter Effects on Perfectly Smooth (111) and (100) Surfaces of Au–Ag Alloys Studied by AES
(68.40 and 68.55; 64.75; 79.20; S1.3)
(a) Vol. **106**, No. 1, p. 133—144 (1988).

N. J. CASTELLANI and D. B. LEROY
*Final State Core Level Shifts in Supported and Free Metal Particles
(68.40; 79.60; S4)
(b) Vol. **148**, No. 1, p. 121—128 (1988).

W. VOGELSBERGER, H.-G. FRITSCHE, and E. MÜLLER
*Size Dependence of Surface Thermodynamic Parameters of Microclusters and the Location of the Gibbs' "Surface of Tension"
(68.40; 64.10; S5.2)
(b) Vol. **148**, No. 1, p. 155—164 (1988).

T. HALICIOGLU, H. O. PAMUK, and S. ERKOC
Interatomic Potentials with Multi-Body Interactions
(63.10; 61.50; 68.40; S5.11)
(b) Vol. **149**, No. 1, p. 81—92 (1988).

Z. ONYSZKIEWICZ
*Statistical Oprator for Thin Solid Films
(68.40)
(b) Vol. **150**, No. 1, p. 109—117 (1988).

68.45. Solid–fluid interface processes

T. N. Mamontova, A. S. Kochemirovskii, and L. V. Pivovarova
Electronic Processes on the Surface of As–Se Chalcogenide Glassy Semiconductors (**Review Article**)
(73.25; 68.45; 72.40; 73.20; S8)
(a) Vol. **107**, No. 1, p. 11—43 (1988).

V. Švorčík, V. Rybka, and V. Myslík
*Photoelectrochemical Etching of n-GaAs and n-InP
(68.45 and 78.70; S7.11; S7.12)
(a) Vol. **106**, No. 1, p. K35—K39 (1988).

B. S. Evseev
Non-Equilibrium Gradient-Zone Crystallization in Semiconductors
(64.70; 66.70; 68.45)
(a) Vol. **109**, No. 1, p. K31—K34 (1988).

R. Zângaro and J. Gasiot
*A Surface Effect in $CaSO_4$:Mn/PVDF Mixtures
(68.45 and 78.60; S11.1)
(a) Vol. **109**, No. 2, p. K109—K113 (1988).

Yu. B. Bolkhovityanov, R. I. Bolkhovityanova, T. A. Gavrilova, and A. E. Dolbak
*Relaxation Kinetics of a Non-Equilibrium Interface of a Multicomponent Liquid Phase–Binary Substrate on the Example of In–Ga–As/InAs System
(68.45; 68.55; S7.12)
(a) Vol. **110**, No. 2, p. 489—494 (1988).

68.48. Solid–solid interfaces
(including bicrystals)
(for grain boundaries, see 61.70)

J. Griesche, R. Enderlein, and D. Schikora
*Orientation and Strain in Heteroepitaxial Growth (**Review Article**)
(61.50 and 68.48; 68.20; S5.11; S7.12; S8.13)
(a) Vol. **109**, No. 1, p. 11—38 (1988).

E. M. Shpilevskii and D. A. Gorbachevskii
*Argon Ion Effect on Solid Phase Reaction in Copper–Tin, Copper–Antimony Films
(61.80 and 68.48; 66.30; S1.3; S4)
(a) Vol. **106**, No. 1, p. K23—K26 (1988).

K. Meinel, M. Klaua, Ch. Ammer, and H. Bethge
*AES Investigations on Low-Temperature Interdiffusion in Thin Au Films on Ag(111) Substrates
(66.30 and 68.48; 68.60; S1.3)
(a) Vol. **106**, No. 2, p. 493—507 (1988).

E. Jároli, B. Pécz, R. Veresegyházy, F. Pászti, T. Lohner, M. Fried, I. Mojzes, and J. Gyulai
*Effect of Ion Beam Mixing on the Evolution of Arsenic from the Au–GaAs System
(61.80 and 68.48; S1.3; S7.12)
(a) Vol. **107**, No. 1, p. K15—K17 (1988).

R. Flagmeyer, K. Lenkeit, T. Baumbach, Yu. O. Kanter, and A. A. Fedorov
*Characterization of $In_xGa_{1-x}As$/GaAs Strained Layer Superlattices by Ion Backscattering-Channeling and X-Ray Diffraction
(68.48; 62.20; S7.12; S7.15)
(a) Vol. **107**, No. 1, p. K19—K24 (1988).

Guoan Cheng, Hengde Li, and Baixin Liu
*Enhanced Adhesion in Metal/Oxide Pairs by Ion Irradiation
(61.80 and 68.48; S1.1; S1.2; S10.1)
(a) Vol. **107**, No. 1, p. K29—K32 (1988).

C. Boulesteix, M. Ben Salem, B. Yangui, Z. Kang, and L. Eyring
*Thickness of Interfaces between Twins, Glide Domains, and Grain Boundaries in Oxides from HREM Studies (Sm_2O_3, Nd_2O_3, Pr_7O_{12}, $Tb_{12}O_{22}$, and $YBa_2Cu_3O_{7-x}$
(68.48; 61.16; S10)
(a) Vol. **107**, No. 2, p. 469—480 (1988).

H.-J. Ou and J. M. Cowley
The Surface Reaction of Pd/MgO Studied by Scanning Reflection Electron Microscopy
(68.20; 68.48; S1; S10)
(a) Vol. **107**, No. 2, p. 719—729 (1988).

I. Manna and S. K. Pabi
*A Resistometric Study of Migrating Boundary Characteristics in Pb–Sn Alloys
(64.75 and 68.48; S4)
(a) Vol. **108**, No. 1, p. K19—K23 (1988).

N. Q. KHANH, M. FRIED, G. BATTISTIG, Z. LACZIK, T. LOHNER, E. JÁROLI, V. SCHILLER, and J. GUYLAI
*Ellipsometric and Ion Backscattering Measurements of the Properties of Silicon-On-Insulator Structure Formed by Thermally Activated Redistribution of High-Dose Ion Implanted Nitrogen
(61.70 and 68.48; 78.65; S5.11)
(a) Vol. **108**, No. 1, p. K35—K40 (1988).

R. MATTHEIS and D. HESSE
*A Reinvestigation of Ni_2Si Thin Film Growth on Si(111) by TEM and RBS. Evidence of the Presence of an Interfacial NiSi Layer prior to Ni Consumption
(68.48 and 68.55; S1.61)
(a) Vol. **109**, No. 1, p. 217—229 (1988).

L. GUOBA, L. PRANEVIČIUS, V. SARGŪNAS, and G. ZUBAUSKAS
*Protective Coatings on GaAs and InP. High-Energy Ion Scattering Analysis
(68.48; 79.20; S7.11; S7.12)
(a) Vol. **109**, No. 1, p. 239—244 (1988).

J. SUÑÉ, I. PLACENCIA, E. FARRÉS, N. BARNIOL, and X. AYMERICH
*On the Si-SiO_2 Interface Roughness in VLSI–MOS Structures
(68.48 and 73.60; S5.11; S10.1)
(a) Vol. **109**, No. 2, p. 479—491 (1988).

K. MEINEL, M. KLAUA, and CH. AMMER
*Determination of Concentration-Dependent Diffusion Coefficients from Thin Film/Substrate Interdiffusion
(66.30 and 68.48)
(a) Vol. **109**, No. 2, p. 525—529 (1988).

K. MEINEL and M. KLAUA
*On Pore Formation in Thin Film/Substrate Inderdiffusion at Low Temperatures
(66.30 and 68.48; 68.60; S1.3)
(a) Vol. **109**, No. 2, p. K97—K100 (1988).

YONG YAN, FUQUAN JI, JUN CHEN, and DUAN FENG
*Imaging of the SOI Interface by High Resolution Electron Microscopy
(68.48; S6; S10.1)
(a) Vol. **110**, No. 1, p. K25—K28 (1988).

E. MA and M.-A. NICOLET
*Pt_2Al_3 Formation on Evaporated and Large-Grained Al Substrates
(68.48 and 68.55; S4)
(a) Vol. **110**, No. 2, p. 509—514 (1988).

H. KRAUTZ, CH. WENZEL, K. BORNKESSEL, and G. BLASEK
*Barrier Behaviour of TiW between Copper and Aluminium
(68.48; S1.2; S4)
(a) Vol. **110**, No. 2, p. K77—K79 (1988).

68.55. Thin film growth, structure, and epitaxy

Z. T. Al-Dhhan, C. A. Hogarth, and N. Riddleston
The Optical Absorption Edge in Thin Amorphous Oxide Films Based on Cerium Dioxide
(78.65; 68.55; S10.15)
(b) Vol. **145**, No. 1, p. 145—149 (1988).

J. G. Fleming and D. A. Stevenson
*Interdiffusion in Mercury Cadmium Telluride Evaluated from Vapor Phase Growth Kinetics
(66.30; 68.55; S8.15)
(a) Vol. **105**, No. 1, p. 77—85 (1988).

K. Ponnuraju, P. R. Vaya, and J. Sobhanadri
*Structural and Electrical Properties of $Pb_{1-x}Sn_xTe$ Films Grown by Molecular Beam Epitaxy
(68.55 and 73.60; S8.15)
(a) Vol. **105**, No. 1, p. 161—169 (1988).

S. Sen, S. Annamalai, and P. K. Acharya
*Study and Characterization of PECVD Oxides
(68.55 and 78.65; S10.1)
(a) Vol. **105**, No. 1, p. 171—176 (1988).

A. E. Rakhshani and J. Varghese
*The Effect of Temperature on Electrodeposition of Cuprous Oxide
(68.55; S10.1)
(a) Vol. **105**, No. 1, p. 183—188 (1988).

J. Griesche and D. Schikora
*Thermodynamic Model of the Gas Transport in a HWE System
(68.55; S8.13)
(a) Vol. **105**, No. 1, p. 189—195 (1988).

T. Baumbach, H.-G. Brühl, U. Pietsch, and H. Terauchi
*Characterization of $Ga_{1-x}Al_xAs/GaAs$ Superlattices and Thin Single Layers by X-Ray Diffraction
(61.10 and 68.55; S7.12; S7.15)
(a) Vol. **105**, No. 1, p. 197—205 (1988).

B. B. Nayak, S. K. Singh, and B. S. Acharya
*Characterization of As-Grown and Iodinated Bi_2S_3 Thin Films on Silicon and Quartz Substrates by XRD
(68.55; S8)
(a) Vol. **105**, No. 2, p. 455—459 (1988).

N. Ganesan and V. Sivaramakrishnan
Variation in Electrical Properties of SnTe Thin Films with Substrate Temperature
(73.60; 68.55; S8)
(a) Vol. **105**, No. 2, p. 461—466 (1988).

S. K. Biswas, S. Chaudhuri, and A. Choudhury
Effects of Heat Treatment on the Optical and Structural Properties of InSe Thin Films
(78.65; 68.55; S8)
(a) Vol. **105**, No. 2, p. 467—475 (1988).

P. R. VAYA, K. R. MURALI, and M. SUNDARAM
*Some Studies one Zinc Phosphide Films Grown by Hot Wall Epitaxy
(68.55; 73.60; 78.65; S7)
(a) Vol. **105**, No. 2, p. 477—484 (1988).

S. PFEFFER and B. SCHUBERT
*The Dependence of Layer Thickness on Final Cell Voltage for Anodic Oxide Films Grown Galvanostatically on $Hg_{1-x}Cd_xTe$ ($x = 0.02, 0.16, 0.20, 0.22, 0.34, 1.00$)
(68.55; S8.15)
(a) Vol. **105**, No. 2, p. K115—K119 (1988).

G. PASTOR, P. TEJEDOR, I. JIMÉNEZ, E. DOMÍNGUEZ, M. TORRES, and J. V. GARCÍA-RAMOS
*Low Pressure Chemical Vapour Deposition Amorphous Silicon Behaviour under Annealing
(61.40 and 68.55; 78.30; S5.11)
(a) Vol. **106**, No. 1, p. 11—16 (1988).

R. KLABES, A. THOMAS, G. KLUGE, W. BEYER, R. GRÖTZSCHEL, and P. SÜPTITZ
*Ion-Beam Induced Silver Doping in the $Ag_2Se/GeSe_2$-Resist System
(61.80 and 68.55; 66.30; S8)
(a) Vol. **106**, No. 1, p. 57—65 (1988).

J. GEORGE, K. S. JOSEPH, B. PRADEEP, and T. I. PALSON
*Reactively Evaporated Films of Indium Sulphide
(68.55 and 78.65; S8)
(a) Vol. **106**, No. 1, p. 123—131 (1988).

K. MEINEL, M. KLAUA, and H. BETHGE
*Segregation and Sputter Effects on Perfectly Smooth (111) and (100) Surfaces of Au–Ag Alloys Studied by AES
(68.40 and 68.55; 64.75; 79.20; S1.3)
(a) Vol. **106**, No. 1, p. 133—144 (1988).

F. LUKEŠ
*Natural Surface Films on GeS
(68.55; 78.65; S8)
(a) Vol. **106**, No. 1, p. K27—K30 (1988).

J. W. TOMM, L. WERNER, D. GENZOW, K. HERRMANN, D. SCHIKORA, and J. GRIESCHE
*Experimental Evidence of Strain-Induced Gap Shifts in PbTe Epitaxial Layers by Photoluminescence
(68.55 and 78.55; S8)
(a) Vol. **106**, No. 2, p. 509—514 (1988).

A. SINGH
*Oxidise–Etch–Oxidise and Anneal Method for Growing High Quality Ultrathin SiO_2
(68.55 and 73.40; S10.1)
(a) Vol. **106**, No. 2, p. K139—K142 (1988).

W. A. MURAD, S. M. AL-SHAMARI, F. H. AL-KHATEEB, and R. H. MISHO
*Band Gap Measurements of Co_3O_4 Thin Films Prepared by the Method of Chemical Spray Pyrolysis
(68.55 and 78.65; S10.1)
(a) Vol. **106**, No. 2, p. K143—K146 (1988).

L. N. Aleksandrov, S. V. Lozovskii, and S. Y. Knyazev
*Silicon Zone Sublimation Regrowth
(68.55; 64.75; 71.55; S5.11)
(a) Vol. **107**, No. 1, p. 213—223 (1988).

M. S. M. Sastry
Post-Deposition Resistance Increase in Island Aluminium Films
(73.60; 68.55; S4)
(a) Vol. **107**, No. 1, p. 225—233 (1988).

G. M. Gusakov, A. A. Komarnitskii, and A. S. Em
*Dynamics of Optical Parameters of Silicon and Germanium Single Crystals during Two-Pulsed Nanosecond Laser Irradiation
(61.80 and 68.55; 64.70; 78.20; S5.11; S5.12)
(a) Vol. **107**, No. 1, p. 261—271 (1988).

B. Tillack, R. Banisch, and H. H. Richter
*Generation and Characterization of Thick Silicon-on-Insulator Films
(68.55; S5.11)
(a) Vol. **107**, No. 1, p. 281—289 (1988).

A. A. A. Hady
Annealing Effects on the Electrical Properties of $Hg_{1-x}Cd_xTe$ Grown by Two Different Techniques
(73.60; 68.55; S8.15)
(a) Vol. **107**, No. 1, p. K25—K28 (1988).

V. E. Borisenko, D. I. Zarovskii, and V. V. Tokarev
*Influence of Argon Implantation on the Formation of Platinum Silicides
(61.70 and 68.55; S1.61)
(a) Vol. **107**, No. 1, p. K33—K35 (1988).

B. Wolf and A. Zehe
*Improved Layer Homogeneity by an Optimized System Configuration in MBE
(68.55)
(a) Vol. **107**, No. 1, p. K37—K42 (1988).

N. K. Flevaris, Th. Karakostas, and J. Stoemenos
*Coherency Strains and Growth of Artificially Modulated Cu–Ni Thin Films
(68.55; S1.2)
(a) Vol. **107**, No. 2, p. 579—587 (1988).

C. Sarrazin, R. J. Gaboriaud, and J. P. Rivière
*TEM Investigations of Ni–Cu Thin Films Coatings Obtained by Multilayer Technique, Coevaporation, and Ion Beam Assisted Deposition
(68.55; 61.16; S1.2)
(a) Vol. **107**, No. 2, p. 867—871 (1988).

H. Rhan and U. Pietsch
*Investigation of Nanometer Layer Heterostructures by X-Ray Grazing Incidence Diffraction
(68.55; 61.10; S7.15)
(a) Vol. **107**, No. 2, p. K93—K98 (1988).

A. Zehe, A. Adán, and M. Mäder
*A Regenerative Switching Device in a Silicon–Earthalkalifluoride Heterostructure Grown by Molecular Beam Epitaxy
(68.55; 73.60; S9.12)
(a) Vol. **107**, No. 2, p. K119—K122 (1988).

Zhigang He, Jiayou Feng, Yudian Fan, and Hengde Li
*In–Se Thin Films: Crystallization and Optical Properties
(68.55 and 78.65; S8)
(a) Vol. **107**, No. 2, p. K123—K128 (1988).

R. M. Cândea, L. P. Birò, N. Dădârlat, G. Borodi, Al. Darabont, P. Fitori, and R. Turcu
*The Influence of the Film History on Some Electrophysical Properties of VE, CAD, and COD PbSe Films
(68.55 and 73.60; S8)
(a) Vol. **108**, No. 1, p. 233—240 (1988).

B. Schumann, R. Flagmeyer, and H. Weinert
Multi-Quantum-Well Structures Prepared by the Flash Evaporation Technique. Growth, Structural, and Optical Characterization
(73.40; 68.55; 78.65; S7.1; S9.15)
(a) Vol. **108**, No. 1, p. 257—264 (1988).

S. Iselborn, B. Schröder, and J. Geiger
*The Influence of Hydrogen Surface Reactions on the Growth of Evaporated a-Si:H Films
(68.55; 78.30; S5.11)
(a) Vol. **108**, No. 1, p. 275—284 (1988).

T. Drüsedau, M. Eckler, and R. Bindemann
*Photoconductive a-Si:H with Dominant Monohydride Bonding Prepared by DC-Magnetron Sputtering
(68.55 and 73.60; 76.30; 78.65; S5.11)
(a) Vol. **108**, No. 1, p. 285—293 (1988).

M. Hada, K. Sumiyama, and Y. Nakamura
*Structure and Magnetic Properties of Nonequilibrium Fe–Ru Alloys Produced by Vapor Quenching
(68.55 and 75.70; 76.80; S1.1)
(a) Vol. **108**, No. 1, p. 295—302 (1988).

A. Popov, A. Bahnev, and S. Dardjonov
*Photoluminescence and EPR Spectra of Fe-Doped $Ga_{1-x}In_xP$ Epitaxial Layers
(68.55 and 78.55; 76.30; S7.15)
(a) Vol. **108**, No. 2, p. 631—636 (1988).

R. Mattheis and D. Hesse
*A Reinvestigation of Ni_2Si Thin Film Growth in Si(111) by TEM and RBS. Evidence of the Presence of an Interfacial NiSi Layer prior to Ni Consumption
(68.48 and 68.55; S1.61)
(a) Vol. **109**, No. 1, p. 217—229 (1988).

I. A. Bachilo, F. F. Komarov, A. P. Novikov, and S. A. Petrov
*Structural Analysis of Silicon Doped with High Doses of C^+ Ions
(61.80 and 68.55; S1.61; S5.11)
(a) Vol. **109**, No. 1, p. 231—238 (1988).

G. T. Baumbach, H. Rhan, and U. Pietsch
*Detection of the Active Layer of $A^{III}B^{V}$ Semiconductor Quantum-Well Structures by High Resolution X-Ray Diffractometry
(61.10 and 68.55; S7.12; S7.15)
(a) Vol. **109**, No. 1, p. K7—K10 (1988).

Fangqing Zhang, Guanghua Chen, Yafei Zhang, and Gong Yu
*IR and ESR Studies of the Structural Properties of Hydrogenated Amorphous Carbon Films
(68.55; 76.30; 78.65; S5)
(a) Vol. **109**, No. 1, p. K39—K44 (1988).

G. K. Padam, G. L. Malhotra, and S. U. M. Rao
*Electron Microscopy and Optical Studies of Chemically Deposited $CuInS_2$ Thin Films
(68.55; 78.65; S7.16)
(a) Vol. **109**, No. 1, p. K45—K49 (1988).

S. Nara
*A Note on the Icosahedral Morphology of Ultrafine Metal Particles
(68.55; 61.70; 73.20; S1.3)
(b) Vol. **149**, No. 2, p. 555—564 (1988).

A. Abdellaoui, L. Martin, and A. Donnadieu
*Structure and Optical Properties of MoO_3 Thin Films Prepared by Chemical Vapor Deposition
(68.55 and 78.65; S10.1)
(a) Vol. **109**, No. 2, p. 455—462 (1988).

M. Anwar and C. A. Hogarth
Optical Properties of Amorphous Thin Films of MoO_3 Deposited by Vacuum Evaporation
(78.65; 68.55; S10.1)
(a) Vol. **109**, No. 2, p. 469—478 (1988).

A. Tempel, M. Mäder, A. Zehe, and R. Grötzschel
*Thermal Mismatch Biased Rhombohedral Structure of Strained Epitaxial CaF_2 Films on Si (111)
(68.55; 65.70; S9.12)
(a) Vol. **109**, No. 2, p. 493—501 (1988).

W. Gawalek, W. Michalke, H. Bruchlos, T. Eick, R. Hergt, and G. Schmidt
*Preparation of High-T_c Y–Ba–Cu–O and Bi–Sr–Ca–Cu–O Thin Films by RF-Magnetron Sputtering
(68.55; 74.70; S10.15)
(a) Vol. **109**, No. 2, p. 503—509 (1988).

S. Chaudhuri, S. K. Biswas, and A. Choudhury
*The Effects of Heat Treatment on the Structural Properties of Amorphous $In_{0.3}Se_{0.7}$ Films
(68.55; S8)
(a) Vol. **109**, No. 2, p. 511—516 (1988).

R. H. Misho, W. A. Murad, G. H. Fatahalah, I. M. Abdul Aziz, and H. M. Al-Doori
*Preparation and Optical Properties of Thin Nickel Oxide Solid Films
(68.55 and 78.65; S10.1)
(a) Vol. **109**, No. 2, p. K101—K104 (1988).

N. ROMEO, A. BOSIO, and V. CANEVARI
*$Zn_{0.15}Cd_{0.80}S$ Thin Films by RF Sputtering in an Ar–H_2 Atmosphere
(68.55 and 73.60; S8.15)
(a) Vol. **109**, No. 2, p. K105—K108 (1988).

D. KASHCHIEV and YU. O. KANTER
*Oscillations of Specular Beam Intensity in Reflection Diffraction from the Surface of a Growing Epitaxial Film. A Theoretical Study
(61.14 and 68.55; S1.2)
(a) Vol. **110**, No. 1, p. 61—76 (1988).

W. ANDRÄ, G. GÖTZ, H. HOBERT, V. MISYUCHENKO, V. A. SAMUILOV, and V. STELMAKH
*IR and RBS Spectroscopy Investigation of Semi-Insulating Phosphorus-Doped Polycrystalline Silicon Layers
(68.55 and 78.65; S5.11)
(a) Vol. **110**, No. 1, p. 181—187 (1988).

K. MEINEL, M. KLAUA, and H. BETHGE
*On Twin and Stacking Fault Formation during the Epitaxial Film Growth of F.C.C. Materials on (111) Substrates
(61.70 and 68.55; S1.3)
(a) Vol. **110**, No. 1, p. 189—196 (1988).

A. HEILMANN, C. HAMANN, and DO NGOC UAN
*Fractalic Growth in Metal-Cluster Polymer Composite Films
(68.55; S1.3)
(a) Vol. **110**, No. 1, p. K29—K31 (1988).

R. KUSZELEWICZ, J. L. OUDAR, R. AZOULAY, J. C. MICHEL, J. BRANDON, and O. EMILE
*All-Epitaxial GaAs/AlAs Nonlinear Etalons. Towards Continuous and Parallel Operation
(68.55 and 78.20; S7.12; S7.15)
(b) Vol. **150**, No. 2, p. 465—470 (1988).

YU. B. BOLKHOVITYANOV, R. I. BOLKHOVITYANOVA, T. A. GAVRILOVA, and A. E. DOLBAK
Relaxation Kinetics of a Non-Equilibrium Interface of a Multicomponent Liquid Phase–Binary Substrate on the Example of In–Ga–As/InAs System
(68.45; 68.55; S7.12)
(a) Vol. **110**, No. 2, p. 489—494 (1988).

E. MA and M.-A. NICOLET
*Pt_2Al_3 Formation on Evaporated and Large-Grained Al Substrates
(68.48 and 68.55; S4)
(a) Vol. **110**, No. 2, p. 509—514 (1988).

68.60. Physical Properties of thin films: nonelectronic

K. Meinel, M. Klaua, Ch. Ammer, and H. Bethge
AES Investigations on Low-Temperature Interdiffusion in Thin Au Films on Ag(111) Substrates
(66.30 and 68.48; 68.60; S1.3)
(a) Vol. **106**, No. 2, p. 493—507 (1988).

K. Löbe and R. Boyn
Dye-Laser 4f–4f-Line Probing of Crystal-Field Fluctuations in ZnS:Sm^{3+} Polycrystalline Films
(71.55; 68.60; 78.55; S8.11)
(b) Vol. **148**, No. 1, p. 263—272 (1988).

K. Meinel and M. Klaua
On Pore Formation in Thin Film/Substrate Interdiffusion at Low Temperatures
(66.30 and 68.48; 68.60; S1.3)
(a) Vol. **109**, No. 2, p. K97—K100 (1988).

B. A. Budkevich, I. A. Ges, S. P. Zhvavyi, G. D. Ivlev, and V. A. Pilipovich
Modification of Electrochromic WO_3 Films by Plused Ultraviolet Laser Irradiation
(61.80 and 73.60; 68.60; S10.1)
(a) Vol. **110**, No. 1, p. 155—164 (1988).

R. Klabes, A. Thomas, and G. Kluge
*Utilization of Ag:$GeSe_2$ Films as Inorganic Positive Resist
(68.60; S8)
(a) Vol. **110**, No. 2, p. K81—K84 (1988).

70. Condensed matter: electronic structure; electrical, magnetic, and optical properties

71. Electronic states

(see also 63. Lattice dynamics, 73. Electronic structure and electrical properties of surfaces, interfaces, and thin films)

71.10. General theories and computational techniques

T. Bhattacharya, A. Manna, and A. K. Ghosh
*Construction of Electronic Charge Density in Solids. Calculation of Atomic Scattering Factors of Elements and Energy of Formation of Alloys
(71.10; 61.55; S1)
(b) Vol. **145**, No. 1, p. 181—186 (1988).

L. C. R. Alfred
*Ion Interaction in Simple Metals
(71.10; 71.25; S2)
(b) Vol. **145**, No. 1, p. 203—208 (1988).

P. Ziesche and R. Kaschner
Rigorous Theorems on Surface Stress and Relaxation Forces at Surfaces and Edges
(68.25 and 68.40; 71.10)
(b) Vol. **145**, No. 1, p. K9—K13 (1988).

S. Romanowski
An Application of the Extended CPA Method to Electron Work Function and Surface Charge Calculation for Cu Single Crystals
(73.30; 71.10; S1.2)
(b) Vol. **145**, No. 2, p. 467—482 (1988).

R. Pérez-Alvarez and H. Rodriguez-Coppola
*Transfer Matrix in 1 D Schrödinger Problems with Constant and Position-Dependent Mass
(71.10 and 73.40)
(b) Vol. **145**, No. 2, p. 493—500 (1988).

R. Pérez-Alvarez, H. Rodriguez-Coppola, J. López-Gondar, and M. Lago-Izquierdo
*Transfer Matrix in the Quasiclassical Approximation with Constant and Position-Dependent Mass. Resonant Tunneling
(71.10 and 73.40)
(b) Vol. **145**, No. 2, p. 501—508 (1988).

S. Satpathy and Z. Pawlowska
*Construction of Bond-Centered Wannier Functions for Silicon Valence Bands
(71.10; 71.25; S5.11)
(b) Vol. **145**, No. 2, p. 555—565 (1988).

E. Kolley and W. Kolley
*Collective Action Formalism on the Keldysh Contour for Disordered and Hubbard-Correlated Electrons
(71.10; 71.45)
(b) Vol. **145**, No. 2, p. 585—596 (1988).

I. M. Mladenov
*Generalization of the Saxon-Hutner Theorem in the Relativistic Domain
(71.10 and 71.25)
(b) Vol. **145**, No. 2, p. K115—K119 (1988).

B. I. CRAIG and P. V. SMITH
*The Periodic LUC-MINDO Method and Its Application to Diamond
(71.10; S5)
(b) Vol. **146**, No. 1, p. 149—160 (1988).

N. A. SHILKOVA and V. P. SHIROKOVSKII
*Calculation of Dipole Transition Matrix Elements in Muffin-Tin Models
(71.10; 71.25)
(b) Vol. **146**, No. 1, p. 173—180 (1988).

F. BROSENS and J. T. DEVREESE
*Frequenz-Dependent Solution of the Time-Dependent Hartree-Fock Equation in the Electron Gas
(71.10 and 71.45)
(b) Vol. **147**, No. 1, p. 173—183 (1988).

J. GANSTER and D. GEISS
*Orientation Distribution Function Average of Second-Rank Material Constant Tensors. Susceptibility, Conductivity, and Extinction
(71.10)
(b) Vol. **147**, No. 1, p. 191—197 (1988).

V. DRCHAL and J. KUDRNOVSKÝ
*Theory of Auger Spectra from Alloys with Partially Filled Narrow Bands
(71.10 and 79.20; S1.2)
(b) Vol. **147**, No. 1, p. 215—222 (1988).

B. I. CRAIG and P. V. SMITH
*Determination of the Bulk Properties of Silicon by the Periodic LUC-MINDO Method
(71.10; 62.20; S5.11)
(b) Vol. **147**, No. 1, p. K31—K35 (1988).

J. GRAEFENSTEIN, I. MERTIG, E. MROSAN, V. N. ANTONOV, and VL. N. ANTONOV
*Calculated Fermi Surface Characteristics of the Noble Metals
(71.10; 71.25; 71.38; S1.3)
(b) Vol. **147**, No. 2, p. 575—582 (1988).

P. RENNERT and NGUYEN VAN HUNG
Curved Wave and Multiple Scattering Corrections in Calculated EXAFS Spectra
(61.10 and 78.70; 71.10)
(b) Vol. **148**, No. 1, p. 49—61 (1988).

R. A. VOLKOV and A. F. CHUIKO
*Finite Motion of Microparticles in Tunnel-Effect Multiquantum Well Structures
(71.10 and 73.40; 71.45)
(b) Vol. **148**, No. 1, p. 107—120 (1988).

E. Heiner
*Nonequilibrium Thermalized Stage of Noble Metals
(71.10; 65.40; 71.38; S1.3)
(b) Vol. **148**, No. 2, p. 599—609 (1988).

F. Marlow and B. Esser
*Exciton Propagation in the Field of a Driven Vibration
(71.10 and 71.35)
(b) Vol. **148**, No. 2, p. 619—627 (1988).

D. B. Tran Thoai and E. Zeitler
*Dynamical Image Potential of a Point Charge Approaching a Metal Sphere
(71.10 and 73.25)
(b) Vol. **149**, No. 1, p. 169—174 (1988).

N. A. Shilkova and V. P. Shirokovskii
*Calculation of Dipole Transition Matrix Elements in Crystals by the Relativistic Green's Function Method
(71.10)
(b) Vol. **149**, No. 1, p. 195—201 (1988).

N. A. Shilkova and V. P. Shirokovskii
*Calculation of Electron Velocity in Crystal ab Initio. Role of Relativistic Effects
(71.10; 71.25)
(b) Vol. **149**, No. 2, p. 571—577 (1988).

R. Ramírez and M. C. Böhm
*Analytic Expressions for the One-Center Elements of Compton Profiles. Slater-Type Orbitals
(71.10 and 78.70)
(b) Vol. **149**, No. 2, p. K135—K140 (1988).

R. V. Vedrinskii, L. A. Bugaev, and I. G. Levin
Single and Multiple Electron Scattering Description in X-Ray Absorption Spectrum Theory and the Problem of Solid Structure Investigation
(78.70; 61.10; 71.10; S9.11; S10.1; S11.1)
(b) Vol. **150**, No. 1, p. 307—314 (1988).

B. Esser and F. Marlow
*On the Theory of Exciton Transport Described by GME
(71.10 and 71.35)
(b) Vol. **150**, No. 1, p. K35—K38 (1988).

71.20. Electronic density of states determinations
(including energy states of liquid semiconductors)
(see also 65.40. Electronic heat capacity)

H. N. NAZARENO, I. FONSECA MORENO, R. M. RIBEIRO-TEIXEIRA, and J. R. IGLESIAS
*Density of States of a Binary Alloy with Short-Range Order in a Bethe Lattice
(71.20)
(b) Vol. **145**, No. 2, p. 525—534 (1988).

W. B. COSTA and E. L. ALBUQUERQUE
*Impurity Binding Energies in Heterostructures
(71.20; 71.55; S7.12; S7.15)
(b) Vol. **145**, No. 2, p. 597—602 (1988).

A. N. TIMOSHEVSKII
*The Peculiarities of the Electronic Structure of the $La_{2-x}Sr_xCuO_4$ Compound
(71.20 and 71.25; S11)
(b) Vol. **146**, No. 1, p. 161—171 (1988).

H. J. FISCHBECK
*A New Electronic Pinning Mechanism for Sliding Charge Density Waves
(71.20; 64.80; 71.38; 71.45)
(b) Vol. **146**, No. 2, p. 555—572 (1988).

G. PETERMANN
*DC and AC Characterization of Grain Boundaries in Float Zone Silicon
(71.20 and 73.40; S5.11)
(a) Vol. **106**, No. 2, p. 535—549 (1988).

H. HE
*Influence of Rings on Electronic Structure in Topologically Disordered Systems
(61.40 and 71.20)
(b) Vol. **147**, No. 1, p. 185—190 (1988).

P. KLEINERT
*Self-Consistent Density of States of a Disordered Electron System in the Cooper Channel
(71.20; 71.50)
(b) Vol. **147**, No. 1, p. 237—242 (1988).

A. F. BARABANOV, R. O. KUZIAN, and L. A. NIKIFOROVA
*Use of CPA in the Recursion Method for Disordered Transition Metal Alloys and Semiconductors
(71.20)
(b) Vol. **147**, No. 1, p. K25—K29 (1988).

I. Ya. Nikiforov and A. B. Kolpachev
*Electronic Structure of Niobium Nitrocarbides
(71.20; 74.70; S1.61; S1.62)
(b) Vol. **148**, No. 1, p. 205—211 (1988).

J. Kudrnovský and V. Drchal
*The Electronic Structure of Palladium–Noble Metal Alloys
(71.20; 71.55; S1.3)
(b) Vol. **148**, No. 1, p. K23—K27 (1988).

E. L. Albuquerque and W. B. Costa
*Impurity Binding Energies in Heterostructures. Two-Band Model
(71.20 and 73.40; S7.12; S7.15)
(b) Vol. **148**, No. 2, p. 543—548 (1988).

V. Nolting and W. Nolting
*Influence of Electron–Magnon Interaction on Electronic and Magnetic Properties of Ferromagnetic 4f Systems
(71.20 and 75.30; 75.10)
(b) Vol. **149**, No. 1, p. 313—325 (1988).

O. Foltin
Oblique Incidence Specular Reflectivity of Uniaxial Semiconductors near Three-Dimensional Critical Points
(78.20; 71.20)
(b) Vol. **149**, No. 1, p. K73—K76 (1988).

F. López-Aguilar, S. Balle, and J. Costa-Quintana
*Dynamic Potentials for Unstable Narrow Band Systems. Application to Intermediate Valence and Heavy Fermion Compounds
(71.20 and 71.25; S1.5; S1.63)
(b) Vol. **149**, No. 2, p. 585—593 (1988).

A. Šimůnek and G. Wiech
Angle-Resolved X-Ray Emission Spectra (ARXES) of Carbon in Potassium Graphite C_8K
(78.70; 71.20; S6)
(b) Vol. **149**, No. 2, p. 765—774 (1988).

71.25. Nonlocalized single-particle electronic states

E. L. NAGAEV
Photoinduced Magnetism and Conduction Electrons in Magnetic Semiconductors
(Review Article)
(75.30 and 78.20; 71.25; 75.60)
(b) Vol. **145**, No. 1, p. 11—64 (1988).

R. ENDERLEIN, DESHENG JIANG, and YINSHENG TANG
On the Mechanisms of Photoreflectance in Multiple Quantum Wells
(78.20; 71.25; 71.35; S7.12; S7.15)
(b) Vol. **145**, No. 1, p. 167—180 (1988).

S. YA. GOROSHCHENKO and I. I. UKRAINSKII
*Two-Dimensional Electron Pair in a Transverse Magnetic Field
(71.25; 72.10; 73.40)
(b) Vol. **145**, No. 1, p. 187—194 (1988).

L. C. R. ALFRED
Ion Interaction in Simple Metals
(71.10; 71.25; S2)
(b) Vol. **145**, No. 1, p. 203—208 (1988).

V. I. GAVRILENKO, N. I. KLYUI, V. G. LITOVCHENKO, and V. E. STRELNITSKII
*Characteristic Features on the Electronic Structure of Carbon Films
(71.25; 78.20; S5)
(b) Vol. **145**, No. 1, p. 209—217 (1988).

K. TAKARABE
*Optical Properties of InS under High Pressure
(71.25; 78.40; S8)
(b) Vol. **145**, No. 1, p. 219—225 (1988).

S. SATPATHY and Z. PAWLOWSKA
Construction of Bond-Centered Wannier Functions for Silicon Valence Bands
(71.10; 71.25; S5.11)
(b) Vol. **145**, No. 2, p. 555—565 (1988).

I. M. MLADENOV
*Generalization of the Saxon-Hutner Theorem in the Relativistic Domain
(71.10 and 71.25)
(b) Vol. **145**, No. 2, p. K115—K119 (1988).

M. BURKERT, R. LINK, N. PUHLMANN, and W. VOLLMANN
*Electron Energy Loss Measurements on Tetracyanoquinodimethane (TCNQ)
(71.25; 71.45; S12)
(b) Vol. **145**, No. 2, p. K121—K123 (1988).

M. de Dios Leyva, J. López Gondar, and J. Sabín del Valle
*Erratum to: Nonparabolicity Parameter in the Conduction Band of GaAs–$Al_xGa_{1-x}As$ Quantum Wells
(71.25; S7.13; S7.15)
(b) Vol. **145**, No. 2, p. K155 (1988).

A. N. Timoshevskii
*The Peculiarities of the Electronic Structure of the $La_{2-x}Sr_xCuO_4$ Compound
(71.20 and 71.25; S11)
(b) Vol. **146**, No. 1, p. 161—171 (1988).

N. A. Shilkova and V. P. Shirokovskii
Calculation of Dipole Transition Matrix Elements in Muffin-Tin Models
(71.10; 71.25)
(b) Vol. **146**, No. 1, p. 173—180 (1988).

M. Hartmann, R. Zimmermann, and H. Stolz
*Carrier Kinetics in a Semiconductor with Light-Induced Gaps
(71.25 and 72.20; 72.10)
(b) Vol. **146**, No. 1, p. 357—369 (1988).

P. Sobkowicz
*Self-Consistent Calculation of the 2D Subband Structure at Grain Boundaries: Some Shortcomings of the Semiclassical Approach
(71.25 and 73.40)
(b) Vol. **146**, No. 1, p. K7—K10 (1988).

E. Majková and P. Duhaj
*XPS Study of Co–Fe–B Metallic Glasses
(71.25 and 79.60; S1.2)
(b) Vol. **146**, No. 1, p. K17—K20 (1988).

A. V. Soldatov, Yu. V. Sukhetskii, A. G. Khasabov, and A. N. Gusatinskii
*X-Ray Spectroscopic and Theoretical Investigation of the Energy Band Structure of SbSI
(71.25 and 78.70; S9)
(b) Vol. **146**, No. 1, p. K21—K24 (1988).

N. A. Gushchina, S. M. Dunaevskii, and V. K. Nikulin
Energy Band Structure of $(GaAs)_m$ $(AlAs)_m$ Semiconductor Superlattices with Ultrathin Layers
(73.40; 71.25; S7.15)
(b) Vol. **146**, No. 2, p. 511—516 (1988).

M. K. Bose, C. Majumdar, A. B. Maity, and A. N. Chakravarti
*On Electron States of Quantum Wells in Ultrathin Films of Wide-Gap Semiconductors in an Oriented Magnetic Field
(71.25 and 73.60; S7.12)
(b) Vol. **146**, No. 2, p. 525—530 (1988).

P. E. VAN CAMP, V. E. VAN DOREN, and J. T. DEVREESE
*Ground State and Electronic Properties of Silicon Carbide and Boron Nitride
(71.25; S6; S7.1)
(b) Vol. **146**, No. 2, p. 573—587 (1988).

M. T. MOHAMMAD and W. A. ABDUL-GHAFOR
Properties of Fluorine-Doped SnO_2 Films Prepared by the Spray Pyrolysis Technique
(78.65; 71.25; S10.1)
(a) Vol. **106**, No. 2, p. 479—484 (1988).

R. PÉREZ-ALVAREZ, J. L. PARRA-SANTIAGO, and P. PAJÓN-SUAREZ
*Cylindrical Quantum Well with Position-Dependent Mass
(71.25 and 73.40; S7.12; S7.15)
(b) Vol. **147**, No. 1, p. 127—133 (1988).

A. R. HASSAN
*Phonon-Assisted Two-Photon Exciton Transitions in Semiconductors
(71.25 and 71.35; S7.11; S9)
(b) Vol. **147**, No. 1, p. 211—214 (1988).

L. E. OLIVEIRA
*On the Static Dielectric Constant and Plasmon Energy of $TiSe_2$ and $TiSe_2$
(71.25; 71.45; S8)
(b) Vol. **147**, No. 1, p. 223—228 (1988).

V. M. RATAG
*Influence of Bond-Angle and Dihedral-Angle Disorders on the Top of the Valence Band of Amorphous Germanium
(71.25; S5.12)
(b) Vol. **147**, No. 1, p. 243—251 (1988).

H. VAN CONG, S. CHARAR, and S. BRUNET
*Band-Gap Narrowing Due to Many-Body Effects in n-Type Degenerate GaAs Crystals
(71.25; S7.12)
(b) Vol. **147**, No. 1, p. 253—260 (1988).

K. R. ALLAKHVERDIEV, N. A. BAKHYSHOV, S. S. GUSEINOV, T. G. MAMEDOV, M. A. NIZAMETDINOVA, and I. K. EFENDIEVA
*Band Gap Variation in Mixed Crystals with TlSe-Type Crystal Structure
(71.25 and 78.40; S8)
(b) Vol. **147**, No. 1, p. K99—K104 (1988).

R. PÉREZ-ALVAREZ and P. PAJÓN-SUÁREZ
*Optical Absorption Coefficient in Cylindrical Quantum Wells
(71.25 and 73.40; S7.12)
(b) Vol. **147**, No. 2, p. 547—552 (1988).

J. GRAEFENSTEIN, I. MERTIG, E. MROSAN, V. N. ANTONOV, and VL. N. ANTONOV
Calculated Fermi Surface Characteristics of the Noble Metals
(71.10; 71.25; 71.38; S1.3)
(b) Vol. **147**, No. 2, p. 575—582 (1988).

S. Handschack, G. Paasch, and G. Gobsch
*Self-Consistent Calculation of the Electronic Structure of an Inversion Layer in a Perpendicular Magnetic Field
(71.25 and 73.40; S7.13)
(b) Vol. **147**, No. 2, p. K125—K129 (1988).

M. K. Bose, C. Majumdar, A. B. Maity, and A. N. Chakravarti
*Effect of a Tilted Magnetic Field on the Size-Quantized Levels of Quantum Wells in Ultrathin Films of Narrow-Gap Semiconductors
(71.25 and 73.60; S7.13)
(b) Vol. **148**, No. 1, p. 165—172 (1988).

E. Dębowska and J. Rudzińska-Girulska
*Electronic Structure of $Pd_{1-x}Ni_x$ Alloys Investigated by Positron Annihilation
(71.25; 71.60; S1.2)
(b) Vol. **148**, No. 1, p. 227—231 (1988).

V. V. Karyagin, I. I. Lyapilin, and V. V. Dyakin
*Thermoelectric Coefficients of a Limited 2D Gas in Quantizing Magnetic Field
(71.25 and 73.40)
(b) Vol. **148**, No. 2, p. 501—513 (1988).

R. Riera, F. Comas, C. Trallero Giner, and S. T. Pavlov
*Electron Raman Scattering in Semiconductor Quantum Wells
(71.25 and 78.30; S7.12; S7.15)
(b) Vol. **148**, No. 2, p. 533—542 (1988).

A. Kumar and I. Singh
*Electronic Structure of the Ge–ZnSe, Ge–GaAs, and Ge–AlAs(110) Interfaces and Superlattices
(71.25 and 73.40; S5.12; S7.12; S8.12)
(b) Vol. **148**, No. 2, p. 549—558 (1988).

R. Sen Gupta and S. Chatterjee
*Electronic Structure of SmH_2
(71.25; S1.4)
(b) Vol. **148**, No. 2, p. 579—584 (1988).

L. K. Mishra and D. N. Tripathy
*A Self-Consistent Method for Evaluating the Effective Mass of Electrons in Metals
(71.25; S2; S4)
(b) Vol. **148**, No. 2, p. 585—592 (1988).

J.-P. Zöllner, G. Paasch, G. Gobsch, and H. Übensee
*Theory for n-Surface Charge Layers in $Hg_{1-x}Cd_xTe$ MIS Structures
(71.25 and 73.40; S8.15)
(b) Vol. **148**, No. 2, p. 611—618 (1988).

H. Nakanishi, M. Isomae, S. Endo, and T. Irie
Piezoresistance in $CdIn_2S_4$
(72.20; 71.25; S8.16)
(b) Vol. **148**, No. 2, p. 673—682 (1988).

M. de Dios Leyva, J. López Gondar, and J. Sabin del Valle
*Interface Boundary Conditions and Nonparabolicity Effects in Quantum Wells
(71.25 and 73.40)
(b) Vol. **148**, No. 2, p. K113—K117 (1988).

A. Dargys
Addendum to "Tunneling in *k*-Space"
(72.10; 71.25; S5.12)
(b) Vol. **148**, No. 2, p. K125—K127 (1988).

V. P. Zhukov, N. I. Medvedeva, D. L. Novikov, and V. A. Gubanov
*The Study of Energy Band Structure and Chemical Bonding in Yttrium Carbide, Yttrium and Strontium Dicarbides by the LMTO and X_α DV Method
(71.25; S1.61)
(b) Vol. **149**, No. 1, p. 175—186 (1988).

B. Kumar and B. Agarwal
*Energy Band Structure of Phenazine N Oxide and Phenazine N Dioxide. Candidate Materials for Organic Photovoltaic Devices
(71.25; 72.20; S12)
(b) Vol. **149**, No. 1, p. 221—229 (1988).

N. A. Shilkova and V. P. Shirokovskii
Calculation of Electron Velocity in Crystal ab Initio. Role of Relativistic Effects
(71.10; 71.25)
(b) Vol. **149**, No. 2, p. 571—577 (1988).

T. Takeuchi
*Electronic Structure of $CuCl_4^{-3}$, $CuCl_4^{-4}$ and $CuCl_4^{-2}$ Clusters
(71.25; S9)
(b) Vol. **149**, No. 2, p. 579—584 (1988).

F. López-Aguilar, S. Balle, and J. Costa-Quintana
*Dynamic Potentials for Unstable Narrow Band Systems. Application to Intermediate Valence and Heavy Fermion Compounds
(71.20 and 71.25; S1.5; S1.63)
(b) Vol. **149**, No. 2, p. 585—593 (1988).

V. G. Khlabutin, V. I. Gridnev, and E. I. Rozum
*Quantum Modulation of Electron Dechanneling
(71.25; S5.11)
(b) Vol. **149**, No. 2, p. K129—K134 (1988).

M. Stordeur, M. Stölzer, H. Sobotta, and V. Riede
*Investigation of the Valence Band Structure of Thermoelectric $(Bi_{1-x}Sb_x)_2Te_3$ Single Crystals
(71.25; 72.20; 72.30; 78.20; S1.63)
(b) Vol. **150**, No. 1, p. 165—176 (1988).

A. Bonnet, A. Conan, M. Morsli, M. Ganne, and M. Tournoux
Electronic and Structural Properties of Tungsten Bronzes
(72.20; 71.25; S10)
(b) Vol. **150**, No. 1, p. 225—235 (1988).

J. Batarūnas, A. Audzijonis, N. Mykolaitienė, and K. Žičkus
*The Electronic Potential of Paraelectric SbSI
(71.25; S9)
(b) Vol. **150**, No. 1, p. K31—K34 (1988).

M. Iwamatsu
*On the Temperature Dependence of the Average Gap of SnTe and PbTe
(71.25; S8)
(b) Vol. **150**, No. 1, p. K39—K44 (1988).

I. N. Uraltsev, E. L. Ivchenko, P. S. Kopev, V. P. Kochereshko, and D. R. Yakovlev
Exciton Parameters and Electron Miniband Structure of GaAs/AlGaAs Superlattices
(71.35 and 78.20; 71.25; S7.12; S7.15)
(b) Vol. **150**, No. 2, p. 673—678 (1988).

L. J. Huang, E. Ma, and B. X. Liu
The Band Structure and the Anomalous Alloying Behavior of Noble Metal Based Systems
(64.75; 71.25; S1.3)
(a) Vol. **110**, No. 2, p. 443—448 (1988).

71.30. Metal–insulator transitions

I. M. Tsidilkovskii, Yu. G. Arapov, M. L. Zvereva, N. B. Brandt, V. A. Kulbachinskii, I. N. Gorbatyuk, and A. I. Elizarov
*The Influence of Compensation on the Hall Effect and Magnetoresistance of n-$Hg_{0.8}Cd_{0.2}Te$
(71.30; 72.20; S8.15)
(b) Vol. **148**, No. 1, p. 197—204 (1988).

W. Kraak, U. Troppenz, R. Herrmann, S. M. Chudinov, and V. A. Kulbachinskii
*Magnetic Freeze-Out and Magnetic-Field-Induced Semiconductor–Semimetal Transition in $Bi_{1-x}Sb_x$ Alloys under High Hydrostatic Pressure
(71.30 and 72.20; S7.13)
(b) Vol. **148**, No. 1, p. 333—347 (1988).

T. Yu. Bilgildeeva and T. A. Polyanskaya
The Quantum Corrections to the Conductivity in Strong Spin–Orbit Coupling Semiconductors
(72.20; 71.30; S7.12)
(b) Vol. **149**, No. 2, p. 649—655 (1988).

71.35. Excitons and related phenomena
(including electron–hole drops)

O. Goede and W. Heimbrodt
Optical Properties of (Zn, Mn) and (Cd, Mn) Chalcogenide Mixed Crystals and Superlattices **(Review Article)**
(78.20; 71.35; 78.55; S8.1)
(b) Vol. **146**, No. 1, p. 11—62 (1988).

R. Enderlein, Desheng Jiang, and Yinsheng Tang
On the Mechanisms of Photoreflectance in Multiple Quantum Wells
(78.20; 71.25; 71.35; S7.12; S7.15)
(b) Vol. **145**, No. 1, p. 167—180 (1988).

F. Beerwerth, D. Fröhlich, and V. Leinweber
*Temperature Dependence of Exciton Linewidth in CsI and CsBr
(71.35; 78.20; S9.11)
(b) Vol. **145**, No. 1, p. 195—201 (1988).

A. V. Sherman
*Temperature Dependence of Exciton Absorption Spectra. A Calculation Based on the Recursion Method
(71.35 and 78.40)
(b) Vol. **145**, No. 1, p. 319—332 (1988).

E. P. Pokatilov, S. I. Beril, V. M. Fomin, V. G. Litovchenko, D. V. Korbutyak, E. G. Lashkevich, and E. V. Mikhailovskaya
*The Size-Quantized States of the Wannier-Mott Exciton in Structures with Superthin Films
(71.35 and 78.65; S9)
(b) Vol. **145**, No. 2, p. 535—544 (1988).

T. Itoh, Y. Iwabuchi, and M. Kataoka
*Study on the Size and Shape of CuCl Microcrystals Embedded in Alkali-Chloride Matrices and Their Correlation with Exciton Confinement
(61.10 and 71.35; 78.20; S9)
(b) Vol. **145**, No. 2, p. 567—577 (1988).

J. Röseler, R. I. Shekhmametev, and J. Neugebauer
*Diamagnetic Shift of Bielectrons in BiI_3
(71.35; 78.20; S9)
(b) Vol. **145**, No. 2, p. 579—584 (1988).

K. Polák, D. J. S. Birch, and M. Nikl
Decay Kinetics of UV Luminescence from Undoped $PbCl_2$ Crystals
(78.55; 71.35; S9)
(b) Vol. **145**, No. 2, p. 741—747 (1988).

B. P. Antonyuk, P. Bussemer, L. Leine, and A. A. Mukhamedov
*Bistability of Molecular Excitons
(71.35 and 78.55)
(b) Vol. **145**, No. 1, p. 759—766 (1988).

A. I. Bobrysheva, S. S. Russu, and V. A. Zaloi
Light Absorption and Hyper-Raman Scattering by Many-Electron Complexes in Quantum Well Structures
(78.20 and 78.30; 71.35; S7.15)
(b) Vol. **146**, No. 1, p. 329—334 (1988).

V. S. Dneprovskii, A. I. Furtichev, V. I. Klimov, E. V. Nazvanova, D. K. Okorokov, and U. V. Vandishev
*Excitons at High Density in CdS and GaSe, and Optical Bistability
(71.35 and 78.20; S8; S8.11)
(b) Vol. **146**, No. 1, p. 341—350 (1988).

V. D. Egorov, P. Flögel, Hoang Xuan Nguyen, and M. Kaschke
*Picosecond Transient Optical Nonlinearities in CdS and CdSSe
(71.35 and 78.20; S8.11; S8.15)
(b) Vol. **146**, No. 1, p. 351—356 (1988).

R. Zimmermann
*Nonlinear Optics and the Mott Transition in Semiconductors
(71.35 and 78.20)
(b) Vol. **146**, No. 1, p. 371—384 (1988).

J. H. Collet and M. Pugnet
Picosecond Plasma Dynamics and All-Optical Data Processing
(78.20; 71.35; S8.12)
(b) Vol. **146**, No. 1, p. 393—401 (1988).

T. Tsuboi
*Energy Level Discreteness of CuCl Electrons Confined to a Three-Dimensional Potential Well
(71.35 and 78.40; S9; S9.12)
(b) Vol. **146**, No. 1, p. K11—K15 (1988).

O. Goede, W. Heimbrodt, M. Lamla, and V. Weinhold
Energy Gap of MnS
(78.40 and 78.55; 71.35; S8.11)
(b) Vol. **146**, No. 1, p. K65—K69 (1988).

T. Itoh, Y. Iwabuchi, and T. Kirihara
*Size-Quantized Excitons in Microcrystals of Cuprous Halides Embedded in Alkali-Halide Matrices
(71.35; 78.55; S9)
(b) Vol. **146**, No. 2, p. 531—543 (1988).

R. Zimmermann
*On the Dynamical Stark Effect of Excitons. The Low-Field Limit
(71.35; 78.20)
(b) Vol. **146**, No. 2, p. 545—554 (1988).

T. Hönig and J. Gutowski
*Stimulated Emission of the I_1—TA(P_M) Band in CdS
(71.35 and 71.55; 78.45; S8.11)
(b) Vol. **146**, No. 2, p. 589—595 (1988).

A. R. Hassan
*Phonon-Assisted Two-Photon Exciton Transitions in Semiconductors
(71.25 and 71.35; S7.11; S9)
(b) Vol. **147**, No. 1, p. 211—214 (1988).

A. V. Egranov and A. I. Nepomnyachikh
*Excitonic Processes and Thermoluminescence in NaCl Crystals Doped with Mg or Mn
(61.70 and 71.35; 78.60; S9.11)
(b) Vol. **147**, No. 1, p. 267—272 (1988).

W. Schäfer, K.-H. Schuldt, and J. Treusch
*Many-Body Effects in Resonantly Excited Dense Exciton Systems
(71.35 and 78.20)
(b) Vol. **147**, No. 2, p. 699—710 (1988).

A. I. Bobrysheva, S. A. Moskalenko, and Yu. M. Shvera
*The Coherent Pairing and Bose-Einstein Condensation of Excitons in Semiconductors
(71.35)
(b) Vol. **147**, No. 2, p. 711—715 (1988).

P. I. Khadzhi, G. D. Shibarshina, A. Kh. Rotaru, and Yu. D. Slavov
*Optical Bistability and Multistability in a System of High Density Excitons and Biexcitons in Semiconductors
(71.35 and 78.20)
(b) Vol. **147**, No. 2, p. 721—725 (1988).

A. A. Klochikhin, B. S. Razbirin, D. K. Nelson, T. Amand, M. Brousseau, J. Collet, A. Cornet, and M. Pugnet
Picosecond Kinetics of Luminescence in Highly Excited Direct-Gap Semiconductors
(78.55; 71.35; S8.11; S8.12)
(b) Vol. **147**, No. 2, p. 727—732 (1988).

G. Manzke and K. Henneberger
*Nonlinear Propagation of Laser Pulses in GaAs–GaAlAs MQWS
(71.35 and 78.20; 73.40; S7.15)
(b) Vol. **147**, No. 2, p. 733—739 (1988).

P. I. Khadzhi and E. S. Kiseleva
The Dispersion Relations of the Nonlinear Surface Waves Due to the Exciton and Biexciton Formation
(78.20; 71.35; 73.40)
(b) Vol. **147**, No. 2, p. 741—745 (1988).

D. Kremp, M. Schlanges, and T. Bornath
*Kinetic Theory of Chemically Reacting Systems. Nonideality Effects in Rate Equations
(71.35; S5.12)
(b) Vol. **147**, No. 2, p. 747—756 (1988).

TH. SCHMIDT, J. KÖHLER, C. KRYSCHI, and D. SCHMID
Sensitized Luminescence of Pure and Doped $NaNO_2$ Single Crystals. III. Excitation Spectroscopy and Time Resolved Trap Fluorescence
(78.55; 71.35; 71.55; S11)
(b) Vol. **147**, No. 2, p. 797—809 (1988).

F. R. KESSLER, J. NIEDER, R. NIES, and E. GAULKE
*Indirect Exciton Excitation and the Oscillatory Magnetophotoconductivity of Germanium
(71.35 and 72.40; 78.20; S5.12)
(b) Vol. **147**, No. 2, p. K143—K148 (1988).

I. CH. AKOPYAN, D. N. GROMOV, A. A. KLOCHIKHIN, B. V. NOVIKOV, and S. G. OGLOBLIN
*Effect of Lattice Disorder on the Excitonic States in Superionic Crystals
(71.35; 78.20; S9)
(b) Vol. **148**, No. 1, p. 185—195 (1988).

V. P. GRIBKOVSKII, L. G. ZIMIN, S. V. GAPONENKO, I. E. MALINOVSKII, P. I. KUZNETSOV, and G. G. YAKUSHCHEVA
*Optical Hysteresis near Excitonic Resonance in Zinc Selenide Monocrystals
(71.35 and 78.20; S8.12)
(b) Vol. **148**, No. 1, p. K65—K69 (1988).

F. MARLOW and B. ESSER
*Exciton Propagation in the Field of a Driven Vibration
(71.10 and 71.35)
(b) Vol. **148**, No. 2, p. 619—627 (1988).

K. DUNPHY and W. W. DULEY
Multiphoton Excitation of the 520 nm Luminescence Band in MgO:Al
(78.55; 71.35; 71.55; S10.1)
(b) Vol. **148**, No. 2, p. 729—735 (1988).

B. M. NITSOVICH, O. V. KRECHIVSKII, and G. M. PESTRYAKOV
*Nonlinearity of Absorption in an Exciton–Phonon System
(71.35 and 78.20)
(b) Vol. **148**, No. 2, p. K165—K168 (1988).

E. JÄGER, M. MOSSA, and K. HEHL
Difference Equations in the Theory of Biphonons
(63.20; 68.30; 71.35)
(b) Vol. **149**, No. 1, p. 111—120 (1988).

YA. M. STRELNIKER
*Theory of Inelastic Low-Energy Electron Scattering with Generation of Surface and Bulk Excitons in Molecular Crystals
(71.35; 73.20)
(b) Vol. **149**, No. 2, p. 603—613 (1988).

M. FLIYOU, M. A. KHAN, and J. RINGEISSEN
*Dynamical Effects on the Binding Energies of Core Excitons in Covalent and Polar Semiconductors
(71.35; S5.11; S5.12; S7.1; S8.1)
(b) Vol. **149**, No. 2, p. 615—621 (1988).

K. P. TCHAKPELE and J. P. ALBERT
*Excitons Bound to Te Impurities in CdS, ZnS, and Their Mixed Compounds with Wurtzite Structure
(71.35 and 71.55; S8.11; S8.15)
(b) Vol. **149**, No. 2, p. 641—648 (1988).

A. V. SHERMAN
*A Recursion Method for Calculating Many-Particle Green's Functions. Resonant Secondary Radiation and Excitation Spectra in the Exciton Region
(71.35 and 78.30)
(b) Vol. **149**, No. 2, p. 725—737 (1988).

R. BALTRAMIEJŪNAS, V. STEPANKEVIČIUS, G. TAMULAITIS, and A. ŽUKAUSKAS
*Investigation of Hot Exciton Interaction in ZnTe Crystals
(71.35; 71.38; 78.55; S8.13)
(b) Vol. **150**, No. 1, p. 139—146 (1988).

F. DUJARDIN, B. STÉBÉ, and G. MUNSCHY
*Binding Energies of Neutral Bound Excitons for Intermediate to High Magnetic Fields
(71.35 and 71.55)
(b) Vol. **150**, No. 1, p. 201—209 (1988).

B. ESSER and F. MARLOW
*On the Theory of Exciton Transport Described by GME
(71.10 and 71.35)
(b) Vol. **150**, No. 1, p. K35—K38 (1988).

S. SCHMITT-RINK
*Coherent Nonlinear Optical Processes in Semiconductors. Physics and Applications
(71.35 and 78.20; S7.12; S7.15; S12)
(b) Vol. **150**, No. 2, p. 349—356 (1988).

B. D. FLUEGEL, J. P. SOKOLOFF, F. JARKA, S. W. KOCH, M. LINDBERG, N. PEYGHAMBARIAN, M. JOFFRE, D. HULIN, A. MIGUS, A. ANTONETTI, C. ELL, L. BANYAI, and H. HAUG
*Measurements of Ultrafast Optical Nonlinearities in Semiconductors
(71.35 and 78.20; S8.11)
(b) Vol. **150**, No. 2, p. 357—363 (1988).

R. ZIMMERMANN and M. HARTMANN
*Resonant and Off-Resonant Light-Matter Interaction in Semiconductors. Kinetics and Transients
(71.35 and 78.20)
(b) Vol. **150**, No. 2, p. 365—378 (1988).

M. LINDBERG and S. W. KOCH
*Theory of the Optical Stark Effect in Semiconductors under Ultrashort-Pulse Excitation
(71.35 and 78.20; S8.11)
(b) Vol. **150**, No. 2, p. 379—385 (1988).

J. M. HVAM, C. DÖRNFELD, and H. SCHWAB
*Optical Nonlinearity and Phase Coherence in CdSe and $CdSe_xS_{1-x}$
(71.35 and 78.20; S8.12; S8.15)
(b) Vol. **150**, No. 2, p. 387—391 (1988).

C. Ell, J. F. Müller, K. El Sayed, L. Banyai, and H. Haug
*Evaluation of the Hartree-Fock Theory of the Excitonic Optical Stark Effect
(71.35 and 78.20)
(b) Vol. **150**, No. 2, p. 393—399 (1988).

S. A. Moskalenko, A. H. Rotaru, and V. A. Zaloi
*Periodic and Stochastic Self-Pulsations in the System of Coherent Excitons, Photons, and Biexcitons
(71.35 and 78.20)
(b) Vol. **150**, No. 2, p. 401—406 (1988).

W. Schäfer, K.-H. Schuldt, and R. Binder
Theory of the Optical Stark Effect in Semiconductors
(78.20; 71.35)
(b) Vol. **150**, No. 2, p. 407—412 (1988).

I. Balslev and A. Stahl
*Coherent Nonlinear Response of Excitons in Semiconductors
(71.35 and 78.20; S7.12; S7.15)
(b) Vol. **150**, No. 2, p. 413—418 (1988).

J. Puls, W. Rudolph, F. Henneberger, and D. Lab
*Femtosecond Studies of Room Temperature Optical Nonlinearities in Wide-Gab II–VI Compounds
(71.35 and 78.20; S8.11; S8.12; S8.15)
(b) Vol. **150**, No. 2, p. 419—423 (1988).

K. Henneberger and F. Jahnke
*Laser Excited Semiconductors: Light-Induced Gaps and Saturation Kinetics
(71.35 and 78.20; S8.11)
(b) Vol. **150**, No. 2, p. 425—430 (1988).

T. Higashimura, T. Iida, and T. Komatsu
Optical Stark Effect in an Exciton–Phonon System
(78.20; 71.35)
(b) Vol. **150**, No. 2, p. 431—436 (1988).

A. L. Ivanov and G. S. Vygovskii
The Exciton Reflection Shifting and Modification in Presence of Intensive Polariton Wave Interacting with LO-Phonons in Polar Semiconductors
(78.20; 71.35; 71.36; S8.11)
(b) Vol. **150**, No. 2, p. 443—446 (1988).

B. M. Ashkinadze, V. V. Belkov, and A. V. Subashiev
Optical and Electrical Bistability Induced by Exciton Ionization Processes
(78.20; 71.35; S5.11)
(b) Vol. **150**, No. 2, p. 533—537 (1988).

B. M. Nitsovich, O. V. Krechivskii, and G. M. Pestryakov
*Exciton–Phonon Mechanism of Optical Bistability
(71.35 and 78.20)
(b) Vol. **150**, No. 2, p. 561—565 (1988).

S. V. Bogdanov and V. G. Lyssenko
Optoelectronic Bistability with Internal Feedback in Cadmium Sulfide
(78.20; 71.35; S8.11)
(b) Vol. **150**, No. 2, p. 593—598 (1988).

A. I. Ekimov and A. L. Efros
*Nonlinear Optics of Semiconductor-Doped Glasses
(71.35 and 78.20; 78.60)
(b) Vol. **150**, No. 2, p. 627—633 (1988).

I. Akai, T. Karasawa, T. Komatsu, and Y. Kaifu
*Large Optical Nonlinearity of the Two-Dimensional Exciton States at a Stacking Fault Interface in BiI_3
(71.35 and 78.20; 71.70; 73.40; S9)
(b) Vol. **150**, No. 2, p. 635—639 (1988).

U. Woggon, F. Henneberger, and M. Müller
Optical Nonlinearities of Semiconductor Microcrystallites
(78.20; 71.35; S8.1; S9.11)
(b) Vol. **150**, No. 2, p. 641—645 (1988).

Th. Peschel
Phenomenological Model for the Optical Properties of Semiconductor-Doped Glasses
(78.20; 71.35; S8.1)
(b) Vol. **150**, No. 2, p. 657—661 (1988).

I. N. Uraltsev, E. L. Ivchenko, P. S. Kopev, V. P. Kochereshko, and D. R. Yakovlev
*Exciton Parameters and Electron Miniband Structure of GaAs/AlGaAs Superlattices
(71.35 and 78.20; 71.25; S7.12; S7.15)
(b) Vol. **150**, No. 2, p. 673—678 (1988).

E. Lach, M. Walther, G. Tränkle, A. Forchel, and G. Weimann
*Optical Investigation of Exciton Screening in GaAs/GaAlAs MQW Structures
(71.35 and 78.20; S7.12; S7.15)
(b) Vol. **150**, No. 2, p. 679—683 (1988).

G. Bongiovanni, J. L. Staehli, and D. Martin
*Many-Body Effects in the Electron–Hole Plasma Confined in GaAs–(GaAl)As Quantum Wells
(71.35 and 78.65; 71.45; S7.12; S7.15)
(b) Vol. **150**, No. 2, p. 685—690 (1988).

K. Henneberger and G. Manzke
*Dynamics of Incoherent Excitonic Optical Nonlinearities in GaAs–GaAlAs MQWS's
(71.35 and 78.20; S7.12; S7.15)
(b) Vol. **150**, No. 2, p. 691—693 (1988).

H. Weinert, V. Latinis, S. Juršenas, and R. Baltramiejunas
*Time-Resolved Photoluminescence of a GaAs/GaAlAs Superlattice
(71.35 and 78.55; S7.12; S7.15)
(b) Vol. **150**, No. 2, p. 699—704 (1988).

V. P. Gribkovskii, L. G. Zimin, S. V. Gaponenko, I. E. Malinovskii, P. I. Kuznetsov, and G. G. Yakushcheva
*Resonatorless Optical Bistability near the Excitonic Resonance of ZnSe
(71.35 and 78.65; S8.12)
(b) Vol. **150**, No. 2, p. 761—764 (1988).

Ch. Spiegelberg, M. Kretzschmar, J. Puls, and F. Henneberger
*Strong Dispersive Nonlinearity in the Excitonic Region of Wide-Gap II–VI Compounds at Room Temperature
(71.35 and 78.20; S8.11)
(b) Vol. **150**, No. 2, p. 769—775 (1988).

D. K. Nelson, A. A. Klochikhin, G. V. Mikhailov, A. G. Panfilov, and B. S. Razbirin
*Nonlinear Effects in the Emission Spectra of High Density Excitons in CdS Crystals
(71.35 and 78.55; S8.11)
(b) Vol. **150**, No. 2, p. 777—781 (1988).

N. Nagasawa
*Optical Nonlinearity Due to the Excitonic System in CuCl
(71.35 and 78.20; S9)
(b) Vol. **150**, No. 2, p. 817—823 (1988).

Nguyen Ba An
*Exciton Nonequilibrium Steady States in Laser-Excited Molecular Crystals
(71.35; 78.20)
(b) Vol. **150**, No. 2, p. 845—850 (1988).

R. P. Seisyan, A. A. Toropov, V. P. Khvostikov, and T. V. Shubina
*Excitonic Electroabsorption in GaAs–GaAlAs Single Quantum Well Waveguide Structures
(71.35 and 78.20; S7.12; S7.15)
(b) Vol. **150**, No. 2, p. 863—866 (1988).

71.36. Polaritons
(including photon–phonon and photon–magnon interactions)

Y. Q. Sheng, E. Recknagel, A. Weidinger, Z. G. Gu, Z. Y. Lai, and Z. C. Zhuang
*Raman Scattering at the CaF_2 Crystal Surface Excited by an Evanescent Optical Wave Field
(71.36 and 73.20; 78.30; S9.12)
(b) Vol. **145**, No. 1, p. 151—156 (1988).

I. I. Burshta, E. F. Venger, Yu. A. Pasechnik, and O. V. Snitko
*Simplest Properties of Guided Wave Polaritons in Three-Layer Structures
(71.36 and 73.20; S6)
(b) Vol. **146**, No. 2, p. 517—524 (1988).

K. Yonashiro, T. Tomoyose, M. Yamashiro, and M. Kobayashi
*Surface Modes of a Type II Semiconductor Superlattice
(71.36 and 71.45; 68.30; 73.40)
(b) Vol. **147**, No. 1, p. 141—148 (1988).

B. M. Nitsovich, O. V. Krekhivskii, G. M. Pestryakov, and G. T. Avanesyan
*Phonon-Induced Splitting of Polariton States in Crystals
(71.36 and 71.38)
(b) Vol. **148**, No. 1, p. K29—K32 (1988).

S. A. Moskalenko, A. H. Rotaru, Yu. M. Shvera, and V. A. Zaloi
*Quantum Fluctuations and Time Structures of High Density Polaritons
(71.36)
(b) Vol. **149**, No. 1, p. 187—194 (1988).

A. A. Maksimov, I. I. Tartakovskii, and T. A. Krivenko
*Determination of Polariton–Phonon Interaction Parameters in Anthracene Crystals
(71.36; 78.30; 78.55; S12)
(b) Vol. **150**, No. 1, p. 125—132 (1988).

A. L. Ivanov and G. S. Vygovskii
The Exciton Reflection Shifting and Modification in Presence of Intensive Polariton Wave Interacting with LO-Phonons in Polar Semiconductors
(78.20; 71.35; 71.36; S8.11)
(b) Vol. **150**, No. 2, p. 443—446 (1988).

71.38. Polarons and electron–phonon interactions
(see also 63.20. phonon–electron interactions in lattices)

E. HEINER, S. L. DRECHSLER, and J. MALEK
*On the Charge Structure of Polarons in Trans-Polyacetylene
(71.38 and 71.50; S12)
(b) Vol. **145**, No. 1, p. 233—240 (1988).

K. LEO
Exact Formula for Energy-Independent Recombination Heating
(72.30; 71.38)
(b) Vol. **145**, No. 1, p. K31—K33 (1988).

F. BROSENS and J. T. DEVREESE
*Rigorous Ground State Energy of a Linearized Model of a Polaron in a Magnetic Field
(71.38)
(b) Vol. **145**, No. 2, p. 517—523 (1988).

H. J. FISCHBECK
A New Electronic Pinning Mechanism for Sliding Charge Density Waves
(71.20; 64.80; 71.38; 71.45)
(b) Vol. **146**, No. 2, p. 555—572 (1988).

A. P. ZHERNOV and T. N. KULAGINA
*The Microcontact Spectrum of Aluminium
(71.38; 72.10; S4)
(b) Vol. **147**, No. 1, p. 149—153 (1988).

E. P. POKATILOV, S. I. BERIL, and V. M. FOMIN
*Image Potentials and Image Forces in the Polaron Theory
(71.38 and 73.40; 71.45)
(b) Vol. **147**, No. 1, p. 163—172 (1988).

V. K. FEDYANIN and V. A. OSIPOV
On the Theory of Linear Diatomic Polymers
(71.45; 71.38; S12.1)
(b) Vol. **147**, No. 1, p. 199—209 (1988).

R. RUCKH and E. SIGMUND
*Large Polarons in the Discrete SSH Model
(71.38)
(b) Vol. **147**, No. 1, p. 229—235 (1988).

J. MALEK, S. L. DRECHSLER, E. HEINER, and R. KAHNT
Solitons and Polarons in Polyyne. Discrete Lattice and Broken Electron–Hole Symmetry Effects
(71.50; 71.38; 71.45; S12.1)
(b) Vol. **147**, No. 1, p. 281—295 (1988).

J. GRAEFENSTEIN, I. MERTIG, E. MROSAN, V. N. ANTONOV, and VL. N. ANTONOV
Calculated Fermi Surface Characteristics of the Noble Metals
(71.10; 71.25; 71.38; S1.3)
(b) Vol. **147**, No. 2, p. 575—582 (1988).

J. S. PAN and H. B. PAN
Size-Quantum Effect of the Energy of a Charge Carrier in a Semiconductor Crystallite
(68.30 and 77.50; 71.38)
(b) Vol. **148**, No. 1, p. 129—141 (1988).

B. M. NITSOVICH, O. V. KREKHIVSKII, G. M. PESTRYAKOV, and G. T. AVANESYAN
*Phonon-Induced Splitting of Polariton States in Crystals
(71.36 and 71.38)
(b) Vol. **148**, No. 1, p. K29—K32 (1988).

E. HEINER
Nonequilibrium Thermalized Stage of Noble Metals
(71.10; 65.40; 71.38; S1.3)
(b) Vol. **148**, No. 2, p. 599—609 (1988).

W. W. RÜHLE and K. LEO
*Carrier Heating in GaAs by Nonradiative Recombination
(71.38; S7.12)
(b) Vol. **149**, No. 1, p. 215—220 (1988).

H. W. STREITWOLF
*Symmetry Properties of the Polaron Lattice
(71.38)
(b) Vol. **149**, No. 1, p. K13—K17 (1988).

H. LEON and F. COMAS
*On the Scattering of Electrons by Confined LO Phonons in a Semiconductor Quantum Well
(71.38 and 73.40; S7.12; S7.15)
(b) Vol. **149**, No. 2, p. 533—542 (1988).

R. BALTRAMIEJŪNAS, V. STEPANKEVIČIUS, G. TAMULAITIS, and A. ŽUKAUSKAS
Investigation of Hot Exciton Interaction in ZnTe Crystals
(71.35; 71.38; 78.55; S8.13)
(b) Vol. **150**, No. 1, p. 139—146 (1988).

H. PUFF and H. W. STREITWOLF
*Optical Absorption and Stability of the Bipolaron Lattice
(71.38; 78.20; S12)
(b) Vol. **150**, No. 1, p. 147—164 (1988).

YAN-MIN LI and LI-YUAN ZHANG
Interplay between Superconductivity and Antiferromagnetism in the Triplet Local-Electron-Pair System
(74.20 and 75.30; 71.38)
(b) Vol. **150**, No. 1, p. 211—224 (1988).

A. A. BUGAEV and B. P. ZAKHARCHENYA
Picosecond Holographic Diagnostic of Light Pulse Absorption in Semiconductors
(73.20 and 78.20; 71.38; S5.11; S7.12)
(b) Vol. **150**, No. 2, p. 891—894 (1988).

M. G. KAPLUNOV, N. D. KUSHCH, and E. B. YAGUBSKII
Optical Properties of the New Organic Superconductor (BEDT-TTF)$_2$ Cu(SCN)$_2$
(78.30; 71.38; S12)
(a) Vol. **110**, No. 2, p. K111—K116 (1988).

71.45. Collective effects

G. Irmer, M. Herms, J. Monecke, and H. Bock
Raman Scattering from LO-Phonon–Plasmon Modes in $GaAs_{1-x}P_x$ Crystals
(63.20 and 78.30; 71.45; S7.15)
(b) Vol. **145**, No. 1, p. K79—K82 (1988).

Tran Thoai Duy Bao
*Plasmon Excitation in Metal Surface Layers by Fast Electrons
(71.45 and 73.20; 61.80; S4)
(b) Vol. **145**, No. 2, p. 509—515 (1988).

E. Kolley and W. Kolley
Collective Action Formalism on the Keldysh Contour for Disordered and Hubbard-Correlated Electrons
(71.10; 71.45)
(b) Vol. **145**, No. 2, p. 585—596 (1988).

M. Burkert, R. Link, N. Puhlmann, and W. Vollmann
Electron Energy Loss Measurements on Tetracyanoquinodimethane (TCNQ)
(71.25; 71.45; S12)
(b) Vol. **145**, No. 2, p. K121—K123 (1988).

D. B. Tran Thoai and E. Zeitler
*Image Potential of an Electrcn in a Gap between Two Metals
(71.45 and 73.20; S1.3; S4)
(b) Vol. **146**, No. 1, p. 137—140 (1988).

K. S. Srivastava, A. Sinha, R. Srivastava, and A. Tandon
*Interaction between Two Surface Excitations in Degenerate Polar Semiconductors
(68.30 and 71.45; S7.1)
(b) Vol. **146**, No. 1, p. 141—147 (1988).

F. Kh. Abdullaev, A. A. Abdumalikov, and E. N. Tsoi
Soliton and Breather Dynamics in a Nematic Liquid Crystal under the Action of Constant and Time-Varying Magnetic Fields
(63.10; 71.45)
(b) Vol. **146**, No. 2, p. 457—466 (1988).

L. N. Kristoforov and A. V. Zolotaryuk
Dynamics of Ionic and Bonding Defects in Quasi-One-Dimensional Hydrogen-Bonded Chains
(61.70 and 63.90; 66.90; 71.45; S12.1)
(b) Vol. **146**, No. 2, p. 487—501 (1988).

H. J. Fischbeck
A New Electronic Pinning Mechanism for Sliding Charge Density Waves
(71.20; 64.80; 71.38; 71.45)
(b) Vol. **146**, No. 2, p. 555—572 (1988).

L. S. Brizhik, Yu. B. Gaididei, A. A. Vakhnenko, and V. A. Vakhnenko
Soliton Generation in Semi-Infinite Molecular Chains
(71.50; 71.45; S12.1)
(b) Vol. **146**, No. 2, p. 605—612 (1988).

A. S. Davydov
*Bisoliton Mechanism of High-Temperature Superconductivity
(71.45 and 74.20; S12.1)
(b) Vol. **146**, No. 2, p. 619—641 (1988).

K. Yonashiro, T. Tomoyose, M. Yamashiro, and M. Kobayashi
*Surface Modes of a Type II Semiconductor Superlattive
(71.36 and 71.45; 68.30; 73.40)
(b) Vol. **147**, No. 1, p. 141—148 (1988).

V. N. Kadantsev, L. N. Lupichev, and A. V. Savin
*Intramolecular Excitation Dynamics in a Thermalized Chain. II. Formation of Autolocalized States in a Chain with Free Ends
(71.45)
(b) Vol. **147**, No. 1, p. 155—161 (1988).

E. P. Pokatilov, S. I. Beril, and V. M. Fomin
Image Potentials and Image Forces in the Polaron Theory
(71.38 and 73.40; 71.45)
(b) Vol. **147**, No. 1, p. 163—172 (1988).

F. Brosens and J. T. Devreese
*Frequency-Dependent Solution of the Time-Dependent Hartree-Fock Equation in the Electron Gas
(71.10 and 71.45)
(b) Vol. **147**, No. 1, p. 173—181 (1988).

V. K. Fedyanin and V. A. Osipov
*On the Theory of Linear Diatomic Polymers
(71.45; 71.38; S12.1)
(b) Vol. **147**, No. 1, p. 199—209 (1988).

L. E. Oliveira
On the Static Dielectric Constant and Plasmon Energy of TiS_2 and $TiSe_2$
(71.25; 71.45; S8)
(b) Vol. **147**, No. 1, p. 223—228 (1988).

J. Malek, S. L. Drechsler, E. Heiner, and R. Kahnt
Solitons and Polarons in Polyyne. Discrete Lattice and Broken Electron–Hole Symmetry Effects
(71.50; 71.38; 71.45; S12.1)
(b) Vol. **147**, No. 1, p. 281—295 (1988).

YU B. GAIDIDEI and V. M. LOKTEV
On a Theory of the Electronic Spectrum and Magnetic Properties of High-T_c Superconductors
(74.20; 71.45; S12.1)
(b) Vol. **147**, No. 1, p. 307—319 (1988).

D. K. PAPADEMITRIOU
*Scattering of X-Rays by Plasmons in Boron
(71.45 and 78.30; S5)
(b) Vol. **147**, No. 1, p. K95—K98 (1988).

NGUYEN VAN TRONG
*Plasma Waves in Degenerate Electron Systems with Random Impurities
(71.45)
(b) Vol. **147**, No. 2, p. 567—573 (1988).

M. F. BURKHARD, H. ROTHARD, and K.-O. E. GROENEVELD
*Single-Electron Deexcitation of Volume Plasmons Induced by Heavy Ions in Thin Solid Foils
(71.45; 79.20; S1.2; S4; S5)
(b) Vol. **147**, No. 2, p. 589—592 (1988).

D. B. TRAN THOAI and E. ZEITLER
Inelastic Scattering of Fast Electrons by Small Metal Particles
(61.80 and 73.20; 71.45; S4)
(a) Vol. **107**, No. 2, p. 791—797 (1988).

R. A. VOLKOV and A. F. CHUIKO
Finite Motion of Microparticles in Tunnel-Effect Multiquantum Well Structures
(71.10 and 73.40; 71.45)
(b) Vol. **148**, No. 1, p. 107—120 (1988).

A. S. DAVYDOV and V. N. ERMAKOV
Stability of a Superconducting Condensate of Bisolitons
(74.20; 71.45)
(b) Vol. **148**, No. 1, p. 305—318 (1988).

B. S. TOŠIĆ, J. P. ŠETRAJČIĆ, and D. LJ. MIRJANIĆ
*The Influence of Phonons on Dielectric Properties of Molecular Crystals
(71.45; 77.20)
(b) Vol. **148**, No. 2, p. 593—598 (1988).

R. BALTRAMIEJŪNAS and A. ŽUKAUSKAS
Manifestation of Hot Phonons in the Luminescence Spectra of Highly Photoexcited $A^{II}B^{VI}$ Crystals
(63.20 and 78.55; 71.45; S8.12)
(b) Vol. **149**, No. 1, p. 337—346 (1988).

H. K. Sy and L. M. Song
*Surface Plasmons in Type II Superlattices
(71.45 and 73.20)
(b) Vol. **149**, No. 2, p. 595—602 (1988).

L. Wendler, R. Haupt, and V. G. Grigoryan
*Intersubband Mixing of Coupled Plasmon–Phonons in Double Heterostructures
(71.45 and 73.40)
(b) Vol. **149**, No. 2, p. K123—K128 (1988).

K. H. Aharonian, H. L. Erknapetian, and D. R. Tilley
*Collective Plasmon Modes in Thin Semiconductor Films Including the Effect of the Image Charges
(71.45; 73.60; S8)
(b) Vol. **150**, No. 1, p. 133—138 (1988).

G. Bongiovanni, J. L. Staehli, and D. Martin
Many-Body Effects in the Electron–Hole Plasma Confined in GaAs–(GaAl)As Quantum Wells
(71.35 and 78.65; 71.45; S7.12; S7.15)
(b) Vol. **150**, No. 2, p. 685—690 (1988).

Lun-Biao Xu, Shou-Fu Cai, and Guo-Xin Xie
Quantum-Statistical Investigation on the Nonlinear Optical Oscillation in Amorphous Semiconductor $GeSe_2$ Films
(78.20 and 78.65; 71.45; S8)
(b) Vol. **150**, No. 2, p. 797—803 (1988).

71.50. Localized single-particle electronic states
(excluding impurities)

E. Heiner, S. L. Drechsler, and J. Malek
*On the Charge Structure of Polarons in Trans-Polyacetylene
(71.38 and 71.50; S12)
(b) Vol. **145**, No. 1, p. 233—240 (1988).

R. A. Evarestov and V. P. Smirnov
Double-Valued Band Representations of Monoclinic and Orthorhombic Space Groups
(61.50; 71.50)
(b) Vol. **145**, No. 2, p. 391—399 (1988).

C. Zwanziger, S. Petrowa, J. Reinhold, P. Streubel, and A. Meisel
*Approximate Calculation of Inner-Shell Binding Energies and Auger Transition Energies of Solid Silicon Compounds
(71.50; 79.20; S5.11; S6)
(b) Vol. **145**, No. 2, p. 603—607 (1988).

G. Dräger, R. Frahm, G. Materlik, and O. Brümmer
On the Multipole Character of the X-Ray Transitions in the Pre-Edge Structure of Fe K Absorption Spectra. An Experimental Study
(78.70; 71.50; S1.1; S10.1)
(b) Vol. **146**, No. 1, p. 287—294 (1988).

L. S. Brizhik, Yu. B. Gaididei, A. A. Vakhnenko, and V. A. Vakhnenko
*Soliton Generation in Semi-Infinite Molecular Chains
(71.50; 71.45; S12.1)
(b) Vol. **146**, No. 2, p. 605—612 (1988).

P. Kleinert
Self-Consistent Density of States of a Disordered Electron System in the Cooper Channel
(71.20; 71.50)
(b) Vol. **147**, No. 1, p. 237—242 (1988).

J. Malek, S. L. Drechsler, E. Heiner, and R. Kahnt
*Solitons and Polarons in Polyyne. Discrete Lattice and Broken Electron–Hole Symmetry Effects
(71.50; 71.38; 71.45; S12.1)
(b) Vol. **147**, No. 1, p. 281—295 (1988).

R. E. Hummel and W. M. Goho
*Optical Properties and Electronic Structure of Copper–Cobalt Alloys
(71.50 and 78.20; S1.2)
(b) Vol. **147**, No. 1, p. 417—420 (1988).

A. Brezini, M. Bendaoud, M. Sebbani, and F. Behilil
*Effects of Electron–Electron Interactions on the Nature of the Eigenstates in Amorphous Materials
(71.50)
(b) Vol. **148**, No. 1, p. 233—242 (1988).

W. Weller and M. Kasner
*Localization in a System of N Coupled Chains. Recursion Method. I. General Approach
(71.50)
(b) Vol. **148**, No. 1, p. 273—287 (1988).

M. Kasner and W. Weller
*Localization in a System of N Coupled Chains — Recursion Method. II. Explicit Calculation for Two Coupled Chains
(71.50)
(b) Vol. **148**, No. 2, p. 635—644 (1988).

71.55. Impurity and defect levels

R. Boyn
4f–4f Luminescence of Rare-Earth Centers in II–VI Compounds **(Review Article)**
(78.55 and 78.60; 71.55; S8.1)
(b) Vol. **148**, No. 1, p. 11—47 (1988).

L. F. Makarenko and L. I. Murin
*Trapping of Minority Carriers in Thermal U^--Donors in n-Si
(71.55; 72.40; S5.11)
(b) Vol. **145**, No. 1, p. 241—253 (1988).

H. Engelmann, W. Gatzweiler, I. Dézsi, and U. Gonser
*A Comparative Study of Impurity Defects in Crystalline and Amorphous $LiNbO_3$
(71.55; 76.80; S11.1)
(a) Vol. **105**, No. 1, p. 219—230 (1988).

K. Žďánský and Le Thanh Binh
*Characterization of DX Centers in GaAlAs Grown by Liquid-Phase Epitaxy
(71.55; S7.15)
(a) Vol. **105**, No. 1, p. K51—K55 (1988).

V. N. Brudnyi and V. V. Peschev
*Electron Traps in n-GaAs Irradiated with High Electron Beam Fluxes at High Temperatures
(61.80 and 71.55; S7.12)
(a) Vol. **105**, No. 1, p. K57—K60 (1988).

W. B. Costa and E. L. Albuquerque
Impurity Binding Energies in Heterostructures
(71.20; 71.55; S7.12; S.7.15)
(b) Vol. **145**, No. 2, p. 597—602 (1988).

K. Kreissl and W. Gehlhoff
*Electron Paramagnetic Resonance of the Mn_4^0 Cluster in Silicon
(71.55 and 76.30; 61.70; S5.11)
(b) Vol. **145**, No. 2, p. 609—617 (1988).

M. Grinberg
*The Substitutional Shallow Donor Size Effect
(71.55; S8.12)
(b) Vol. **145**, No. 2, p. 619—624 (1988).

N. Russo, M. Toscano, V. Barone, and C. Minichino
*Model Clusters and Electronic Characteristics of Deep-Level Impurities in Silicon
(71.55; S5.11)
(b) Vol. **145**, No. 2, p. K125—K130 (1988).

W. DĄBROWSKI
*Investigation of Deep Impurities in Schottky Diodes on High-Resistivity Silicon
(71.55; 73.40; S5.11)
(a) Vol. **105**, No. 2, p. 511—520 (1988).

R. FORNARI and L. DOZSA
*Deep Level Characterization in Semi-Insulating LEC Gallium Arsenide
(71.55; 72.40; 78.50; S7.12)
(a) Vol. **105**, No. 2, p. 521—530 (1988).

I. V. MARCHISHIN, V. N. OVSYUK, and S. B. SEVASTIANOV
*Deep Level Profiling Using an Admittance Spectroscopy Method
(71.55; 73.40)
(a) Vol. **106**, No. 1, p. 153—166 (1988).

J. HŘEBÍČEK, V. ČECH, and A. BRABLEC
*Determination of Trap Concentrations and Energy Levels in Insulators and Semiconductors from Steady-State Space-Charge-Limited Currents
(71.55; 72.20)
(a) Vol. **106**, No. 1, p. 167—172 (1988).

K. KAWABATA, T. SEIYAMA, S. OKUDA, and Y. SHONO
*Thermally Stimulated Exo-Electron Emission and Thermally Stimulated Luminescence of Irradiated α-Al_2O_3 Powder
(71.55 and 79.75; 78.60; S10.1)
(a) Vol. **106**, No. 1, p. 181—190 (1988).

D. S. DOMANEVSKII, S. V. ZHOKHOVETS, and M. V. PROKOPENYA
Peculiarities of Radiative Recombination in Gallium Arsenide Doped with Shallow Donors and Acceptors
(78.55; 71.55; S7.12)
(a) Vol. **106**, No. 1, p. 249—260 (1988).

T. HÖNIG and J. GUTOWSKI
*Stimulated Emission of the I_1—TA(P_M) Band in CdS
(71.35 and 71.55; 78.45; S8.11)
(b) Vol. **146**, No. 2, p. 589—595 (1988).

V. DRCHAL and J. KUDRNOVSKÝ
*Impurity Auger Spectra. Host Crystal with Unfilled Valence Band
(71.55 and 79.20; S1)
(b) Vol. **146**, No. 2, p. 597—604 (1988).

I. V. MARCHISHIN, V. N. OVSYUK, and S. B. SEVASTIANOV
*Investigation of Radiation Defects in Hyperabrupt p–n Junctions
(71.55 and 73.40; 61.80)
(a) Vol. **106**, No. 2, p. 557—568 (1988).

G. KOSCHEK, H. LAKNER, and E. KUBALEK
On the Origin of Cathodoluminescence Contrast Phenomena in Semiinsulating GaAs
(78.60; 71.55; S7.12)
(a) Vol. **106**, No. 2, p. 651—658 (1988).

L. F. Makarenko
*On the Cross Section of Electron Capture into the Deep State of Thermal U^- Donors in Si
(71.55; 72.40; S5.11)
(a) Vol. **106**, No. 2, p. K153—K155 (1988).

E. M. Dianov, V. O. Sokolov, and V. B. Sulimov
*Theory of H(I) Centers in Vitreous Silicon Dioxide
(71.55; S10.1)
(b) Vol. **147**, No. 1, p. 261—266 (1988).

S. G. Lyapin, M. I. Eremets, A. M. Shirokov, R. Herrmann, H.-U. Müller, and F. Ludwig
*Submillimetre Magnetospectroscopy of Tellurium under Hydrostatic Pressure
(71.55 and 78.50; 72.40; 76.40; S5)
(b) Vol. **147**, No. 1, p. 297—306 (1988).

L. N. Aleksandrov, S. V. Lozovskii, and S. Y. Knyazev
Silicon Zone Sublimation Regrowth
(68.55; 64.75; 71.55; S5.11)
(a) Vol. **107**, No. 1, p. 213—223 (1988).

F. S. Nasredinov, V. F. Masterov, Ch. S. Saidov, P. P. Seregin, N. N. Troitskaya, and H. U. Tschirner
*Nature of the Electrical Inactivity of Tin and Iron Impurity Atoms in In_2Te_3 and Ga_2Te_3
(71.55; 76.80; S8)
(a) Vol. **107**, No. 1. p. 291—298 (1988).

B. Rheinländer, G. Grummt, J. Kováč, H. Haefner, and R. Heilmann
Minority-Carrier Generation and Recombination in 1.35 μm InGaAsP/InP Double Heterostructure Diodes
(73.40; 71.55; 72.40; 78.60; S7.11; S7.15)
(a) Vol. **107**, No. 1, p. 405—418 (1988).

V. V. Kosarev, V. V. Popov, V. S. Vekshina, and N. I. Pepik
*Intrinsic and Deep Centre Charge Carriers in Indium Antimonide
(71.55; S7.13)
(a) Vol. **107**, No. 1, p. K43—K48 (1988).

O. V. Kulikova, L. L. Kulyuk, N. M. Nartya, S. I. Radautsan, E. V. Russu, A. V. Siminel, and E. E. Strumban
*On the Possibility of Formation of Associated Impurity Centers in Semi-Insulating Indium Phosphide Doped by Oxygen
(71.55; 72.40; S7.11)
(a) Vol. **107**, No. 1, p. K57—K60 (1988).

J. Misiewicz
*The Low Temperature Photoluminescence in Zn_3P_2
(71.55 and 78.55; S7)
(a) Vol. **107**, No. 1, p. K65—K68 (1988).

C. Ruiz-Mejía
*About the F Center in Cesium Halides
(71.55; 61.70; S9.11)
(b) Vol. **147**, No. 2, p. 593—600 (1988).

Th. Schmidt, J. Köhler, C. Kryschi, and D. Schmid
Sensitized Luminescence of Pure and Doped $NaNO_2$ Single Crystals. III. Excitation Spectroscopy and Time Resolved Trap Fluorescence
(78.55; 71.35; 71.55; S11)
(b) Vol. **147**, No. 2, p. 797—809 (1988).

P. Jungwirth and L. Skála
*Exact Solution of the Schrödinger Equation for the Infinite Linear Chain with a Single Trap
(71.55)
(b) Vol. **147**, No. 2, p. K149—K152 (1988).

N. G. Romanov, V. V. Dyakonov, V. A. Vetrov, and P. G. Baranov
*EPR Detected via Magnetic Circular Dichroism of Optical Absorption in Irradiated Alkali Halide Crystals Doped with Cadmium
(71.55 and 76.30; S9.11)
(b) Vol. **147**, No. 2, p. K171—K174 (1988).

G. I. Bersuker
*Dynamics in the System of Dipolar Impurity Centers in Crystals
(63.20 and 71.55; S11)
(b) Vol. **148**, No. 1, p. 243—249 (1988).

S. Shimanuki
*Radiative and Non-Radiative Processes of Localized Impurity Centers in Alkali Halides: $KBr:Ag^-$
(71.55 and 78.55; S9.11)
(b) Vol. **148**, No. 1, p. 251—262 (1988).

K. Löbe and R. Boyn
*Dye-Laser 4f–4f-Line Probing of Crystal-Field Fluctuations in $ZnS:Sm^{3+}$ Polycrystalline Films
(71.55; 68.60; 78.55; S8.11)
(b) Vol. **148**, No. 1, p. 263—272 (1988).

J. Kudrnovský and V. Drchal
The Electronic Structure of Palladium–Noble Metal Alloys
(71.20; 71.55; S1.3)
(b) Vol. **148**, No. 1, p. K23—K27 (1988).

V. A. Gubanov, A. F. Zatsepin, V. S. Kortov, S. P. Freidman, and G. B. Cherlov
*Electronic Structure of Intrinsic Defects in Amorphous GeO_2
(71.55; S10.1)
(b) Vol. **148**, No. 1, p. K33—K36 (1988).

V. I. Kuznetsov, P. F. Lugakov, and V. V. Lukyanitsa
Influence of Irradiation Temperature on Radiation Defect Accumulation in Silicon
(61.80; 71.55; S5.11)
(a) Vol. **108**, No. 1, p. 143—148 (1988).

V. V. Litvinov, G. V. Palchik, and V. I. Urenev
*Extrinsic Self-Trapping and Negative *U* of Oxygen-Related Thermal Donors in Germanium
(71.55; 61.70; S5.12)
(a) Vol. **108**, No. 1, p. 311—321 (1988).

A. Serpi and J. P. Zielinger
On the Photoconductivity Relaxation in $ZnIn_2S_4$
(72.40; 71.55; S8.16)
(a) Vol. **108**, No. 1, p. 351—362 (1988).

A. Opanowicz
*Comments on the Interpretation of Thermally Stimulated Luminescence and Conductivity in Condutive ZnSe Crystals
(71.55; 78.60; S8.12)
(a) Vol. **108**, No. 1, p. K47—K51 (1988).

S. Shigetomi, T. Ikari, Y. Koga, and S. Shigetomi
*Electrical Properties of Cd-Doped p-InSe
(71.55 and 72.80; S8)
(a) Vol. **108**, No. 7, p. K53—K57 (1988).

S. I. Radautsan, I. M. Tiginyanu, and N. B. Pyshnaya
*Electrical Properties of InP:Fe Single Crystals Implanted by Phosphorus Ions
(71.55 and 72.80; S7.11)
(a) Vol. **108**, No. 1, p. K59—K61 (1988).

J. Calas, S. M. Abdel Maksoud, and M. Averous
*Energy Distribution of Trapping Centers in Vitreous $GeSe_3$ as Obtained from Transit Time Measurements in the Temperature Range 60 to 300 K
(71.55; 72.40; S8)
(b) Vol. **148**, No. 2, p. 629—634 (1988).

R. Callens, F. Callens, P. Matthys, and E. Boesman
*EPR of a New O_2^- Centre in KCl
(71.55 and 76.30; S9.11)
(b) Vol. **148**, No. 2, p. 683—688 (1988).

V. Thomas, J. Barrau, M. Brousseau, K. Kirouni, and G. Couloumiers
The Luminescence of Palladium-Doped Silicon
(78.55; 71.55; S5.11)
(b) Vol. **148**, No. 2, p. 723—727 (1988).

K. Dunphy and W. W. Duley
Multiphoton Excitation of the 520 nm Luminescence Band in MgO:Al
(78.55; 71.35; 71.55; S10.1)
(b) Vol. **148**, No. 2, p. 729—735 (1988).

V. B. Shikin and N. I. Shikina
*Charged Dislocations in Semiconductors
(61.70 and 71.55; S5.11; S5.12)
(a) Vol. **108**, No. 2, p. 669—681 (1988).

G. Koschek, H. Lakner, and E. Kubalek
*On the Improvement of the Semiinsulating GaAs Substrate Homogeneity Due to Annealing
(71.55; 78.60; S7.12)
(a) Vol. **108**, No. 2, p. 683—688 (1988).

W. Siegel, H. Witte, G. Kühnel, H. A. Schneider, and T. Flade
*Thermal Activation Energies of Undoped Semi-Insulating LEC GaAs. Correlation of TDH and TSC Results
(71.55 and 72.20; S7.12)
(a) Vol. **108**, No. 2, p. 689—697 (1988).

P. S. Gladkov and K. B. Ozanyan
*Photoconductivity in Bulk and Epitaxial GaAs:V
(71.55 and 72.40; 61.70; S7.12)
(a) Vol. **108**, No. 2, p. K125—K130 (1988).

Z. Dreger, J. Kalinowski, I. Davoli, S. Stizza, and S. Feliziani
*Thermoluminescence Study of Charge Traps in Polycrystalline Layers of Phenanthrene
(71.55 and 78.60; S12)
(b) Vol. **149**, No. 1, p. 363—370 (1988).

G. R. Allan, H. A. MacKenzie, J. J. Hunter, and B. S. Wherrett
The Identification of Urbach, Acceptor Impurity, and Phonon-Assisted Components of the Optical Absorption Band Tail of n-Type InSb
(78.20; 71.55; S7.13)
(b) Vol. **149**, No. 1, p. 383—389 (1988).

S. Muramatsu
*Electron–Lattice Interaction Parameters Determined from Absorption and MCD Spectra of the F Centers in CsCl and CsBr
(71.55 and 78.50; 61.70; S9.11; S9.12)
(b) Vol. **149**, No. 1, p. K19—K23 (1988).

S. N. Martynov and S. B. Petrov
*Single-Ion Anisotropy of Exchange-Coupled Mn^{2+} Ions in $(CH_3)_4NCdCl_3$ and $CsCdCl_3$
(71.55 and 76.30; S9)
(b) Vol. **149**, No. 1, p. K41—K44 (1988).

H. WIECZOREK and W. FUHS
Transient Photocurrents in a-Si:H Diodes: Effects of Deep Trapping
(72.40 and 73.40; 61.40; 71.55; S5.11)
(a) Vol. **109**, No. 1, p. 245—253 (1988).

V. PANDIAN and V. KUMAR
*Silver Related Deep Levels in Silicon
(71.55; S5.11)
(a) Vol. **109**, No. 1, p. 273—278 (1988).

K. SCHMALZ, F.-G. KIRSCHT, and K. TITTELBACH-HELMRICH
*DLTS Study of Deep Level Defects in Cz n-Si Due to Heat Treatment at 600 to 900 °C
(71.55; S5.11)
(a) Vol. **109**, No. 1, p. 279—294 (1988).

G. VÖLKEL, A PÖPPL, and B. VOIGTSBERGER
*Investigation of the Oxygen Vacancy Balance in ZnO Ceramics by Means of EPR
(71.55; 76.30; S10.1)
(a) Vol. **109**, No. 1, p. 295—299 (1988).

E. A. ZHILIŃSKAYA, V. N. LAZUKIN, E. A. BYCHKOV, and S. B. ROSENKOV
*Correlation between the Structures of Ge–S Glasses and that of Impurity Mn^{2+} Complexes
(61.60 and 71.55; 76.30; S8.16)
(b) Vol. **149**, No. 2, p. 427—433 (1988).

E. RADZHABOV
*Reaction of OH^- and F Centers in Sodium Halides
(61.70 and 71.55; 78.50; S9.11)
(b) Vol. **149**, No. 2, p. 441—446 (1988).

J. P. JOUART and G. MARY
*Up-Conversion from Er^{3+}–Er^{3+} Pairs in CdF_2 Crystals
(71.55; 78.50; 78.55; S9.12)
(b) Vol. **149**, No. 2, p. 633—639 (1988).

K. P. TCHAKPELE and J. P. ALBERT
*Excitons Bound to Te Impurities in CdS, ZnS, and Their Mixed Compounds with Wurtzite Structure
(71.35 and 71.55; S8.11; S8.15)
(b) Vol. **149**, No. 2, p. 641—648 (1988).

C. A. LONDOS and J. GRAMMATIKAKIS
Notes on the Carbon-Associated Deep Level Complex in Irradiated Silicon
(61.80; 71.55; S5.11)
(a) Vol. **109**, No. 2, p. 421—426 (1988).

J. KRYNICKI and H. RZEWUSKI
*Deep Electronic Levels in Carbon-Implanted Silicon
(61.70 and 71.55; S5.11)
(a) Vol. **109**, No. 2, p. 531—536 (1988).

V. M. Babich, N. P. Baran, A. A. Bugai, A. A. Konchits, V. B. Kovalchuk, V. M. Maksimenko, and B. D. Shanina
*Electrical and Paramagnetic Properties of Thermodonors-II in Silicon. Discussion of a Model
(71.55; 72.80; 76.30; S5.11)
(a) Vol. **109**, No. 2, p. 537—547 (1988).

S. Kishida, K. Matsuura, H. Mori, Y. Mizuguchi, and I. Tsurumi
Temperature Dependence of the 2.5 eV Emission Band in Se-Treated ZnSe Crystals
(78.55; 71.55; S8.12)
(a) Vol. **109**, No. 2, p. 617—623 (1988).

P. Krispin, N. T. Nghi, N. T. Q. Hai, D. D. Khang, and D. A. Tuan
*Defects in GaP Green Light-Emitting Diodes
(71.55; 73.40; 85; S7.11)
(a) Vol. **109**, No. 2, p. K115—K119 (1988).

K.-P. Möllmann
*Additional Deep Level in Strongly Indium-Doped Lead Tin Telluride
(71.55; 72.80; S8)
(a) Vol. **109**, No. 2, p. K121—K124 (1988).

N. T. Bagraev and I. S. Polovtsev
*Reactions for Iron Centers in Silicon under Optical Pumping
(61.70 and 71.55; 72.40; 76.30; 76.60; S5.11)
(b) Vol. **150**, No. 1, p. 19—24 (1988).

W. Ulrici, K. Friedland, L. Eaves, D. P. Halliday, and C. A. Payling
*Optical and Electrical Studies of GaP:Ti
(71.55; 72.20; 78.50; 78.55; S7.11)
(b) Vol. **150**, No. 1, p. 177—190 (1988).

K. Löbe and R. Boyn
*Sm^{3+} in ZnS on Lattice Sites and Wurtzite-Type Interstitial Sites?
(61.70 and 71.55; 71.70; 78.55; S8.11)
(b) Vol. **150**, No. 1, p. 191—200 (1988).

F. Dujardin, B. Stébé, and G. Munschy
*Binding Energies of Neutral Bound Excitons for Intermediate to High Magnetic Fields
(71.35 and 71.55)
(b) Vol. **150**, No. 1, p. 201—209 (1988).

N. I. Tarbaev, J. Schreiber, and G. A. Shepelskii
*Physical Properties of $A^{II}B^{VI}$ Semiconductor Crystals after Plastic Deformation at Low Temperature
(61.70 and 71.55; 72.40; 78.50; 78.55; S8.1)
(a) Vol. **110**, No. 1, p. 97—106 (1988).

T. V. Torchinskaya, A. A. Shmatov, and M. K. Sheinkman
*Recombination-Enhanced Transformation of Deep Centers in Red Light-Emitting AlGaAs Diodes
(71.55; 73.40; 78.55; S7.15)
(a) Vol. **110**, No. 1, p. 213—220 (1988).

D. S. Domanevskii, S. V. Zhokhovets, and M. V. Prokopenya
*The Valence Band Tail State Filling in Heavily Doped and Compensated Crystals of Gallium Arsenide
(71.55; 78.55; S7.12)
(a) Vol. **110**, No. 1, p. 221—226 (1988).

A. G. J. Green, B. Ray, I. V. F. Viney, and J. W. Brightwell
*Luminescence Studies of CaS Phosphors Doped with Cerium and Copper
(71.55 and 78.55; 76.30; 78.60; S8)
(a) Vol. **110**, No. 1, p. 269—276 (1988).

E. Krause
*Effects of Growth and Annealing Conditions on Deep Level Emissions in CVD Poly-ZnSe Layers
(71.55 and 78.60; S8.12)
(a) Vol. **110**. No. 1, p. K33—K37 (1988).

J. K. Požela, Z. N. Tamaševičienė, A. V. Tamaševičius, J. K. Ulbikas, and G. V. Bandurkina
Self-Generated Chaos in n-Ge After Electron Irradiation
(72.20; 71.55; S5.12)
(a) Vol. **110**, No. 2, p. 555—563 (1988).

S. M. Wasim and J. G. Albornóz
Electrical and Optical Properties of n- and p-Type $CuInTe_2$
(72.20 and 78.20; 71.55; S8.16)
(a) Vol. **110**, No. 2, p. 575—583 (1988).

71.60. Positron states
(see also 78.70. Positron annihilation)

S. B. Shrivastava, V. K. Gupta, and K. S. Bhadoriya
*Positronium at Metal Surfaces
(71.60 and 73.20; S1.2; S1.3)
(b) Vol. **147**, No. 2, p. 553—557 (1988).

E. Dębowska and J. Rudzińska-Girulska
Electronic Structure of $Pd_{1-x}Ni_x$ Alloys Investigated by Positron Annihilation
(71.25; 71.60; S1.2)
(b) Vol. **148**, No. 1, p. 227—231 (1988).

71.70. Level splitting and interactions

(see 75.10. — in magnetic phenomena, 75.30. Exchange and superexchange interactions, 73.20. Electronic surface states)

YI-YANG ZHOU
*Theoretical Study of the Optical Absorption, Mössbauer, and EPR Spectra of Fe^{2+} Ion in Orthopyroxene
(71.70; 76.30; 76.80; 78.30; S11)
(b) Vol. **145**, No. 1, p. 227—232 (1988).

WAN-LUN YU and J. L. RAO
*Pressure Dependence of the EPR Parameters of Cs_3CoCl_5 Crystal
(71.70; 76.30; S9.16)
(b) Vol. **145**, No. 1, p. 255—259 (1988).

E. A. GOREMYCHKIN, E. MÜHLE, and I. NATKANIEC
*Refinement of the Crystal Field Parameters of $PrCu_5$ by Inelastic Neutron Scattering
(71.70; S1.4)
(b) Vol. **145**, No. 1, p. K15—K19 (1988).

HUANG MING
*High-Order Perturbation Formula for the EPR Trigonal Symmetric g_i-Value of the d^8 Ion
(71.70; S9.16)
(b) Vol. **145**, No. 1, p. K21—K25 (1988).

TIESONG ZHAO, YONG ZHU, and HANMIN JIN
Estimation of the Crystal Field at R Sites in $R_2Fe_{14}B$ Compounds
(75.30; 71.70; S1.1; S1.4)
(b) Vol. **145**, No. 2, p. 679—686 (1988).

WEN-CHEN ZHENG
Determination of the Local Thermal Expansion Coefficient in the Vicinity of Cr^{3+} Ions Associated with Vacancies in MgO
(61.70 and 65.70; 71.70; S10.1)
(b) Vol. **145**, No. 2, p. K99—K102 (1988).

A. P. ELISEEV, A. M. YURKIN, and E. G. SAMOILOVA
Cr^{3+} Energy Levels in Alexandrite
(78.55; 71.70; S10)
(a) Vol. **105**, No. 2, p. K169—K173 (1988).

D. NEOGY and T. PUROHIT
*The Behavior of Active Centers in a Laser Host. A Crystal Field Investigation on Dy^{3+} in LaF_3 Single Crystal
(71.70; 78.50; S9)
(b) Vol. **146**, No. 1, p. 181—187 (1988).

A. K. MUKHERJEE and D. NEOGY
*Magnetism and Crystal Field Analysis of Gadolinium Bromate Enneahydrate Crystal
(71.70 and 75.50; S11)
(b) Vol. **146**, No. 1, p. K45—K50 (1988).

D. Neogy, A. K. Mukherjee, and T. Purohit
*Calculation of Electrostatic Crystal Field Parameters of Rare Earth Hydrated Single Crystals
(71.70; S12)
(a) Vol. **106**, No. 1, p. 173—180 (1988).

S. V. Melnichuk, V. M. Kramar, and K. D. Tovstyuk
*Vibronic Interaction in the Absorption Spectra of Vanadium-Doped $A^{II}B^{VI}$ Semiconductors
(71.70 and 78.50; S8.1)
(b) Vol. **146**, No. 2, p. 613—617 (1988).

Yi-Yang Zhou
*A Theoretical Study of EPR Spectra in Cu^{2+}- and Mn^{2+}-Doped $CaCd(CH_3COO)_4 \cdot 6\,H_2O$
(71.70 and 76.30; S11)
(b) Vol. **147**, No. 1, p. 273—280 (1988).

A. E. Szukiel, P. J. Markowski, and K. Durczewski
Collective Excitation on the Electrical Resistivity of Singlet Ground State Paramagnets of Metallic Conduction
(72.15; 71.70; S1.4)
(b) Vol. **147**, No. 1, p. 321—334 (1988).

M. Budzyński, M. Subotowicz, H. Niezgoda, H. Spustek, W. Tańska-Krupa, and R. Wasiewicz
The Hyperfine Fields and Their Pressure Derivatives in $(Y_{0.9}Hf_{0.1})Fe_2$ and $(Zr_{0.9}Hf_{0.1})Fe_2$
(76.80; 71.70; S1.1)
(b) Vol. **147**, No. 2, p. 685—689 (1988).

B. M. Anghelov
*Spectroscopic Properties of $LiNdP_4O_{12}$
(71.70 and 78.50; S10)
(b) Vol. **148**, No. 1, p. K83—K88 (1988).

B. F. Minaev, Yu. A. Serebrennikov, and H. D. Rempel
*Non-Equilibrium Spin Polarization of the Si-Sl Centre in Silicon Induced by Spin–Orbit Coupling
(71.70 anf 76.30; S5.11)
(b) Vol. **148**, No. 2, p. 689—698 (1988).

J. L. Rao, R. M. Krishna, and S. V. J. Lakshman
*Crystal Field Energy Levels, EPR Parameters, and Magnetic Susceptibility of the Co^{2+} Ion in Cs_3ZnCl_5
(71.70; 76.30; 78.50; S9.12)
(b) Vol. **148**, No. 2, p. K119—K123 (1988).

A. G. Makhanek, V. S. Korolkov, and L. A. Yuguryan
*Analysis of the Intensity and Polarization Dependence of Two-Photon Absorption in the Channel $E_{3/2}(4f)$–$E_{3/2}(5d)$ for Ce^{3+} in CaF_2
(71.70; 78.50; S9.12)
(b) Vol. **149**, No. 1, p. 231—240 (1988).

J. Nandi and D. Neogy
*Magnetic Behavior and Crystal Field of Pr^{3+} in Praseodymium Selenate Octahydrate
(71.70 and 75.30; 75.40; S11)
(b) Vol. **149**, No. 1, p. 275—281 (1988).

A. J. de Vries, J. P. M. van Vliet, and G, Blasse
*Mechanism of Concentration Quenching of Gd^{3+} and Eu^{3+} Emission in Perovskite-Like Aluminates
(71.70 and 78.55; S11.1)
(b) Vol. **149**, No. 1, p. 391—401 (1988).

P. Matthys, F. Callens, G. Vekeman, and E. Boesman
**g*-Tensor Analysis of Sulfur Centresin Alkali Halides
(71.70 and 76.30; S9.11)
(b) Vol. **149**, No. 1, p. K25—K30 (1988).

R. B. Prabhu and R. B. Tangsali
*Self-Consistent Calculations in the Theory of Mix-Valence
(71.70)
(b) Vol. **149**, No. 2, p. 623—632 (1988).

Wen-Chen Zheng
On the Equations of State of $ZnSiF_6 \cdot 6\,H_2O:Ni^{2+}$ Crystal Determined from the Pressure Dependence of EPR Spectra
(64.30; 71.70; 76.30; S11)
(b) Vol. **149**, No. 2, p. K105—K109 (1988).

K. Löbe and R. Boyn
Sm^{3+} in ZnS on Lattice Sites and Wurtzite-Type Interstitial Sites?
(61.70 and 71.55; 71.70; 78.55; S8.11)
(b) Vol. **150**, No. 1, p. 191—200 (1988).

I. Akai, T. Karasawa, T. Komatsu, and Y. Kaifu
Large Optical Nonlinearity of the Two-Dimensional Exciton States at a Stacking Fault Interface in BiI_3
(71.35 and 78.20; 71.70; 73.40; S9)
(b) Vol. **150**, No. 2, p. 635—639 (1988).

72. Electronic transport in condensed matter
(for surfaces, interfaces, and thin films, see 73)

72.10. Theory of electronic transport; scattering mechanism

S. Ya. Goroshchenko and I. I. Ukrainskii
Two-Dimensional Electron Pair in a Transverse Magnetic Field
(71.25; 72.10; 73.40)
(b) Vol. **145**, No. 1, p. 187—194 (1988).

D. H. Jha
*Calculation of Electrical Resistivity of Some Liquid Metals by the Mode-Coupling Method
(72.10; 72.15; S1.1; S1.2; S1.3; S1.4)
(b) Vol. **145**, No. 1, p. 277—282 (1988).

F. Berz
*Effect of Barrier Thickness on the Quantum Mechanical Transmission across Parabolic Barriers
(72.10)
(b) Vol. **145**, No. 1, p. K27—K30 (1988).

Yu. A. Firsov, E. P. Nakhmedov, V. N. Prigodin, and W. Weller
*Low-Temperature Conductivity of a Weakly Disordered, Strongly Anisotropic Two-Dimensional System with Electron–Electron Interaction. Application to Germanium Bicrystals and to Surface Superlattices
(72.10; 73.40; S5.12)
(b) Vol. **145**, No. 2, p. 625—635 (1988).

M. Hartmann, R. Zimmermann, and H. Stolz
Carrier Kinetics in a Semiconductor with Light-Induced Gaps
(71.25 and 72.20; 72.10)
(b) Vol. **146**, No. 1, p. 357—369 (1988).

H. Glaeske and M. Schubert
*Green's Function Approach in Nonequilibrium Charge Carriers in Direct Gap Semiconductors Generated by Short Light Pulses
(72.10; 78.20)
(b) Vol. **146**, No. 1, p. 385—391 (1988).

R. Grill
*Comparison of Two Methods to Solve the Boltzmann Equation in Polar Semiconductors
(72.10; 72.20)
(b) Vol. **146**, No. 1, p. K35—K38 (1988).

Yu. B. Gaididei and I. I. Yakimenko
*On Trigger Properties of Molecular Organizates. III. Nonlinear Waves in Stratified Molecular Structures
(72.10; 73.40)
(b) Vol. **146**, No. 2, p. 643—654 (1988).

E. Heiner
*Evolution Equations for Highly Excited Direct Semiconductors in the Athermal Stage
(72.10; 72.20)
(b) Vol. **146**, No. 2, p. 655—660 (1988).

T. Satow, H. Ishikawa, and O. Uemura
*Semiconductivity in the Liquid Cu–TlTe System
(72.10 and 72.80; 75.30; S1.63)
(b) Vol. **146**, No. 2, p. 661—669 (1988).

V. A. Shenderovskii, P. M. Tomchuk, P. N. Gorley, N. P. Gavaleshko, and I. I. Tarasyuk
*The Effect of Structural Imperfections on Macroscopic Properties of Tellurium-Type Anisotropic Semiconductors
(72.10; 72.20; S5)
(b) Vol. **146**, No. 2, p. 671—681 (1988).

A. P. Zhernov and T. N. Kulagina
The Microcontact Spectrum of Aluminium
(71.38; 72.10; S4)
(b) Vol. **147**, No. 1, p. 149—153 (1988).

R. Chicón
*Conductivity and Dielectric Constant of Interacting Anderson Localized Systems at $T = 0$
(72.10)
(b) Vol. **147**, No. 1, p. 335—341 (1988).

S. Godoy and S. Fujita
*Exact Solutions of Few First Passage Problems in Linear Correlated Walks. Applications to the Electron Trapping in a Semiconductor
(72.10)
(b) Vol. **147**, No. 2, p. 649—660 (1988).

C. A. C. Sequeira and J. M. B. Fernandes Diniz
*Transport Equations in Common Solid State Electrochemical Situations
(66.30 and 72.10; S10)
(a) Vol. **108**, No. 1, p. 155—162 (1988).

Z. Kletowski
*Resistance of Some Lanthanum and Lutetium Compounds of the Type (La, Lu)Me_3
(72.10; 72.15; S1.4)
(a) Vol. **108**, No. 1, p. 363—367 (1988).

A. Dargys
*Addendum to "Tunneling in *k*-Space"
(72.10; 71.25; S5.12)
(b) Vol. **148**, No. 2, p. K125—K127 (1988).

K. Král, Z. Khás, and B. Hejda
*Hot Phonon Generation Rate with Electron Energy Collision Broadening in Polar Semiconductors
(72.10; S7.12)
(b) Vol. **148**, No. 2, p. K129—K131 (1988).

E. Heiner
*Evolution Equations for Highly Excited Direct-Semiconductors in the Athermal Stage — Entropy and Entropy Production
(72.10; 72.20)
(b) Vol. **149**, No. 2, p. K153—K156 (1988).

F. Khater and M. El-Hiti
*Temperature Coefficient of the Strain Coefficient of Electrical Resistivity of Double-Layer Thin Metallic Films
(72.10 and 73.60; S1.3)
(a) Vol. **109**, No. 2, p. 517—523 (1988).

W. Hergert and S. Hildebrandt
Unified Theoretical Description of EBIC, LBIC, CL, and PL Experiments. Transient Analysis
(78.55 and 78.60; 72.10)
(a) Vol. **109**, No. 2, p. 625—633 (1988).

F. Ludwig, H.-U. Müller, and R. Herrmann
Effective Scattering Times of Inversion Electrons in InSb Grain Boundaries Deduced from Cyclotron Resonance Measurements
(73.40 and 76.40; 72.10; S7.13)
(b) Vol. **150**, No. 1, p. 119—123 (1988).

72.15. Electronic conduction in metals and alloys

D. N. Jha
Calculation of Electrical Resistivity of Some Liquid Metals by the Mode-Coupling Method
(72.10; 72.15; S1.1; S1.2; S1.3; S1.4)
(b) Vol. **145**, No. 1, p. 277—282 (1988).

W. Schiller and D. Schwitalla
*The Influence of Electron–Electron Interaction on the Temperature Dependence of the Electrical Resistivity of Disordered Alloys
(72.15)
(b) Vol. **145**, No. 1, p. K35—K40 (1988).

E. Kokmeijer, E. Huizer, B. J. Thijsse, and A. van den Beukel
*Resistivity Changes during Structural Relaxation of Amorphous $Fe_{40}Ni_{40}B_{20}$
(72.15; 61.40; S1.1)
(a) Vol. **105**, No. 1, p. 235—242 (1988).

N. Kapoor and A. N. Nigam
*Influence of Hydrogen Diffusion on Thermal Conductivity of Nickel
(66.30 and 72.15; S1.2)
(a) Vol. **105**, No. 2, p. 403—409 (1988).

P. Svoboda and P. Vašek
Low Temperature Thermal Conductivity of Metallic Glasses $Fe_xNi_{80-x}B_{20}$
(66.70; 72.15; S1.1)
(a) Vol. **106**, No. 1, p. 115—121 (1988).

L. VATSKICHEV and M. VATSKICHEVA
*Relations between Some Galvanomagnetic and Magnetic Properties of Thin Ferromagnetic Films
(72.15 and 75.70; S1.2; S1.4)
(a) Vol. **106**, No. 1, p. K75—K78 (1988).

Z. HENKIE, P. J. MARKOWSKI, R. HORYŃ, Z. BUKOWSKI, and J. KLAMUT
*The Thermopower Sign Inversion for the High T_c Superconductor $YBa_2Cu_3O_{7-\delta}$
(72.15 and 74.70; S10.15)
(b) Vol. **146**, No. 2, p. K131—K134 (1988).

A. E. SZUKIEL, P. J. MARKOWSKI, and K. DURCZEWSKI
*Collective Excitation Effects on the Electrical Resistivity of Singlet Ground State Paramagnets of Metallic Conduction
(72.15; 71.70; S1.4)
(b) Vol. **147**, No. 1, p. 321—334 (1988).

K. K. MUKHERJEE, S. M. RAFIQUE, M. MEHAR KHAN, and S. BAGCHI
*Energetic Approach to the Electrical Resistivity of Liquid Magnesium
(72.15; S3)
(b) Vol. **147**, No. 1, p. K51—K55 (1988).

S. OHTAKI and I. HASHIMOTO
Numerical Evaluation for Stage II_A Recovery in Cold-Worked Aluminium
(61.70; 62.20; 72.15; S4)
(a) Vol. **107**, No. 1, p. 123—130 (1988).

P. KWAPULIŃSKI, J. RASEK, and Z. GIERAK
*Scattering of Conductivity Electrons on Grain Boundaries in Metals
(72.15; S1.1; S1.2)
(a) Vol. **107**, No. 1, p. 299—304 (1988).

F. SACHSLEHNER
*The Influence of Precipitates on the Low-Field Hall Coefficient of Cu–Be 2
(72.15; S1.2)
(a) Vol. **107**, No. 1, p. 305—313 (1988).

A. W. SHEIKH and M. EL-FAZANI
*A Phenomenological Model for the Magnetic Resistivity of NM–TM Spin Glasses
(72.15; 75.30; S1.2; S1.3)
(b) Vol. **147**, No. 2, p. 601—612 (1988).

M. D. TIWARI, G. THUMMES, and V. KUKERMANN
*Deviations from Matthiessen's Rule in Ag–Au Alloys
(72.15; S1.3)
(b) Vol. **147**, No. 2, p. 661—675 (1988).

F. M. MUNTYANU, V. G. KISTOL, and I. A. POPOV
*Quantum Oscillations of Transport Phenomena of H-Holes in $Bi_{1-x}Sb_x$ Alloys ($x \geqq 0.25$)
(72.15; S4)
(b) Vol. **148**, No. 1, p. K37—K40 (1988).

M. HEYTING, B. J. THIJSSE, A. VAN DEN BEUKEL
*Change of Electrical Resistance during Crystallization of Icosahedral $Al_{86}Mn_{14}$
(61.50 and 72.15; S4)
(a) Vol. **108**, No. 1, p. 97—101 (1988).

F. KHATER and M. EL-HITI
Strain Coefficients of Electrical Resistance of Double-Layer Thin Metallic Films
(73.60; 72.15; S1.3)
(a) Vol. **108**, No. 1, p. 241—249 (1988).

S. ARAJS and T. A. JENKINS
*Electrical Resistivity of Some Directionally Solidified Nickel-Base Eutectic Composites
(72.15; S1.2)
(a) Vol. **108**, No. 1, p. 343—349 (1988).

Z. KLETOWSKI
Resistance of Some Lanthanum and Lutetium Compounds of the Type (La, Lu) Me_3
(72.10; 72.15; S1.4)
(a) Vol. **108**, No. 1, p. 363—367 (1988).

A. RAOUFI and S. ARAJS
*Stress Dependence of Magnetoresistance in Crystalline and Amorphous Ferromagnets
(72.15; 75.80; S1.1)
(a) Vol. **108**, No. 1, p. 369—374 (1988).

YU. V. ZAKHAROV, YU. I. MANKOV, and L. S. TITOV
*Electric Conductivity of a Metallic Ferromagnet and Magnetization Process
(72.15; 75.60; S1.1)
(b) Vol. **149**, No. 1, p. 241—248 (1988).

CH. SEIDEL
*On the Pinning of Spin-Density Waves by Impurities
(72.15 and 75.30; S12.1)
(b) Vol. **149**, No. 1, p. 327—335 (1988).

72.20. Conductivity phenomena in semiconductors and insulators
(for nonelectronic conduction, see 66.70)

F. Pomer, X. Bonet, A. Segura, and A. Chevy
*Electrical Conductivity Anisotropy in Tin-Doped n-Type Indium Selenide
(72.20; S8)
(b) Vol. **145**, No. 1, p. 261—268 (1988).

D. K. Gautam, W. S. Khokle, and K. B. Garg
*Transition Probability of Impact Ionization by Holes in Silicon
(72.20; S5.11)
(b) Vol. **145**, No. 1, p. 269—275 (1988).

C. A. C. Sequeira and J. M. B. Fernandez Diniz
*Conduction Processes in Oxide Solid Electrolytes
(66.30 and 72.20; 61.70; S10)
(a) Vol. **105**, No. 1, p. 123—130 (1988).

A. I. Eatha, A. A. Ghani, and E. El Faramawy
*Effect of Sintering Temperature on the Electrical Conductivity and Thermoelectric Power of $CuFe_2O_4$
(72.20; S11.2)
(a) Vol. **105**, No. 1, p. 231—233 (1988).

P. Höschl, P. Moravec, V. Prosser, V. Szöcs, and R. Grill
*Galvanomagnetic Properties of p-$Hg_{1-x}Cd_xTe$
(72.20; S8.15)
(b) Vol. **145**, No. 2, p. 637—648 (1988).

G. F. Leal Ferreira and O. N. Oliveira, Jr.
*Hole Transport in Corona-Charged 12 μm Sheldahl FEP Samples Interpreted through a Field-Dependent Schubweg Model
(72.20; S12)
(a) Vol. **105**, No. 2, p. 531—539 (1988).

Ll. Closas and J. Fernández Rubio
*An EBIC Model with Extended Generation Profile of Electron–Hole Pairs
(72.20 and 73.40)
(a) Vol. **105**, No. 2, p. 541—548 (1988).

G. Irmer and W. Siegel
*Carrier Concentration Profiles across GaP Grain Boundaries Observed by Raman Measurements
(61.70 and 72.20; 78.30; S7.11)
(a) Vol. **105**, No. 2, p. 549—553 (1988).

M. A. Hassan, M. N. Khan, and C. A. Hogarth
*Electrical and Optical Properties of V_2O_5–GeO_2–NiO Glasses
(72.20 and 78.20; 78.30; S10.15)
(a) Vol. **105**, No. 2, p. 609—616 (1988).

H. L. Cui, X. L. Lei, and N. J. M. Horing
*Hot Electron Magnetophonon Resonances in Conduction and Noise for a Nondegenerate Semiconductor
(72.20 and 72.70)
(b) Vol. **146**, No. 1, p. 189—197 (1988).

R. Rentzsch, K. J. Friedland, and A. N. Ionov
*Negative Magnetoresistance of Neutron-Transmutation-Doped Gallium Arsenide at Variable-Range Hopping
(72.20; 72.60; S7.12)
(b) Vol. **146**, No. 1, p. 199—206 (1988).

M. Hartmann, R. Zimmermann, and H. Stolz
*Carrier Kinetics in a Semiconductor with Light-Induced Gaps
(71.25 and 72.20; 72.10)
(b) Vol. **146**, No. 1, p. 357—369 (1988).

R. Grill
Comparison of Two Methods to Solve the Boltzmann Equation in Polar Semiconductors
(72.10; 72.20)
(b) Vol. **146**, No. 1, p. K35—K38 (1988).

J. Hřebíček, V. Čech, and A. Brablec
Determination of Trap Concentrations and Energy Levels in Insulators and Semiconductors from Steady-State Space-Charge-Limited Currents
(71.55; 72.20)
(a) Vol. **106**, No. 1, p. 167—172 (1988).

E. Heiner
Evolution Equations for Highly Excited Direct Semiconductors in the Athermal Stage
(72.10; 72.20)
(b) Vol. **146**, No. 2, p. 655—660 (1988).

V. A. Shenderovskii, P. M. Tomchuk, P. N. Gorley, N. P. Gavaleshko, and I. I. Tarasyuk
The Effect of Structural Imperfections on Macroscopic Properties of Tellurium-Type Anisotropic Semiconductors
(72.10; 72.20; S5)
(b) Vol. **146**, No. 2, p. 671—681 (1988).

O. B. Alekseevskii, S. A. Vorobiev, and N. S. Kostyukov
*Radiation-Induced Conductivity of Dielectric Solids Irradiated by Intensive Fluxes of Fast Neutrons
(61.80 and 72.20)
(b) Vol. **146**, No. 2, p. K109—K113 (1988).

A. V. Andreeva, A. V. Bazhenov, N. A. Bulenkov, and A. A. Firsova
Structural and Photoelectrical Properties of a Third-Order Twin Boundary ($\Sigma = 27$) in Silicon
(61.55; 61.70; 72.20; S5.11)
(a) Vol. **106**, No. 2, p. 351—362 (1988).

C. B. Steele and J. Beynon
Thermopower and DC Electrical Conductivity in Au/SiO Thin Cermet Films
(73.60; 72.20; S10.1)
(a) Vol. **106**, No. 2, p. 515—523 (1988).

D. BOSE and B. R. NAG
*High-Field Electron Transport in GaAs and Ga(0.47)In(0.53)As Rectangular Quantum Wells
(72.20; S7.12; S7.15)
(a) Vol. **106**, No. 2, p. 569—575 (1988).

E. SUSI, L. PASSARI, M. MERLI, and M. C. CAROTTA
*Determination of Minority Carriers Mobility in Silicon from Stationary and Transient Lifetime Measurements
(72.20; S5.11)
(a) Vol. **106**, No. 2, p. 583—587 (1988).

G. P. TRIBERIS, N. GUSKOS, and G. KIRIAKIDIS
*New Aspects on the Electronic Conduction in Semi-Insulating $CdF_2:Gd^{3+}$ Single Crystals
(72.20; S9.12)
(b) Vol. **147**, No. 1, p. 349—353 (1988).

S. GHOSH and A. N. CHAKRAVARTI
*Effect of Broadening of Tail States on the Einstein Relation in Heavily Doped Compensated Semiconductors
(72.20; S7.13)
(b) Vol. **147**, No. 1, p. 355—360 (1988).

A. GRAJA, G. SEKRETARCZYK, W. PUKACKI, S. NEŠPŮREK and O. RYBA
*Conduction-Electron Spin Resonance and Thermoelectric Power of the Mixed-Valence Tetrathiafulvalenium Chloride
(72.20 and 76.30; S12)
(b) Vol. **147**, No. 1, p. 401—408 (1988).

P. CZARNECKI, B. IDZIKOWSKI, and A. WRZECIONO
*Magnetoresistance of $Pr_xFe_{80-x}B_{20}$ and $Ho_xCo_{70-x}B_{30}$ Amorphous Alloys in Magnetic Fields up to 22 T
(72.20; S1.2)
(b) Vol. **147**, No. 1, p. K47—K50 (1988).

V. ČÁPEK
*Cancellation of Local Shifts of the Chemical Potential in Phonon-Assisted Hopping
(72.20)
(b) Vol. **147**, No. 1, p. K57—K61 (1988).

K. I. WYSOKIŃSKI and W. BRENIG
*On the Hopping Thermopower of 2D Systems in Strong Magnetic Fields
(72.20)
(b) Vol. **147**, No. 1, p. K67—K70 (1988).

D. M. ZAYACHUK, P. M. STARIK, and V. A. SHENDEROVSKII
The Effect of Nonstoichiometric Tellurium on the Properties of Lead and Tin Tellurides
(61.70; 72.20; 75.20; S8.13)
(a) Vol. **107**, No. 1, p. 95—99 (1988).

H. ABE and Y. NAKAMURA
*The Effect of Cu Doping on the DC Conductivity and Thermoelectric Power of As_2Se_3-Based Glasses
(72.20; S8)
(a) Vol. **107**, No. 1, p. 315—319 (1988).

D. WRUCK and A. KNAUER
*Doping Inhomogeneities and Compensation in n-Type LEC InP Wafers
(72.20; 78.50; S7.11)
(a) Vol. **107**, No. 1, p. 321—328 (1988).

A. GALDIKAS, S. GREBINSKII, and S. MICKEVIČIUS
*Piezoresistance Anisotropy of the Ferromagnetic Semiconductor $HgCr_2Se_4$
(72.20; S8.16)
(a) Vol. **107**, No. 1, p. K53—K55 (1988).

M. KOYANO, S. ŌHARA, H. NEGISHI, M. SASAKI, M. INOUE, M. NOMURA, and H. FUJIWARA
*Effect of Pressure on Electrical Resistivity and Crystal Structure of Quasi-Two-Dimensional Mo_4O_{11}
(61.60 and 72.20; S10)
(b) Vol. **147**, No. 2, p. 559—565 (1988).

E. M. KOGAN and M. I. AUSLENDER
*Anderson Localization in Ferromagnetic Semiconductors Due to Spin Disorder. I. Narrow Conduction Band
(72.20; 72.60; S8.16)
(b) Vol. **147**, No. 2, p. 613—620 (1988).

P. HÖSCHL, P. MORAVEC, E. BELAS, J. FRANC, and R. GRILL
*Thermoelectric Properties of p-$Hg_{1-x}Cd_xTe$
(72.20; S8.15)
(b) Vol. **147**, No. 2, p. 621—628 (1988).

B. N. NARAHARI ACHAR and J. R. FERGUSON
*Quantized Damping of Helicons in a GaAs/AlGaAs Superlattice
(72.20 and 73.60; S7.12; S7.15)
(b) Vol. **147**, No. 2, p. K137—K141 (1988).

I. M. TSIDILKOVSKII, YU. G. ARAPOV, M. L. ZVEREVA, N. B. BRANDT, V. A. KULBACHINSKII, I. N. GORBATYUK, and A. I. ELIZAROV
The Influence of Compensation on the Hall Effect and Magnetoresistance of n-$Hg_{0.8}Cd_{0.2}Te$
(71.30; 72.20; S8.15)
(b) Vol. **148**, No. 1, p. 197—204 (1988).

M. I. AUSLENDER, E. M. KOGAN, and S. V. TRETYAKOV
*Anderson Localization in Ferromagnetic Semiconductors Due to Spin Disorder. II. Wide Conduction Band
(72.20; 72.60; S8.1; S10)
(b) Vol. **148**, No. 1, p. 289—295 (1988).

P. HÖSCHL, P. MORAVEC, R. GRILL, and V. ČÁPEK
*Weak Impurity Metallic Behaviour of p-$Hg_{1-x}Cd_xTe$
(72.20; S8.15)
(b) Vol. **148**, No. 1, p. 319—324 (1988).

W. KRAAK, U. TROPPENZ, R. HERRMANN, S. M. CHUDINOV, and V. A. KULBACHINSKII
*Magnetic Freeze-Out and Magnetic-Field-Induced Semiconductor–Semimetal Transition in $Bi_{1-x}Sb_x$ Alloys under High Hydrostatic Pressure
(71.30 and 72.20; S7.13)
(b) Vol. **148**, No. 1, p. 333—347 (1988).

A. A. Samokhvalov, N. A. Viglin, B. A. Gizhevskii, T. I. Arbuzova, and N. M. Chebotaev
*Magnetic Polarons in EuO
(72.20 and 75.30; 76.30; S10.1)
(b) Vol. **148**, No. 1, p. 361—367 (1988).

D. Chattopadhyay, P. C. Rakshit, and C. K. Sarkar
*Small-Signal AC Mobility of Two-Dimensional Hot Electrons in Semiconductor Quantum Wells at Low Temperatures
(72.20 and 73.40; S7.12)
(b) Vol. **148**, No. 1, p. K17—K22 (1988).

R. Keiper, R. Nolte, and O. Ziep
*Some Peculiarities of Charge Transport in the Quantum Hall Regime
(72.20)
(b) Vol. **148**, No. 1, p. K41—K45 (1988).

H. Kahnt, F. Schirrmeister, and A. Feltz
Model of Photostimulated Diffusion of Ag in Amorphous Chalcogenides
(66.30; 72.20; S8)
(a) Vol. **108**, No. 1, p. 149—153 (1988).

O. A. Golikova, M. M. Mezdrogina, N. A. Feoktistov, K. L. Sorokina, M. M. Kazanin, P. M. Karageorgy-Alkalaev, and A. Yu. Leiderman
*Injection and Recombination Currents in a-Si:H Structures
(72.20 and 73.40; S5.11)
(a) Vol. **108**, No. 1, p. 323—329 (1988).

K. P. Ghatak and M. Mondal
*On the Einstein Relation in n-Channel Inversion Layers on Ternary Semiconductors under Magnetic Quantization
(72.20; S8.15)
(b) Vol. **148**, No. 2, p. 645—652 (1988).

J. H. Haritonidis and D. S. Kyriakos
*Magnetoresistance Behaviour in Orthorhombic Crystals
(72.20)
(b) Vol. **148**, No. 2, p. 653—658 (1988).

M. Inoue, S. Ōhara, S. Horisaka, M. Koyano, and H. Negishi
*Transport Properties of Quasi-Two-Dimensional Mo_4O_{11} Crystals
(72.20; S10)
(b) Vol. **148**, No. 2, p. 659—671 (1988).

H. Nakanishi, M. Isomae, S. Endo, and T. Irie
*Piezoresistance in $CdIn_2S_4$
(72.20; 71.25; S8.16)
(b) Vol. **148**, No. 2, p. 673—682 (1988).

E. Susi, A. Poggi, G. Lulli, C. Monteverdi, and M. Merli
Lifetime and Crystal Order in Annealed CZ Silicon
(61.70; 72.20; S5.11)
(a) Vol. **108**, No. 2, p. 503—508 (1988).

W. SIEGEL, H. WITTE, G. KÜHNEL, H. A. SCHNEIDER, and T. FLADE
*Thermal Activation Energies of Undoped Semi-Insulating LEC GaAs. Correlation of TDH and TSC Results
(71.55 and 72.20; S7.12)
(a) Vol. **108**, No. 2, p. 689—697 (1988).

M. MOLDOVANOVA, R. ASSENOV, and L. PARTHIER
*Carrier Mobility in PbTe(I)
(72.20; S8)
(a) Vol. **108**, No. 2, p. 699—704 (1988).

H. KRAUSE, H.-P. BÄR, and H. SCHULZE
*Correlations between Current–Voltage Characteristics and Visible Defect Channels in Thin SiO_2 Films after Field Stress
(61.70 and 72.20; S10.1)
(a) Vol. **108**, No. 2, p. K131—K133 (1988).

H.-P. BÄR, H. KRAUSE, and H. SCHULZE
*On the Nature of Structural Defects in SiO_2 Films Visible with the Aid of the Liquid Crystal Technique
(61.70 and 72.20; S10.1)
(a) Vol. **108**, No. 2, p. K135—K137 (1988).

N. KRISTIANPOLLER, B. TRIEMAN, R. CHEN, and Y. KIRSH
The Application of Thermally Stimulated Processes to the Study of Defects in Perovskite Type Fluorides
(61.70; 72.20; 78.60; S9.16)
(b) Vol. **149**, No. 1, p. 45—54 (1988).

V. M. YARTSEV
*Optical Properties of Quasi-1D *n*-Merized Molecular Semiconductors
(72.20 and 78.20; S12.1)
(b) Vol. **149**, No. 1, p. 157—167 (1988).

B. KUMAR and B. AGARWAL
Energy Band Structure of Phenazine N Oxide and Phenazine N Dioxide. Candidate Materials for Organic Photovoltaic Devices
(71.25; 72.20; S12)
(b) Vol. **149**, No. 1, p. 221—229 (1988).

H. ISHIKAWA, Y. KAMEDA, O. UEMURA, and T. SATOW
*Electronic Properties of Liquid Tl–$GeTl_2Se_4$ and Tl–$SnTl_2Se_4$ Systems
(72.20 and 75.30; S1.63; S8.16)
(b) Vol. **149**, No. 1, p. 257—266 (1988).

T. YU. BILGILDEEVA and T. A. POLYANSKAYA
*The Quantum Corrections to the Conductivity in Strong Spin–Orbit Coupling Semiconductors
(72.20; 71.30; S7.12)
(b) Vol. **149**, No. 2, p. 649—655 (1988).

B. MITRA and K. P. GHATAK
*On the Einstein Relation in Semiconductor Superlattices under Magnetic Quantization
(72.20 and 73.40; S7.12; S7.15)
(b) Vol. **149**, No. 2, p. K117—K122 (1988).

K. Unger
*Reversible Formulae to Approximate Fermi Integrals
(72.20)
(b) Vol. **149**, No. 2, p. K141—K144 (1988).

B. Pödör
*Transport Coefficients for Non-Polar Optical Phonon Scattering
(72.20)
(b) Vol. **149**, No. 2, p. K149—K151 (1988).

E. Heiner
Evolution Equations for Highly Excited Direct-Semiconductors in the Athermal Stage — Entropy and Entropy Production
(72.10; 72.20)
(b) Vol. **149**, No. 2, p. K153—K156 (1988).

M. Stordeur, M. Stölzer, H. Sobotta, and V. Riede
Investigation of the Valence Band Structure of Thermoelectric $(Bi_{1-x}Sb_x)_2Te_3$ Single Crystals
(71.25; 72.20; 72.30; 78.20; S1.63)
(b) Vol. **150**, No. 1, p. 165—176 (1988).

W. Ulrici, K. Friedland, L. Eaves, D. P. Halliday, and C. A. Payling
Optical and Electrical Studies of GaP:Ti
(71.55; 72.20; 78.50; 78.55; S7.11)
(b) Vol. **150**, No. 1, p. 177—190 (1988).

A. Bonnet, A. Conan, M. Morsli, M. Ganne, and M. Tournoux
*Electronic and Structural Properties of Tungsten Bronzes
(72.20; 71.25; S10)
(b) Vol. **150**, No. 1, p. 225—235 (1988).

V. A. Kulbachinskii, N. B. Brandt, P. A. Cheremnykh, S. A. Azou, J. Horak, and P. Lošťák
*Magnetoresistance and Hall Effect in Bi_2Te_3 ⟨Sn⟩ in Ultrahigh Magnetic Fields and under Pressure
(72.20; S8)
(b) Vol. **150**, No. 1, p. 237—243 (1988).

B. I. Smirnov and I. A. Smirnov
Defect Structure and Physico-Mechanical Properties of SmS Single Crystals
(61.70; 62.20; 72.20; S8)
(a) Vol. **110**, No. 1, p. 89—96 (1988).

H. Bergner, V. Brückner, M. Lenzner, and R. Strobel
*Investigation of the Field-Dependent Carrier Mobility in GaAs by Picosecond Photoconductivity Measurements
(72.20; 72.40; S7.12)
(b) Vol. **150**, No. 2, p. 885—889 (1988).

J. K. Požela, Z. N. Tamaševičienė, A. V. Tamaševičius, J. K. Ulbikas, and G. V. Bandurkina
*Self-Generated Chaos in n-Ge After Electron Irradiation
(72.20; 71.55; S5.12)
(a) Vol. **110**, No. 2, p. 555—563 (1988).

J. Pellegrino and J. M. Galligan
*The Photoplastic Effect in II–VI Compounds. Mercury Cadmium Telluride
(72.20; 61.70; S8.15)
(a) Vol. **110**, No. 2, p. 565—573 (1988).

S. M. Wasim and J. G. Albornóz
*Electrical and Optical Properties of n- and p-Type $CuInTe_2$
(72.20 and 78.20; 71.55; S8.16)
(a) Vol. **110**, No. 2, p. 575—583 (1988).

F. Pomer and J. Navasquillo
*A Method for Measuring the Resistivity of a Layered Semiconductor Perpendicular to the Layers
(72.20)
(a) Vol. **110**, No. 2, p. 585—592 (1988).

R. Świetlik, G. Mihàly, and P. Delhaès
*Microwave Conductivity and Thermoelectric Power of Some Ternary Salts of TCNQ and Iodine
(72.20; S12.1)
(a) Vol. **110**, No. 2, p. 593—599 (1988).

72.30. High-frequency effects; plasma effects

K. Leo
*Exact Formula for Energy-Independent Recombination Heating
(72.30; 71.38)
(b) Vol. **145**, No. 1, p. K31—K33 (1988).

M. Mondal and K. P. Ghatak
*Theoretical Analysis of the Plasma Frequency in Bismuth in the Presence of a Quantizing Magnetic Field
(72.30; S4)
(b) Vol. **146**, No. 2, p. K97—K102 (1988).

M. Stordeur, M. Stölzer, H. Sobotta, and V. Riede
Investigation of the Valence Band Structure of Thermoelectric $(Bi_{1-x}Sb_x)_2Te_3$ Single Crystals
(71.25; 72.20; 72.30; 78.20; S1.63)
(b) Vol. **150**, No. 1, p. 165—176 (1988).

A. A. Bugaev and B. P. Zakharchenya
*Picosecond Holographic Diagnostic of Light Pulse Absorption in Semiconductors
(72.30 and 78.20; 71.38; S5.11; S7.12)
(b) Vol. **150**, No. 2, p. 891—894 (1988).

72.40. Photoconduction and photovoltaic effects; photodielectric effects

T. N. Mamontova, A. S. Kochemirovskii, and L. V. Pivovarova
Electronic Processes on the Surface of As–Se Chalcogenide Glassy Semiconductors
(Review Article)
(73.25; 68.45; 72.40; 73.20; S8)
(a) Vol. **107**, No. 1, p. 11—43 (1988).

L. F. Makarenko and L. I. Murin
Trapping of Minority Carriers in Thermal U^--Donors in n-Si
(71.55; 72.40; S5.11)
(b) Vol. **145**, No. 1, p. 241—253 (1988).

R. Fornari and L. Dozsa
Deep Level Characterization in Semi-Insulating LEC Gallium Arsenide
(71.55; 72.40; 78.50; S7.12)
(a) Vol. **105**, No. 2, p. 521—530 (1988).

P. D. Pester and T. Wilson
An Analysis of the Photocurrent Induced in a Semi-Infinite Schottky Barrier by a Time Varying Focussed Laser Beam
(73.30 and 85; 72.40)
(a) Vol. **105**, No. 2, p. 649—660 (1988).

A. Opanowicz and B. Kóscielniak-Mucha
*Determination of the Electron Lifetime in p-$CuInS_2$ from a Photovoltaic Study of In/p-$CuInS_2$ Schottky Barrier Diodes
(72.40 and 73.30; S8.16)
(a) Vol. **105**, No. 2, p. K135—K139 (1988).

R. Sommerfeldt, L. Holtmann, E. Krätzig, and B. C. Grabmaier
Influence of Mg Doping and Composition on the Light-Induced Charge Transport in $LiNbO_3$
(61.70; 72.40; 78.20; S11.1)
(a) Vol. **106**, No. 1, p. 89—98 (1988).

E. Z. Katsnelson and A. G. Karoza
On the Irreversibility of the Processes during Photomagnetization of Ferrites
(78.20; 72.40; S11.2)
(a) Vol. **106**, No. 1, p. 277—282 (1988).

J. Szatkowski and K. Sierański
*Influence of Interface States on the Electrical Properties of Mg–Zn_3P_2 Junctions
(72.40 and 73.30; S7)
(a) Vol. **106**, No. 1, p. K31—K34 (1988).

G. Kluge and J. Schmal
*On the Application of the Ordinary and Photoconductivity Meyer-Neldel Rule to Amorphous $Ge_{30}Se_{60}Te_{10}$ Films
(72.40; S8.16)
(a) Vol. **106**, No. 1, p. K47—K52 (1988).

J. Szatkowski and K. Sierański
Electrical Properties of Mg–Zn_3P_2 Contacts
(73.30; 72.40; S7.1)
(a) Vol. **106**, No. 2, p. 473—477 (1988).

J. Szatkowski, K. Sierański, and J. Misiewicz
*Characteristics of Mg–Zn_3P_2:Cd Contacts
(72.40 and 73.30; S7)
(a) Vol. **106**, No. 2, p. 551—555 (1988).

P. D. Pester and T. Wilson
*Time-Dependent Information Depth in Optical Probing of Semiconductor Materials and Devices
(72.40 and 73.25)
(a) Vol. **106**, No. 2, p. 577—582 (1988).

L. F. Makarenko
On the Cross Section of Electron Capture into the Deep State of Thermal U^- Donors in Si
(71.55; 72.40; S5.11)
(a) Vol. **106**, No. 2, p. K153—K155 (1988).

F. Januschewski, H.-J. Erzgräber, and W. Füssel
Modelling the Local Damage of Short-Channel MOSFETs Due to Hot Electron Injection Using Results from Photoinjection Measurements on MOS Capacitors
(73.40 and 85; 72.40; S5.11)
(a) Vol. **106**, No. 2, p. K215—K220 (1988).

S. G. Lyapin, M. I. Eremets, A. M. Shirokov, R. Herrmann, H.-U. Müller, and F. Ludwig
Submillimetre Magnetospectroscopy of Tellurium under Hydrostatic Pressure
(71.55 and 78.50; 72.40; 76.40; S5)
(b) Vol. **147**, No. 1, p. 297—306 (1988).

A. F. Vavrek and K. K. Christova
X-Ray Polarization and Depolarization Phenomena in $Bi_{12}SiO_{20}$
(77.30 and 78.70; 72.40; S10)
(a) Vol. **107**, No. 1, p. 379—386 (1988).

B. Rheinländer, G. Grummt, J. Kováč, H. Haefner, and R. Heilmann
Minority-Carrier Generation and Recombination in 1.35 μm InGaAsP/InP Double Heterostructure Diodes
(73.40; 71.55; 72.40; 78.60; S7.11; S7.15)
(a) Vol. **107**, No. 1, p. 405—418 (1988).

T. K. Kasumov, F. I. Mamedov, and I. K. Gasymov
*Negative Photoconductivity in $AgGaSe_2$ Single Crystals Due to the Negative Photogalvanic Effect
(72.40; S8.16)
(a) Vol. **107**, No. 1, p. K49—K52 (1988).

O. V. Kulikova, L. L. Kulyuk, N. M. Nartya, S. I. Radautsan, E. V. Russu, A. V. Siminel, and E. E. Strumban
On the Possibility of Formation of Associated Impurity Centers in Semi-Insulating Indium Phosphide Doped by Oxygen
(71.55; 72.40; S7.11)
(a) Vol. **107**, No. 1, p. K57—K60 (1988).

V. K. Goyal, R. Pal, and K. Sen
On the Photogenerated Current in Polycrystalline Silicon MIS Solar Cells
(73.40 and 85; 72.40; S5.11)
(a) Vol. **107**, No. 1, p. K83—K87 (1988).

J. Calas, S. M. Abdel-Maksoud, and M. Averous
*Photocarrier Drift Mobility in Bulk Vitreous $GeSe_3$ in the Temperature Range 60 to 300 K
(72.40; S8)
(b) Vol. **147**, No. 2, p. 629—634 (1988).

F. R. Kessler, J. Nieder, R. Nies, and E. Gaulke
*Indirect Exciton Excitation and the Oscillatory Magnetophotoconductivity of Germanium
(71.35 and 72.40; 78.20; S5.12)
(b) Vol. **147**, No. 2, p. K143—K148 (1988).

T. X. Hoai, T. H. Nhung, D. X. Thanh, and N. D. Hung
*Experimental Study of the Photomagnetoelectric Effect near the Threshold of the Stimulated Emission in Semiconductors
(72.40 and 78.55; S8.12)
(b) Vol. **147**, No. 2, p. K165—K170 (1988).

V. A. Berezovets, W. Braune, N. Kubicki, and A. O. Smirnov
Investigation of Interface Peculiarities in Anodized p-InSb MOS Structures
(73.40; 72.40; 73.25; S7.13)
(a) Vol. **108**, No. 1, p. 303—309 (1988).

A. Serpi and J. P. Zielinger
*On the Photoconductivity Relaxation in $ZnIn_2S_4$
(72.40; 71.55; S8.16)
(a) Vol. **108**, No. 1, p. 351—362 (1988).

J. Calas, S. M. Abdel Maksoud, and M. Averous
Energy Distribution of Trapping Centers in Vitreous $GeSe_3$ as Obtained from Transit Time Measurements in the Temperature Range 60 to 300 K
(71.55; 72.40; S8)
(b) Vol. **148**, No. 2, p. 629—634 (1988).

A. G. Kyazymzade, A. A. Akhmedov, S. Kh. Khalilov, and Ya. A. Bairamov
*Simulation of Residual Photoconductivity at a Metal–Semiconductor Contact
(72.40 and 73.40; S8)
(a) Vol. **108**, No. 2, p. 619—623 (1988).

J. A. de Saja and K. Tanaka
Photoelectrochemical Cells with p-Type Poly(3-Methylthiophene)
(73.40; 72.40; 78.40; S12)
(a) Vol. **108**, No. 2, p. K109—K114 (1988).

T. K. Bandyopadhyay, S. B. Chakraborty, and S. R. Chaudhuri
Effect of Surface Treatment on the Photoanodic Behaviour of Thin Film n-CdSe
(73.60; 72.40; S8.12)
(a) Vol. **108**, No. 2, p. K119—K123 (1988).

P. S. Gladkov and K. B. Ozanyan
*Photoconductivity in Bulk and Epitaxial GaAs:V
(71.55 and 72.40; 61.70; S7.12)
(a) Vol. **108**, No. 2, p. K125—K130 (1988).

M. Igalson, K. Plankiewicz, and R. Trykozko
On the Photoresponse of Cu-Treated Cu_2S/CdS Thin Film Solar Cells
(73.40 and 85; 72.40; S8.11)
(a) Vol. **108**, No. 2, p. K169—K171 (1988).

H. Wieczorek and W. Fuhs
*Transient Photocurrents in a-Si:H Diodes: Effects of Deep Trapping
(72.40 and 73.40; 61.40; 71.55; S5.11)
(a) Vol. **109**, No. 1, p. 245—253 (1988).

E. D. Arama, S. I. Radautsan, I. M. Tiginyanu, V. F. Zhitar, A. Turković, B. Petrović, B. Etlinger, and N. B. Urli
*Photoconductivity and Luminescence Spectra of $ZnIn_2S_4$ Crystals Irradiated by γ-Quanta
(72.40 and 78.55; S8.16)
(a) Vol. **109**, No. 1, p. K55—K57 (1988).

K.-P. Möllmann
*Light Quenching of Photoconductivity in Indium-Doped Lead Tin Telluride
(72.40; S8.16)
(b) Vol. **149**, No. 2, p. 663—667 (1988).

N. T. Bagraev and I. S. Polovtsev
Reactions for Iron Centers in Silicon under Optical Pumping
(61.70 and 71.55; 72.40; 76.30; 76.60; S5.11)
(b) Vol. **150**, No. 1, p. 19—24 (1988).

N. I. Tarbaev, J. Schreiber, and G. A. Shepelskii
Physical Properties of $A^{II}B^{VI}$ Semiconductor Crystals after Plastic Deformation at Low Temperature
(61.70 and 71.55; 72.40; 78.50; 78.55; S8.1)
(a) Vol. **110**, No. 1, p. 97—106 (1988).

N. A. Moldovyan, S. I. Radautsan, and V. Z. Chebotaru
*Photoconductivity, Cathodoluminescence, and Optical Absorption of $CdIn_2S_2Se_2$ Single Crystals
(72.40 and 78.40; 78.60; S8.16)
(a) Vol. **110**, No. 1, p. 277—282 (1988).

R. Srivastava, V. M. Pathak, and V. V. Rao
*Flat Band Potential of As-Grown n-$MoSe_2$ Single Crystal Electrodes
(72.40 and 73.40; S8)
(a) Vol. **110**, No. 1, p. 293—299 (1988).

B. S. Ryvkin
Semiconductor Cells with N-Type Photocurrent–Voltage Characteristic: A New Class of Nonlinear Optical Elements
(78.20; 72.40; S7.12)
(b) Vol. **150**, No. 2, p. 567—574 (1988).

V. N. Astratov, A. V. Ilinskii, and A. S. Furman
Photoinduced Charge Dynamics in Photorefractive Semiconductors
(78.20; 72.40; S10)
(b) Vol. **150**, No. 2, p. 611—615 (1988).

A. Pishchev and N. Kristoffel
*Coherent Second-Order Photoinduced Effects in Noncentrosymmetric Semiconductors
(72.40 and 78.20)
(b) Vol. **150**, No. 2, p. 617—621 (1988).

H. Bergner, V. Brückner, M. Lenzner, and R. Strobel
Investigation of the Field-Dependent Carrier Mobility in GaAs by Picosecond Photoconductivity Measurements
(72.20; 72.40; S7.12)
(b) Vol. **150**, No. 2, p. 885—889 (1988).

G. A. Medvedkin, Yu. V. Rud, and M. A. Tairov
Fundamental Optical Absorption Edge in $MnGa_2Te_4$ Single Crystals
(78.20; 72.40; S8.16)
(a) Vol. **110**, No. 2, p. 631—643 (1988).

72.60. Mixed conductivity and conductivity transitions

M. Hidaka, H. Fujii, Y. Horie, and S. Mase
*Superconductor–Semiconductor Transition in $BaPb_{1-x}Bi_xO_3$
(72.60 and 74.10; S10.15)
(b) Vol. **145**, No. 2, p. 649—658 (1988).

R. Rentzsch, K. J. Friedland, and A. N. Ionov
Negative Magnetoresistance of Neutron-Transmutation-Doped Gallium Arsenide at Variable-Range Hopping
(72.20; 72.60; S7.12)
(b) Vol. **146**, No. 1, p. 199—206 (1988).

E. M. Kogan and M. I. Auslender
Anderson Localization in Ferromagnetic Semiconductors Due to Spin Disorder. I. Narrow Conduction Band
(72.20; 72.60; S8.16)
(b) Vol. **147**, No. 2, p. 613—620 (1988).

M. I. Auslender, E. M. Kogan, and S. V. Tretyakov
Anderson Localization in Ferromagnetic Semiconductors Due to Spin Disorder. II. Wide Conduction Band
(72.20; 72.60; S8.1; S10)
(b) Vol. **148**, No. 1, p. 289—295 (1988).

72.70. Noise processes and phenomena

H. L. Cui, X. L. Lei, and N. J. M. Horing
*Hot Electron Magnetophonon Resonances in Conduction and Noise for a Nondegenerate Semiconductor
(72.20 and 72.70)
(b) Vol. **146**, No. 1, p. 189—197 (1988).

72.80. Conductivity of specific semiconductors and insulators

V. M. Korol
Sodium-Ion Implantation into Silicon **(Review Article)**
(61.80; 66.30; 72.80; 73.40; S5.11)
(a) Vol. **110**, No. 1, p. 9—34 (1988).

A. Kuczkowski
*The Electrical Properties of Polyester Polymer-Quinoline Salt of Tetracyanoquinodimethane Composites
(72.80; S12)
(a) Vol. **105**, No. 1, p. K61—K65 (1988).

S. A. Eshragih, S. Kianian, B. Ostrom, O. M. Stafsudd, and A. L. Gentile
*The Electrical and Optical Properties of n-Type $HgIn_2Te_4$
(72.80 and 78.20; S8.16)
(a) Vol. **105**, No. 2, p. 563—566 (1988).

I. Gyúró, B. Pödör, V. V. Popov, and R. V. Parfenev
*Hall Effect Studies on p-Type GaSb Grown under Mcirogravity Conditions
(72.80; S7.13)
(a) Vol. **105**, No. 2, p. K129—K133 (1988).

V. N. Brudnyi and I. V. Kamenskaya
*The Electrical Properties and Fermi Level Pinning in Proton-Irradiated GaSb
(61.80 and 72.80; S7.13)
(a) Vol. **105**, No. 2, p. K141—K144 (1988).

M. N. Khan, M. A. Hassan, and C. A. Hogarth
*The Electronic and Optical Properties of Germanium Tellurite Glasses Containing Various Transition Metal Oxides
(72.80; 76.30; 78.30; S10.15)
(a) Vol. **106**, No. 1, p. 191—200 (1988).

Yu. Dmitriev, V. Koshkin, and U. Ulmanis
Defocusing of Atomic Knock-Ons in Crystals with Low Local Symmetry of Structural Elements
(61.80; 72.80; S8)
(a) Vol. **106**, No. 1, p. K7—K11 (1988).

T. Satow, H. Ishikawa, and O. Uemura
*Semiconductivity in the Liquid Cu–TlTe System
(72.10 and 72.80; 75.30; S1.63)
(b) Vol. **146**, No. 2, p. 661—669 (1988).

G. Dlubek, A. Dlubek, R. Krause, O. Brümmer, K. Friedland, and R. Rentzsch
Point Defects in GaAs Studied by Correlated Positron Lifetime, Optical, and Electrical Measurements. I. Native Point Defects and Their Complexes in As-Grown GaAs
(61.70; 72.80; 78.55; 78.70; S7.12)
(a) Vol. **106**, No. 2, p. 419—432 (1988).

W. Sienicki
*Investigation of Tungsten Disulphide and Diselenide Intercalated by Thallium
(72.80; 61.70; S1.63)
(a) Vol. **106**, No. 2, p. K157—K159 (1988).

A. Henry, K. Saminadayar, J. L. Pautrat, and N. Magnea
The Kinetics of the Formation of the Thermal Donors in Silicon. Effect of Various Parameters
(61.70; 72.80; 78.30; S5.11)
(a) Vol. **107**, No. 1, p. 101—110 (1988).

J. Horak, Z. Starý, and J. Klikorka
Relations between Structure, Bonding, and Nature of Point Defects in Layered Crystals of Tetradymite Structure
(61.50 and 61.70; 72.80; S8; S8.15)
(b) Vol. **147**, No. 2, p. 501—510 (1988).

G. I. Gashimov, A. G. Rustamov, A. A. Mustafaev, and A. D. Namazov
*Kinetic Properties of the $Fe_{1-x}Zn_xCr_2S_4$ ($0 \leqq x < 1$) System
(72.80; S8)
(a) Vol. **107**, No. 2, p. K129—K133 (1988).

Qing Gu and Xi-Mao Bao
Effect of the Structure of Laser Crystallized Silicon-on-Insulator on Impurity Diffusion
(61.50; 66.30; 72.80; S5.11)
(a) Vol. **108**, No. 1, p. 81—87 (1988).

R. Piasecki, M. Ząbkowska-Wacławek, and W. Wacławek
*An Explanation of Electrical Conductivity Data for Copper Phthalocyanine–Acetylene Carbon Black Mixtures
(72.80; S12)
(a) Vol. **108**, No. 1, p. 331—335 (1988).

S. Shigetomi, T. Ikari, Y. Koga, and S. Shigetomi
*Electrical Properties of Cd-Doped p-InSe
(71.55 and 72.80; S8)
(a) Vol. **108**, No. 1, p. K53—K57 (1988).

S. I. Radautsan, I. M. Tiginyanu, and N. B. Pyshnaya
*Electrical Properties of InP:Fe Single Crystals Implanted by Phosphorus Ions
(71.55 and 72.80; S7.11)
(a) Vol. **108**, No. 1, p. K59—K61 (1988).

I. A. Aksenov, L. A. Makovetskaya, and V. A. Savchuk
*Electrical Properties of Cadmium and Zinc Doped $CuAlS_2$
(72.80; S8.16)
(a) Vol. **108**, No. 1, p. K63—K65 (1988).

G. A. Bordovskii, N. A. Savinova, and L. N. Seregina
*Electrical Activity Peculiarities of Tin and Iron Impurities in the Semiconducting Glass $Ge_{28.5}Pb_{15}S_{56.5}$
(72.80 and 76.80; S8)
(a) Vol. **109**, No. 1, p. K51—K53 (1988).

V. M. Babich, N. P. Baran, A. A. Bugai, A. A. Konchits, V. B. Kovalchuk, V. M. Maksimenko, and B. D. Shanina
Electrical and Paramagnetic Properties of Thermodonors-II in Silicon. Discussion of a Model
(71.55; 72.80; 76.30; S5.11)
(a) Vol. **109**, No. 2, p. 537—547 (1988).

A. M. Karguppikar and A. G. Vedeshwar
*Electrical and Optical Properties of Natural Iron Pyrite (FeS_2)
(72.80 and 78.50; S1.63)
(a) Vol. **109**, No. 2, p. 549—558 (1988).

K.-P. Möllmann
Additional Deep Level in Strongly Indium-Doped Lead Tin Telluride
(71.55; 72.80; S8)
(a) Vol. **109**, No. 2, p. K121—K124 (1988).

G. M. Hegde
*Effect of the Electrode on the Electrical Conduction Process in Organic Semiconductors
(72.80 and 73.40; S12)
(a) Vol. **109**, No. 2, p. K125—K127 (1988).

P. F. Lugakov and T. A. Lukashevich
Peculiarities of Radiation Defect Accumulation under High-Flux γ-Ray and Electron Irradiation of Silicon
(61.80; 72.80; S5.11)
(a) Vol. **110**, No. 2, p. 403—408 (1988).

73. Electronic structure and electrical properties of surfaces, interfaces, and thin films

73.20. Electronic surface states

(for emission and impact phenomena, see 79)

T. N. Mamontova, A. S. Kochemirovskii, and L. V. Pivovarova
Electronic Processes on the Surface of As–Se Chalcogenide Glassy Semiconductors **(Review Article)**
(73.25; 68.45; 72.40; 73.20; S8)
(a) Vol. **107**, No. 1, p. 11—43 (1988).

Y. Q. Sheng, E. Recknagel, A. Weidinger, Z. G. Gu, Z. Y. Lai, and Z. C. Zhuang
*Raman Scattering at the CaF_2 Crystal Surface Excited by an Evanescent Optical Wave Field
(71.36 and 73.20; 78.30; S9.12)
(b) Vol. **145**, No. 1, p. 151—156 (1988).

Tran Thoai Duy Bao
*Plasmon Excitation in Metal Surface Layers by Fast Electrons
(71.45 and 73.20; 61.80; S4)
(b) Vol. **145**, No. 2, p. 509—515 (1988).

H. von Berlepsch and M. Pinnow
Thermally Stimulated Surface Potential Decay in HDPE and the Cross-Over Phenomenon
(73.40; 73.20; S12)
(a) Vol. **105**, No. 2, p. 485—492 (1988).

D. B. Tran Thoai and E. Zeitler
*Image Potential of an Electron in a Gap between Two Metals
(71.45 and 73.20; S1.3; S4)
(b) Vol. **146**, No. 1, p. 137—140 (1988).

I. I. Burshta, Ė. F. Venger, Yu. A. Pasechnik, and O. V. Snitko
*Simplest Properties of Guided Wave Polaritons in Three-Layer Structures
(71.36 and 73.20; S6)
(b) Vol. **146**, No. 2, p. 517—524 (1988).

S. B. Shrivastava, V. K. Gupta, and K. S. Bhadoriya
*Positronium at Metal Surfaces
(71.60 and 73.20; S1.2; S1.3)
(b) Vol. **147**, No. 2, p. 553—557 (1988).

D. B. Tran Thoai and E. Zeitler
*Inelastic Scattering of Fast Electrons by Small Metal Particles
(61.80 and 73.20; 71.45; S4)
(a) Vol. **107**, No. 2, p. 791—797 (1988).

S. Nara
A Note on the Icosahedral Morphology of Ultrafine Metal Particles
(68.55; 61.70; 73.20; S1.3)
(b) Vol. **149**, No. 2, p. 555—564 (1988).

H. K. Sy and L. M. Song
*Surface Plasmons in Type II Superlattices
(71.45 and 73.20)
(b) Vol. **149**, No. 2, p. 595—602 (1988).

Ya. M. Strelniker
Theory of Inelastic Low-Energy Electron Scattering with Generation of Surface and Bulk Excitons in Molecular Crystals
(71.35; 73.20)
(b) Vol. **149**, No. 2, p. 603—613 (1988).

73.25. Surface conductivity

T. N. Mamontova, A. S. Kochemirovskii, and L. V. Pivovarova
*Electronic Processes on the Surface of As–Se Chalcogenide Glassy Semiconductors **(Review Article)**
(73.25; 68.45; 72.40; 73.20; S8)
(a) Vol. **107**, No. 1, p. 11—43 (1988).

G. Beister, J. Maege, and G. Richter
*Identification of Surface Influences on the Current–Voltage Characteristic of Narrow Oxide Stripe Laser Diodes
(73.25; 78.45; S7.15)
(a) Vol. **105**, No. 1, p. K45—K49 (1988).

P. D. Pester and T. Wilson
*Time-Dependent Information Depth in Optical Probing of Semiconductor Materials and Devices
(72.40 and 73.25)
(a) Vol. **106**, No. 2, p. 577—582 (1988).

G. Nachtwei, D. Schulze, G. Gobsch, G. Paasch, W. Kraak, H. Krüger, and R. Herrmann
*Calculations of Magnetotransport Properties of InSb Bicrystals with Several Populated Q2D Subbands
(73.25 and 73.40; S7.13)
(b) Vol. **148**, No. 1, p. 349—359 (1988).

V. A. Berezovets, W. Braune, N. Kubicki, and A. O. Smirnov
Investigation of Interface Peculiarities in Anodized p-InSb MOS Structures
(73.40; 72.40; 73.25; S7.13)
(a) Vol. **108**, No. 1, p. 303—309 (1988).

C. Manfredotti, F. Fizzotti, and G. Amato
*Frenkel Effect in SIPOS Films
(73.25; S5.11)
(a) Vol. **108**, No. 1, p. K25—K30 (1988).

M. Chybicki, J. Liwo, and K. Trzebiatowski
*Modification of the Electrical Surface Conductivity of Lead–Silicate Glasses by H_2^+ Ion Bombardment
(73.25; S10.15)
(a) Vol. **108**, No. 2, p. K105—K107 (1988).

D. B. Tran Thoai and E. Zeitler
*Dynamical Image Potential of a Point Charge Approaching a Metal Sphere
(71.10 and 73.25)
(b) Vol. **149**, No. 1, p. 169—174 (1988).

73.30. Surface double layers, Schottky barriers, and work functions

I. T. Yakubov, A. G. Khrapak, V. V. Pogosov, and S. A. Trigger
Thermodynamical Characteristics of Liquid Metal Droplets
(68.10; 73.30; S2)
(b) Vol. **145**, No. 2, p. 455—466 (1988).

S. Romanowski
*An Application of the Extended CPA Method to Electron Work Function and Surface Charge Calculation for Cu Single Crystals
(73.30; 71.10; S1.2)
(b) Vol. **145**, No. 2, p. 467—482 (1988).

P. D. Pester and T. Wilson
*An Analysis of the Photocurrent Induced in a Semi-Infinite Schottky Barrier by a Time Varying Focussed Laser Beam
(73.30 and 85; 72.40)
(a) Vol. **105**, No. 2, p. 649—660 (1988).

T. Chot, N. N. Xuan, and L. Q. Minh
*Barrier Heights and Interface States of Metal/2,4,7-Trinitro-9-Fluorenone Thin Film Contacts
(73.30; S12)
(a) Vol. **105**, No. 2, p. K121—K124 (1988).

A. Opanowicz and B. Kościelniak-Mucha
*Determination of the Electron Lifetime in p-$CuInS_2$ from a Photovoltaic Study of In/p-$CuInS_2$ Barrier Diodes
(72.40 and 73.30; S8.16)
(a) Vol. **105**, No. 2, p. K135—K139 (1988).

D. Kindl and J. Toušková
*Charge Transport Study in Thin Film Au–CdTe Schottky Diodes
(73.30 and 85; 73.60; S8.13)
(a) Vol. **106**, No. 1, p. 297—304 (1988).

J. Szatkowski and K. Sierański
*Influence of Interface States on the Electrical Properties of Mg–Zn_3P_2 Junctions
(72.40 and 73.30; S7)
(a) Vol. **106**, No. 1, p. K31—K34 (1988).

J. Szatkowski and K. Sierański
*Electrical Properties of Mg–Zn_3P_2 Contacts
(73.30; 72.40; S7.1)
(a) Vol. **106**, No. 2, p. 473—477 (1988).

W. Possart and I. Müller
The Electric Double Layer at the Polymer–Metal Contact. A Phenomenological Model
(73.40; 73.30; S12)
(a) Vol. **106**, No. 2, p. 525—534 (1988).

J. Szatkowski, K. Sierański, and J. Misiewicz
*Characteristics of Mg–Zn_3P_2:Cd Contacts
(72.40 and 73.30; S7)
(a) Vol. **106**, No. 2, p. 551—555 (1988).

A. Opanowicz, B. Kościelniak-Mucha, and G. Boroś
*Electrical and Photovoltaic Properties of Au/n-$CuInSe_2$ Schottky Barrier Diodes
(73.30 and 85; S8.16)
(a) Vol. **106**, No. 2, p. K197—K201 (1988).

P. R. Vaya and J. Majhi
*Effect of Surface Treatment on Al–Si Schottky Structures
(73.30 and 85; S5.11)
(a) Vol. **106**, No. 2, p. K209—K213 (1988).

H. Müller, Ch. Opitz, S. Romanowski, and L. Skála
*On the Size Dependence of the Contact Potential and Its Consequences
(73.30; 73.40)
(b) Vol. **148**, No. 1, p. K11—K15 (1988).

Sh. M. Efendiev, V. E. Bagiev, R. A. Aliev, and E. R. Mustafaev
Nonlinear *I–U* Characteristics and Photoelectret State in Sillenite-Type Crystals
(73.40 and 85; 73.30; S11)
(a) Vol. **109**, No. 1, p. 345—357 (1988).

73.40. Interfaces

H. Pagnia and N. Sotnik
*Bistable Switching in Electroformed Metal–Insulator–Metal Devices **(Review Article)**
(73.40 and 85; S10)
(a) Vol. **108**, No. 1, p. 11—65 (1988).

V. M. Korol
Sodium-Ion Implantation into Silicon **(Review Article)**
(61.80; 66.30; 72.80; 73.40; S5.11)
(a) Vol. **110**, No. 1, p. 9—34 (1988).

R. Herrmann, W. Kraak, S. Handschack, Th. Schurig, D. Kusnick, and B. Schnackenburg
*Magnetotransport Properties and Subband Structure of the Two-Dimensional Electron Gas in the Inversion Layer of InSb Bicrystals under Hydrostatic Pressure
(73.40; S7.13)
(b) Vol. **145**, No. 1, p. 157—166 (1988).

S. Ya. Goroshchenko and I. I. Ukrainskii
Two-Dimensional Electron Pair in a Transverse Magnetic Field
(71.25; 72.10; 73.40)
(b) Vol. **145**, No. 1, p. 187—194 (1988).

Guanghua Chen, Guosheng Sun, and Fangqing Zhang
*ESR in P/I Heterojunctions Based on a-Si:H
(73.40 and 76.30; S5.11)
(a) Vol. **105**, No. 1, p. K41—K43 (1988).

P. Apell and A. Tagliacozzo
*Single Electron Tunneling
(73.40)
(b) Vol. **145**, No. 2, p. 483—491 (1988).

R. Pérez-Alvarez and H. Rodriguez-Coppola
*Transfer Matrix in 1D Schrödinger Problems with Constant and Position-Dependent Mass
(71.10 and 73.40)
(b) Vol. **145**, No. 2, p. 493—500 (1988).

R. Pérez-Alvarez, H. Rodriguez-Coppola, J. López-Gondar, and M. Lago-Izquierdo
*Transfer Matrix in the Quasiclassical Approximation with Constant and Position-Dependent Mass. Resonant Tunneling
(71.10 and 73.40)
(b) Vol. **145**, No. 2, p. 501—508 (1988).

Yu. A. Firsov, E. P. Nakhmedov, V. N. Prigodin, and W. Weller
Low-Temperature Conductivity of a Weakly Disordered, Strongly Anisotropic Two-Dimensional System with Electron–Electron Interaction. Application to Germanium Bicrystals and to Surface Superlattices
(72.10; 73.40; S5.12)
(b) Vol. **145**, No. 2, p. 625—635 (1988).

S. Ya. Goroshchenko and I. I. Ukrainskii
*On the Effective Interaction of Quasi-Two-Dimensional Electrons in a Transverse Magnetic Field
(73.40)
(b) Vol. **145**, No. 2, p. K107—K110 (1988).

P. H. Jiang, Y. T. Zhu, D. Z. Sun, and Y. P. Zeng
*Modulation-Doped $Al_xGa_{1-x}As$/GaAs Heterostructures with Parallel Conducting Layer in $Al_xGa_{1-x}As$
(73.40; S7.12; S7.15)
(b) Vol. **145**, No. 2, p. K111—K114 (1988).

H. von Berlepsch and M. Pinnow
*Thermally Stimulated Surface Potential Decay in HDPE and the Cross-Over Phenomenon
(73.40; 73.20; S12)
(a) Vol. **105**, No. 2, p. 485—492 (1988).

W. Dąbrowski
Investigation of Deep Impurities in Schottky Diodes on High-Resistivity Silicon
(71.55; 73.40; S5.11)
(a) Vol. **105**, No. 2, p. 511—520 (1988).

Ll. Closas and J. Fernández Rubio
*An EBIC Model with Extended Generation Profile of Electron–Hole Pairs
(72.20 and 73.40)
(a) Vol. **105**, No. 2, p. 541—548 (1988).

R. Stegmann, B. Jacobs, H. Kamleh, M. Albani, M. Heider, and F. Monzer
*Carrier Injection at Graded Heterojunctions
(73.40 and 85; 78.60; S7.15)
(a) Vol. **105**, No. 2, p. 661—670 (1988).

P. Sobkowicz
*Self-Consistent Calculation of the 2D Subband Structure at Grain Boundaries: Some Shortcomings of the Semiclassical Approach
(71.25 and 73.40)
(b) Vol. **146**, No. 1, p. K7—K10 (1988).

I. V. Marchishin, V. N. Ovsyuk, and S. B. Sevastianov
Deep Level Profiling Using an Admittance Spectroscopy Method
(71.55; 73.40)
(a) Vol. **106**, No. 1, p. 153—166 (1988).

B. Rheinländer, R. Heilmann, and G. Oelgart
Determination of Temperature-Dependent Carrier Losses in 1.3 μm InGaAsP/InP Double-Heterostructures
(78.45 and 85; 73.40; S7.11; S7.16)
(a) Vol. **106**, No. 1, p. K113—K116 (1988).

N. A. Gushchina, S. M. Dunaevskii, and V. K. Nikulin
*Energy Band Structure of $(GaAs)_m$ $(AlAs)_m$ Semiconductor Superlattices with Ultrathin Layers
(73.40; 71.25; S7.15)
(b) Vol. **146**, No. 2, p. 511—516 (1988).

Yu. B. Gaididei and I. I. Yakimenko
On Trigger Properties of Molecular Organizates. III. Nonlinear Waves in Stratified Molecular Structures
(72.10; 73.40)
(b) Vol. **146**, No. 2, p. 643—654 (1988).

J. J. Wnuk
*Evaluation of the Eliashberg Function from Normal Metal Tunneling Data
(73.40)
(b) Vol. **146**, No. 2, p. K91—K96 (1988).

W. Possart and I. Müller
*The Electric Double Layer at the Polymer–Metal Contact. A Phenomenological Model
(73.40; 73.30; S12)
(a) Vol. **106**, No. 2, p. 525—534 (1988).

G. Petermann
*DC and AC Characterization of Grain Boundaries in Float Zone Silicon
(71.20 and 73.40; S5.11)
(a) Vol. **106**, No. 2, p. 535—549 (1988).

I. V. Marchishin, V. N. Ovsyuk, and S. B. Sevastianov
*Investigation of Radiation Defects in Hyperabrupt p–n Junctions
(71.55 and 73.40; 61.80)
(a) Vol. **106**, No. 2, p. 557—568 (1988).

E. Pinčík, B. Maťátko, J. Bartoš, I. Thurzo, M. Grendel, V. Nádaždy, M. Zubeková, and M. Morvic
*Some Properties of MIS Structures Prepared by Plasma Oxidation of Al Layers on GaAs
(73.40 and 85; S7.12)
(a) Vol. **106**, No. 2, p. 659—667 (1988).

A. P. Kovchavtsev, G. L. Kurishev, K. O. Postnikov, R. A. Sokolov, and I. M. Subbotin
*Inelastic Electron Tunneling Spectroscopy of Si MIS Structures with Ultrathin Thermal Silicon Nitride and Thermal Silica
(73.40 and 85; S5.11)
(a) Vol. **106**, No. 2, p. 669—674 (1988).

A. Singh
*Oxidise–Etch–Oxidise and Anneal Method for Growing High Quality Ultrathin SiO_2
(68.55 and 73.40; S10.1)
(a) Vol. **106**, No. 2, p. K139—K142 (1988).

A. Srivastava
*Effect of Polysilicon–Silicon Interface Recombination Velocity on Current Transport Properties of Poly-Emitter Transistors
(73.40 and 85; S5.11)
(a) Vol. **106**, No. 2, p. K203—K207 (1988).

F. Januschewski, H.-J. Erzgräber, and W. Füssel
*Modelling the Local Damage of Short-Channel MOSFETs Due to Hot Electron Injection Using Results from Photoinjection Measurements on MOS Capacitors
(73.40 and 85; 72.40; S5.11)
(a) Vol. **106**, No. 2, p. K215—K220 (1988).

R. Pérez-Alvarez, J. L. Parra-Santiago, and P. Pajón-Suarez
*Cylindrical Quantum Well with Position-Dependent Mass
(71.25 and 73.40; S7.12; S7.15)
(b) Vol. **147**, No. 1, p. 127—133 (1988).

C. K. Sarkar, S. Kundu, and P. K. Basu
*Cyclotron Resonance Linewidth in a Two-Dimensional Electron Gas
(73.40 and 76.40; S7.12)
(b) Vol. **147**, No. 1, p. 135—140 (1988).

K. Yonashiro, T. Tomoyose, M. Yamashiro, and M. Kobayashi
Surface Modes of a Type II Semiconductor Superlattice
(71.36 and 71.45; 68.30; 73.40)
(b) Vol. **147**, No. 1, p. 141—148 (1988).

E. P. Pokatilov, S. I. Beril, and V. M. Fomin
*Image Potentials and Image Forces in the Polaron Theory
(71.38 and 73.40; 71.45)
(b) Vol. **147**, No. 1, p. 163—172 (1988).

A. Chiari, M. Colocci, F. Fermi, Yuzhang Li, R. Querzoli, A. Vinattieri, and Weihua Zhuang
Temperature Dependence of the Photoluminescence in GaAs–GaAlAs Multiple Quantum Well Structures
(78.55; 73.40; S7.15)
(b) Vol. **147**, No. 1, p. 421—429 (1988).

M. Hrobár and I. Travěnec
*Time-Domain Spectroscopy. Method for Investigation of YSZ
(73.40; 77.55; S10.15)
(a) Vol. **107**, No. 1, p. 235—238 (1988).

B. Cabon and G. Ghibaudo
*Degradation of Submicron MOSFETs after Aging
(73.40 and 85)
(a) Vol. **107**, No. 1, p. 393—404 (1988).

B. Rheinländer, G. Grummt, J. Kováč, H. Haefner, and R. Heilmann
*Minority-Carrier Generation and Recombination in 1.35 μm InGaAsP/InP Double Heterostructure Diodes
(73.40; 71.55; 72.40; 78.60; S7.11; S7.15)
(a) Vol. **107**, No. 1, p. 405—418 (1988).

G. Grummt, H. Haefner, G. Oelgart, W. Thielemann, R. Pickenhain, and B. Jacobs
*Properties of Red Light Emitting (Al, Ga)As Double Heterostructure Diodes
(73.40 and 78.60; 85; S7.12)
(a) Vol. **107**, No. 1, p. 419—428 (1988).

H. Ohyama and K. Nemoto
*Effect of Lattice Defects in the Collector Region of npn Si Transistors on the Degradation of h_{FE}
(61.80 and 73.40; S5.11)
(a) Vol. **107**, No. 1, p. 429—439 (1988).

V. K. Goyal, R. Pal, and K. Sen
*On the Photogenerated Current in Polycrystalline Silicon MIS Solar Cells
(73.40 and 85; 72.40; S5.11)
(a) Vol. **107**, No. 1, p. K83—K87 (1988).

R. Pérez-Alvarez and P. Pajón-Suárez
*Optical Absorption Coefficient in Cylindrical Quantum Wells
(71.25 and 73.40; S7.12)
(b) Vol. **147**, No. 2, p. 547—552 (1988).

G. Manzke and K. Henneberger
Nonlinear Propagation of Laser Pulses in GaAs–GaAlAs MQWS
(71.35 and 78.20; 73.40; S7.15)
(b) Vol. **147**, No. 2, p. 733—739 (1988).

P. I. Khadzhi and E. S. Kiseleva
The Dispersion Relations of the Nonlinear Surface Waves Due to the Exciton and Biexciton Formation
(78.20; 71.35; 73.40)
(b) Vol. **147**, No. 2, p. 741—745 (1988).

H. Übensee, G. Paasch, J.-P. Zöllner, and S. Handschack
*Modified Thomas-Fermi Approximation for Accumulation Layers in MIS Structures
(73.40; S7.1)
(b) Vol. **147**, No. 2, p. 823—837 (1988).

H. Yamamoto
*Well Width Determination in Rectangular Double-Barrier Quantum-Well Structures
(73.40)
(b) Vol. **147**, No. 2, p. K115—K118 (1988).

S. HANDSCHACK, G. PAASCH, and G. GOBSCH
*Self-Consistent Calculation of the Electronic Structure of an Inversion Layer in a Perpendicular Magnetic Field
(71.25 and 73.40; S7.13)
(b) Vol. **147**, No. 2, p. K125—K129 (1988).

G. M. SHMELEV, A. T. ZHELEZNYAK, and G. I. TSURKAN
*Interface Screening Breakdown of the Coulomb Field by Strong Electromagnetic Waves in the Presence of a Constant Magnetic Field
(73.40)
(b) Vol. **147**, No. 2, p. K131—K135 (1988).

R. A. VOLKOV and A. F. CHUIKO
*Finite Motion of Microparticles in Tunnel-Effect Multiquantum Well Structures
(71.10 and 73.40; 71.45)
(b) Vol. **148**, No. 1, p. 107—120 (1988).

F. LUDWIG, H.-U. MÜLLER, and R. HERRMANN
Magneto-Optical Investigations on p-InSb Bicrystals in the Submillimeter Wavelength Range
(76.40 and 78.20; 73.40; S7.13)
(b) Vol. **148**, No. 1, p. 143—153 (1988).

R. ENDERLEIN, F. BECHSTEDT, and G. GERECKE
The Complete Set of Polar Optic Phonon Modes of Superlattices. Confined Bulk and Interface Modes
(63.20 and 68.30; 73.40)
(b) Vol. **148**, No. 1, p. 173—183 (1988).

G. NACHTWEI, D. SCHULZE, G. GOBSCH, G. PAASCH, W. KRAAK, H. KRÜGER, and R. HERRMANN
*Calculations of Magnetotransport Properties of InSb Bicrystals with Several Populated Q2D Subbands
(73.25 and 73.40; S7.13)
(b) Vol. **148**, No. 1, p. 349—359 (1988).

H. ÜBENSEE, F. PAASCH, D. BUNDFUSS, and J.-P. ZÖLLNER
*Electrons in MIS Inversion and Accumulation Layers. Exchange and Correlation Effects within a Modified Thomas-Fermi Theory
(73.40; S5.11; S7.1)
(b) Vol. **148**, No. 1, p. 421—434 (1988).

H. MÜLLER, CH. OPITZ, S. ROMANOWSKI, and L. SKÁLA
On the Size Dependence of the Contact Potential and Its Consequences
(73.30; 73.40)
(b) Vol. **148**, No. 1, p. K11—K15 (1988).

D. CHATTOPADHYAY, P. C. RAKSHIT, and C. K. SARKAR
*Small-Signal AC Mobility of Two-Dimensional Hot Electrons in Semiconductor Quantum Wells at Low Temperatures
(72.20 and 73.40; S7.12)
(b) Vol. **148**, No. 1, p. K17—K22 (1988).

B. Schumann, R. Flagmeyer, and H. Weinert
*Multi-Quantum-Well Structures Prepared by the Flasch Evaporation Technique. Growth, Structural, and Optical Characterization
(73.40; 68.55; 78.65; S7.1; S9.15)
(a) Vol. **108**, No. 1, p. 257—264 (1988).

V. A. Berezovets, W. Braune, N. Kubicki, and A. O. Smirnov
*Investigation of Interface Peculiarities in Anodized p-InSb MOS Structures
(73.40; 72.40; 73.25; S7.13)
(a) Vol. **108**, No. 1, p. 303—309 (1988).

O. A. Golikova, M. M. Mezdrogina, N. A. Feoktistov, K. L. Sorokina, M. M. Kazanin, P. M. Karageorgy-Alkalaev, and A. Yu. Leiderman
*Injection and Recombination Currents in a-Si:H Structures
(72.20 and 73.40; S5.11)
(a) Vol. **108**, No. 1, p. 323—329 (1988).

U. Henniger, B. Schmidt, R. Tempel, and H. Moeller
*Temperature Distribution in Square-Shaped Semiconductor Chips
(73.40 and 85)
(a) Vol. **108**, No. 1, p. 469—478 (1988).

B. Lepley, A. Bath, C. Carabatos-Nedelec, G. J. Papaioannou, P. C. Euthymiou, and S. Ravelet
*Determination by Optical DLTS of the Distribution of States near the Valence Band of Plasma Oxidized n-Type InP
(73.40; S7.11)
(a) Vol. **108**, No. 1, p. K41—K46 (1988).

V. V. Karyagin, I. I. Lyapilin, and V. V. Dyakin
*Thermoelectric Coefficients of a Limited 2D Gas in Quantizing Magnetic Field
(71.25 and 73.40)
(b) Vol. **148**, No. 2, p. 501—513 (1988).

E. L. Albuquerque and W. B. Costa
*Impurity Binding Energies in Heterostructures. Two-Band Model
(71.20 and 73.40; S7.12; S7.15)
(b) Vol. **148**, No. 2, p. 543—548 (1988).

A. Kumar and I. Singh
*Electronic Structure of the Ge–ZnSe, Ge–GaAs, and Ge–AlAs(110) Interfaces and Superlattices
(71.25 and 73.40; S5.12; S7.12; S8.12)
(b) Vol. **148**, No. 2, p. 549—558 (1988).

W. Kraak, G. Nachtwei, R. Herrmann, and M. Gliński
*Properties of the Quantum Hall Effect of the Two-Dimensional Electron Gas in the n-Inversion Layer of InSb Grain Boundaries under High Hydrostatic Pressure
(73.40; S7.13)
(b) Vol. **148**, No. 2, p. 567—578 (1988).

J.-P. Zöllner, G. Paasch, G. Gobsch, and H. Übensee
*Theory for n-Surface Charge Layers in $Hg_{1-x}Cd_xTe$ MIS Structures
(71.25 and 73.40; S8.15)
(b) Vol. **148**, No. 2, p. 611—618 (1988).

M. DE DIOS LEYVA, J. LÓPEZ GONDAR, and J. SABIN DEL VALLE
*Interface Boundary Conditions and Nonparabolicity Effects in Quantum Wells
(71.25 and 73.40)
(b) Vol. **148**, No. 2, p. K113—K117 (1988).

P. J. RATCLIFFE and R. A. COLLINS
Thermal Oxidation of Ion Implanted Copper
(61.70; 73.40; S1.2)
(a) Vol. **108**, No. 2, p. 537—543 (1988).

A. G. KYAZYMZADE, A. A. AKHMEDOV, S. KH. KHALILOV, and YA. A. BAIRAMOV
*Simulation of Residual Photoconductivity at a Metal–Semiconductor Contact
(72.40 and 73.40; S8)
(a) Vol. **108**, No. 2, p. 619—623 (1988).

E. G. SALMAN and V. N. VERTOPRAKHOV
*Thermally Stimulated Depolarization Current Controlled by Surface Charge Change
(73.40 and 77.30; 77.55; S10.1)
(a) Vol. **108**, No. 2, p. 625—630 (1988).

YA. D. LEBEDEV, M. V. FOK, and M. D. SHAGALOV
*On the Mechanism of Polarized Electroluminescence from Light-Emitting Structures Based on GaN
(73.40 and 78.60; S7.1)
(a) Vol. **108**, No. 2, p. 657—667 (1988).

R. AGRAWAL, R. GOPAL, R. DWIVEDI, and S. K. SRIVASTAVA
*Analytical Modelling of a MOSFET with Non-Uniform Impurity Profile
(73.40 and 85)
(a) Vol. **108**, No. 2, p. 803—811 (1988).

G. GRUMMT
*Calculation of the *C–U* Profiles of Graded Heterojunctions
(73.40)
(a) Vol. **108**, No. 2, p. 813—820 (1988).

J. A. DE SAJA and K. TANAKA
*Photoelectrochemical Cells with p-Type Poly(3-Methylthiophene)
(73.40; 72.40; 78.40; S12)
(a) Vol. **108**, No. 2, p. K109—K114 (1988).

V. N. KATERINCHUK, Z. D. KOVALYUK, and V. A. MANASSON
*On a Mechanism of Photo-E.M.F. Formation in SIS Structures
(73.40; 85; S8; S10.1)
(a) Vol. **108**, No. 2, p. K115—K118 (1988).

M. IGALSON, K. PLANKIEWICZ, and R. TRYKOZKO
*On the Photoresponse of Cu-Treated Cu_2S/CdS Thin Film Solar Cells
(73.40 and 85; 72.40; S8.11)
(a) Vol. **108**, No. 2, p. K169—K171 (1988).

D. A. ARONOV, R. KABULOV, and YU. M. YUABOV
*The Current Characteristics of p–i–n and p^+–i–p^+ Structures Based on Hydrogenated Amorphous Silicon at Various Temperatures and Excitation Levels
(73.40; 73.60; S5.11)
(a) Vol. **109**, No. 1, p. 211—216 (1988).

H. WIECZOREK and W. FUHS
*Transient Photocurrents in a-Si:H Diodes: Effects of Deep Trapping
(72.40 and 73.40; 61.40; 71.55; S5.11)
(a) Vol. **109**, No. 1, p. 245—253 (1988).

F. KOCH, W. HERGERT, G. OELGART, and N. PUHLMANN
*Determination of Semiconductor Parameters by Electron Beam Induced Current and Cathodoluminescence Measurements
(73.40; 78.60; S7.12)
(a) Vol. **109**, No. 1, p. 261—272 (1988).

SH. M. EFENDIEV, V. E. BAGIEV, R. A. ALIEV, and E. R. MUSTAFAEV
*Nonlinear *I–U* Characteristics and Photoelectret State in Sillenite-Type Crystals
(73.40 and 85; 73.30; S11)
(a) Vol. **109**, No. 1, p. 345—357 (1988).

J. P. BANERJEE, S. P. PATI, and S. K. ROY
*Computer Simulation Experiment on the mm-Wave Properties of Indium Phosphide Double Drift Impatts
(73.40 and 85; S7.11)
(a) Vol. **109**, No. 1, p. 359—364 (1988).

B. SANDOW, E. NEBAUER, and H. STRUSNY
*Contact Resistance Reduction by Implanted n^+-Layers at the Ni/AuGe–GaAs System
(73.40; S7.12)
(a) Vol. **109**, No. 1, p. K35—K37 (1988).

H. LEON and F. COMAS
*On the Scattering of Electrons by Confined LO Phonons in a Semiconductor Quantum Well
(71.38 and 73.40; S7.12; S7.15)
(b) Vol. **149**, No. 2, p. 533—542 (1988).

CHR. PRATSCH and M. SUHRKE
*Low-Temperature Electron Transport in Quasi-Two-Dimensional Systems
(73.40; S7.12; S7.15)
(b) Vol. **149**, No. 2, p. 547—553 (1988).

B. MITRA and K. P. GHATAK
*On the Einstein Relation in Semiconductor Superlattices under Magnetic Quantization
(72.20 and 73.40; S7.12; S7.15)
(b) Vol. **149**, No. 2, p. K117—K122 (1988).

L. WENDLER, R. HAUPT, and V. G. GRIGORYAN
*Intersubband Mixing of Coupled Plasmon–Phonons in Double Heterostructures
(71.45 and 73.40)
(b) Vol. **149**, No. 2, p. K123—K128 (1988).

V. S. Postnikov, B. I. Sysoev, A. V. Budanov, N. N. Bezryadin, Yu. K. Shlyk, and B. L. Agapov
*Heterostructures on the Basis of Indium Arsenide with Semi-Insulating $A_2^{III}B_2^{VI}$ Compound Layers
(73.40; S7.12; S8)
(a) Vol. **109**, No. 2, p. 463—467 (1988).

P. Krispin, N. T. Nghi, N. T. Q. Hai, D. D. Khang, and D. A. Tuan
Defects in GaP Green Light-Emitting Diodes
(71.55; 73.40; 85; S7.11)
(a) Vol. **109**, No. 2, p. K115—K119 (1988).

G. M. Hegde
*Effect of the Electrode on the Electrical Conduction Process in Organic Semiconductors
(72.80 and 73.40; S12)
(a) Vol. **109**, No. 2, p. K125—K127 (1988).

F. Ludwig, H.-U. Müller, and R. Herrmann
*Effective Scattering Times of Inversion Electrons in InSb Grain Boundaries Deduced from Cyclotron Resonance Measurements
(73.40 and 76.40; 72.10; S7.13)
(b) Vol. **150**, No. 1, p. 119—123 (1988).

V. A. Kulbachinskii and D. Yu. Rodichev
*Peculiarities of the Localization of Electrons in GaAs–$Ga_{1-x}Al_xAs$ Heterostructures
(73.40; S7.15)
(b) Vol. **150**, No. 1, p. K25—K29 (1988).

W. Possart and I. Müller
*The Estimation of the Contact Potential Difference from Contact Charging between Polymer and Metal. New Aspects
(73.40; S12)
(a) Vol. **110**, No. 1, p. 205—211 (1988).

T. V. Torchinskaya, A. A. Shmatov, and M. K. Sheinkman
Recombination-Enhanced Transformation of Deep Centers in Red Light-Emitting AlGaAs Diodes
(71.55; 73.40; 78.55; S7.15)
(a) Vol. **110**, No. 1, p. 213—220 (1988).

R. Srivastava, V. M. Pathak, and V. V. Rao
*Flat Band Potential of As-Grown n-$MoSe_2$ Single Crystal Electrodes
(72.40 and 73.40; S8)
(a) Vol. **110**, No. 1, p. 293—299 (1988).

H. Ohyama and K. Nemoto
*Process of Defect Formation in the Collector Region of an Electron-Irradiated npn Si Transistor
(61.80 and 73.40; 85; S5.11)
(a) Vol. **110**, No. 1, p. 301—308 (1988).

I. Akai, T. Karasawa, T. Komatsu, and Y. Kaifu
Large Optical Nonlinearity of the Two-Dimensional Exciton States at a Stacking Fault Interface in BI_3
(71.35 and 78.20; 71.70; 73.40; S9)
(b) Vol. **150**, No. 2, p. 635—639 (1988).

H. Uchiki, T. Kobayashi, and E. Tokunaga
*Carrier Diffusion and Trapping by Quantum Wells
(73.40 and 78.55; S7.12; S7.15)
(b) Vol. **150**, No. 2, p. 667—672 (1988).

F. V. Karpushko
*Fast and Strong Nonlinearity of Thin-Film Semiconductor Interferometers Due to Intrinsic Photo- and Electrooptical Effects of the Semiconductor Interlayer
(73.40 and 78.20; S7.12; S8.12)
(b) Vol. **150**, No. 2, p. 791—796 (1988).

Er-wang Mao, Wen-qin Zhao, Hong-rui Zhang, Ai-zhen Li, Jian-min Chen, and Guo-ping Fang
*The Influence of Strain and Dislocations on Transport Properties of GaAs/Si Strained-Layer Heterojunctions
(73.40; 78.30; S5.11; S7.12)
(a) Vol. **110**, No. 2, p. 515—520 (1988).

D. V. Morgan, H. Thomas, W. T. Anderson, P. Thompson, A. Christou, and D. J. Diskett
*High Temperature Metallisation for GaAs Device Processing
(61.80 and 73.40; S1.61; S4; S7.12)
(a) Vol. **110**, No. 2, p. 531—536 (1988).

H. Krause and H.-P. Bär
*Charge Injection into SiO_2 Films at Fields between 1 and 3 MV cm^{-1} after Electrical Stress
(73.40; 73.60; S10.1)
(a) Vol. **110**, No. 2, p. 537—547 (1988).

G. O. Müller, R. Mach, B. Selle, and G. Schulz
*Measuring on Thin Film Electroluminescent Devices
(73.40 and 78.60; 85; S8.11)
(a) Vol. **110**, No. 2, p. 657—669 (1988).

R. S. Gupta, C. Jagadish, G. S. Chilana, and G. P. Srivastava
*A Method to Determine Surface Doping and Substrate Doping Profile of n-Channel MOSFETs
(73.40 and 85)
(a) Vol. **110**, No. 2, p. 671—675 (1988).

H. Ohyama and K. Nemoto
*Recovery Mechanism of Lattice Defects Formed in the Collector Region for Electron-Irradiated npn Si Transistors
(61.70 and 73.40; 85; S5.11)
(a) Vol. **110**, No. 2, p. 677—686 (1988).

G. Grummt, J. Toušek, and B. Tryzna
*Transients in $p^+\pi n^+$ Photodiodes
(73.40 and 85)
(a) Vol. **110**, No. 2, p. 687—695 (1988).

73.60. Electronic properties of thin films

K. Ponnuraju, P. R. Vaya, and J. Sobhanadri
*Structural and Electrical Properties of $Pb_{1-x}Sn_xTe$ Films Grown by Molecular Beam Epitaxy
(68.55 and 73.60; S8.15)
(a) Vol. **105**, No. 1, p. 161—169 (1988).

I. J. Jabr, J. D. Meyer, K. A. Al-Saleh, and N. S. Saleh
Ion-Beam Mixing of Te/Au. Metastable Phase Formation
(61.80 and 64.75; 73.60; S1.3; S5)
(a) Vol. **105**, No. 1, p. 177—181 (1988).

N. Ganesan and V. Sivaramakrishnan
*Variation in Electrical Properties of SnTe Thin Films with Substrate Temperature
(73.60; 68.55; S8)
(a) Vol. **105**, No. 2, p. 461—466 (1988).

P. R. Vaya, K. R. Murali, and M. Sundaram
Some Studies on Zinc Phosphide Films Grown by Hot Wall Epitaxy
(68.55; 73.60; 78.65; S7)
(a) Vol. **105**, No. 2, p. 477—484 (1988).

R. Gilde, H. J. van Daal, and H. H. Brongersma
*Electromigration and Coagulation in Thin Films of Pure Indium and Pure Tin
(73.60; S4)
(a) Vol. **105**, No. 2, p. 493—499 (1988).

T. Luciński and J. Baszyński
*The Influence of Two Amorphous Phases on Transport Properties of Amorphous $Fe_{1-x}Si_x$ Films
(73.60; 75.70; S1.61)
(a) Vol. **105**, No. 2, p. 501—509 (1988).

N. Clark and M. Segnini
*The Electrical Conductance of Indium Thin Films Immediately after Being Deposited
(73.60; S4)
(a) Vol. **105**, No. 2, p. K125—K128 (1988).

K. Rajanna and S. Mohan
*Response of Thin Film Strain Gauge to Tensile and Compressive Strains by Cantilever Technique
(62.20 and 73.60; S1.2)
(a) Vol. **105**, No. 2, p. K181—K184 (1988).

M. Pattabi, M. S. M. Sastry, and V. Sivaramakrishnan
*Studies on the Stability of Discontinuous Silver Films with Overlayers of Al_2O_3 and SiO_2
(73.60; S1.3; S10.1)
(a) Vol. **106**, No. 1, p. 145—152 (1988).

D. Kindl and J. Toušková
Charge Transport Study in Thin Film Au–CdTe Schottky Diodes
(73.30 and 85; 73.60; S8.13)
(a) Vol. **106**, No. 1, p. 297—304 (1988).

N. Sasi, C. Balasubramanian, and Sa. K. Narayandass
*Electrical Conduction in Germanium Dioxide Thin Films
(73.60; S10.1)
(a) Vol. **106**, No. 1, p. K41—K45 (1988).

M. K. Bose, C. Majumdar, A. B. Maity, and A. N. Chakravarti
*On Electron States of Quantum Wells in Ultrathin Films of Wide-Gap Semiconductors in an Oriented Magnetic Field
(71.25 and 73.60; S7.12)
(b) Vol. **146**, No. 2, p. 525—530 (1988).

M. A. Angadi and R. Angadi
*Electrical Properties of MnSb Films
(73.60; 75.70; S7)
(a) Vol. **106**, No. 2, p. 485—491 (1988).

C. B. Steele and J. Beynon
*Thermopower and DC Electrical Conductivity in Au/SiO Thin Cermet Films
(73.60; 72.20; S10.1)
(a) Vol. **106**, No. 2, p. 515—523 (1988).

N. N. Dinh, V. T. Bich, N. H. Hoang, T. X. Hoai, V. N. Quynh, and N. V. Chanh
*Vacuum Annealing Effects on Structural and Electrical Properties of Reactively Evaporated Indium Tin Oxide
(73.60; 78.65; S10.1)
(a) Vol. **106**, No. 2, p. K147—K151 (1988).

S. I. Pokutnyi
*Energy Transfer of Electron Excitation: Kinetics of Impurity Luminescence Decay in Thin Films
(73.60 and 78.55)
(b) Vol. **147**, No. 1, p. K21—K24 (1988).

M. S. M. Sastry
*Post-Deposition Resistance Increase in Island Aluminium Films
(73.60; 68.55; S4)
(a) Vol. **107**, No. 1, p. 225—233 (1988).

A. J. Abu El-Haija, K. A. Al-Saleh, N. A. Halim, J. M. Khalifeh, and N. S. Saleh
*Ion Beam Induced Mixing of the Cu/Si System Using Electrical Resistivity and RBS Measurements
(61.80 and 73.60; S1.2; S5.11)
(a) Vol. **107**, No. 1, p. 253—260 (1988).

K. Konan, G. Galibert, and J. Calas
Silver Photodissolution and Optical Properties of $GeSe_3$ Vitreous Glass Thin Films
(66.30 and 78.65; 73.60; S8)
(a) Vol. **107**, No. 1, p. 273—279 (1988).

A. A. A. Hady
*Annealing Effects on the Electrical Properties of $Hg_{1-x}Cd_xTe$ Grown by Two Different Techniques
(73.60; 68.55; S8.15)
(a) Vol. **107**, No. 1, p. K25—K28 (1988).

A. Urbaniak-Kucharczyk
*The Magnetic Contribution to the Electrical Resistivity in Some Thin Antiferromagnetic Films
(73.60 and 75.70)
(b) Vol. **147**, No. 2, p. K119—K124 (1988).

B. N. Narahari Achar and J. R. Ferguson
*Quantized Damping of Helicons in a GaAs/AlGaAs Superlattice
(72.20 and 73.60; S7.12; S7.15)
(b) Vol. **147**, No. 2, p. K137—K141 (1988).

J. Ledzion
*Electrical Conductivity of Thin Metallic Films
(73.60; S1.3)
(a) Vol. **107**, No. 2, p. K115—K118 (1988).

A. Zehe, A. Adán, and M. Mäder
A Regenerative Switching Device in a Silicon–Earthalkalifluoride Heterostructure Grown by Molecular Beam Epitaxy
(68.55; 73.60; S9.12)
(a) Vol. **107**, No. 2, p. K119—K122 (1988).

M. K. Bose, C. Majumdar, A. B. Maity, and A. N. Chakravarti
*Effect of a Tilted Magnetic Field on the Size-Quantized Levels of Quantum Wells in Ultrathin Films of Narrow-Gap Semiconductors
(71.25 and 73.60; S7.13)
(b) Vol. **148**, No. 1, p. 165—172 (1988).

R. M. Cândea, L. P. Biró, N. Dădârlat, G. Borodi, Al. Darabont, P. Fitori, and R. Turcu
*The Influence of the Film History on Some Electrophysical Properties of VE, CAD, and COD PbSe Films
(68.55 and 73.60; S8)
(a) Vol. **108**, No. 1, p. 233—240 (1988).

F. Khater and M. El-Hiti
*Strain Coefficients of Electrical Resistance of Double-Layer Thin Metallic Films
(73.60; 72.15; S1.3)
(a) Vol. **108**, No. 1, p. 241—249 (1988).

H.-J. Fitting and D. Hecht
*Secondary Electron Field Emission
(73.60 and 79.20; S10.1)
(a) Vol. **108**, No. 1, p. 265—273 (1988).

T. Drüsedau, M. Eckler, and R. Bindemann
*Photoconductive a-Si:H with Dominant Monohydride Bonding Prepared by DC-Magnetron Sputtering
(68.55 and 73.60; 76.30; 78.65; S5.11)
(a) Vol. **108**, No. 1, p. 285—293 (1988).

C. Majumdar, A. B. Maity, and A. N. Chakravarti
Quantized Photoemission from Quantum Wells in Ultrathin Films of Wide-Gap Semiconductors in a Tilted Magnetic Field
(79.60; 73.60; S7.12)
(b) Vol. **148**, No. 2, p. 559—565 (1988).

D. Dădârlat, R. M. Cândea, R. Turcu, L. P. Biró, I. I. Zasavitskii, M. V. Valeiko, and A. P. Shotov
*Size Effects in Polycrystalline PbSe Films Obtained by Chemical Deposition
(73.60 and 78.65; S8)
(a) Vol. **108**, No. 2, p. 637—641 (1988).

V. Parizek and T. A. Chuong
*Influence of In on Transport Properties and Phase Transition in $Pb_{1-x}Sn_xTe$ Thin Films
(73.60; 64.70; S8.16)
(a) Vol. **108**, No. 2, p. 643—650 (1988).

T. K. Bandyopadhyay, S. B. Chakraborty, and S. R. Chaudhuri
*Effect of Surface Treatment on the Photoanodic Behaviour of Thin Film n-CdSe
(73.60; 72.40; S8.12)
(a) Vol. **108**, No. 2, p. K119—K123 (1988).

P. Lobotka and F. Gömöry
*The Complex AC Susceptibility of Superconducting Y–Ba–CuO Thin Film and Bulk Samples
(73.60 and 75.30; 74.70; S10.15)
(a) Vol. **109**, No. 1, p. 205—210 (1988).

D. A. Aronov, R. Kabulov, and Yu. M. Yuabov
The Current Characteristics of p–i–n and p^+–i–p^+ Structures Based on Hydrogenated Amorphous Silicon at Various Temperatures and Excitation Levels
(73.40; 73.60; S5.11)
(a) Vol. **109**, No. 1, p. 211—216 (1988).

S. M. Javed Akhtar and D. Ristau
*Mechanism of Voltage Generation in Bi Films Due to Laser Irradiation
(73.60; 79.20; S4)
(a) Vol. **109**, No. 1, p. 255—259 (1988).

A. Urbaniak-Kucharczyk
*Electrical Resistivity Due to Electron-Phonon Scattering in Thin Gadolinum Films
(73.60; S1.4)
(b) Vol. **149**, No. 2, p. 543—546 (1988).

A. B. Maity, C. Majumdar, and A. N. Chakravarti
*Effect of Magneto-Size Quantization on Electric-Field-Aided Photoemission from Quantum Wells in Ultrathin Films of Wide-Gap Semiconductors
(73.60 and 79.60; S7.12)
(b) Vol. **149**, No. 2, p. 565—570 (1988).

J. Suñé, I. Placencia, E. Farrés, N. Barniol, and X. Aymerich
*On the Si–SiO_2 Interface Roughness in VLSI–MOS Structures
(68.48 and 73.60; S5.11; S10.1)
(a) Vol. **109**, No. 2, p. 479—491 (1988).

F. Khater and M. El-Hiti
*Temperature Coefficient of the Strain Coefficient of Electrical Resistivity of Double-Layer Thin Metallic Films
(72.10 and 73.60; S1.3)
(a) Vol. **109**, No. 2, p. 517—523 (1988).

N. Romeo, A. Bosio, and V. Canevari
*$Zn_{0.15}Cd_{0.85}S$ Thin Films by RF Sputtering in an Ar–H_2 Atmosphere
(68.55 and 73.60; S8.15)
(a) Vol. **109**, No. 2, p. K105—K108 (1988).

K. H. Aharonian, H. L. Erknapetian, and D. R. Tilley
Collective Plasmon Modes in Thin Semiconductor Films Including the Effect of the Image Charges
(71.45; 73.60; S8)
(b) Vol. **150**, No. 1, p. 133—138 (1988).

B. A. Budkevich, I. A. Ges, S. P. Zhvavyi, G. D. Ivlev, and V. A. Pilipovich
*Modification of Electrochromic WO_3 Films by Pulsed Ultraviolet Laser Irradiation
(61.80 and 73.60; 68.60; S10.1)
(a) Vol. **110**, No. 1, p. 155—164 (1988).

B. L. Evans and G. Y. Nasser
*The DC Conductivity of Carbon Films
(73.60; S5)
(a) Vol. **110**, No. 1, p. 165—179 (1988).

F. Koch, R. Mitdank, G. Oelgart, and C. Chlupsa
*Excess Carrier Recombination in Indirect-Gap $GaAs_{1-x}P_x$:N
(73.60 and 78.60; S7.15)
(a) Vol. **110**, No. 1, p. 197—204 (1988).

Z. T. Al-Dhhan and C. A. Hogarth
*DC Conduction in Thin Films of SiO/In_3O_3 before and after Electroforming
(73.60; S10.1)
(a) Vol. **110**, No. 1, p. K39—K44 (1988).

J. C. Bernède, G. Safoula, A. Ameziane, and P. Burgaud
*Experimental Study of Oxygen Effect on the Variation of Revistivity with Temperature for Polycrystalline Selenium Thin Films
(73.60; S5; S8)
(a) Vol. **110**, No. 2, p. 521—530 (1988).

H. Krause and H.-P. Bär
Charge Injection into SiO_2 Films at Fields between 1 and 3 MV cm^{-1} after Electrical Stress
(73.40; 73.60; S10.1)
(a) Vol. **110**, No. 2, p. 537—547 (1988).

G. Horváth and J. Bánkuti
*Resistivity Increase in Thin Conducting Films Considering the Size Effect
(73.60)
(a) Vol. **110**, No. 2, p. 549—554 (1988).

74. Superconductivity

74.10. Occurrence, critical temperature

M. Hidaka, H. Fujii, Y. Horie, and S. Mase
*Superconductor–Semiconductor Transition in $BaPb_{1-x}Bi_xO_3$
(72.60 and 74.10; S10.15)
(b) Vol. **145**, No. 2, p. 649—658 (1988).

B. Fellmuth and D. Elefant
*Influence of Residual Resistance Measurements on the Superconducting Transition of Ultra-High-Purity Niobium
(74.10; S1)
(b) Vol. **146**, No. 1, p. 207—216 (1988).

K. B. Garg, R. K. Singhal, K. V. R. Rao, U. Chandra, H. S. Chauhan, J. Singh, K. S. Jerath, and D. C. Jain
*XANES Study of Copper Mixed Valence and Mechanism of Superconductivity in $YBa_2Cu_3O_{6.5+\delta}$ System
(74.10; 78.70; S11)
(b) Vol. **147**, No. 1, p. 343—348 (1988).

P. Kleinert
*The Influence of Disorder on Superconductivity
(74.10; 74.20)
(b) Vol. **148**, No. 1, p. 325—331 (1988).

74.20. Theory

A. S. Davydov
*Bisoliton Mechanism of High-Temperature Superconductivity
(71.45 and 74.20; S12.1)
(b) Vol. **146**, No. 2, p. 619—641 (1988).

Lianping Wang, Guangtian Zou, and Lizhong Wang
*Superconductivity of the Weak-Coupling Extended Hubbard Model
(74.20)
(b) Vol. **146**, No. 2, p. K115—K118 (1988).

G. Giaquinta, R. Fazio, G. Falci, and M. Apostol
*Flux-Flow Resistance, Vortex Depairing, and Temperature Dependence of the Ginzburg-Landau Parameter in Dirty Quasi-2D Superconductors
(74.20)
(b) Vol. **146**, No. 2, p. K125—K129 (1988).

Yu. B. Gaididei and V. M. Loktev
*On a Theory of the Electronic Spectrum and Magnetic Properties of High-T_c Superconductors
(74.20; 71.45; S12.1)
(b) Vol. **147**, No. 1, p. 307—319 (1988).

T. K. Kopeć
*Charge Density Waves in Systems with Condensed Local Pairs and Superconductivity in $La_{2-x}M_xCuO_4$ Compounds
(74.20; 10.15)
(b) Vol. **147**, No. 1, p. K37—K42 (1988).

Nguyen Van Hieu, Ha Vinh Tan, Nguyen Toan Thang, and Nguyen Ai Viet
*Electron Pairing in Isotropic Superconductors with Spin-Dependent Effective Electron–Electron Interaction
(74.20)
(b) Vol. **147**, No. 1, p. K63—K66 (1988).

K. Svozil
*Heavy Fermion Superconductivity via Kondo-Type Pairing
(74.20)
(b) Vol. **147**, No. 2, p. 635—647 (1988).

V. Z. Lozovsky, A. B. Nazarenko, and B. I. Khudik
*Theory of Superconductors with Strong Coupling in the Bogolyubov Method
(74.20)
(b) Vol. **148**, No. 1, p. 297—303 (1988).

A. S. Davydov and V. N. Ermakov
*Stability of a Superconducting Condensate of Bisolitons
(74.20; 71.45)
(b) Vol. **148**, No. 1, p. 305—318 (1988).

P. Kleinert
The Influence of Disorder on Superconductivity
(74.10; 74.20)
(b) Vol. **148**, No. 1, p. 325—331 (1988).

G. Ciobanu, I. Corcotoi, and A. B. Fazakas
*Theory of the Superconductivity in Two Energy Band Systems
(74.20)
(b) Vol. **149**, No. 1, p. 249—256 (1988).

E. Krüger
*Condition for Superconductivity
(74.20)
(b) Vol. **149**, No. 2, p. 669—673 (1988).

Yan-Min Li and Li-Yuan Zhang
*Interplay between Superconductivity and Antiferromagnetism in the Triplet Local-Electron-Pair System
(74.20 and 75.30; 71.38)
(b) Vol. **150**, No. 1, p. 211—224 (1988).

74.30. General properties

J. Sosnowski
*Influence of Magnetic History on Flux-Cutting Electric Fields for an Arbitrary Pinning Force Model
(74.30)
(a) Vol. **105**, No. 2, p. 555—562 (1988).

Z. Frait, D. Fraitová, E. Pollert, and L. Půst
*Observation of the Meissner-Ochsenfeld Field of the High-Temperature Superconductor Y–Ba–Cu–O by Means of FREPR
(74.30 and 76.30; S10.15)
(b) Vol. **146**, No. 2, p. K119—K123 (1988).

B. Lippold, H. Schmiedel, and H. Börner
*Evidence of Frozen Magnetic Flux in $Y_1Ba_2Cu_3O_{7-x}$ Ceramics
(74.30; 74.70; S10.15)
(b) Vol. **147**, No. 1, p. K43—K45 (1988).

74.50. Tunneling phenomena, Josephson effect, and proximity effects

F. Kh. Abdullaev and A. A. Abdumalikov
*Emission of Electromagnetic Waves by a Breather
(74.50)
(b) Vol. **145**, No. 2, p. K131—K134 (1988).

F. Müller and H.-J. Köhler
*Microwave-Induced Constant Voltage Steps at 0.1 Volt from a Series Array of Josephson Tunnel Junctions
(74.50; S4)
(a) Vol. **106**, No. 2, p. K165—K168 (1988).

74.60. Type-II superconductivity

T. Krzyszton and P. Wróbel
*Single Vortex Line in a Heavy Fermion Superconductor with Uniaxial Anisotropy
(74.60)
(b) Vol. **145**, No. 1, p. K41—K45 (1988).

A. V. Gurevich and V. T. Kovachev
*Influence of the Surface Irregularity on the Bean-Livingston Barrier in Type II Superconductors
(74.60)
(b) Vol. **145**, No. 1, p. K47—K50 (1988).

L. Schimmele, H. Kronmüller, and H. Teichler
*Intrinsic Pinning in Superconductors with Extremely Small Coherence Lengths
(74.60)
(b) Vol. **147**, No. 1, p. 361—372 (1988).

A. G. Saif
*Behaviour and Hysteresis Losses of Irreversible Type-II Superconducting Cylindrical Shells
(74.60)
(b) Vol. **149**, No. 2, p. 657—661 (1988).

74.70. Superconducting materials

R. HERRMANN, N. KUBICKI, H.-U. MÜLLER, H. DWELK, N. PRUSS, A. KRAPF, and L. ROTHKIRCH
*Effect of Fluorine on the Superconductivity of $YBa_2Cu_3O_{7-\delta}$
(74.70; S10.15)
(b) Vol. **146**, No. 1, p. K25—K28 (1988).

G. KRABBES, M. RITSCHEL, H. VINZELBERG, E. WOLF, H. ESCHRIG, W. HOLZHÄUSER, D. ELEFANT, and U. WIESNER
*Relationship between Oxygen Chemical Potential, Annealing Conditions, and Superconducting Properties of $YBa_2Cu_3O_x$
(74.70; S10.15)
(b) Vol. **146**, No. 1, p. K29—K33 (1988).

B. R. SINGH, H. S. KOTHARI, O. P. DAGA, J. K. SNGH, G. S. T. RAO, J. D. JAIN, and W. S. KHOKLE
*X-Ray Diffraction and EDS Analysis of $Y_1Ba_2Cu_3O_{9-y}$ Superconducting Compound
(61.50 and 74.70; S10.15)
(a) Vol. **106**, No. 1, p. K53—K56 (1988).

K. BENTE and D. SIEBERT
*EPR Investigations on High Temperature Superconducting $YBa_2Cu_3O_{7-x}$
(74.70 and 76.30; S10.15)
(a) Vol. **106**, No. 1, p. K57—K59 (1988).

J. PETZELT, S. KAMBA, S. PAČESOVÁ, E. POLLERT, J. ŠRÁMEK, O. SMRČKOVÁ, and D. SÝKOROVÁ
Infrared Reflectivity of High-T_c $YBa_2Cu_3O_{7-x}$ and Related Ceramics
(63.20 and 78.30; 74.70; S10.15)
(b) Vol. **146**, No. 2, p. 743—755 (1988).

K. ROGACKI, C. SUŁKOWSKI, A. J. ZALESKI, Z. BUKOWSKI, B. GREŃ, R. HORYŃ, E. TROJNAR, and J. KLAMUT
*Upper Critical Fields of High-T_c Superconductor $YBa_2Cu_3O_{7-\delta}$: Possibility of 350 T
(74.70; S10.15)
(b) Vol. **146**, No. 2, p. K103—K107 (1988).

Z. HENKIE, P. J. MARKOWSKI, R. HORYŃ, Z. BUKOWSKI, and J. KLAMUT
*The Thermopower Sign Inversion for the High T_c Superconductor $YBa_2Cu_3O_{7-\delta}$
(72.15 and 74.70; S10.15)
(b) Vol. **146**, No. 2, p. K131—K134 (1988).

H. PFEIFFER
*On Oxygen Ordering in the High Temperature Superconductor Y–Ba–Cu–O
(64.70 and 74.70; S10.15)
(a) Vol. **106**, No. 2, p. K161—K164 (1988).

B. LIPPOLD, H. SCHMIEDEL, and H. BÖRNER
Evidence if Frozen Magnetic Flux in $Y_1Ba_2Cu_3O_{7-x}$ Ceramics
(74.30; 74.70; S10.15)
(b) Vol. **147**, No. 1, p. K43—K45 (1988).

Y. H. LIANG, J. CHEN, L. C. WANG, Y. YIAN, X. Q. PAN, M. WEI, G. Q. WANG, X. JING, S. Y. DING, D. FENG, and Q. R. ZHANG
*A HREM Study on the Defects in Y–Ba–Cu–O Superconductors
(61.60 and 74.70; S10.15)
(a) Vol. **107**, No. 1, p. 63—72 (1988).

S. X. DOU, A. J. BOURDILLON, X. Y. SUN, H. K. LIU, J. P. ZHOU, N. SAVVIDES, C. C. SORRELL, and K. E. EASTERLING
*Dependence of the Superconducting Transition Temperature on Radii of Alkali and Alkaline Earth Dopants in $Y_1Ba_2Cu_3O_{7-x}$
(74.70; S10.15)
(b) Vol. **147**, No. 2, p. K153—K157 (1988).

R. HERRMANN, G. NACHTWEI, H. KRÜGER, P. SVOBODA, C. BREITLOW, R. HOLZBERGER, H. DWELK, and A. JACOBI
*Measurements of the Upper Critical Fields of Single Phase $YBa_2Cu_3O_{7-\delta}$
(74.70; S10.15)
(b) Vol. **147**, No. 2, p. K159—K163 (1988).

L. A. BURSILL and XU DONG FAN
Surface Structure of the Ceramic Superconductor $YBa_2Cu_3O_7$
(68.20; 61.16; 74.70; S10)
(a) Vol. **107**, No. 2, p. 503—510 (1988).

H. W. ZANDBERGEN and G. THOMAS
HREM Investigations of Defects in Sintered $YBa_2Cu_3O_{7-\delta}$
(61.16 and 61.70; 74.70; S10.15)
(a) Vol. **107**, No. 2, p. 825—843 (1988).

O. SMRČKOVÁ, D. SÝKOROVÁ, J. DOMINEC, L. SMRČKA, and P. VAŠEK
*Resistivity and Magnetic Properties of Bi–Sr–Ca–Cu–O Superconductor
(74.70; S10.15)
(a) Vol. **107**, No. 2, p. K135—K138 (1988).

B. B. BOIKO, F. P. KORSHUNOV, G. V. GATALSKII, A. I. AKIMOV, V. I. GATALSKAYA, S. E. DEMYANOV, and E. K. STRIBUK
*Radiation Effect on the Superconductivity in the Y–Sm–Ba–Cu–O Ceramic System
(74.70; S10.15)
(a) Vol. **107**, No. 2, p. K139—K144 (1988).

K. V. RAO, D.-X. CHEN, H. SZYMCZAK, M. BARAN, P. PRZYSŁUPSKI, and R. PUŹNIAK
*On Magnetic Properties of the Superconductor $Y_1Ba_2Cu_3O_x$
(74.70 and 75.60; S10.15)
(a) Vol. **107**, No. 2, p. K145—K150 (1988).

I. YA. NIKIFOROV and A. B. KOLPACHEV
Electronic Structure of Niobium Nitrocarbides
(71.20; 74.70; S1.61; S1.62)
(b) Vol. **148**, No. 1, p. 205—211 (1988).

Changyong Sun, Wenhui Su, Hongjian Liu, Jianshi Zhou, Fatang Wang, and Yupu Zhang
*Magnetic and Superconducting Properties of $REBa_2Cu_3O_{9-y}$ (RE = Sm, Eu, Gd, Dy, Ho, Er, Tm)
(74.70; 75.20; S10.15)
(a) Vol. **108**, No. 1, p. 337—342 (1988).

Hong Zhang, Xiao-gan Wang, and Yao-xian Fu
Study of the Movement of Oxygen Vacancies in the Orthorhombic Phase of $YBa_2Cu_3O_{7-x}$ by Positron Doppler Broadening Spectroscopy
(61.70; 74.70; 78.70; S10.15)
(a) Vol. **109**, No. 1, p. 135—139 (1988).

P. Lobotka and F. Gömöry
The Complex AC Susceptibility of Superconducting Y–Ba–CuO Thin Film and Bulk Samples
(73.60 and 75.30; 74.70; S10.15)
(a) Vol. **109**, No. 1, p. 205—210 (1988).

Yuhuan Xu, Zhongrong Li, Chaorui Li, Xianduan Lin, Wu Li, Wanming Lao, Hong Wang, and Guowei Chen
*The Internal Strain Effect on T_c in the Y–Ba–Al–Cu Oxide Superconductor
(74.70; S10.15)
(a) Vol. **109**, No. 1, p. 301—306 (1988).

E. M. Gololobov, A. S. Gurevich, E. L. Mager, Z. V. Mezhevich, N. A. Prytkova, Zh. M. Tomilo, D. M. Turtsevich, and N. M. Shimanskaya
*Orthorhombic Lattice Distortion of $YBa_2Cu_3O_{7-y}$ and the Superconducting Transition Temperature
(61.70 and 74.70; S10.15)
(b) Vol. **149**, No. 2, p. K145—K148 (1988).

N. Guskos, Th. Leventouri, Ch. Trikalinos, and M. Calamiotou
*EPR Spectra of Cu^{2+} Ions in Orthorhombic Complexes of the $YBa_2Cu_3O_x$ Compound
(74.70 and 76.30; S10.15)
(b) Vol. **149**, No. 2, p. K157—K162 (1988).

W. Gawalek, W. Michalke, H. Bruchlos, T. Eick, R. Hergt, and G. Schmidt
Preparation of High-T_c Y–Ba–Cu–O and Bi–Sr–Ca–Cu–O Thin Films by RF-Magnetron Sputtering
(68.55; 74.70; S10.15)
(a) Vol. **109**, No. 2, p. 503—509 (1988).

Xianyi Zhou, Huilin Jiang, Qirui Zhang, Guoqing Pan, and Qiwei Long (C. W. Lung)
*Positron Trapping in $YBa_2Cu_3O_x$ Superconductors
(74.70 and 78.70; S10.15)
(a) Vol. **109**, No. 2, p. K129—K132 (1988).

Hong Zhang, Xiao-Gang Wang, and Yao-Xian Fu
*Positron Annihilation in Thermally Quenched $YBa_2Cu_3O_{7-x}$
(74.70 and 78.70; S10.15)
(a) Vol. **109**, No. 2, p. K133—K138 (1988).

J. K. Liang, S. S. Xie, Y. L. Zhang, G. C. Che, G. H. Rao, J. Q. Huang, and X. R. Cheng
*Crystal Structure and Superconductivity of the $Tl_{2+x}Ba_2Ca_{2-x}Cu_3O_{10}$ Compound
(61.60 and 74.70; S11)
(a) Vol. **110**, No. 1, p. 55—60 (1988).

C. S. Sunandana
*On the Normal-State Electrical Resistivity of $YBa_2Cu_3O_7$ and Other Oxide Superconductors
(74.70; S10.15)
(a) Vol. **110**, No. 2, p. K85—K88 (1988).

Huimin Shao, Huaqin Wang, Yuanfu Hsia, Rongchuan Liu, and Xin Jin
*Tetragonal-to-Orthorhombic Phase Transition of Fe-Substituted $YBa_2Cu_3O_7$
(64.70 and 74.70; S10.15)
(a) Vol. **110**, No. 2, p. K89—K91 (1988).

A. K. Pradhan, P. C. Jana, and B. K. Roul
*Bismuth Strontium Calcium Copper Oxide High-T_c Superconductors from Nitrate Solutions
(74.70; S10.15)
(a) Vol. **110**, No. 2, p. K93—K96 (1988).

75. Magnetic properties and materials

75.10. General theory and models of magnetic ordering

(see also 71.25. Nonlocalized single-particle electronic states, 71.70j Level splitting and interactions)

A. B. Fazakas, I. Corcoţoi, and G. Ciobanu

*Influence of Non-Magnetic Impurities on the Ferromagnetic Phase Transition and Its Critical Properties
(75.10 and 75.30)
(b) Vol. **145**, No. 1, p. 283—289 (1988).

G. Wiatrowski, J. Mielnicki, and T. Balcerzak

*The Magnetic Susceptibility of Diluted Ferromagnetic Thin Films (Ising Model, $S = 1/2$)
(75.10 and 75.70)
(b) Vol. **145**, No. 1, p. 299—308 (1988).

V. Urumov

*The Critical Temperature of a Spin-One Ising Model on a Rectangular Lattice
(75.10)
(b) Vol. **145**, No. 1, p. K59—K64 (1988).

J. M. Wesselinowa

*On the Theory of the Broad Spin Response in Uranium Compounds
(75.10; 63.10; S1.5)
(b) Vol. **145**, No. 1, p. K65—K69 (1988).

A. A. Kazakov and S. M. Kleshchev

*Parametric Green Function Method in the Heisenberg Model
(75.10; 75.30; 75.40; S1.4)
(b) Vol. **145**, No. 2, p. 659—667 (1988).

A. V. Makhankov and V. G. Makhankov

*Spin Coherent States, Holstein-Primakoff Transformations for Heisenberg Spin Chain Models, and Status of the Landau-Lifshits Equation
(75.10)
(b) Vol. **145**, No. 2, p. 669—678 (1988).

S. L. Talim and A. S. T. Pires

*Soliton Energy in a Quantum Ferromagnet with Two Anisotropies
(75.10)
(b) Vol. **145**, No. 2, p. 711—718 (1988).

M. Tanaka and Y. Fujiwara

*Correlation Identities for the Spin-One Ising Model Including Biquadratic Interactions
(75.10)
(b) Vol. **145**, No. 2, p. K141—K144 (1988).

G. G. Bokuchava, L. L. Buishvili, and A. I. Tugushi
*Indirect Nuclear Spin Interaction in Ferromagnets. Highest Approximation
(75.10; 75.50)
(b) Vol. **146**, No. 1, p. 217—221 (1988).

C. Z. Yang and Z. Y. Li
*Mean Field Renormalization Group for Site-Bond Diluted Ising Ferromagnets
(75.10; 75.50)
(b) Vol. **146**, No. 1, p. 223—228 (1988).

B. V. da Costa and A. S. T. Pires
*Solitons in TMMC. Transfer Matrix Results and Phenomenological Interpretation
(75.10; 75.40)
(b) Vol. **146**, No. 1, p. 229—239 (1988).

Siming Zheng
*The Curie Temperature of the Heisenberg Ferromagnet of Mixed Spins with Uniaxial Anisotropy
(75.10)
(b) Vol. **146**, No. 1, p. K39—K44 (1988).

J. M. Wesselinowa
*Influence of the Band Occupation on the Electronic Damping in the s–f Model
(75.10)
(b) Vol. **146**, No. 1, p. K55—K59 (1988).

W. Korneta and Z. Pytel
*The Local Correlation Lengths in Ising Models with Free Surfaces
(75.10 and 75.70)
(b) Vol. **146**, No. 2, p. 697—706 (1988).

V. P. Silin and A. Z. Solontsov
*Spin-Fluctuation Theory of Magneto-Volume Effects in Itinerant Elecron Magnets
(75.10; 75.50)
(b) Vol. **147**, No. 1, p. 373—381 (1988).

R. J. Jelitto
*Model Calculations for Spin-Density-Wave Instabilities in Itinerant Magnets
(75.10)
(b) Vol. **147**, No. 1, p. 391—400 (1988).

B. V. Costa and P. R. Silva
*Monte Carlo Simulation of the Site Diluted Ising Model on the Rectangular Lattice
(75.10)
(b) Vol. **147**, No. 1, p. K79—K82 (1988).

A. ZIEGLER
*A Simple and Accurate Estimate of the Curie Temperature for the Heisenberg Model with Spin $S = 1/2$
(75.10)
(b) Vol. **147**, No. 1, p. K83—K86 (1988).

S. J. GŁADYSZ
*On the Curie Temperature of Ferromagnetic Semiconductors
(75.10; 75.50)
(b) Vol. **147**, No. 1, p. K87—K90 (1988).

WEI-JI SONG and ZHEN-YA LI
*Surface Phase Diagrams of the Site-Bond Diluted Surface Ordering Ising Ferromagnet with a Transverse Field
(75.10; 75.30)
(b) Vol. **148**, No. 1, p. 369—376 (1988).

A. ZAGÓRSKI and T. PURTAK
*An Infinite Ising Superlattice: A Simple Theory
(75.10)
(b) Vol. **148**, No. 1, p. K57—K60 (1988).

J. R. FALEIRO FERREIRA
*A Variational Linear Approach for the Free Energy Applied to the Ising Model
(75.10)
(b) Vol. **148**, No. 2, p. 709—714 (1988).

C. Z. YANG and Z. Y. LI
*Site-Random Ising Spin Glass
(75.10 and 75.50)
(b) Vol. **148**, No. 2, p. K133—K137 (1988).

K. I. GROZDEV
*Effect of Annealed Bond Impurities on the Phase Transition Temperature in a Two-Dimensional Ising Lattice
(75.10)
(b) Vol. **148**, No. 2, p. K143—K146 (1988).

I. A. GOICHUK, V. V. KUKHTIN, and E. G. PETROV
Orientational Vibrations of Two-Dimensional Dipole Lattice on Support
(63.20; 75.10)
(b) Vol. **149**, No. 1, p. 55—61 (1988).

L. ŠAMAJ
*A Theory of Higher-Order Correlations. I. Ising Model with Spin 1/2
(75.10)
(b) Vol. **149**, No. 1, p. 291—300 (1988).

V. Nolting and W. Nolting
Influence of Electron–Magnon Interaction on Electronic and Magnetic Properties of Ferromagnetic 4f Systems
(71.20 and 75.30; 75.10)
(b) Vol. **149**, No. 1, p. 313—325 (1988).

L. Šamaj
*A Theory of Higher-Order Correlations. II. Generalization to the Ising Model with Arbitrary Spin and to the Blume-Capel Model
(75.10 and 75.40)
(b) Vol. **149**, No. 2, p. 675—683 (1988).

A. Bobák, J. Karaba, and Ľ. Tóth
*A Correlated Effective Field Theory for the Ising Model: Two-Spin Cluster Approximation
(75.10)
(b) Vol. **149**, No. 2, p. K167—K171 (1988).

T. Kaneyoshi, E. F. Sarmento, and I. P. Fittipaldi
*An Effective-Field Theory of Mixed Ising Spin Systems in Transverse Fields
(75.10 and 75.40)
(b) Vol. **150**, No. 1, p. 261—272 (1988).

J. R. Faleiro Ferreira
*An Improved Mean Field Approximation
(75.10)
(b) Vol. **150**, No. 1, p. 281—288 (1988).

J. R. Faleiro Ferreira
*A Cluster Variational Approach to the Two-Dimensional Ising Model with Alternated Interactions along One Direction
(75.10)
(b) Vol. **150**, No. 1, p. K51—K54 (1988).

V. Ilkovič
*The Application of the Differential Operator Method to the Kinetic Ising Model
(75.10)
(b) Vol. **150**, No. 1, p. K55—K56 (1988).

75.20. Diamagnetism and paramagnetism

D. M. Zayachuk, P. M. Starik, and V. A. Shenderovskii
The Effect of Nonstoichiometric Tellurium on the Properties of Lead and Tin Tellurides
(61.70; 72.20; 75.20; S8.13)
(a) Vol. **107**, No. 1, p. 95—99 (1988).

M. Mondal and K. P. Ghatak
*Effect of Crossed Oscillatory Electric and Quantizing Magnetic Fields on the Magnetic Susceptibilities of the Electrons in Degenerated Semiconductors
(75.20)
(b) Vol. **147**, No. 2, p. K179—K183 (1988).

Changyong Sun, Wenhui Su, Hongjian Liu, Jianshi Zhou, Fatang Wang, and Yupu Zhang
Magnetic and Superconducting Properties of $REBa_2Cu_3O_{9-y}$ (RE = Sm, Eu, Gd, Dy, Ho, Er, Tm)
(74.70; 75.20; S10.15)
(a) Vol. **108**, No. 1, p. 337—342 (1988).

75.25. Spin arrangements in magnetically ordered materials
(neutron studies, etc.)

T. Tsuboi
*Light Scattering by Magnetic Polaritons in the Ferromagnet K_2CuF_4
(75.25 and 76.50; S9.16)
(b) Vol. **145**, No. 2, p. K145—K148 (1988).

A. Mayr, W. Paul, and H. G. Kahle
$Er_2O_2SO_4$ — An Ising System with Six Magnetic Phases
(75.50; 75.25; 75.30; S11.1)
(b) Vol. **147**, No. 1, p. 383—390 (1988).

M. Błaszak and B. Fechner
*Π- and 2Π-Magnetic Solitons in Uniform Stochastic Fields
(75.25; 75.30)
(b) Vol. **148**, No. 2, p. K147—K150 (1988).

W. Lottermoser and H. Fuess
*Magnetic Structure of the Orthosilicates Mn_2SiO_4 and Co_2SiO_4
(75.25; S11.1)
(a) Vol. **109**, No. 2, p. 589—595 (1988).

75.30. Magnetically ordered materials, other intrinsic properties
(for critical point efects, see 75.40)

E. L. NAGAEV
*Photoinduced Magnetism and Conduction Electrons in Magnetic Semiconductors (**Review Article**)
(75.30 and 78.20; 71.25; 75.60)
(b) Vol. **145**, No. 1, p. 11—64 (1988).

A. B. FAZAKAS, I. CORCOŢOI, and G. CIOBANU
*Influence of Non-Magnetic Impurities on the Ferromagnetic Phase Transition and Its Critical Properties
(75.10 and 75.30)
(b) Vol. **145**, No. 1, p. 283—289 (1988).

A. A. KAZAKOV and S. M. KLESHCHEV
*Two-Mode Approximation in the Low-Temperature Region. Biquadratic Exchange and Single-Ion Anisotropy
(75.30)
(b) Vol. **145**, No. 1, p. 291—297 (1988).

A. A. HIRSCH and T. MASSALHA
*Magnetic Metastable States in the Critical Transitions of Nickel and Iron
(75.30; 76.80; S1.1; S1.2)
(a) Vol. **105**, No. 1, p. 249—262 (1988).

A. EXTREMERA
*The Magnetic Behaviour of $NaHg_4$ Peritectic
(75.30; 61.55; S1.3)
(a) Vol. **105**, No. 1, p. 281—284 (1988).

F. LECCABUE, O. ARÉS MUZIO, and J. L. SÁNCHEZ LLAMAZARES
*Shape Anisotropy Effect in Barium Ferrite Interacting Submicronic Particles
(75.30; S11.2)
(a) Vol. **105**, No. 1, p. 285—290 (1988).

A. A. KAZAKOV and S. M. KLESHCHEV
Parametric Green Function Method in the Heisenberg Model
(75.10; 75.30; 75.40; S1.4)
(b) Vol. **145**, No. 2, p. 659—667 (1988).

TIESONG ZHAO, YONG ZHU, and HANMIN JIN
*Estimation of the Crystal Field at R Sites in $R_2F_{14}B$ Compounds
(75.30; 71.70; S1.1; S1.4)
(b) Vol. **145**, No. 2, p. 679—686 (1988).

D. SKRZYPEK, K. BIEŁAS-BORGIEŁ, and Z. CELIŃSKI
*Investigations on Magnetically Diluted $KNi_xMg_{1-x}F_3$ Perovskite Solid Solution
(75.30 and 76.30; S9)
(a) Vol. **105**, No. 2, p. 567—572 (1988).

A. APOSTOLOV, M. MIKHOV, N. SCHELUDKO, P. TCHOLAKOV, and V. SKUMRIEV
*The Effect of Zr-Substitution of Co on the Intersublattice Exchange in $Gd_2Co_{17-x}Zr_x$
(75.30; S1.4)
(a) Vol. **105**, No. 2, p. K145—K149 (1988).

J. SZYMAŃSKI and H. COFTA
*Versatility of the Two-Ion Quartic Spin Model of Spontaneous Metamagnetism
(75.30)
(b) Vol. **146**, No. 1, p. 241—251 (1988).

T. KANEYOSHI
*Phase Diagrams of an Amorphous Ferromagnetic Insulator with a Fractional Effective Coordination Number
(75.30; 75.50)
(b) Vol. **146**, No. 1, p. 253—262 (1988).

F. G. BASS, N. N. NASONOV, and O. V. NAUMENKO
*On the Radiation of Spin Waves Caused by Forced Motion of Bloch Walls in Alternating Magnetic Field
(75.30)
(b) Vol. **146**, No. 1, p. K61—K63 (1988).

T. T. DUNG, N. P. THUY, N. M. HONG, and T. D. HIEN
*Anomaly in the Magnetic Properties of the YCo_4B Compound
(75.30; S1.2; S1.4)
(a) Vol. **106**, No. 1, p. 201—206 (1988).

T. SATOW, H. ISHIKAWA, and O. UEMURA
Semiconductivity in the Liquid Cu–TlTe System
(72.10 and 72.80; 75.30; S1.63)
(b) Vol. **146**, No. 2, p. 661—669 (1988).

W. PRYSTASZ
*Quantum Tunneling between AF and SF or Equivalent States in a Uniaxial Antiferromagnet
(75.30)
(b) Vol. **146**, No. 2, p. 707—716 (1988).

S. MUKHOPADHYAY and I. CHATTERJEE
*Tricritical Behaviour in Correlated Effective Field Theory
(75.30 and 75.40)
(b) Vol. **146**, No. 2, p. 717—721 (1988).

F. G. BASS, N. N. NASONOV, and O. V. NAUMENKO
*Solitary Electromagnetic Spin Waves in a Ferrite Waveguide
(75.30; S11.2)
(b) Vol. **146**, No. 2, p. K145—K148 (1988).

A. I. GUSEV, A. A. REMPEL, and V. N. LIPATNIKOV
Magnetic Susceptibility and Atomic Ordering in Tantalum Carbide
(64.60; 75.30; S1.61)
(a) Vol. **106**, No. 2, p. 459—466 (1988).

A. MAYR, W. PAUL, and H. G. KAHLE
$Er_2O_2SO_4$ — An Ising System with Six Magnetic Phases
(75.50; 75.25; 75.30; S11.1)
(b) Vol. **147**, No. 1, p. 383—390 (1988).

A. URBANIAK-KUCHARCZYK
*The Temperature Dependence of the Spin Autocorrelation Time in Thin Antiferromagnetic Films
(75.30 and 78.70)
(b) Vol. **147**, No. 1, p. K71—K74 (1988).

L. G. SHPINKOVA and A. A. SOROKIN
*Magnetic Quenching of the Electric Quadrupole Interaction of ^{181}Ta in Ferromagnetic Laves Phases AFe_2 (A = Zr, Y, Lu)
(75.30 and 76.80; S1.1)
(b) Vol. **147**, No. 1, p. K75—K78 (1988).

A. W. SHEIKH and M. EL-FAZANI
A Phenomenological Model for the Magnetic Resistivity of NM–TM Spin Glasses
(72.15; 75.30; S1.2; S1.3)
(b) Vol. **147**, No. 2, p. 601—612 (1988).

K. BÄRNER
*On the Electronic States of MnAs
(75.30; S1.62)
(b) Vol. **147**, No. 2, p. K189—K194 (1988).

A. A. SAMOKHVALOV, N. A. VIGLIN, B. A. GIZHEVSKII, T. I. ARBUZOVA, and N. M. CHEBOTAEV
*Magnetic Polarons in EuO
(72.20 and 75.30; 76.30; S10.1)
(b) Vol. **148**, No. 1, p. 361—367 (1988).

WEI-JI SONG and ZHEN-YA LI
Surface Phase Diagrams of the Site-Bond Diluted Surface Ordering Ising Ferromagnet with a Transverse Field
(75.10; 75.30)
(b) Vol. **148**, No. 1, p. 369—376 (1988).

H. POLAT and S. EKMEKÇI
*Phase Diagram for the Simple Antiferromagnetic Ising Model Using Constant Coupling and Cluster Variation Approximations
(75.30 and 75.40)
(b) Vol. **148**, No. 1, p. 377—386 (1988).

T. HATTORI, T. IDOGAKI, and N. URYÛ
*Magnetic Susceptibility of the Tetrameric Complex Compound of Ruthenium Pentafluoride RuF_5
(75.30; S9)
(b) Vol. **148**, No. 1, p. K47—K51 (1988).

K.-D. DURST, H. KRONMÜLLER, and W. ERVENS
*Investigations of the Magnetic Properties and Demagnetization Processes of an Extremely High Coercive $Sm(Co, Cu, Fe, Zr)_{7.6}$ Permanent Magnet. I. Determination of Intrinsic Magnetic Material Parameters
(75.30; 75.50; 75.60; S1.2)
(a) Vol. **108**, No. 1, p. 403—416 (1988).

E. N. Bulgakov
*Phase Diagram of the Quantum Magnet with Competing Couplings
(75.30)
(b) Vol. **148**, No. 2, p. K139—K141 (1988).

M. Błaszak and B. Fechner
Π- and 2Π-Magnetic Solitons in Uniform Stochastic Fields
(75.25; 75.30)
(b) Vol. **148**, No. 2, p. K147—K150 (1988).

M. Quintero, P. Grima, J. E. Avon, G. Lamarche, and J. C. Woolley
Phase Diagram, Optical Energy Gap, and Magnetic Susceptibility of $(CuIn)_{1-z}Mn_{2z}Se_2$ Alloys
(64.75; 75.30; S8.16)
(a) Vol. **108**, No. 2, p. 599—606 (1988).

K.-D. Durst, H. Kronmüller, and W. Ervens
*Investigation of the Magnetic Properties and Demagnetization Processes of an Extremely High Coercive $Sm(Co, Cu, Fe, Zr)_{7.6}$ Permanent Magnet. II. The Coercivity Mechanism
(75.30; 75.50; 75.60; S1.1; S1.2)
(a) Vol. **108**, No. 2, p. 705—719 (1988).

H. Ishikawa, Y. Kameda, O. Uemura, and T. Satow
*Electronic Properties of Liquid $Tl-GeTl_2Se_4$ and $Tl-SnTl_2Se_4$ Systems
(72.20 and 75.30; S1.63; S8.16)
(b) Vol. **149**, No. 1, p. 257—266 (1988).

S. S. Aplesnin
*Monte-Carlo Study of Two-Dimensional Antiferromagnets with Competing Anisotropies
(75.30)
(b) Vol. **149**, No. 1, p. 267—273 (1988).

J. Nandi and D. Neogy
*Magnetic Behaviour and Crystal Field of Pr^{3+} in Praseodymium Selenate Octahydrate
(71.70 and 75.30; 75.40; S11)
(b) Vol. **149**, No. 1, p. 275—281 (1988).

I. A. Vakarchuk and I. F. Margolych
*Magnon Spectrum of Two-Component Amorphous Ferromagnets
(75.30 and 75.50)
(b) Vol. **149**, No. 1, p. 301—312 (1988).

V. Nolting and W. Nolting
*Influence of Electron–Magnon Interaction on Electronic and Magnetic Properties of Ferromagnetic 4f Systems
(71.20 and 75.30; 75.10)
(b) Vol. **149**, No. 1, p. 313—325 (1988).

Ch. Seidel
*On the Pinning of Spin-Density Waves by Impurities
(72.15 and 75.30; S12.1)
(b) Vol. **149**, No. 1, p. 327—335 (1988).

V. V. Gorbach
*On a Possible Multimagnon Generation Mechanism under Exciton Light Absorption in Antiferromagnetics
(75.30; 76.50)
(b) Vol. **149**, No. 1, p. K49—K54 (1988).

P. Lobotka and F. Gömöry
*The Complex AC Susceptibility of Superconducting Y–Ba–CuO Thin Film and Bulk Samples
(73.60 and 75.30; 74.70; S10.15)
(a) Vol. **109**, No. 1, p. 205—210 (1988).

S. K. Asadov, E. A. Zavadskii, Yu. D. Zavorotnev, and B. M. Todris
*Low-Temperature Magnetic Transformations in Ferrous Fluorsilicate Induced by Pressure
(75.30; S11)
(a) Vol. **109**, No. 1, p. 307—312 (1988).

J. L. Sánchez Llamazares, S. García García, S. Díaz Castañón, and N. Suarez Almodovar
*On the Effect of Al Addition on the Phase Composition and Magnetic Properties of Nd–Fe–B Permanent Magnet Alloys
(75.30; 61.60; S1.1)
(a) Vol. **109**, No. 2, p. 569—573 (1988).

R. Höhne
*A Study of Induced Magnetic Anisotropy in Gallium–Iron Ferrites above Room Temperature
(75.30 and 75.50; S11.2)
(a) Vol. **109**, No. 2, p. 575—587 (1988).

Yan-Min Li and Li-Yuan Zhang
*Interplay between Superconductivity and Antiferromagnetism in the Triplet Local-Electron-Pair System
(74.20 and 75.30; 71.38)
(b) Vol. **150**, No. 1, p. 211—224 (1988).

A. W. Sheikh and M. El-Fazani
*A New Maximum in the Spin–Spin Resistivity of NM-TM Spin Glasses
(75.30)
(b) Vol. **150**, No. 1, p. K45—K49 (1988).

P. Stefányi, A. Orendáčová, and A. Feher
*The Influence of a Magnetic Field on the Heat Capacity of $CsDy(MoO_4)_2$
(75.30 and 75.40; S11)
(b) Vol. **150**, No. 1, p. K63—K66 (1988).

A. Kowalczyk and A. Wrzeciono
*Magnetic and Crystallographic Properties of $R_2Fe_{14-x}Cr_xB$ Compounds (R = Y, Nd, and Gd)
(75.30; 61.55; S1.1; S1.2)
(a) Vol. **110**, No. 1, p. 241—245 (1988).

75.40. Critical-point effects, specific heats, short-range order
(see also 65.40. Heat capacities)

A. A. Kazakov and S. M. Kleshchev
Parametric Green Function Method in the Heisenberg Model
(75.10; 75.30; 75.40; S1.4)
(b) Vol. **145**, No. 2, p. 659—667 (1988).

J. Schreiber and D. Schulz
*Magnetic Relaxation in the Heisenberg Spin Glass
(75.40)
(b) Vol. **145**, No. 2, p. 695—709 (1988).

B. V. da Costa and A. S. T. Pires
Solitons in TMMC. Transfer Matrix Results and Phenomenological Interpretation
(75.10; 75.40)
(b) Vol. **146**, No. 1, p. 229—239 (1988).

S. Mukhopadhyay and I. Chatterjee
*Tricritical Behaviour in Correlated Effective Field Theory
(75.30 and 75.40)
(b) Vol. **146**, No. 2, p. 717—721 (1988).

A. Czopnik, H. Mädge, and B. Staliński
*Anomalous Strong Divergence of the Specific Heat of $ErIn_3$ at the Antiferromagnetic Transition
(75.40; S1.4)
(a) Vol. **107**, No. 2, p. K151—K155 (1988).

H. Polat and S. Ekmekçi
*Phase Diagram for the Simple Antiferromagnetic Ising Model Using Constant Coupling and Cluster Variation Approximations
(75.30 and 75.40)
(b) Vol. **148**, No. 1, p. 377—386 (1988).

M. Funke and S. Trimper
*Low Temperature Behaviour of the Random Field Ising Model
(75.40; 64.60)
(b) Vol. **148**, No. 2, p. K151—K156 (1988).

J. Nandi and D. Neogy
Magnetic Behaviour and Crystal Field of Pr^{3+} in Praseodymium Selenate Octahydrate
(71.70 and 75.30; 75.40; S11)
(b) Vol. **149**, No. 1, p. 275—281 (1988).

L. Šamaj
*A Theory of Higher-Order Correlations. II. Generalization to the Ising Model with Arbitrary Spin and to the Blume-Capel Model
(75.10 and 75.40)
(b) Vol. **149**, No. 2, p. 675—683 (1988).

W. Reich
*Thermodynamic Functions for a Three-Dimensional Ferromagnet within Niemeijer and van Leeuven's Method
(75.40)
(b) Vol. **149**, No. 2, p. K163—K165 (1988).

T. Kaneyoshi, E. F. Sarmento, and I. P. Fittipaldi
*An Effective-Field Theory of Mixed Ising Spin Systems in Transverse Fields
(75.10 and 75.40)
(b) Vol. **150**, No. 1, p. 261—272 (1988).

A. Duda
*The Critical Temperature of the Axial Next-Nearest-Neighbour Ising Thin Film
(75.40 and 75.70)
(b) Vol. **150**, No. 1, p. 273—280 (1988).

P. Stefányi, A. Orendáčová, and A. Feher
*The Influence of a Magnetic Field on the Heat Capacity of $CsDy(MoO_4)_2$
(75.30 and 75.40; S11)
(b) Vol. **150**, No. 1, p. K63—K66 (1988).

J.-W. Schünemann, K. Bärner, H. Fjellvag, A. Kjekshus, J. Eigemann, E. Gmelin, and U. Sondermann
Classification of Martensitic-Like Transformations in $Cr_xMn_{1-x}As_{1-y}P_y$ Crystals
(64.70; 75.40; S7.15)
(a) Vol. **110**, No. 1, p. 141—154 (1988).

A. Czopnik, H. Mädge, R. Pott, and B. Staliński
*Thermal Properties of $NdIn_3$
(75.40; S1.4)
(a) Vol. **110**, No. 2, p. 601—614 (1988).

75.50. Studies of specific magnetic materials

J. L. Sánchez Llamazares, S. García García, O. Arés Muzio, and S. Díaz Castañón
*Mössbauer and Thermomagnetic Analysis of Al-Containing Nd–Fe–B Permanent Magnet Alloys
(75.50; 76.80; S1.1)
(a) Vol. **105**, No. 1, p. 243—247 (1988).

M. Takahashi, F. Sato, and T. Miyazaki
*Magnetic Anisotropy and Magnetostriction in Fe–Nd Alloys
(75.50; 75.80; S1.1)
(a) Vol. **105**, No. 1, p. 263—273 (1988).

Ż. Kąkol, H. Figiel, J. Chmist, and A. Kołodziejczyk
*The Influence of Manganese on Magnetic Properties of $RE_2(Co_{1-x}Mn_x)_{17}$, RE = Y, Gd Compounds
(75.50; S1.2; S1.4)
(b) Vol. **145**, No. 2, p. 687—693 (1988).

A. C. Bódi and I. Ileo
*The Effect of Neutron Irradiation on the Cation Distribution in Iron-Deficient Ni–Zn Ferrites
(75.50; S11.2)
(a) Vol. **105**, No. 2, p. 573—577 (1988).

Yan-Zhong Zhang
*Magnetic Instability of Metallic Glass $(Fe_{0.1}Ni_{0.33}Co_{0.55}Cr_{0.02})_{78}Si_8B_{14}$. I. Reversible Disaccommodation Phenomenon with Respect to Demagnetization
(75.50; S1.1; S1.2)
(a) Vol. **105**, No. 2, p. 579—587 (1988).

G. G. Bokuchava, L. L. Buishvili, and A. I. Tugushi
Indirect Nuclear Spin Interaction in Ferromagnets. Highest Approximation
(75.10; 75.50)
(b) Vol. **146**, No. 1, p. 217—221 (1988).

C. Z. Yang and Z. Y. Li
Mean Field Renormalization Group for Site-Bond Diluted Ising Ferromagnets
(75.10; 75.50)
(b) Vol. **146**, No. 1, p. 223—228 (1988).

T. Kaneyoshi
Phase Diagrams of an Amorphous Ferromagnetic Insulator with a Fractional Effective Coordination Number
(75.30; 75.50)
(b) Vol. **146**, No. 1, p. 253—262 (1988).

A. K. Mukherjee and D. Neogy
*Magnetism and Crystal Field Analysis of Gadolinium Bromate Enneahydrate Crystal
(71.70 and 75.50; S11)
(b) Vol. **146**, No. 1, p. K45—K50 (1988).

M. Pawłowska, W. Bronowska, and W. Wojciechowski
*Anomalies in the Magnetic Susceptibility of $RbMnCl_3$ and $RbMnCl_3:Sm^{3+}$
(75.50; S9)
(b) Vol. **146**, No. 1, p. K51—K54 (1988).

YAN-ZHONG ZHANG
*Magnetic Instability of Metallic Glass $(Fe_{0.1}Ni_{0.33}Co_{0.55}Cr_{0.02})_{78}Si_8B_{14}$. II. Irreversible Magnetic Permeability Decay with Respect to Demagnetization
(75.50; S1.1; S1.2)
(a) Vol. **106**, No. 1, p. 207—215 (1988).

Y. SRINIVASA RAO and C. S. SUNANDANA
*Characterization of Diamagnetically Substituted (Na^+, Te^{4+}, Cl^-) Lithium Ferrite
(75.50; 61.60; S11.2)
(a) Vol. **106**, No. 1, p. 217—225 (1988).

Y. F. LI, L. HEDMAN, and Ö. RAPP
*Magnetic Susceptibility of Amorphous $Cu_{1-x}Y_x$ Alloys
(75.50; S1.2; S1.4)
(a) Vol. **106**, No. 1, p. 233—237 (1988).

S. ROTH and G. STEPHANI
*Influence of the Melt Temperature on the Formation of Soft Magnetic Properties of an Amorphous $Fe_{40}Ni_{40}P_{14}B_6$ Alloy
(75.50; 75.60; S1.1; S1.2)
(a) Vol. **106**, No. 1, p. K61—K63 (1988).

S. LIGENZA, B. PALUCHOWSKA, and M. KONWICKI
*Neutron Diffraction Studies of 0.2 Titanium Substituted Lithium Ferrite
(75.50; S11.2)
(a) Vol. **106**, No. 1, p. K71—K74 (1988).

F. YASSIN, V. CHRISTOPH, and L. JAHN
*On the Determination of the Curie Temperature of Anisotropic Ferromagnets
(75.50; S10)
(a) Vol. **106**, No. 1, p. K79—K82 (1988).

V. P. SILIN and A. Z. SOLONTSOV
Spin-Fluctuation Theory of Magneto-Volume Effects in Itinerant Electron Magnets
(75.10; 75.50)
(b) Vol. **147**, No. 1, p. 373—381 (1988).

A. MAYR, W. PAUL, and H. G. KAHLE
*$Er_2O_2SO_4$ — An Ising System with Six Magnetic Phases
(75.50; 75.25; 75.30; S11.1)
(b) Vol. **147**, No. 1, p. 383—390 (1988).

S. J. GŁADYSZ
On the Curie Temperature of Ferromagnetic Semiconductors
(75.10; 75.50)
(b) Vol. **147**, No. 1, p. K87—K90 (1988).

F. SATO, S. ISHIO, and T. MIYAZAKI
*Soft Magnetic Properties of Rapidly Quenched Fe–Al–M (M = Si, B, Ge) Alloys
(75.50; 75.60; S1.1)
(a) Vol. **107**, No. 1, p. 355—363 (1988).

A. Kowalczyk, P. Stefański, and A. Wrzeciono
*Magnetic Properties of $Pr_2Fe_{14-x-y}Co_xCr_yB$ Compounds
(75.50; S1.1)
(a) Vol. **107**, No. 1, p. K61—K64 (1988).

H. Krikor, P. Nagels, J. Vansummeren, and A. van den Bosch
*Static Magnetic Susceptibilities of Poly(phenylethynylcopper) Doped with Iodine
(75.50; S12)
(a) Vol. **107**, No. 2, p. 639—646 (1988).

K.-D. Durst, H. Kronmüller, and W. Ervens
Investigations of the Magnetic Properties and Demagnetization Processes of an Extremely High Coercive $Sm(Co, Cu, Fe, Zr)_{7.6}$ Permanent Magnet. I. Determination of Intrinsic Magnetic Material Parameters
(75.30; 75.50; 75.60; S1.2)
(a) Vol. **108**, No. 1, p. 403—416 (1988).

J. Gwiazda, E. Mariańska, and W. Zych
The Influence of Transition Metal Substitution and $B_{3/2}$ Values for Some Amorphous Alloys
(76.80; 75.50; S1.1)
(a) Vol. **108**, No. 1, p. K67—K72 (1988).

C. Z. Yang and Z. Y. Li
*Site-Random Ising Spin Glass
(75.10 and 75.50)
(b) Vol. **148**, No. 2, p. K133—K137 (1988).

P. V. Reddy
High-Temperature Elastic Behaviour of Mn–Mg Mixed Ferrites
(62.20; 75.50; S11.2)
(a) Vol. **108**, No. 2, p. 607—611 (1988).

K.-D. Durst, H. Kronmüller, and W. Ervens
Investigations of the Magnetic Properties and Demagnetization Processes of an Extremely High Coercive $Sm(Co, Cu, Fe, Zr)_{7.6}$ Permanent Magnet. II. The Coercivity Mechanism
(75.30; 75.50; 75.60; S1.1; S1.2)
(a) Vol. **108**, No. 2, p. 705—719 (1988).

C. de Francisco, J. M. Muñoz, J. Ayala, and J. I. Iñiguez
*Magnetic Disaccommodation in Ferrous Zinc Ferrites
(75.50; S11.2)
(a) Vol. **108**, No. 2, p. 721—731 (1988).

I. Škorvánek, B. Idzikowski, A. Zentko, and E. Mosiniewicz-Szablewska
*Influence of Neutron Irradiation on the Magnetic Properties of FeNiCrMoSiB Amorphous Alloys
(75.50; 61.80; S1.1; S1.2)
(a) Vol. **108**, No. 2, p. 747—751 (1988).

I. A. Vakarchuk and I. F. Margolych
*Magnon Spectrum of Two-Component Amorphous Ferromagnets
(75.30 and 75.50)
(b) Vol. **149**, No. 1, p. 301—312 (1988).

S. Ligenza and M. Konwicki
*Cluster Structure of the Li–Ti–Zn Ferrite System Studied by Neutron Scattering
(75.50; 61.12; S11.2)
(a) Vol. **109**, No. 1, p. 319—327 (1988).

M. Zelený, F. Zounová, and A. V. Andreev
*Magnetic Properties of UCuGe
(75.50; S1.5)
(a) Vol. **109**, No. 1, p. K67—K69 (1988).

J. Świerczek and S. Szymura
*Temperature Dependence of Magnetization in Amorphous Co–Fe–Si–B Alloys
(75.50; S1.1; S1.2)
(a) Vol. **109**, No. 2, p. 559—567 (1988).

R. Höhne
*A Study of Induced Magnetic Anisotropy in Gallium–Iron Ferrites above Room Temperature
(75.30 and 75.50; S11.2)
(a) Vol. **109**, No. 2, p. 575—587 (1988).

J. Kuriata, L. Sadłowski, B. Bojanowski, J. Walczak, M. Kurzawa, and J. Pichet
The Magnetic Resonance of $Fe_2V_{4-x}Mo_xO_{13}$ Type Phase
(76.30; 75.50; S11)
(a) Vol. **109**, No. 2, p. K139—K143 (1988).

T. Kaneyoshi
*Magnetic Properties of a Surface with an Amorphous Layer
(75.50)
(b) Vol. **150**, No. 1, p. 297—306 (1988).

A. Schröder and M. Feller-Kniepmeier
The Ferromagnetic Diffusion Anomaly of the Diffusion System Iron–Antimony
(66.30; 75.50; S1.1)
(a) Vol. **110**, No. 1, p. 107—121 (1988).

E. A. Eivazov, A. F. Safarov, and A. O. Ismailov
Thermal Properties of $Co_{1-x}Cu_xCr_2S_4$ in the Magnetic Phase Transition Region
(65.70; 75.50; S8)
(a) Vol. **110**, No. 1, p. K45—K49 (1988).

E. Z. Katsnelson, A. G. Karoza, and L. A. Meleshchenko
*Photoinduced Magnetization Effect on IR Spectra of Ferrites
(75.50 and 78.30; S11.2)
(a) Vol. **110**, No. 1, p. K51—K56 (1988).

C. H. Westphal and C. C. Becerra
*Magnetic Transitions in $Mn_{1-x}Ni_x(HCOO)_2 \cdot 2\,H_2O$
(75.50; S1.2; S11)
(a) Vol. **110**, No. 2, p. K97—K100 (1988).

75.60. Domain effects, magnetization curves, and hysteresis

E. L. NAGAEV
Photoinduced Magnetism and Conduction Electrons in Magnetic Semiconductors **(Review Article)**
(75.30 and 78.20; 71.25; 75.60)
(b) Vol. **145**, No. 1, p. 11—64 (1988).

W. APPEL
*Structure and Energy of 180° Domain Walls in H.C.P. Cobalt as Functions of the Temperature
(75.60 and 75.70; S1.2)
(a) Vol. **105**, No. 1, p. K67—K69 (1988).

W. APPEL and W. ANDRÄ
The Atomic Layer Model as an Instrument in Calculating the Magnetization Reversal in Continuous Thin Films
(75.70; 75.60)
(a) Vol. **105**, No. 2, p. 589—596 (1988).

R. A. KOSIŃSKI
*On the Stabilization of a Vertical Bloch Line Pair in a Domain Wall in an Ion-Implanted Garnet Film
(75.60; S11.2)
(a) Vol. **106**, No. 1, p. 227—231 (1988).

J. EHLERT, F. K. HÜBNER, and W. SPERBER
*Micromagnetics of Cylindrical Particles
(75.60)
(a) Vol. **106**, No. 1, p. 239—248 (1988).

S. ROTH and G. STEPHANI
Influence of the Melt Temperature on the Formation of Soft Magnetic Properties of an Amorphous $Fe_{40}Ni_{40}P_{14}B_6$ Alloy
(75.50; 75.60; S1.1; S1.2)
(a) Vol. **106**, No. 1, p. K61—K63 (1988).

A. H. WAFIK and S. A. MAZEN
*Effect of Mg–Fe Replacement in the System $Mg_xZn_{0.3}Fe_{2.7-x}O_{4\pm\delta}$ on Barkhausen Jumps
(75.60; S11.2)
(a) Vol. **106**, No. 1, p. K65—K69 (1988).

M. GUYOT, T. MERCERON, V. CAGAN, and A. MESSEKHER
*Mobility and/or Damping of the Domain Wall. Comparison between Experimental and Theoretical Values in Ferrimagnets
(75.60; S11.2)
(a) Vol. **106**, No. 2, p. 595—612 (1988).

S. RAM
Magnetic and Magnetic Resonance Studies on Oxygen-Deficient $SrFeO_{3-y}$ Perovskites
(61.40; 75.60; 76.30; S11.2)
(a) Vol. **107**, No. 1, p. 55—62 (1988).

J. F. Fuertes and M. Tejedor
*Wrinkle-Like Domain Structures on Imperfect Iron Whiskers
(75.60; S1.1)
(a) Vol. **107**, No. 1, p. 339—346 (1988).

F. Sato, S. Ishio, and T. Miyazaki
Soft Magnetic Properties of Rapidly Quenched Fe–Al–M (M = Si, B, Ge) Alloys
(75.50; 75.60; S1.1)
(a) Vol. **107**, No. 1, p. 355—363 (1988).

K. V. Rao, D.-X. Chen, H. Szymczak, M. Baran, P. Przysłupski, and R. Puźniak
*On Magnetic Properties of the Superconductor $Y_1Ba_2Cu_3O_x$
(74.70 and 75.60; S10.15)
(a) Vol. **107**, No. 2, p. K145—K150 (1988).

A. H. Wafik and G. C. Hallam
*Effect of Tensile Strain on Barkhausen Jumps in Amorphous Metal Ribbons of $Fe_{80}B_{20}$
(75.60; S1.1)
(a) Vol. **107**, No. 2, p. K157—K161 (1988).

B. Idzikowski and A. Wrzeciono
*Properties of Amorphous $Sm_xFe_{80-x}B_{20}$ Ribbons
(75.60; 75.80; S1.1; S1.4)
(a) Vol. **108**, No. 1, p. 375—380 (1988).

M. A. Fahim and A. R. Ali
*On the Influence of Copper Atoms on Domain Wall Energy and Pinning Force in Ni–Cu Alloys
(75.60; S1.2)
(a) Vol. **108**, No. 1, p. 381—386 (1988).

U. Hartmann
*Observation of a New Domain Nucleation Process in Soft Magnetic Materials
(75.60; S1.1)
(a) Vol. **108**, No. 1, p. 387—401 (1988).

K.-D. Durst, H. Kronmüller, and W. Ervens
Investigations of the Magnetic Properties and Demagnetization Processes of an Extremely High Coercive $Sm(Co, Cu, Fe, Zr)_{7.6}$ Permanent Magnet. I. Determination of Intrinsic Magnetic Material Parameters
(75.30; 75.50; 75.60; S1.2)
(a) Vol. **108**, No. 1, p. 403—416 (1988).

D. C. Jiles
*Variation of the Magnetic Properties of AISI 4140 Steels with Plastic Strain
(75.60; 62.20; S1.1)
(a) Vol. **108**, No. 1, p. 417—429 (1988).

K.-D. Durst, H. Kronmüller, and W. Ervens
Investigations of the Magnetic Properties and Demagnetization Processes of an Extremely High Coercive $Sm(Co, Cu, Fe, Zr)_{7.6}$ Permanent Magnet. II. The Coercivity Mechanism
(75.30; 75.50; 75.60; S1.1; S1.2)
(a) Vol. **108**, No. 2, p. 705—719 (1988).

YU. V. ZAKHAROV, YU. I. MANKOV, and L. S. TITOV
Electric Conductivity of a Metallic Ferromagnet and Magnetization Process
(72.15; 75.60; S1.1)
(b) Vol. **149**, No. 1, p. 241—248 (1988).

A. R. ALI and M. A. FAHIM
*On the Influence of Internal Stress and Magnetizing Frequency on Domain Wall Motion in High-Purity Ni
(75.60; S1.2)
(a) Vol. **109**, No. 1, p. 313—318 (1988).

DE-GANG ZHANG and BO-ZANG LI
*The Stability of the Curling Mode of an Infinite Cylinder
(75.60)
(b) Vol. **149**, No. 2, p. K173—K176 (1988).

F. YASSIN and V. CHRISTOPH
*Hysteresis Loops of Fine Particle Magnets of the Type NdFeB
(75.60; S1.1)
(a) Vol. **109**, No. 2, p. K145—K148 (1988).

A. UNKROTH and W. ANDRÄ
*Coercivity of Ion Implanted Cobalt Films
(75.60; S1.2)
(a) Vol. **109**, No. 2, p. K149—K152 (1988).

K. ELK and L. JAHN
*On the Coercivity of Texturized Uniaxial Permanent Magnets
(75.60)
(b) Vol. **150**, No. 1, p. 289—295 (1988).

K. POLAŃSKI, I. ZASADA, K. DOLECKI, and E. MISIAK
*The Structural Phase Transition in Cobalt Monocrystals Investigated by Means of SEM
(64.70 and 75.60; S1.2)
(a) Vol. **110**, No. 1, p. 231—240 (1988).

U. HARTMANN
*Static Magnetization Irreversibilities in Soft Ferromagnetic Media of High Crystalline Perfection
(75.60; S1.1)
(a) Vol. **110**, No. 1, p. 247—257 (1988).

D. I. SEMENTSOV and R. M. GRECHISHKIN
Light Diffraction by Stripe Domain Structures in Magnetic Crystals
(78.20; 75.60)
(a) Vol. **110**, No. 1, p. 259—267 (1988).

75.70. Magnetic films and plates

G. Wiatrowski, J. Mielnicki, and T. Balcerzak
*The Magnetic Susceptibility of Diluted Ferromagnetic Thin Films (Ising Model, $S = 1/2$)
(75.10 and 75.70)
(b) Vol. **145**, No. 1, p. 299—308 (1988).

J. Barnaś
*Magnetostatic Modes in Three-Layered Superlattices
(75.70; 76.50)
(b) Vol. **145**, No. 1, p. K55—K58 (1988).

K. Sumiyama and Y. Nakamura
*Nonequilibrium $Fe_{1-x}Co_x$ Alloys Produced by Vapor Quenching
(75.70; 61.10; 76.80; S1.1)
(a) Vol. **105**, No. 1, p. 291—295 (1988).

W. Appel
*Structure and Energy of 180° Domain Walls in H.C.P. Cobalt as Functions of the Temperature
(75.60 and 75.70; S1.2)
(a) Vol. **105**, No. 1, p. K67—K69 (1988).

T. Luciński and J. Baszyński
The Influence of Two Amorphous Phases on Transport Properties of Amorphous $Fe_{1-x}Si_x$ Films
(73.60; 75.70; S1.61)
(a) Vol. **105**, No. 2, p. 501—509 (1988).

W. Appel and W. Andrä
*The Atomic Layer Model as an Instrument in Calculating the Magnetization Reversal in Continuous Thin Films
(75.70; 75.60)
(a) Vol. **105**, No. 2, p. 589—596 (1988).

L. Vatskichev and M. Vatskicheva
*Relations between Some Galvanomagnetic and Magnetic Properties of Thin Ferromagnetic Films
(72.15 and 75.70; S1.2; S1.4)
(a) Vol. **106**, No. 1, p. K75—K78 (1988).

W. Korneta and Z. Pytel
*The Local Correlation Lengths in Ising Models with Free Surfaces
(75.10 and 75.70)
(b) Vol. **146**, No. 2, p. 697—706 (1988).

M. A. Angadi and R. Angadi
Electrical Properties of MnSb Films
(73.60; 75.70; S7)
(a) Vol. **106**, No. 2, p. 485—491 (1988).

A. Unkroth, W. Andrä, and G. Götz
*Composition Dependence of Magnetization of Ion Implanted Cobalt Layers
(75.70; 61.80; S1.2; S.61; S1.62)
(a) Vol. **106**, No. 2, p. 589—594 (1988).

J. M. Alameda, M. C. Contreras, F. Carmona, and F. López
*Analysis of Magnetic Anisotropy and Magnetization Dispersion in Polycrystalline Films Evaporated at Low Incidence Angles ($\alpha < 60°$)
(75.70; S1.2)
(a) Vol. **107**, No. 1, p. 329—337 (1988).

A. Urbaniak-Kucharczyk
*The Magnetic Contribution to the Electrical Resistivity in Some Thin Antiferromagnetic Films
(73.60 and 75.70)
(b) Vol. **147**, No. 2, p. K119—K124 (1988).

M. Hada, K. Sumiyama, and Y. Nakamura
*Structure and Magnetic Properties of Nonequilibrium Fe–Ru Alloys Produced by Vapor Quenching
(68.55 and 75.70; 76.80; S1.1)
(a) Vol. **108**, No. 1, p. 295—302 (1988).

A. Zagòrski and M. Olko
*Magnetization in Films with Modulated Surfaces
(75.70)
(a) Vol. **109**, No. 1, p. K63—K65 (1988).

K. Schröder and H. Hejase
*Magnetism in Thin Chromium Films
(75.70; S1.2)
(b) Vol. **149**, No. 2, p. 685—690 (1988).

A. Duda
*The Critical Temperature of the Axial Next-Nearest-Neighbour Ising Thin Film
(75.40 and 75.70)
(b) Vol. **150**, No. 1, p. 273—280 (1988).

L. Smardz and J. Baszyński
*Average Magnetization of Compositionally Modulated Cu/Ni Films
(75.70; S1.2)
(a) Vol. **110**, No. 1, p. 227—230 (1988).

75.80. Magnetomechanical and magnetoelectric effects, magnetostriction

For galvanomagnetic effects, see 72.15 an 72.20
For magneto-optical effects, see 78.20

M. Takahashi, F. Sato, and T. Miyazaki
Magnetic Anisotropy and Magnetostriction in Fe–Nd Alloys
(75.50; 75.80; S1.1)
(a) Vol. **105**, No. 1, p. 263—273 (1988).

J. P. Allemand, F. Fouquet, J. Perez and B. de Guillebon
Evolution of the Mechanical and Micromechanical Properties as a Function of Quench Rate in a Metallic Fe–Cr–C–P–Si Glass
(62.20; 61.40; 75.80; S1.1)
(a) Vol. **105**, No. 2, p. 427—439 (1988).

H. K. Lachowicz, A. Siemko, N. Moser, A. Forkl, and H. Kronmüller
*Direct Evidence for Stress-Dependent Saturation Magnetostriction in Metallic Glasses
(75.80; 78.20; S1.1)
(a) Vol. **105**, No. 2, p. 597--600 (1988).

S. U. Jen and C. J. Weng
Thermal Expansion of $(FeM)_{80}B_{14}Si_6$ Metallic Glasses (M = V, Cr, Mn)
(65.70; 75.80; S1.61)
(a) Vol. **106**, No. 1, p. 99—105 (1988).

A. Węgrzyn, K. Kułakowski, and Z. Obuszko
*Magnetostriction and Magnetization of Fe-Rich $(Fe_{1-x}Ni_x)_{90}Zr_{10}$ Amorphous Alloys
(75.80; S1.1; S1.2)
(a) Vol. **106**, No. 2, p. K169—K172 (1988).

A. Raoufi and S. Arajs
Stress Dependence of Magnetoresistance in Crystalline and Amorphous Ferromagnets
(72.15; 75.80; S1.1)
(a) Vol. **108**, No. 1, p. 369—374 (1988).

B. Idzikowski and A. Wrzeciono
Properties of Amorphous $Sm_xFe_{80-x}B_{20}$ Ribbons
(75.60; 75.80; S1.1; S1.4)
(a) Vol. **108**, No. 1, p. 375—380 (1988).

M. Cinal and S. Olszewski
*Quantum-Statistical Approach to Elastic Constants in Metals Influenced by Magnetization
(75.80; 62.20; S1; S1.1; S1.2)
(b) Vol. **148**, No. 2, p. 699—707 (1988).

L. Kraus
*On the Stress Dependence of the Saturation Magnetostriction in Amorphous Alloys
(75.80)
(a) Vol. **109**, No. 1, p. K71—K74 (1988).

76. Magnetic resonances and relaxation in condensed matter; Mössbauer effect

76.20. General theory of resonances and relaxation

E. R. Butkevich and R. Kh. Sabirov
*On Anisotropy of Nuclear Spin-Diffusion Coefficient
(76.20)
(b) Vol. **146**, No. 2, p. 683—689 (1988).

J. Łuczka and E. Piotrowski
*Effective Relaxation Time of a Dipole System in an Alternating Field
(76.20; 77.40)
(b) Vol. **149**, No. 1, p. K35—K39 (1988).

76.30. Electron paramagnetic resonance and relaxation

Yi-Yang Zhou
Theoretical Study of the Optical Absorption, Mössbauer, and EPR Spectra of Fe^{2+} Ion in Orthopyroxene
(71.70; 76.30; 76.80; 78.30; S11)
(b) Vol. **145**, No. 1, p. 227—232 (1988).

Wan-Lun Yu and J. L. Rao
Pressure Dependence of the EPR Parameters of Cs_3CoCl_5 Crystal
(71.70; 76.30; S9.16)
(b) Vol. **145**, No. 1, p. 255—259 (1988).

M. Umar, M. I. Haque, and R. J. Singh
*EPR Study of Mn^{2+} Doped Single Crystal of α-$Ca(HCOO)_2$
(76.30; S12)
(b) Vol. **145**, No. 1, p. K51—K54 (1988).

Guanghua Chen, Guosheng Sun, and Fangqing Zhang
*ESR in P/I Heterojunctions Based on a-Si:H
(73.40 and 76.30; S5.11)
(a) Vol. **105**, No. 1, p. K41—K43 (1988).

M. V. Chernitskii, A. E. Nikiforov, and A. I. Krotkii
*Microscopic Mechanisms of Paramagnetic Probe–Lattice Interaction for the System $RbCdF_3:Cr^{3+}$ near the Structural Phase Transition
(64.60 and 76.30; S9.16)
(b) Vol. **145**, No. 2, p. 425—434 (1988).

J. Kreissl and W. Gehlhoff
*Electron Paramagnetic Resonance of the Mn_4^0 Cluster in Silicon
(71.55 and 76.30; 61.70; S5.11)
(b) Vol. **145**, No. 2, p. 609—617 (1988).

S. V. Nistor, A. Bouwen, and D. Schoemaker
*ESR of Pb^{3+} Centers in Alkali Chlorides
(61.70 and 76.30; S9.11)
(b) Vol. **145**, No. 2, p. K135—K139 (1988).

D. Skrzypek, K. Bualas-Borgieł, and Z. Celiński
*Investigations on Magnetically Diluted $KNi_xMg_{1-x}F_3$ Perovskite Solid Solution
(75.30 and 76.30; S9)
(a) Vol. **105**, No. 2, p. 567—572 (1988).

I. M. Askerov, A. O. Mekhrabov, G. K. Aslanov, B. G. Tagiev, and S. M. Nakhmetov
*Transition-Metal Electron States in Imperfect Ga_2S_3 Crystals
(76.30 and 76.80; S8)
(a) Vol. **105**, No. 2, p. K151—K154 (1988).

S. Dhanuskodi and N. Hariharan
Radiation Damage of Bisglycine Cadmium Chloride Single Crystals. An EPR Study
(61.80; 76.30; S12)
(a) Vol. **106**, No. 1, p. 67—72 (1988).

M. N. Khan, M. A. Hassan, and C. A. Hogarth
The Electronic and Optical Properties of Germanium Tellurite Glasses Containing Various Transition Metal Oxides
(72.80; 76.30; 78.30; S10.15)
(a) Vol. **106**, No. 1, p. 191—200 (1988).

K. Bente and D. Siebert
*EPR Investigations on High Temperature Superconducting $YBa_2Cu_3O_{7-x}$
(74.70 and 76.30; S10.15)
(a) Vol. **106**, No. 1, p. K57—K59 (1988).

S. N. Martynov
*Single-Ion Anisotropy and ESR Linewidth in Quasi-Low-Dimensional Heisenberg Magnets
(76.30)
(b) Vol. **146**, No. 2, p. 691—696 (1988).

Z. Frait, D. Fraitová, E. Pollert, and L. Půst
*Observation of the Meissner-Ochsenfeld Field of the High-Temperature Superconductor Y–Ba–Cu–O by Means of FREPR
(74.30 and 76.30; S10.15)
(b) Vol. **146**, No. 2, p. K119—K123 (1988).

A. A. Alybakov, V. A. Gubanova, K. Kudabaev, and K. Sharsheev
*Optical Absorption and EPR Spectra of Cu^{2+} Ions in $LiKSO_4$ Single Crystals
(76.30 and 78.50; S11.1)
(b) Vol. **146**, No. 2, p. K135—K139 (1988).

P. Matthys, F. Callens, and E. Boesman
*An Algorithm for the Fitting of EPR Steady-State Saturation Curves
(76.30)
(b) Vol. **146**, No. 2, p. K149—K151 (1988).

Yi-Yang Zhou
*A Theoretical Study of EPR Spectra in Cu^{2+}- and Mn^{2+}-Doped $CaCd(CH_3COO)_4 \cdot 6\,H_2O$
(71.70 and 76.30; S11)
(b) Vol. **147**, No. 1, p. 273—280 (1988).

A. Graja, G. Sekretarczyk, W. Pukacki, S. Nešpůrek, and O. Ryba
*Conduction-Electron Spin Resonance and Thermoelectric Power of the Mixed-Valence Tetrathiafulvalenium Chloride
(72.20 and 76.30; S12)
(b) Vol. **147**, No. 1, p. 401—408 (1988).

M. Kumeda, M. Kiwaki, and T. Shimizu
Temperature Dependence of the E′ Center Creation in Silica Glasses
(61.70; 76.30; S10.1)
(b) Vol. **147**, No. 1, p. K1—K4 (1988).

I. N. Kravtsova
*Angular Dependences of EPR Linewidth of Paramagnetic Centres with Quadrupole Moments
(76.30)
(b) Vol. **147**, No. 1, p. K91—K94 (1988).

S. Ram
Magnetic and Magnetic Resonance Studies on Oxygen-Deficient $SrFeO_{3-y}$ Perovskites
(61.40; 75.60; 76.30; S11.2)
(a) Vol. **107**, No. 1, p. 55—62 (1988).

S. K. Misra and C. Rudowicz
*Effect of Monoclinic Symmetry on the EPR Spectra of Gd^{3+}-Doped Hydrated Single Crystals of Rare-Earth Trichlorides
(76.30; S11)
(b) Vol. **147**, No. 2, p. 677—684 (1988).

A. B. Vassilikou-Dova and G. Lehmann
*EPR of V^{4+} in Apophyllite
(61.70 and 76.30; S11)
(b) Vol. **147**, No. 2, p. 691—697 (1988).

N. G. Romanov, V. V. Dyakonov, V. A. Vetrov, and P. G. Baranov
*EPR Detected via Magnetic Circular Dichroism of Optical Absorption in Irradiated Alkali Halide Crystals Doped with Cadmium
(71.55 and 76.30; S9.11)
(b) Vol. **147**, No. 2, p. K171—K174 (1988).

A. A. Romanyukha, Yu. N. Shvachko, A. V. Skripov, and A. P. Stepanov
**Conduction Electron Spin Resonance in Organic Conductors α- and β-$(ET)_2IBr_2$
(76.30; S12)
(b) Vol. **147**, No. 2, p. K185—K188 (1988).

A. A. SAMOKHVALOV, N. A. VIGLIN, B. A. GIZHEVSKII, T. I. ARBUZOVA, and N. M. CHEBOTAEV
Magnetic Polarons in EuO
(72.20 and 75.30; 76.30; S10.1)
(b) Vol. **148**, No. 1, p. 361—367 (1988).

K. BIELECKI
*A High Pressure EPR Study of Cr^{3+} Doped Deuterated Potassium Aluminium Alum
(76.30; 62.50; S11.1)
(b) Vol. **148**, No. 1, p. K53—K56 (1988).

V. A. IVANSHIN, I. N. KURKIN, and G. VÖLKEL
*The Influence of Optical Modes on the Spin-Lattice Relaxation of Paramagnetic Centres in Irradiated Rochelle Salt Single Crystals
(76.30 and 77.80; S11.1)
(b) Vol. **148**, No. 1, p. K61—K63 (1988).

T. DRÜSEDAU, M. ECKLER, and R. BINDEMANN
Photoconductive a-Si:H with Dominant Monohydride Bonding Prepared by DC-Magnetron Sputtering
(68.55 and 73.60; 76.30; 78.65; S5.11)
(a) Vol. **108**, No. 1, p. 285—293 (1988).

R. CALLENS, F. CALLENS, P. MATTHYS, and E. BOESMAN
*EPR of a New O_2^- Centre in KCl
(71.55 and 76.30; S9.11)
(b) Vol. **148**, No. 2, p. 683—688 (1988).

B. F. MINAEV, YU. A. SEREBRENNIKOV, and H. D. REMPEL
*Non-Equilibrium Spin Polarization of the Si-Sl Centre in Silicon Induced by Spin-Orbit Coupling
(71.70 and 76.30; S5.11)
(b) Vol. **148**, No. 2, p. 689—698 (1988).

J. L. RAO, R. M. KRISHNA, and S. V. J. LAKSHMAN
Crystal Field Energy Levels, EPR Parameters, and Magnetic Susceptibility of the Co^{2+} Ion in Cs_3ZnCl_5
(71.70; 76.30; 78.50; S9.12)
(b) Vol. **148**, No. 2, p. K119—K123 (1988).

V. K. JAIN and V. S. YADAV
*Estimation of T_1 for Co(II) from the EPR Linewidths of Mn(II) in Dibarium Cobalt Formate Tetrahydrate Single Crystals
(76.30; S12)
(b) Vol. **148**, No. 2, p. K157—K160 (1988).

A. POPOV, A. BAHNEV, and S. DARDJONOV
Photoluminescence and EPR Spectra of Fe-Doped $Ga_{1-x}In_xP$ Epitaxial Layers
(68.55 and 78.55; 76.30; S7.15)
(a) Vol. **108**, No. 2, p. 631—636 (1988).

F. X. CABAÑAS and C. F. SCHWERDTFEGER
*Temperature Dependence of the g-Tensor of N-Methyl-N-Ethyl Morpholinium Tetracyanoquinodimenthane
(76.30; S12)
(a) Vol. **108**, No. 2, p. 733—737 (1988).

FANGQING ZHANG, GUOSHENG SUN, and GUANGHUA CHEN
*ESR and IR Studies on P-Doped a-Si:H/a-Si:H Junctions
(76.30 and 78.30; S5.11)
(a) Vol. **108**, No. 2, p. K139—K143 (1988).

P. MATTHYS, F. CALLENS, G. VEKEMAN, and E. BOESMAN
**g*-Tensor Analysis of Sulfur Centres in Alkali Halides
(71.70 and 76.30; S9.11)
(b) Vol. **149**, No. 1, p. K25—K30 (1988).

A. G. BADALYAN, T. I. MAKSIMOVA, and J. ROSA
*EPR of MnO_4^{2-} Centres in Alkali Halide Crystals
(76.30; 78.50; S9.11)
(b) Vol. **149**, No. 1, p. K31—K34 (1988).

S. N. MARTYNOV and S. B. PETROV
*Single-Ion Anisotropy of Exchange-Coupled Mn^{2+} Ions in $(CH_3)_4NCdCl_3$ and $CsCdCl_3$
(71.55 and 76.30; S9)
(b) Vol. **149**, No. 1, p. K41—K44 (1988).

M. TANAKA
*EPR Linewidth of Two-Dimensional Diluted Magnetic Systems
(76.30; S9.15)
(b) Vol. **149**, No. 1, p. K45—K48 (1988).

V. A. KHRAMTSOV, V. N. LOMASOV, YA. YA. PILKEVICH, M. P. VLASENKO, and L. S. VLASENKO
Accumulation and Transformation of Radiation Defects in Silicon under Different Doses and Intensities of Electron Irradiation
(61.80; 76.30; S5.11)
(a) Vol. **109**, No. 1, p. 127—134 (1988).

J. MINGE and T. KRAJEWSKI
EPR Study of SO_4^- Radicals in $K_4LiH_3(SO_4)_4$ Crystals
(64.70; 76.30; S11.1)
(a) Vol. **109**, No. 1, p. 193—198 (1988).

G. VÖLKEL, A. PÖPPL, and B. VOIGTSBERGER
Investigation of the Oxygen Vacancy Balance in ZnO Ceramics by Means of EPR
(71.55; 76.30; S10.1)
(a) Vol. **109**, No. 1, p. 295—299 (1988).

FANGQING ZHANG, GUANGHUA CHEN, YAFEI ZHANG, and GONG YU
IR and ESR Studies of the Structural Properties of Hydrogenated Amorphous Carbon Films
(68.55; 76.30; 78.65; S5)
(a) Vol. **109**, No. 1, p. K39—K44 (1988).

E. A. ZHILINSKAYA, V. N. LAZUKIN, E. A. BYCHKOV, and S. B. ROSENKOV
Correlation between the Structures of Ge–S Glasses and that of Impurity Mn^{2+} Complexes
(61.60 and 71.55; 76.30; S8.16)
(b) Vol. **149**, No. 2, p. 427—433 (1988).

S. A. MARSHALL, C. YU, and Y. N. ZHANG
*Comments on the EPR Spectrum of Tetravalent Chromium in Ruby
(76.30; S10.1)
(b) Vol. **149**, No. 2, p. 691—695 (1988).

WEN-CHEN ZHENG
On the Equations of State of $ZnSiF_6 \cdot 6\,H_2O:Ni^{2+}$ Crystal Determined from the Pressure Dependence of EPR Spectra
(64.30; 71.70; 76.30; S11)
(b) Vol. **149**, No. 2, p. K105—K109 (1988).

N. GUSKOS, TH. LEVENTOURI, CH. TRIKALINOS, and M. CALAMIOTOU
*EPR Spectra of Cu^{2+} Ions on Orthorhombic Complexes of the $YBa_2Cu_3O_x$ Compound
(74.70 and 76.30; S10.15)
(b) Vol. **149**, No. 2, p. K157—K162 (1988).

V. M. BABICH, N. P. BARAN, A. A. BUGAI, A. A. KONCHITS, V. B. KOVALCHUK, V. M. MAKSIMENKO, and B. D. SHANINA
Electrical and Paramagnetic Properties of Thermodonors-II in Silicon. Discussion of a Model
(71.55; 72.80; 76.30; S5.11)
(a) Vol. **109**, No. 2, p. 537—547 (1988).

J. KURIATA, L. SADŁOWSKI, B. BOJANOWSKI, J. WALCZAK, M. KURZAWA, and J. PICHET
*The Magnetic Resonance of $Fe_2V_{4-x}Mo_xO_{13}$ Type Phase
(76.30; 75.50; S11)
(a) Vol. **109**, No. 2, p. K139—K143 (1988).

N. T. BAGRAEV and I. S. POLOVTSEV
Reactions for Iron Centers in Silicon under Optical Pumping
(61.70 and 71.55; 72.40; 76.30; 76.60; S5.11)
(b) Vol. **150**, No. 1, p. 19—24 (1988).

H. R. ASATRYAN
*ESR Study of Mo^{3+} Containing Mixed Yttrium–Scandium–Aluminium Garnet Single Crystals
(76.30; S11.2)
(b) Vol. **150**, No. 1, p. 253—259 (1988).

A. HAMAÏDIA
Colour Centres in Alkali Metal Doped SrF_2
(78.20 and 78.50; 61.70; 76.30; S9.12)
(b) Vol. **150**, No. 1, p. 323—336 (1988).

J. KREISSL, W. ULRICI, and W. GEHLHOFF
*Electron Paramagnetic Resonance of Ti^{3+} in GaP
(76.30; S7.11)
(b) Vol. **150**, No. 1, p. K57—K61 (1988).

A. G. J. GREEN, B. RAY, I. V. F. VINEY, and J. W. BRIGHTWELL
Luminescence Studies of CaS Phosphors Doped with Cerium and Copper
(71.55 and 78.55; 76.30; 78.60; S8)
(a) Vol. **110**, No. 1, p. 269—276 (1988).

A. STESMANS and G. DE VOS
*MgO Powder Containing Low Concentrations of Isotopically Pure $^{52}Cr^{3+}$ Ions. Its Application as ESR Marker
(76.30; 61.70; S10.1)
(a) Vol. **110**, No. 2, p. 615—629 (1988).

A. N. GURYANOV, E. M. DIANOV, V. M. KIM, V. M. MASHINSKII, V. B. NEUSTRUEV, V. A. TIKHOMIROV, and V. F. KHOPIN
UV and γ-Induced Paramagnetic Colour Centres in Germanium-Doped Silica Glass
(78.50; 76.30; S10.15)
(a) Vol. **110**, No. 2, p. K107—K110 (1988).

76.40. Diamagnetic and cyclotron resonances

C. K. Sarkar, S. Kundu, and P. K. Basu
*Cyclotron Resonance Linewidth in a Two-Dimensional Electron Gas
(73.40 and 76.40; S7.12)
(b) Vol. **147**, No. 1, p. 135—140 (1988).

S. G. Lyapin, M. I. Eremets, A. M. Shirokov, R. Herrmann, H.-U. Müller, and F. Ludwig
Submillimetre Magnetospectroscopy of Tellurium under Hydrostatic Pressure
(71.55 and 78.50; 72.40; 76.40; S5)
(b) Vol. **147**, No. 1, p. 297—306 (1988).

F. Ludwig, H.-U. Müller, and R. Herrmann
*Magneto-Optical Investigations on p-InSb Bicrystals in the Submillimeter Wavelength Range
(76.40 and 78.20; 73.40; S7.13)
(b) Vol. **148**, No. 1, p. 143—153 (1988).

F. Ludwig, H.-U. Müller, and R. Herrmann
*Effective Scattering Times of Inversion Electrons in InSb Grain Boundaries Deduced from Cyclotron Resonance Measurements
(73.40 and 76.40; 72.10; S7.13)
(b) Vol. **150**, No. 1, p. 119—123 (1988).

76.50. Ferromagnetic, antiferromagnetic, and ferrimagnetic resonances; spin-wave resonance
(see also 75.30. Spin waves)

J. Barnaś
Magnetostatic Modes in Three-Layered Superlattices
(75.70; 76.50)
(b) Vol. **145**, No. 1, p. K55—K58 (1988).

T. Tsuboi
*Light Scattering by Magnetic Polaritons in the Ferromagnet K_2CuF_4
(75.25 and 76.50; S9.16)
(b) Vol. **145**, No. 2, p. K145—K148 (1988).

V. V. Gorbach
On a Possible Multimagnon Generation Mechanism under Exciton Light Absorption in Antiferromagnetics
(75.30; 76.50)
(b) Vol. **149**, No. 1, p. K49—K54 (1988).

76.60. Nuclear magnetic resonance and relaxation

A. A. Sukhovskii, I. P. Aleksandrova, Yu. N. Moskvich, and O. V. Rozanov
Phase Diagram of the Monoclinic and Orthorhombic Ammonium Hydrogen Selenate
(64.70; 76.60; S11.1)
(a) Vol. **105**, No. 1, p. 145—152 (1988).

T. J. Bastow and S. N. Stuart
*NMR Study of the Zinc Chalcogenides (ZnX, X = O, S, Se, Te)
(76.60; S8.1)
(b) Vol. **145**, No. 2, p. 719—728 (1988).

M. E. Gouvea and A. S. T. Pires
*Spin Dynamics in the Classical Easy-Plane Antiferromagnet: Proton Spin-Lattice Relaxation in TMMC
(76.60; S12.1)
(b) Vol. **146**, No. 1, p. 263—268 (1988).

S. F. Sagnowski, S. A. Hodorowicz, and B. Borzecka-Prokop
*^{1}H and ^{19}F Magnetic Resonance Study on Molecular Dynamics in Nickel(II) Hexaammino-Tetrafluoroborate
(76.60; S11)
(a) Vol. **107**, No. 1, p. 347—353 (1988).

R. G. Barnes, D. R. Torgeson, and P. J. Bray
*On NMR Powder Spectrum Simulation for Nuclei with $I > 1/2$
(76.60)
(b) Vol. **147**, No. 2, p. K175—K178 (1988).

M. Maćkowiak and P. Kozioł
*Effect of Pressure on the Symmetric Hydrogen Bond in $(CCl_3COO)_2HK$
(76.60; 62.50; S11)
(a) Vol. **108**, No. 2, p. 739—745 (1988).

I. O. Troyanchuk, L. A. Bashkirov, A. A. Shemyakov, and V. K. Prokopenko
*NMR-Spectroscopy Study of $Ca(Cu_{3-x}Mn_x)Mn_4O_{12}$ Perovskites
(76.60; S10.15)
(a) Vol. **109**, No. 1, p. K59—K61 (1988).

M. Mertig, V. G. Luppov, and Yu. K. Pilipenko
*Nuclear Polarization of Atomic Hydrogen
(76.60; 67.90)
(b) Vol. **149**, No. 2, p. 697—709 (1988).

N. T. Bagraev and I. S. Polovtsev
Reactions for Iron Centers in Silicon under Optical Pumping
(61.70 and 71.55; 72.40; 76.30; 76.60; S5.11)
(b) Vol. **150**, No. 1, p. 19—24 (1988).

R. C. Zamar and A. H. Brunetti
*Uniaxial-Stress Dependence of the NQR Frequency of ^{35}Cl in Paradichlorobenzene
(76.60; S12)
(b) Vol. **150**, No. 1, p. 245—252 (1988).

76.70. Magnetic double resonances and cross effects

M. Kiknadze, R. Mirianashvili, and T. Sanadze
*Nature of Optically Induced Trigonal Centers of Tm^{2+} in BaF_2
(76.70; S9.12)
(b) Vol. **146**, No. 2, p. K141—K143 (1988).

76.80. Mössbauer effect; other γ-ray spectroscopy

Yi-Yang Zhou
Theoretical Study of the Optical Absorption, Mössbauer, and EPR Spectra of Fe^{2+} Ion in Orthopyroxene
(71.70; 76.30; 76.80; 78.30; S11)
(b) Vol. **145**, No. 1, p. 227—232 (1988).

P. W. Mauriz and E. L. Albuquerque
Calculation of the Debye-Waller Factor for Mössbauer Surface Atoms: Anisotropic Effects
(63.20; 76.80)
(b) Vol. **145**, No. 1, p. K5—K8 (1988).

H. Engelmann, W. Gatzweiler, I. Dézsi, and U. Gonser
A Comparative Study of Impurity Defects in Crystalline and Amorphous $LiNbO_3$
(71.55; 76.80; S11.1)
(a) Vol. **105**, No. 1, p. 219—230 (1988).

J. L. Sánchez Llamazares, S. García García, O. Arés Muzio, and S. Díaz Castañón
Mössbauer and Thermomagnetic Analysis of Al-Containing Nd–Fe–B Permanent Magnet Alloys
(75.50; 76.80; S1.1)
(a) Vol. **105**, No. 1, p. 243—247 (1988).

A. A. Hirsch and T. Massalha
Magnetic Metastable States in the Critical Transitions of Nickel and Iron
(75.30; 76.80; S1.1; S1.2)
(a) Vol. **105**, No. 1, p. 249—262 (1988).

S. N. Gupta and H. C. Verma
*Studies of $Zr_{0.8}Fe_{0.2}$ Alloys Using Mössbauer Spectroscopy
(76.80; S1; S1.1)
(a) Vol. **105**, No. 1, p. 275—280 (1988).

K. Sumiyama and Y. Nakamura
Nonequilibrium $Fe_{1-x}Co_x$ Alloys Produced by Vapor Quenching
(75.70; 61.10; 76.80; S1.1)
(a) Vol. **105**, No. 1, p. 291—295 (1988).

I. M. Askerov, A. O. Mekhrabov, G. K. Aslanov, B. G. Tagiev, and S. M. Nakhmetov
*Transition-Metal Electron States in Imperfect Ga_2S_3 Crystals
(76.30 and 76.80; S8)
(a) Vol. **105**, No. 2, p. K151—K154 (1988).

V. Angelov, Ts. Bonchev, V. Rusanov, and I. Mandjukov
*On the Probability of Recoilless Absorption f' and Reemission f'' in Some Fe-Containing Compounds
(76.80; S1.1)
(b) Vol. **146**, No. 1, p. 269—277 (1988).

D. Baltrūnas, D. M. Zayachuk, S. Motiejūnas, and P. M. Starik
*Direct Observation of Giant Deviations from Stoichiometry in Single Crystalline $Pb_{1-x}Sn_xTe$ by Means of the Mössbauer Effect on the ^{119}Sn Nuclei
(76.80; S8.15)
(a) Vol. **106**, No. 2, p. K173—K175 (1988).

L. G. Shpinkova and A. A. Sorokin
*Magnetic Quenching of the Electric Quadrupole Interaction of ^{181}Ta in Ferromagnetic Laves Phases AFe_2 (A = Zr, Y, Lu)
(75.30 and 76.80; S1.1)
(b) Vol. **147**, No. 1, p. K75—K78 (1988).

F. S. Nasredinov, V. F. Masterov, Ch. S. Saidov, P. P. Seregin, N. N. Troitskaya, and H. U. Tschirner
Nature of the Electrical Inactivity of Tin and Iron Impurity Atoms in In_2Te_3 and Ga_2Te_3
(71.55; 76.80; S8)
(a) Vol. **107**, No. 1, p. 291—298 (1988).

M. Budzyński, M. Subotowicz, H. Niezgoda, H. Spustek, W. Tańska-Krupa, and R. Wasiewicz
*The Hyperfine Fields and Their Pressure Derivatives in $(Y_{0.9}Hf_{0.1})Fe_2$ and $(Zr_{0.9}Hf_{0.1})Fe_2$
(76.80; 71.70; S1.1)
(b) Vol. **147**, No. 2, p. 685—689 (1988).

J. Chojcan and M. Szuszkiewicz
*The Mössbauer Effect and Positron Annihilation Studies of Chromium–Iron Alloys
(76.80 and 78.70; S1.1)
(b) Vol. **147**, No. 2, p. K201—K204 (1988).

K. Melzer and A. Martin
*A Mössbauer Study of the Reaction between Barium Carbonate and Iron(III) Oxide
(76.80; S11.2)
(a) Vol. **107**, No. 2, p. K163—K168 (1988).

D. Baltrūnas, B. Vengalis, S. Motiejūnas, and N. Shiktorov
Effect of a Deviation from Stoichiometry on the Mössbauer Parameters of $Sn_{0.8}Ge_{0.2}Te$ during Phase Transition
(64.70; 76.80; S8.15)
(a) Vol. **108**, No. 1, p. 197—204 (1988).

M. Hada, K. Sumiyama, and Y. Nakamura
Structure and Magnetic Properties of Nonequilibrium Fe–Ru Alloys Produced by Vapor Quenching
(68.55 and 75.70; 76.80; S1.1)
(a) Vol. **108**, No. 1, p. 295—302 (1988).

J. Gwiazda, E. Mariańska, and W. Zych
*The Influence of Transition Metal Substitution and $B_{3/2}$ Values for Some Amorphous Alloys
(76.80; 75.50; S1.1)
(a) Vol. **108**, No. 1, p. K67—K72 (1988).

S. Motiejūnas, D. Baltrūnas, V. I. Garasim, and P. M. Starik
*A Correlation between Ligand Electronegativity and the Value of the Mössbauer Isomer Shift of Stannous Chalcogenides
(76.80; S8)
(b) Vol. **148**, No. 2, p. K161—K164 (1988).

M. Haas, V. Hizhnyakov, E. Realo, and J. Jõgi
*Nuclear Polariton in the Mössbauer Absorber
(76.80)
(b) Vol. **149**, No. 1, p. 283—290 (1988).

G. A. Bordovskii, N. A. Savinova, and L. N. Seregina
*Electrical Activity Pecularities of Tin and Iron Impurities in the Semiconducting Glass $Ge_{28.5}Pb_{15}S_{56.5}$
(72.80 and 76.80; S8)
(a) Vol. **109**, No. 1, p. K51—K53 (1988).

A. A. Opalenko and A. A. Kornilova
*Anisotropy of the Mössbauer Effect in β-Sn under Pressure and the Lattice Dynamics
(63.20 and 76.80; 62.50; S4)
(b) Vol. **150**, No. 1, p. 25—29 (1988).

77. Dielectric properties and materials
(for conductivity phenomena, see 72.20 and 72.80)

77.20. Permittivity

E. D. Yakushkin and V. N. Anisimova
Phase Transition in Monoclinic KD_2PO_4 Crystal
(64.70; 77.20; S11.1)
(a) Vol. **105**, No. 1, p. 139—143 (1988).

S. S. Sastry, G. Satyanandam, A. Subrahmanyam, and V. R. K. Murthy
*Dielectric Properties of $HgCl_2$:2 KCl:H_2O Single Crystals in the Radio Frequency Region
(77.20 and 77.40; S9)
(a) Vol. **105**, No. 1, p. K71—K75 (1988).

K. S. Aleksandrov, S. V. Melnikova, I. N. Flerov, A. D. Vasilev, A. I. Kruglik, and I. T. Kokov
Phase Transitions in Cs_2CdI_4 Single Crystals
(64.70; 61.60; 65.40; 77.20; 78.20; S9.16)
(a) Vol. **105**, No. 2, 441—446 (1988).

T. R. Middya, D. Roy, and A. N. Basu
*Multiple-Scattering-Theory Calculation of Elastic and Dielectric Properties of Alkali-Halide Solid Solutions
(62.20 and 77.20; S9.11)
(b) Vol. **146**, No. 1, p. 117—124 (1988).

H. Kresse and W. Weissflog
*Strong Antiparallel Correlation in Liquid Crystalline Esters with Lateral Groups
(77.20; 77.40; S12)
(a) Vol. **106**, No. 1, p. K89—K91 (1988).

A. De and K. V. Rao
*Dielectric Properties of Synthetic Quartz Crystals Irradiated with γ-Rays or X-Rays under a High Electric Field
(77.20; 77.40; S10.1)
(a) Vol. **106**, No. 1, p. K101—K105 (1988).

Om Parkash, Ch. Durga Prasad, and Devendra Kumar
*Dielectric Properties of Lanthanum Lead Cobalt Titanium Oxide $(La_{1-x}Pb_x)$ $(Co_{1-x}Ti_x)O_3$ $(0.50 \leqq x \leqq 0.90)$
(77.20; and 77.40; 77.80; S10.15)
(a) Vol. **106**, No. 2, p. 627—633 (1988).

D. Hans, A. A. S. Sangachin, and M. Kumar
Elastic, Dielectric, and Photoelastic Constants of Alkali Halide Crystals
(62.20; 63.90; 77.20; S9.11)
(b) Vol. **147**, No. 2, p. 529—536 (1988).

A. De and K. V. Rao
*Dielectric Properties of CaF_2 Single Crystals Irradiated with X-Rays under a High Electric Field or γ-Rays
(77.20 and 77.40; S9.12)
(a) Vol. **108**, No. 1, p. 457—462 (1988).

B. S. Tošić, J. P. Šetrajčić, and D. Lj. Mirjanić
The Influence of Phonons on Dielectric Properties of Molecular Crystals
(71.45; 77.20)
(b) Vol. **148**, No. 2, p. 593—598 (1988).

I. Kosacki and K. Zalibowski
*Polarisation Phenomena in PbF_2 Superionic Crystals
(77.20 and 77.40; S9)
(a) Vol. **108**, No. 2, p. 765—770 (1988).

A. V. Zaretskii, V. F. Petrenko, and V. A. Chesnakov
*The Protonic Conductivity of Heavily KOH-Doped Ice
(66.90 and 77.20; 64.70; S10)
(a) Vol. **109**, No. 2, p. 373—381 (1988).

R. Mizeris, J. Grigas, V. Samulionis, V. Skritski, A. I. Baranov, and L. A. Shuvalov
Microwave and Ultrasonic Investigations of Superionic Phase Transitions in $CsDSO_4$ and $CsDSeO_4$
(64.70; 62.65; 77.20; S11.1)
(a) Vol. **110**, No. 2, p. 429—436 (1988).

77.30. Polarization and depolarization effects

C. Vijayan and Y. V. G. S. Murti
Ionic Thermocurrents Due to Z-Centres in $KCl:Gd^{2+}$
(61.70; 77.30; 77.40; 78.50; S9.11)
(a) Vol. **105**, No. 2, p. 397—402 (1988).

R. Vila, A. Ibarra, and M. Jiménez de Castro
*Improvements of Prakash's Method for Analyzing ITC Peaks
(77.30 and 77.40)
(a) Vol. **105**, No. 2, p. 601—607 (1988).

J. Prakash and F. Fischer
*TSD in Pressed Pellets of $NaCl:HCO_2^-$
(77.30; 61.70; S9.11)
(b) Vol. **146**, No. 1, p. K71—K73 (1988).

J. Prakash and A. K. Nishad
*On the Analysis of the Ionic Thermocurrent Spectrum
(77.30; S9.11)
(a) Vol. **106**, No. 2, p. K177—K179 (1988).

A. F. Vavrek and K. K. Christova
*X-Ray Polarization and Depolarization Phenomena in $Bi_{12}SiO_{20}$
(77.30 and 78.70; 72.40; S10)
(a) Vol. **107**, No. 1, p. 379—386 (1988).

E. G. Salman and V. N. Vertoprakhov
*Thermally Stimulated Depolarization Current Controlled by Surface Charge Change
(73.40 and 77.30; 77.55; S10.1)
(a) Vol. **108**, No. 2, p. 625—630 (1988).

M. P. Madan and M. N. Sharma
*On the Change in Energy of Free Ions in Alkali Halides
(77.30; S9.11)
(b) Vol. **149**, No. 1, p. K55—K59 (1988).

V. R. K. Murthy and R. Raman
*Effect of Thickness and Length of the Sample on the Dielectric Constant (ε') and Loss (ε'') Measurements in the Microwave Cavity Perturbation Technique
(77.30 and 77.40)
(a) Vol. **109**, No. 1, p. K75—K78 (1988).

C. Christodoulides
*Effective Temperature of Polarization in Measurements of Thermally Stimulated Depolarization Currents
(77.30)
(a) Vol. **109**, No. 2, p. 611—616 (1988).

77.40. Dielectric loss and relaxation

S. S. Sastry, G. Satyanandam, A. Subrahmanyam, and V. R. K. Murthy
*Dielectric Properties of $HgCl_2$:2 KCl:H_2O Single Crystals in the Radio Frequency Region
(77.20 and 77.40; S9)
(a) Vol. **105**, No. 1, p. K71—K75 (1988).

C. Vijayan and Y. V. G. S. Murti
Ionic Thermocurrents Due to Z-Centres in KCl:Gd^{2+}
(61.70; 77.30; 77.40; 78.50; S9.11)
(a) Vol. **105**, No. 2, p. 397—402 (1988).

R. Vila, A. Ibarra, and M. Jiménez de Castro
*Improvements of Prakash's Method for Analyzing ITC Peaks
(77.30 and 77.40)
(a) Vol. **105**, No. 2, p. 601—607 (1988).

H. Kresse and W. Weissflog
Strong Antiparallel Correlation in Liquid Crystalline Esters with Lateral Groups
(77.20; 77.40; S12)
(a) Vol. **106**, No. 1, p. K89—K91 (1988).

A. De and K. V. Rao
Dielectric Properties of Synthetic Quartz Crystals Irradiated with γ-Rays or X-Rays under a High Electric Field
(77.20; 77.40; S10.1)
(a) Vol. **106**, No. 1, p. K101—K105 (1988).

Om Parkash, Ch. Durga Prasad, and Devendra Kumar
*Dielectric Properties of Lanthanum Lead Cobalt Titanium Oxide $(La_{1-x}Pb_x)(Co_{1-x}Ti_x)O_3$ $(0.50 \leqq x \leqq 0.90)$
(77.20 and 77.40; 77.80; S10.15)
(a) Vol. **106**, No. 2, p. 627—633 (1988).

A. E. Aliev, A. Sh. Akramov, L. N. Fershtat, and P. K. Khabibullaev
Mechanism of a Superion Phase Transition in α-$LiIO_3$
(64.60; 77.40; S11.1)
(a) Vol. **108**, No. 1, p. 189—196 (1988).

A. De and K. V. Rao
*Dielectric Properties of CaF_2 Single Crystals Irradiated with X-Rays under a High Electric Field or γ-Rays
(77.20 and 77.40; S9.12)
(a) Vol. **108**, No. 1, p. 457—462 (1988).

I. Kosacki and K. Ząlibowski
*Polarisation Phenomena in PbF_2 Superionic Crystals
(77.20 and 77.40; S9)
(a) Vol. **108**, No. 2, p. 765—770 (1988).

J. ŁUCZKA and E. PIOTROWSKI
Effective Relaxation Time of a Dipole System in an Alternating Field
(76.20; 77.40)
(b) Vol. **149**, No. 1, p. K35—K39 (1988).

V. R. K. MURTHY and R. RAMAN
*Effect of Thickness and Length of the Sample on the Dielectric Constant (ε') and Loss (ε") Measurements in the Microwave Cavity Perturbation Technique
(77.30 and 77.40)
(a) Vol. **109**, No. 1, p. K75—K78 (1988).

77.50. Dielectric breakdown and space-charge effects

J. S. PAN and H. B. PAN
*Size-Quantum Effect of the Energy of a Charge Carrier in a Semiconductor Crystallite
(68.30 and 77.50; 71.38)
(b) Vol. **148**, No. 1, p. 129—141 (1988).

G. JEANDEL, A. WARIN, and G. MORLOT
*Dielectric Constant and ITC Measurements of $BaF_2:Dy^{3+}$
(66.30 and 77.50; S9.12)
(a) Vol. **108**, No. 1, p. 183—187 (1988).

77.55. Dielectric thin films

M. HROBÁR and I. TRAVĚNEC
Time-Domain Spectroscopy. Method for Investigation of YSZ
(73.40; 77.55; S10.15)
(a) Vol. **107**, No. 1, p. 235—238 (1988).

E. G. SALMAN and V. N. VERTOPRAKHOV
Thermally Stimulated Depolarization Current Controlled by Surface Charge Change
(73.40 and 77.30; 77.55; S10.1)
(a) Vol. **108**, No. 2, p. 625—630 (1988).

77.80. Ferroelectricity and antiferroelectricity

N. Kristoffel and P. Konsin
*Vibronic Theory of Structural Phase Transitions and Diplacive Ferroelectrics **(Review Article)**
(63.75 and 77.80; 64.70; S8; S11)
(b) Vol. **149**, No. 1, p. 11—40 (1988).

Om Parkash, Ch. Durga Prasad, and Devendra Kumar
Dielectric Properties of Lanthanum Lead Cobalt Titanium Oxide $(La_{1-x}Pb_x)(Co_{1-x}Ti_x)O_3$
$(0.50 \leqq x \leqq 0.90)$
(77.20 and 77.40; 77.80; S10.15)
(a) Vol. **106**, No. 2, p. 627—633 (1988).

M. D. Fontana, A. Ridah, and G. E. Kugel
*Frequency Dependence of the Soft Mode in $KNbO_3$
(77.80 and 78.30; 63.20; S11.1)
(b) Vol. **147**, No. 1, p. 441—447 (1988).

V. D. Sajfert, Lj. D. Mašković, R. P. Djajić-Jovanović, and U. F. Kozmidis-Luburić
Phase Transition in 1D KDP-Type Ferroelectrics
(64.60; 77.80; S11.1)
(b) Vol. **148**, No. 1, p. 81—90 (1988).

V. A. Ivanshin, I. N. Kurkin, and G. Völkel
*The Influence of Optical Modes on the Spin-Lattice Relaxation of Paramagnetic Centres in Irradiated Rochelle Salt Single Crystals
(76.30 and 77.80; S11.1)
(b) Vol. **148**, No. 1, p. K61—K63 (1988).

R. Kawashima and T. Matsuda
*Temperature Dependence of the Dielectric Constant in Samarium Nitrate Crystals
(77.80; S11.1)
(a) Vol. **108**, No. 1, p. K73—K75 (1988).

M. Marvan, V. Dvořák, and M. Glogarová
*Dielectric Dispersion Due to the Electrocaloric Effect in Incommensurate Systems
(77.80)
(b) Vol. **149**, No. 1, p. K61—K65 (1988).

Jun Chen and Duan Feng
TEM Study of Phases and Domains in $NaNbO_3$ at Room Temperature
(64.70; 77.80; S10)
(a) Vol. **109**, No. 1, p. 171—185 (1988).

K. Biedrzycki
*Thermally Stimulated Electron Emission from Triglycine Sulphate Crystals Excited by AC Electric Field
(77.80 and 79.40; S12)
(a) Vol. **109**, No. 1, p. K79—K83 (1988).

78. Optical properties and condensed-matter spectroscopy and other interactions of matter with particles and radiation

(for phonon spectra, see 63)

78.20. Optical properties and materials

E. L. Nagaev
*Photoinduced Magnetism and Conduction Electrons in Magnetic Semiconductors (**Review Article**)
(75.30 and 78.20; 71.25; 75.60)
(b) Vol. **145**, No. 1, p. 11—64 (1988).

C. S. Sunandana
*Physical Applications of Photoacoustic Spectroscopy (**Review Article**)
(78.20)
(a) Vol. **105**, No. 1, p. 11—43 (1988).

O. Goede and W. Heimbrodt
*Optical Properties of (Zn, Mn) and (Cd, Mn) Chalcogenide Mixed Crystals and Superlattices **Review Article**)
(78.20; 71.35; 78.55; S8.1)
(b) Vol. **146**, No. 1, p. 11—62 (1988).

R. Enderlein, Desheng Jiang, and Yinsheng Tang
*On the Mechanisms of Photoreflectance in Multiple Quantum Wells
(78.20; 71.25; 71.35; S7.12; S7.15)
(b) Vol. **145**, No. 1, p. 167—180 (1988).

F. Beerwerth, D. Fröhlich, and V. Leinweber
Temperature Dependence of Excition Linewidth in CsI and CsBr
(71.35; 78.20; S9.11)
(b) Vol. **145**, No. 1, p. 195—201 (1988).

V. I. Gavrilenko, N. I. Klyui, V. G. Litovchenko, and V. E. Strelnitskii
Characteristic Features of the Electronic Structure of Carbon Films
(71.25; 78.20; S5)
(b) Vol. **145**, No. 1, p. 209—217 (1988).

E. K. Martynchuk and F. F. Sizov
*The Faraday Effect Approach to a Study of $A^{IV}B^{VI}$ Narrow-Gap Semiconductor Band Structure
(78.20; S8; S8.15)
(b) Vol. **145**, No. 1, p. 309—318 (1988).

P. Aghamkar, P. Sen, and P. K. Sen
*Effect of Doping on Parametric Amplification in Piezoelectric Semiconductors
(78.20; S7.13)
(b) Vol. **145**, No. 1, p. 343—349 (1988).

F. Henneberger, H. Rossmann, and A. Schülzgen
*Mirrorless Dispersive Optical Bistability at the Brewster Angle Using a Nonlinear Prism
(78.20; S8.12)
(b) Vol. **145**, No. 1, p. K83—K88 (1988).

T. Bremer, W. Heiland, A. Klekamp, and E. Krätzig
SIMS Investigations of Iron Profiles in $LiNbO_3$:Ti Waveguides
(66.30 and 79.20; 78.20; S11.1)
(a) Vol. **105**, No. 1, p. K17—K20 (1988).

T. Itoh, Y. Iwabuchi, and M. Kataoka
Study on the Size and Shape of CuCl Microcrystals Embedded in Alkali-Chloride Matrices and Their Correlation with Exciton Confinement
(61.10 and 71.35; 78.20; S9)
(b) Vol. **145**, No. 2, p. 567—577 (1988).

J. Röseler, R. I. Shekhmametev, and J. Neugebauer
Diamagnetic Shift of Bielectrons in BiI_3
(71.35; 78.20; S9)
(b) Vol. **145**, No. 2, p. 579—584 (1988).

B. Pilawa, G. Hess, H. G. Kahle, and A. Kasten
*Linear Birefringence Investigations of the Jahn-Teller Distorted Phase in the Mixed Crystal Systems $(Tb_x, Tm_{1-x})VO_4$ and $(Tb_x, Yb_{1-x})VO_4$
(78.20; 64.60; S11.1)
(b) Vol. **145**, No. 2, p. 729—739 (1988).

K. S. Aleksandrov, S. V. Melnikova, I. N. Flerov, A. D. Vasilev, A. I. Kruglik, and I. T. Kokov
Phase Transitions in Cs_2CdI_4 Single Crystals
(64.70; 61.60; 65.40; 77.20; 78.20; S9.10)
(a) Vol. **105**, No. 2, p. 441—446 (1988).

S. A. Eshraghi, S. Kianian, B. Ostrom, O. M. Stafsudd, and A. L. Gentile
*The Electrical and Optical Properties of n-Type $HgIn_2Te_4$
(72.80 and 78.20; S8.16)
(a) Vol. **105**, No. 2, p. 563—566 (1988).

H. K. Lachowicz, A. Siemko, N. Moser, A. Forkl, and H. Kronmüller
Direct Evidence for Stress-Dependent Saturation Magnetostriction in Metallic Glasses
(75.80; 78.20; S1.1)
(a) Vol. **105**, No. 2, p. 597—600 (1988).

M. A. Hassan, M. N. Khan, and C. A. Hogarth
*Electrical and Optical Properties of V_2O_5–GeO_2–NiO Glasses
(72.20 and 78.20; 78.30; S10.15)
(a) Vol. **105**, No. 2, p. 609—616 (1988).

P. Royer, J. P. Goudonnet, T. Inagaki, G. Chabrier, and E. T. Arakawa
*Photoacoustic Study of the Optical Absorption of Oblate Silver Spheroids in Attenuated-Total-Reflection Geometry
(78.20; 78.65; S1.3)
(a) Vol. **105**, No. 2, p. 617—625 (1988).

F. R. Kessler, U. Barkow, R. Nies, and N. Unzner
*Optical Profile Analysis of Diffused and Ion-Implanted Silicon
(78.20; 61.70; 66.30; 78.50; S5.11)
(a) Vol. **105**, No. 2, p. 627—635 (1988).

A. M. Elkorashy
*Temperature Dependence of the Two-Dimensional Optical Energy Gap for Germanium Selenide Single Crystals
(78.20; S8)
(b) Vol. **146**, No. 1, p. 279—285 (1988).

V. A. Kochelap and V. N. Sokolov
*Resonatorless Dissipative Optical Bistability. Influence of External Fields and Sample Thickness
(78.20)
(b) Vol. **146**, No. 1, p. 311—317 (1988).

V. A. Kochelap, N. R. Kulish, M. P. Lisitsa, N. I. Malysh, and V. N. Sokolov
*Band Edge Absorption Saturation Dynamics of Semiconductors
(78.20; S8.12)
(b) Vol. **146**, No. 1, p. 319—327 (1988).

A. I. Bobrysheva, S. S. Russu, and V. A. Zaloi
*Light Absorption and Hyper-Raman Scattering by Many-Electron Complexes in Quantum Well Structures
(78.20 and 78.30; 71.35; S7.15)
(b) Vol. **146**, No. 1, p. 329—334 (1988).

G. Jungk
*Optical Bistability in Composite Media
(78.20)
(b) Vol. **146**, No. 1, p. 335—340 (1988).

V. S. Dneprovskii, A. I. Furtichev, V. I. Klimov, E. V. Nazvanova, D. K. Okorokov, and U. V. Vandishev
*Excitons at High Density in CdS and GaSe, and Optical Bistability
(71.35 and 78.20; S8; S8.11)
(b) Vol. **146**, No. 1, p. 341—350 (1988).

V. D. Egorov, P. Flögel, Hoang Xuan Nguyen, and M. Kaschke
*Picosecond Transient Optical Nonlinearities in CdS and CdSSe
(71.35 and 78.20; S8.11; S8.15)
(b) Vol. **146**, No. 1, p. 351—356 (1988).

R. Zimmermann
*Nonlinear Optics and the Mott Transition in Semiconductors
(71.35 and 78.20)
(b) Vol. **146**, No. 1, p. 371—384 (1988).

H. Glaeske and M. Schubert
Green's Function Approach to Nonequilibrium Charge Carriers in Direct Gap Semiconductors Generated by Short Light Pulses
(72.10; 78.20)
(b) Vol. **146**, No. 1, p. 385—391 (1988).

J. H. Collet and M. Pugnet
*Picosecond Plasma Dynamics and All-Optical Data Processing
(78.20; 71.35; S8.12)
(b) Vol. **146**, No. 1, p. 393—401 (1988).

R. Sommerfeldt, L. Holtmann, E. Krätzig, and B. C. Grabmaier
Influence of Mg Doping and Composition on the Light-Induced Charge Transport in $LiNbO_3$
(61.70; 72.40; 78.20; S11.1)
(a) Vol. **106**, No. 1, p. 89—98 (1988).

E. Z. Katsnelson and A. G. Karoza
*On the Irreversibility of the Processes during Photomagnetization of Ferrites
(78.20; 72.40; S11.2)
(a) Vol. **106**, No. 1, p. 277—282 (1988).

G. C. Bhar and N. P. Ghosh
*Broadband Tuning in an Infrared Parametric Oscillator
(78.20; S7.16; S8.12; S8.16)
(a) Vol. **106**, No. 1, p. K93—K96 (1988).

R. Zimmermann
On the Dynamical Stark Effect of Excitons. The Low-Field Limit
(71.35; 78.20)
(b) Vol. **146**, No. 2, p. 545—554 (1988).

A. K. Abass, N. A. Al-Liabi, and W. A. Taha
*Optical Properties of Bromine-Doped SnO_2 Coatings for Solar Applications
(78.20; S10)
(a) Vol. **106**, No. 2, p. 613—618 (1988).

P. Lošťák, R. Novotný, J. Navrátil, and J. Šrámková
*Optical Properties of Gallium-Doped Bi_2Te_3 Single Crystals
(78.20; 78.30; S8)
(a) Vol. **106**, No. 2, p. 619—625 (1988).

R. E. Hummel, W. M. Goho, and V. Gerold
*The Influence of Precipitation on the Optical Properties and Electronic Structure of Copper–Cobalt Alloys
(61.55 and 78.20; S1.2)
(a) Vol. **106**, No. 2, p. 641—649 (1988).

R. Bukowski and Z. Kleszczewski
*Non-Linear Acousto-Optical Interaction at Deviation from Bragg Angle
(78.20)
(b) Vol. **147**, No. 1, p. 409—416 (1988).

R. E. Hummel and W. M. Goho
*Optical Properties and Electronic Structure of Copper–Cobalt Alloys
(71.50 and 78.20; S1.2)
(b) Vol. **147**, No. 1, p. 417—420 (1988).

G. M. Gusakov, A. A. Komarnitskii, and A. S. Em
Dynamics of Optical Parameters of Silicon and Germanium Single Crystals during Two-Pulsed Nanosecond Laser Irradiation
(61.80 and 68.55; 64.70; 78.20; S5.11; S5.12)
(a) Vol. **107**, No. 1, p. 261—271 (1988).

D. K. Ghosh, U. Chatterjee, and L. K. Samanta
*On the Band Gap Dependence of Refractive Indices of Some Quaternary III–V and II–VI Compounds of Device Interest
(78.20; S7.15; S8.15)
(a) Vol. **107**, No. 1, p. K79—K81 (1988).

W. Schäfer, K.-H. Schuldt, and J. Treusch
*Many-Body Effects in Resonantly Excited Dense Exciton Systems
(71.35 and 78.20)
(b) Vol. **147**, No. 2, p. 699—710 (1988).

P. I. Khadzhi, G. D. Shibarshina, A Kh. Rotaru, and Yu. D. Slavov
*Optical Bistability and Multistability in a System of High Density Excitons and Biexcitons in Semiconductors
(71.35 and 78.20)
(b) Vol. **147**, No. 2, p. 721—725 (1988).

G. Manzke and K. Henneberger
*Nonlinear Propagation of Laser Pulses in GaAs–GaAlAs MQWS
(71.35 and 78.20; 73.40; S7.15)
(b) Vol. **147**, No. 2, p. 733—739 (1988).

P. I. Khadzhi and E. S. Kiseleva
*The Dispersion Relations of the Nonlinear Surface Waves Due to the Exciton and Biexciton Formation
(78.20; 71.35; 73.40)
(b) Vol. **147**, No. 2, p. 741—745 (1988).

F. R. Kessler, J. Nieder, R. Nies, and E. Gaulke
Indirect Exciton Excitation and the Oscillatory Magnetophotoconductivity of Germanium
(71.35 and 72.40; 78.20; S5.12)
(b) Vol. **147**, No. 2, p. K143—K148 (1988).

G. Dolino, F. Mogeon, and P. Bastie
Birefringence and γ-Ray Study of the Irreversible Behaviour of the Incommensurate Phase of Quartz
(64.60; 78.20; S10.1)
(a) Vol. **107**, No. 2, p. 559—570 (1988).

J. A. Kalomiros and J. Spyridelis
*Determination of the Optical Constants *n*, *k* of $Zn_3In_2S_6$ and $Zn_5In_2S_8$ from Transmission Measurements
(78.20; S8.16)
(a) Vol. **107**, No. 2, p. 633—637 (1988).

F. Ludwig, H.-U. Müller, and R. Herrmann
*Magneto-Optical Investigations on p-InSb Bicrystals in the Submillimeter Wavelength Range
(76.40 and 78.20; 73.40; S7.13)
(b) Vol. **148**, No. 1, p. 143—153 (1988).

I. CH. AKOPYAN, D. N. GROMOV, A. A. KLOCHIKHIN, B. V. NOVIKOV, and S. G. OGLOBLIN
Effect of Lattice Disorder on the Excitonic States in Superionic Crystals
(71.35; 78.20; S9)
(b) Vol. **148**, No. 1, p. 185—195 (1988).

LIANG-YAO CHEN and D. W. LYNCH
*The Optical Properties of $AuAl_2$ and $PtAl_2$
(78.20; S1.3; S4)
(b) Vol. **148**, No. 1, p. 387—394 (1988).

V. FRITZSCHE
*Calculation of Matrix Elements for Optical Transitions
(78.20; S1.2)
(b) Vol. **148**, No. 1, p. 395—399 (1988).

H. X. NGUYEN, V. D. EGOROV, A. HARENDT, and E. V. NAZVANOVA
*Transverse Spatial Resolution of Thermo-Optical Hysteresis in CdS
(78.20; S8.11)
(b) Vol. **148**, No. 1, p. 407—412 (1988).

V. P. GRIBKOVSKII, L. G. ZIMIN, S. V. GAPONENKO, I. E. MALINOVSKII, P. I. KUZNETSOV, and G. G. YAKUSHCHEVA
*Optical Hysteresis near Excitonic Resonance in Zinc Selenide Monocrystals
(71.35 and 78.20; S8.12)
(b) Vol. **148**, No. 1, p. K65—K69 (1988).

P. M. FAUCHET and K. GZARA
*Determination of Carrier–Carrier and Carrier–Phonon Relaxation Times from Ultrafast Photoinduced Absorption in Amorphous Semiconductors
(78.20)
(b) Vol. **148**, No. 1, p. K71—K75 (1988).

O. GOEDE, W. HEIMBRODT, and TH. KÖPP
*Photoacoustic Spectroscopy of CdTe and (Hg, Cd)Te
(78.20; S8.13)
(a) Vol. **108**, No. 1, p. 443—447 (1988).

H. W. JEON, K. M. KIM, H. K. KIM, H. L. PARK, and C. H. CHUNG
*Determination of x in $Hg_{1-x}Zn_xTe$ Single Crystals by the Temperature Coefficient of the Energy Gap
(78.20; S8.15)
(a) Vol. **108**, No. 1, p. K77—K79 (1988).

B. M. NITSOVICH, O. V. KRECHIVSKII, and G. M. PESTRYAKOV
*Nonlinearity of Absorption in an Exciton–Phonon System
(71.35 and 78.20)
(b) Vol. **148**, No. 2, p. K165—K168 (1988).

J. LIPAVIČIUS and A. AUDZIJONIS
*Optical Birefringence in $Sn_2P_2S_6$
(78.20; S8.16)
(b) Vol. **148**, No. 2, p. K169—K171 (1988).

S. Lakshmi Raghavan, N. Ramesh, and K. Srinivasan
*Electro-Optical Effect in $KCl:Cd^{2+}$ and $KCl:Sr^{2+}$
(78.20; S9.11)
(a) Vol. **108**, No. 2, p. 771—774 (1988).

K. N. Madhusoodanan and J. Philip
*Optical Energy Gap and Thermal Diffusivity of Ge–Se Semiconducting Glasses
(78.20; S5)
(a) Vol. **108**, No. 2, p. 775—782 (1988).

V. M. Yartsev
*Optical Properties of Quasi-1D *n*-Merized Molecular Semiconductors
(72.20 and 78.20; S12.1)
(b) Vol. **149**, No. 1, p. 157—167 (1988).

S. S. Kotsur and I. V. Stasyuk
*The Microscopic Approach to the Description of the Optical Activity in Dielectric Crystals
(78.20)
(b) Vol. **149**, No. 1, p. 347—354 (1988).

G. R. Allan, H. A. MacKenzie, J. J. Hunter, and B. S. Wherrett
*The Identification of Urbach, Acceptor Impurity, and Phonon-Assisted Components of the Optical Absorption Band Tail of n-Type InSb
(78.20; 71.55; S7.13)
(b) Vol. **149**, No. 1, p. 383—389 (1988).

O. Foltin
*Oblique Incidence Specular Reflectivity of Uniaxial Semiconductors near Three-Dimensional Critical Points
(78.20; 71.20)
(b) Vol. **149**, No. 1, p. K73—K76 (1988).

U. Dahms, M. Burkert, M. Starke, H. J. Höhne, R. Link, and C. Hamann
*Electron Energy Loss Measurements on Metal-Free Phthalocyanine
(78.20; S12)
(b) Vol. **149**, No. 1, p. K85—K88 (1988).

M. Kranjčec, B. Čelustka, B. Etlinger, and D. Desnica
*The Indirect Allowed Optical Transitions in $(Ga_{0.3}In_{0.7})_2Se_3$
(78.20; S8.16)
(a) Vol. **109**, No. 1, p. 329—336 (1988).

A. Hamaïdia and A. Hachimi
*Colour Centres in Alkali Metal Doped CaF_2
(78.20 and 78.50; 61.70; 78.55; S9.12)
(b) Vol. **149**, No. 2, p. 711—724 (1988).

P. P. Paskov, L. I. Pavlov, and P. A. Atanasov
*Nonlinear Refraction Due to Dynamical Burstein-Moss Effect in Lead Chalcogenides
(78.20; S8.15)
(b) Vol. **149**, No. 2, p. 739—746 (1988).

A. M. Elkorashy
*Optical Constants of Germanium Selenide Single Crystals in the Transparency Region
(78.20; S8)
(b) Vol. **149**, No. 2, p. 747—758 (1988).

B. Hönerlage, M. Frindi, and J. Miletic
*Time Dependence of Wave-Mixing Processes under Pulsed Resonant Excitation in Three-Level Systems
(78.20; S9)
(b) Vol. **149**, No. 2, p. 775—784 (1988).

R. Capelletti, R. Cywiński, M. Manfredi, and H. Opyrchał
*Influence of Temperature on the Aggregation Processes of the $NaCl:Eu^{2+}$ System
(78.20; S9.11)
(a) Vol. **109**, No. 2, p. 603—610 (1988).

H. Puff and H. W. Streitwolf
Optical Absorption and Stability of the Bipolaron Lattice
(71.38; 78.20; S12)
(b) Vol. **150**, No. 1, p. 147—164 (1988).

M. Stordeur, M. Stölzer, H. Sobotta, and V. Riede
Investigation of the Valence Band Structure of Thermoelectric $(Bi_{1-x}Sb_x)_2Te_3$ Single Crystals
(71.25; 72.20; 72.30; 78.20; S1.63)
(b) Vol. **150**, No. 1, p. 165—176 (1988).

A. Hamaïdia
*Colour Centres in Alkali Metal Doped SrF_2
(78.20 and 78.50; 61.70; 76.30; S9.12)
(b) Vol. **150**, No. 1, p. 323—336 (1988).

D. I. Sementsov and R. M. Grechishkin
*Light Diffraction by Stripe Domain Structures in Magnetic Crystals
(78.20; 75.60)
(a) Vol. **110**, No. 1, p. 259—267 (1988).

S. Schmitt-Rink
*Coherent Nonlinear Optical Processes in Semiconductors. Physics and Applications
(71.35 and 78.20; S7.12; S7.15; S12)
(b) Vol. **150**, No. 2, p. 349—356 (1988).

B. D. Flügel, J. P. Sokoloff, F. Jarka, S. W. Koch, M. Lindberg, N. Peyghambarian, M. Joffre, D. Hulin, A. Migus, A. Antonetti, C. Ell, L. Banyai, and H. Haug
*Measurements of Ultrafast Optical Nonlinearities in Semiconductors
(71.35 and 78.20; S8.11)
(b) Vol. **150**, No. 2, p. 357—363 (1988).

R. Zimmermann and M. Hartmann
*Resonant and Off-Resonant Light-Matter Interaction in Semiconductors. Kinetics and Transients
(71.35 and 78.20)
(b) Vol. **150**, No. 2, p. 365—378 (1988).

M. LINDBERG and S. W. KOCH
*Theory of the Optical Stark Effect in Semiconductors under Ultrashort-Pulse Excitation
(71.35 and 78.20; S8.11)
(b) Vol. **150**, No. 2, p. 379—385 (1988).

J. M. HVAM, C. DÖRNFELD, and H. SCHWAB
*Optical Nonlinearity and Phase Coherence in CdSe and $CdSe_xS_{1-x}$
(71.35 and 78.20; S8.12; S8.15)
(b) Vol. **150**, No. 2, p. 387—391 (1988).

C. ELL, J. F. MÜLLER, K. EL SAYED, L. BANYAI, and H. HAUG
*Evaluation of the Hartree-Fock Theory of the Excitonic Optical Stark Effect
(71.35 and 78.20)
(b) Vol. **150**, No. 2, p. 393—399 (1988).

S. A. MOSKALENKO, A. H. ROTARU, and V. A. ZALOI
*Periodic and Stochastic Self-Pulsations in the System of Coherent Excitons, Photons, and Biexcitons
(71.35 and 78.20)
(b) Vol. **150**, No. 2, p. 401—406 (1988).

W. SCHÄFER, K.-H. SCHULDT, and R. BINDER
*Theory of the Optical Stark Effect in Semiconductors
(78.20; 71.35)
(b) Vol. **150**, No. 2, p. 407—412 (1988).

I. BALSLEV and A. STAHL
*Coherent Nonlinear Response of Excitons in Semiconductors
(71.35 and 78.20; S7.12; S7.15)
(b) Vol. **150**, No. 2, p. 413—418 (1988).

J. PULS, W. RUDOLPH, F. HENNEBERGER, and D. LAB
*Femtosecond Studies of Room Temperature Optical Nonlinearities in Wide-Gap II–VI Compounds
(71.35 and 78.20; S8.11; S8.12; S8.15)
(b) Vol. **150**, No. 2, p. 419—423 (1988).

K. HENNEBERGER and F. JAHNKE
*Laser Excited Semiconductors: Light-Induced Gaps and Saturation Kinetics
(71.35 and 78.20; S8.11)
(b) Vol. **150**, No. 2, p. 425—430 (1988).

T. HIGASHIMURA, T. IIDA, and T. KOMATSU
*Optical Stark Effect in an Exciton–Phonon System
(78.20; 71.35)
(b) Vol. **150**, No. 2, p. 431—436 (1988).

R. MICKEVIČIUS and A. REKLAITIS
*Subpicosecond Dynamics of the Nonequilibrium Electron–Phonon System in GaAs
(78.20; S7.12)
(b) Vol. **150**, No. 2, p. 437—441 (1988).

A. L. Ivanov and G. S. Vygovskii
*The Exciton Reflection Shifting and Modification in Presence of Intensive Polariton Wave Interacting with LO-Phonons in Polar Semiconductors
(78.20; 71.35; 71.36; S8.11)
(b) Vol. **150**, No. 2, p. 443—446 (1988).

N. Streibl and M. E. Prise
*Digital Optics: Architecture and Systems Requirements
(78.20)
(b) Vol. **150**, No. 2, p. 447—454 (1988).

G. V. Sinitsyn
*Digital Circuits for All-Optical Computing Based on Thin-Film Semiconductor Interferometers
(78.20; 78.65; S8.1)
(b) Vol. **150**, No. 2, p. 455—464 (1988).

R. Kuszelewicz, J. L. Oudar, R. Azoulay, J. C. Michel, J. Brandon, and O. Emile
*All-Epitaxial GaAs/AlAs Nonlinear Etalons. Towards Continuous and Parallel Operation
(68.55 and 78.20; S7.12; S7.15)
(b) Vol. **150**, No. 2, p. 465—470 (1988).

A. A. Borshch and V. I. Volkov
*Cavity-Less Optical Bistability in Semiconductors
(78.20; S8.11)
(b) Vol. **150**, No. 2, p. 471—475 (1988).

A. Daunois, J.-Y. Bigot, J. Oberle, N. Cherkaoui Eddeqaqi, M. Wegener, A. Witt, and C. Klingshirn
*Spectral Investigation of the Switching Dynamics in CdS Absorptive Bistable Devices
(78.20; S8.11)
(b) Vol. **150**, No. 2, p. 477—481 (1988).

G. Jungk
*Remarks on Optical Nonlinearity and Bistability in Composite Media
(78.20)
(b) Vol. **150**, No. 2, p. 483—488 (1988).

V. A. Kochelap, A. V. Kuznetsov, and V. N. Sokolov
*Resonatorless Dissipative Optical Bistability of Electronic Origin in Direct-Gap Semiconductors
(78.20; S8.11)
(b) Vol. **150**, No. 2, p. 489—493 (1988).

A. Schülzgen, F. Henneberger, and H. Rossmann
*Cavityless Dispersive Bistability and Optical Adressing Using a Nonlinear Prism
(78.20; S8.12)
(b) Vol. **150**, No. 2, p. 495—499 (1988).

S. E. Esipov and V. A. Stadnik
*Optical Bistability and Formation of Localized and Moving High Absorption Domains in ZnSe and ZnCdS
(78.20; S8.12; S8.15)
(b) Vol. **150**, No. 2, p. 501—505 (1988).

S. P. Apanasevich and A. V. Lyakhnovich
*Transverse Effects in Optical Bistability for Transportation of Light Information Signals
(78.20)
(b) Vol. **150**, No. 2, p. 507—512 (1988).

P. I. Khadzhi, G. D. Shibarshina, A. Kh. Rotaru, and Yu. D. Slavov
*The Theory of Nonlinear Slab Multistable Transmission
(78.20)
(b) Vol. **150**, No. 2, p. 513—518 (1988).

H. X. Nguyen and V. D. Egorov
*Spatial Resolution of Dynamic Thermo-Optical Hysteresis in CdS
(78.20; S8.11)
(b) Vol. **150**, No. 2, p. 519—525 (1988).

B. M. Ashkinadze, V. V. Belkov, and A. V. Subashiev
*Optical and Electrical Bistability Induced by Exciton Ionization Processes
(78.20; 71.35; S5.11)
(b) Vol. **150**, No. 2, p. 533—537 (1988).

A. V. Gainer and G. I. Surdutovich
*New Mechanism of Intrinsic Bistability of Nonlinear Slugged Structures under the Action of Repetitive Pulse Excitation
(78.20)
(b) Vol. **150**, No. 2, p. 539—543 (1988).

N. N. Rosanov, A. V. Fedorov, and G. V. Khodova
*Effects of Spatial Distributivity in Semiconductor Optical Bistable Systems
(78.20)
(b) Vol. **150**, No. 2, p. 545—555 (1988).

A. M. Brodin, S. A. Boiko, M. Ya, Valakh, and G. G. Tarasov
*Effect of Optical Feedback on Self-Induced Change of Light Polarization in Cubic Crystals with Directions Instable for Polarization
(78.20; 78.50; S9.11)
(b) Vol. **150**, No. 2, p. 557—560 (1988).

B. M. Nitsovich, O. V. Krechivskii, and G. M. Pestryakov
*Exciton–Phonon Mechanism of Optical Bistability
(71.35 and 78.20)
(b) Vol. **150**, No. 2, p. 561—565 (1988).

B. S. Ryvkin
*Semiconductor Cells with N-Type Photocurrent–Voltage Characteristic: A New Class of Nonlinear Optical Elements
(78.20; 72.40; S7.12)
(b) Vol. **150**, No. 2, p. 567—574 (1988).

F. Šrobár
*Analysis of Absorptive Bistability in a Two-Section Semiconductor Injection Laser
(78.20 and 78.45)
(b) Vol. **150**, No. 2, p. 581—585 (1988).

R. Müller
*Amplitude and Frequency Modulation of Light Pulses in Semiconductor Fabry-Perot Amplifiers
(78.20)
(b) Vol. **150**, No. 2, p. 587—592 (1988).

S. V. Bogdanov and V. G. Lyssenko
*Optoelectronic Bistability with Internal Feedback in Cadmium Sulfide
(78.20; 71.35; S8.11)
(b) Vol. **150**, No. 2, p. 593—598 (1988).

D. M. Butusov, G. G. Gotsadze, V. R. Larionov, and B. S. Ryvkin
*On the Possibility of the Optical Connection between SEED's within One Array
(78.20; 85; S7.12)
(b) Vol. **150**, No. 2, p. 599—603 (1988).

V. L. Velichanskii, A. V. Gainer, V. I. Saptsov, V. A. Sautenkov, V. F. Seregin, V. F. Solodkov, and G. I. Surdutovich
*Quasisoliton Mechanism of Generation of Bistable Injection Laser with External Cavity
(78.20; 78.45)
(b) Vol. **150**, No. 2, p. 605—609 (1988).

V. N. Astratov, A. V. Ilinskii, and A. S. Furman
*Photoinduced Charge Dynamics in Photorefractive Semiconductors
(78.20; 72.40; S10)
(b) Vol. **150**, No. 2, p. 611—615 (1988).

A. Pishchev and N. Kristoffel
*Coherent Second-Order Photoinduced Effects in Noncentrosymmetric Semiconductors
(72.40 and 78.20)
(b) Vol. **150**, No. 2, p. 617—621 (1988).

Xiaobo Zhang and I. Flint
*The Optical Bistability in the Active and Absorptive Tandem Semiconductor Lasers
(78.20)
(b) Vol. **150**, No. 2, p. 623—626 (1988).

A. I. Ekimov and A. L. Efros
*Nonlinear Optics of Semiconductor-Doped Glasses
(71.35 and 78.20; 78.60)
(b) Vol. **150**, No. 2, p. 627—633 (1988).

I. Akai, T. Karasawa, T. Komatsu, and Y. Kaifu
*Large Optical Nonlinearity of the Two-Dimensional Exciton States at a Stacking Fault Interface in BiI_3
(71.35 and 78.20; 71.70; 73.40; S9)
(b) Vol. **150**, No. 2, p. 635—639 (1988).

U. Woggon, F. Henneberger, and M. Müller
*Optical Nonlinearities of Semiconductor Microcrystallites
(78.20; 71.35; S8.1; S9.11)
(b) Vol. **150**, No. 2, p. 641—645 (1988).

V. P. Gribkovskii, V. A. Zyulkov, A. E. Kazachenko, and S. A. Tikhomirov
*Optical Nonlinearity of Semiconductor Microcrystals CdS_xSe_{1-x} under the Action of Picosecond and Nanosecond Laser Pulses
(78.20; S8.15)
(b) Vol. **150**, No. 2, p. 647—652 (1988).

Th. Peschel
*Phenomenological Model for the Optical Properties of Semiconductor-Doped Glasses
(78.20; 71.35; S8.1)
(b) Vol. **150**, No. 2, p. 657—661 (1988).

Zhao Feng, Han Li, Zhang Yun, and Fei Hao Sheng
*Laser-Induced Transient Grating in Semiconductor-Doped Glasses
(78.20; S8.15)
(b) Vol. **150**, No. 2, p. 663—666 (1988).

I. N. Uraltsev, E. L. Ivchenko, P. S. Kopev, V. P. Kochereshko, and D. R. Yakovlev
*Exciton Parameters and Electron Miniband Structure of GaAs/AlGaAs Superlattices
(71.35 and 78.20; 71.25; S7.12; S7.15)
(b) Vol. **150**, No. 2, p. 673—678 (1988).

E. Lach, M. Walther, G. Tränkle, A. Forchel, and G. Weimann
*Optical Investigation of Exciton Screening in GaAs/GaAlAs MQW Structures
(71.35 and 78.20; S7.12; S7.15)
(b) Vol. **150**, No. 2, p. 679—683 (1988).

K. Henneberger and G. Manzke
*Dynamics of Incoherent Excitonic Optical Nonlinearities in GaAs–GaAlAs MQWS's
(71.35 and 78.20; S7.12; S7.15)
(b) Vol. **150**, No. 2, p. 691—693 (1988).

V. K. Kononenko
*Nonlinear Absorption in Quantum-Size Heterostructures
(78.20; S7.12)
(b) Vol. **150**, No. 2, p. 695—698 (1988).

H. J. Eichler, T. Brand, M. Glotz, and B. Smandek
*Optical Nonlinearity and Bistability in Silicon
(78.20; 85; S5.11)
(b) Vol. **150**, No. 2, p. 705—718 (1988).

G. R. Allan, P. L. Chua, J. J. Hunter, H. A. MacKenzie, C. R. Pidgeon, and A. C. Walker
*Dynamical Nonlinearity and Bistability of Narrow Gap Semiconductors
(78.20; S7.13)
(b) Vol. **150**, No. 2, p. 719—727 (1988).

P. P. Paskov, L. I. Pavlov, P. A. Atanasov, D. B. Kushev, and N. N. Zheleva
*Optical Nonlinearity and Fast Switching in PbSnTe Fabry-Perot Etalon
(78.20; S8.15)
(b) Vol. **150**, No. 2, p. 729—733 (1988).

I. Rückmann, J. Kolenda, F. Henneberger, and M. Petrauskas
*ps-Laser Induced Transmission and Absorption in GaInAsP Epitaxial Layers
(78.20; 78.65; S7.15)
(b) Vol. **150**, No. 2, p. 735—742 (1988).

A. Galeckas, V. Netiksis, M. Petrauskas, and J. Vaitkus
*Reflectivity and Dynamic Gratings in Implanted Si Induced by Picosecond Laser Pulses
(78.20; 61.80; S5.11)
(b) Vol. **150**, No. 2, p. 743—748 (1988).

H.-E. Swoboda, F. A. Majumder, R. Renner, C. Weber, M. Sence, Lu Jie, G. Noll, E. O. Göbel, J. Vaitkus, and C. Klingshirn
*Nonlinear Optical Properties of the System $CdS_{1-x}Se_x$
(78.20; 78.55; S8.15)
(b) Vol. **150**, No. 2, p. 749—759 (1988).

F. Yang, G. Xiong, and X. Xu
*Mechanism of Optical Nonlinearity in CdS Due to Increasing Absorption at Room Temperature
(78.20; S8.11)
(b) Vol. **150**, No. 2, p. 765—768 (1988).

Ch. Spiegelberg, M. Kretzschmar, J. Puls, and F. Henneberger
*Strong Dispersive Nonlinearity in the Excitonic Region of Wide-Gap II–VI Compounds at Room Temperature
(71.35 and 78.20; S8.11)
(b) Vol. **150**, No. 2, p. 769—775 (1988).

I. Jánossy
*Nonlinear Optical Effects in Amorphous Semiconductors
(78.20; 78.65; S8)
(b) Vol. **150**, No. 2, p. 783—789 (1988).

F. V. Karpushko
*Fast and Strong Nonlinearity of Thin-Film Semiconductor Interferometers Due to Intrinsic Photo- and Electrooptical Effects of the Semiconductor Interlayer
(73.40 and 78.20; S7.12; S8.12)
(b) Vol. **150**, No. 2, p. 791—796 (1988).

Lun-Biao Xu, Shou-Fu Cai, and Guo-Xin Xie
*Quantum-Statistical Investigation on the Nonlinear Optical Oscillation in Amorphous Semiconductor $GeSe_2$ Films
(78.20 and 78.65; 71.45; S8)
(b) Vol. **150**, No. 2, p. 797—803 (1988).

K. Ferencz and R. Szipöcs
*Nonlinear Optical Behaviour of Absorbing CdS_xSe_{1-x} Interference Filters
(78.20; S8.15)
(b) Vol. **150**, No. 2, p. 811—815 (1988).

N. Nagasawa
*Optical Nonlinearity Due to the Excitonic System in CuCl
(71.35 and 78.20; S9)
(b) Vol. **150**, No. 2, p. 817—823 (1988).

R. Lévy, B. Hönerlage, and J. B. Grun
*Time-Resolved Light-Induced Grating Spectroscopy in Semiconductors
(78.20; S9)
(b) Vol. **150**, No. 2, p. 825—832 (1988).

T. Hönig and J. Gutowski
*Optical Nonlinearity and Bistability in the Bound Exciton Energy Range of CdS
(78.20; S8.11)
(b) Vol. **150**, No. 2, p. 833—838 (1988).

V. S. Dneprovskii, A. I. Furtichev, V. I. Klimov, Shen Li, E. V. Nazvanova, D. K. Okorokov, and U. V. Vandishev
*Nonlinear Optical Properties of CdS and Optical Bistability
(78.20; S8.11)
(b) Vol. **150**, No. 2, p. 839—844 (1988).

Nguyen Ba An
Exciton Nonequilibrium Steady States in Laser-Excited Molecular Crystals
(71.35; 78.20)
(b) Vol. **150**, No. 2, p. 845—850 (1988).

W. J. Tomlinson
*Nonlinear Phenomena in Single-Mode Optical Fibers
(78.20; 78.30; 78.35)
(b) Vol. **150**, No. 2, p. 851—862 (1988).

R. P. Seisyan, A. A. Toropov, V. P. Khvostikov, and T. V. Shubina
*Excitonic Electroabsorption in GaAs–GaAlAs Single Quantum Well Waveguide Structures
(71.35 and 78.20; S7.12; S7.15)
(b) Vol. **150**, No. 2, p. 863—866 (1988).

U. Trutschel, M. Golz, and M. Abraham
*Optical Bistability in Two-Dimensional Waveguide Structures
(78.20; S7.13)
(b) Vol. **150**, No. 2, p. 867—871 (1988).

P. Dannberg, T. Possner, A. Bräuer, and U. Bartuch
*Experimental Investigations of Optical Nonlinearities in Semiconductor-Doped Glass Waveguides
(78.20)
(b) Vol. **150**, No. 2, p. 873—877 (1988).

K. Jarašiūnas and J. Vaitkus
*Transient Gratings in Metrology of Semiconductor Parameters and Optoelectronic Devices
(78.20; S5.11; S7.12; S7.13; S8.12)
(b) Vol. **150**, No. 2, p. 879—884 (1988).

A. A. Bugaev and B. P. Zakharchenya
*Picosecond Holographic Diagnostic of Light Pulse Absorption in Semiconductors
(72.30 and 78.20; 71.38; S5.11; S7.12)
(b) Vol. **150**, No. 2, p. 891—894 (1988).

H. Kalt, A. L. Smirl, and T. F. Boggess
*Picosecond Optical Nonlinearities and Stimulated Emission in Indirect $Al_xGa_{1-x}As$
(78.20 and 78.45; 78.55; S7.15)
(b) Vol. **150**, No. 2, p. 895—900 (1988).

M. P. Lisitsa, M. Ya. Valakh, G. G. Tarasov, S. A. Boiko, A. M. Brodin, A. M. Shpak, R. Voska, and I. Feldvari
*Nonlinear Polarization Spectroscopy of Impure Cubic Crystals
(78.20; 78.50; S9.11)
(b) Vol. **150**, No. 2, p. 907—911 (1988).

V. A. Ganshin, Yu. N. Korkishko, and T. V. Morozova
*Properties of Proton Exchanged Optical Waveguiding Layers in $LiNbO_3$ and $LiTaO_3$
(61.70 and 78.20; 61.10; S11.1)
(a) Vol. **110**, No. 2, p. 397—402 (1988).

S. M. Wasim and J. G. Albornóz
*Electrical and Optical Properties of n- and p-Type $CuInTe_2$
(72.20 and 78.20; 71.55; S8.16)
(a) Vol. **110**, No. 2, p. 575—583 (1988).

G. A. Medvedkin, Yu. V. Rud, and M. A. Tairov
*Fundamental Optical Absorption Edge in $MnGa_2Te_4$ Single Crystals
(78.20; 72.40; S8.16)
(a) Vol. **110**, No. 2, p. 631—643 (1988).

A. Ghosh
*Fundamental Absorption Edge in Bismuth–Vanadate Glasses
(78.20; S10.15)
(a) Vol. **110**, No. 2, p. 651—656 (1988).

78.30. Infrared and Raman spectra and scattering

I. V. Bodnar, G. F. Smirnova, T. V. Smirnova, Yu. A. Aleshchenko, and L. K. Vodopyanov
Lattice Vibration of $CuGaSe_2$, $CuInSe_2$ Ternary Compounds and $CuGa_xIn_{1-x}Se_2$ Solid Solutions
(63.20; 78.30; S8.16)
(b) Vol. **145**, No. 1, p. 117—122 (1988).

Y. Q. Sheng, E. Recknagel, A. Weidinger, Z. G. Gu, Z. Y. Lai, and Z. C. Zhuang
Raman Scattering at the CaF_2 Crystal Surface Excited by an Evanescent Optical Wave Field
(71.36 and 73.20; 78.30; S9.12)
(b) Vol. **145**, No. 1, p. 151—156 (1988).

Yi-Yang Zhou
Theoretical Study of the Optical Absorption, Mössbauer, and EPR Spectra of Fe^{2+} Ion in Orthopyroxene
(71.70; 76.30; 76.80; 78.30; S11)
(b) Vol. **145**, No. 1, p. 227—232 (1988).

S. Mochizuki
*Infrared Absorption in Nickel Oxide Microcrystals
(78.30; S10.1)
(b) Vol. **145**, No. 1, p. K75—K78 (1988).

G. Irmer, M. Herms, J. Monecke, and H. Bock
*Raman Scattering from LO-Phonon-Plasmon Modes in $GaAs_{1-x}P_x$ Crytals
(63.20 and 78.30; 71.45; S7.15)
(b) Vol. **145**, No. 1, p. K79—K82 (1988).

J. Tatarkiewicz and A M. Witowski
Hydrogen Vibrations in ZnTe
(61.70 and 63.20; 78.30; S8.13)
(b) Vol. **145**, No. 2, p. K95—K98 (1988).

T. D. Ibragimov and N. T. Mamedov
Two-Phonon Absorption Spectra and Phase Transition in $TlGaSe_2$
(63.20 and 64.70; 78.30; S8.16)
(b) Vol. **145**, No. 2, p. K103—K105 (1988).

V. M. Yartsev and C. S. Jacobsen
*Infrared Properties of the Organic Semiconductor $MEM(TCNQ)_2$ for Polarization Perpendicular to the TCNQ Stacks
(78.30; S12)
(b) Vol. **145**, No. 2, p. K149—K153 (1988).

G. Irmer and W. Siegel
Carrier Concentration Profiles across GaP Grain Boundaries Observed by Raman Measurements
(61.70 and 72.20; 78.30; S7.11)
(a) Vol. **105**, No. 2, p. 549—553 (1988).

M. A. Hassan, M. N. Khan, and C. A. Hogarth
Electrical and Optical Properties of V_2O_5–GeO_2–NiO Glasses
(72.20 and 78.20; 78.30; S10.15)
(a) Vol. **105**, No. 2, p. 609—616 (1988).

M. Salagram
*Infrared Spectrum of VO^{2+} Entity in $K_2C_2O_4 \cdot H_2O$ Crystals
(78.30; S11)
(a) Vol. **105**, No. 2, p. K161—K164 (1988).

S. R. Wilk and F. Lüty
*Spectroscopic Determination of the Hindered Rotational Behavior of CN^- Defects in Alkali Halides
(78.30; 78.50; S9.11)
(b) Vol. **146**, No. 1, p. 303—310 (1988).

A. I. Bobrysheva, S. S. Russu, and V. A. Zaloi
*Light Absorption and Hyper-Raman Scattering by Many-Electron Complexes in Quantum Well Structures
(78.20 and 78.30; 71.35; S7.15)
(b) Vol. **146**, No. 1, p. 329—334 (1988).

L. Jansa, R. Novotný, P. Lošťák, and J. Klikorka
*Optical Properties of n-Type $Bi_{2-x}In_xTe_3$ Single Crystals
(78.30; S8.16)
(b) Vol. **146**, No. 1, p. K75—K80 (1988).

G. Pastor, P. Tejedor, I. Jiménez, E. Domínguez, M. Torres, and J. V. García-Ramos
Low Pressure Chemical Vapour Deposition Amorphous Silicon Behaviour under Annealing
(61.40 and 68.55; 78.30; S5.11)
(a) Vol. **106**, No. 1, p. 11—16 (1988).

M. N. Khan, M. A. Hassan, and C. A. Hogarth
The Electronic and Optical Properties of Germanium Tellurite Glasses Containing Various Transition Metal Oxides
(72.80; 76.30; 78.30; S10.15)
(a) Vol. **106**, No. 1, p. 191—200 (1988).

I. I. Reshina
*Raman Study of Disorder in Laser-Annealed III–V Semiconductors
(78.30; 63.50; S7.11; S7.12)
(a) Vol. **106**, No. 1, p. 261—270 (1988).

P. P. Lottici
A MREI Model for the Zincblende-Like Phonons in the Mixed Defect Chalcopyrite $ZnGa_2(S_xSe_{1-x})_4$
(63.20; 78.30; S8.16)
(b) Vol. **146**, No. 2, p. 503—510 (1988).

J. Petzelt, S. Kamba, S. Pačesová, E. Pollert, J. Šrámek, O. Smrčková, and D. Sýkorová
*Infrared Reflectivity of High-T_c $YBa_2Cu_3O_{7-x}$ and Related Ceramics
(63.20 and 78.30; 74.40; S10.15)
(b) Vol. **146**, No. 2, p. 743—755 (1988).

A. ANEDDA, G. BONGIOVANNI, and E. FORTIN
*Raman Scattering from Vibrational Modes in Layered $Cd_xHg_{1-x}I_2$ Solid Solution
(78.30; 63.20; S9.15)
(b) Vol. **146**, No. 2, p. 757—765 (1988).

E. JAHNE, O. GOEDE, and V. WEINHOLD
*Far Infrared Reflectivity of Tetrahedrally and Octahedrally Coordinated MnS
(63.20 and 78.30; S8.11)
(b) Vol. **146**, No. 2, p. K157—K160 (1988).

B. H. BAIRAMOV, V. V. TOPOROV, N. V. AGRINSKAYA, E. A. SAMEDOV, G. IRMER, and J. MONECKE
*Raman Scattering by Coupled LO-Phonon-Plasmon Modes in CdTe
(63.20 and 78.30; S8.13)
(b) Vol. **146**, No. 2, p. K161—K165 (1988).

P. LOŠTÁK, R. NOVOTNÝ, J. NAVRÁTIL, and J. ŠRÁMKOVÁ
Optical Properties of Gallium-Doped Bi_2Te_3 Single Crystals
(78.20; 78.30; S8)
(a) Vol. **106**, No. 2, p. 619—625 (1988).

M. SALAGRAM, K. SEETHARAM, P. SIDDAMBARY, and T. S. N. MURTHY
*Infrared Characterisation of the Oxyanion in Kainite
(78.30; S11.1)
(a) Vol. **106**, No. 2, p. K185—K190 (1988).

C. ZALDO and E. OROZCO
*Raman Spectroscopy of Sr Phases in Potassium Halides
(78.30; S9.11)
(b) Vol. **147**, No. 1, p. 431—440 (1988).

M. D. FONTANA, A. RIDAH, and G. E. KUGEL
*Frequency Dependence of the Soft Mode in $KNbO_3$
(77.80 and 78.30; 63.20; S11.1)
(b) Vol. **147**, No. 1, p. 441—447 (1988).

D. K. PAPADEMITRIOU
*Scattering of X-Rays by Plasmons in Boron
(71.45 and 78.30; S5)
(b) Vol. **147**, No. 1, p. K95—K98 (1988).

A. HENRY, K. SAMINADAYAR, J. L. PAUTRAT, and N. MAGNEA
The Kinetics of the Formation of the Thermal Donors in Silicon. Effect of Various Parameters
(61.70; 72.80; 78.30; S5.11)
(a) Vol. **107**, No. 1, p. 101—110 (1988).

S. E. SARKISOV and A. A. KAMINSKII
*Optical Phonon Spectroscopy of Heterovalent Disordered $Ca_3(Nb, Ga)_2 \cdot Ga_3O_{12}$ Crystals with Garnet Structure
(78.30; S11.2)
(a) Vol. **107**, No. 1, p. 365—371 (1988).

O. V. Kulikova, L. L. Kulyuk, S. I. Radautsan, S. A. Ratseev, E. E. Strumban, V. E. Tezlevan, and V. I. Tsitsanu
*Influence of Defect Generation Processes in $CdIn_2S_4$ Single Crystals on the Photoluminescence and Raman Scattering Spectra
(78.30 and 78.55; S8.16)
(a) Vol. **107**, No. 1, p. 373—377 (1988).

A. Röseler
*Measurement of Surface Polaritons by Spectroscopic Infrared Ellipsometry
(78.30)
(a) Vol. **107**, No. 1, p. K69—K73 (1988).

G. Eidžiūnas, J. Kavaliauskas, G. Krivaitė, and A. Šileika
*Plasma Reflectivity Spectra of Inhomogeneous $Cd_{0.2}Hg_{0.8}Te$ Crystals
(78.30; S8.15)
(a) Vol. **107**, No. 1, p. K75—K78 (1988).

J. Giordano et J. L. Sauvajol
*Etude de l'évolution avec la température des modes actifs en diffusion Raman du ZnF_2 (Study of the Temperature Dependence of the Raman Active Modes of ZnF_2)
(63.20 and 78.30; S9)
(b) Vol. **147**, No. 2, p. 537—545 (1988).

I. Ya. Fugol, A. M. Ratner, L. V. Khashchina, and I. M. Pritula
*Infrared Absorption and Anharmonicity of Crystalline Nitrogen
(78.30; 63.20; S5.2)
(b) Vol. **147**, No. 2, p. 757—766 (1988).

W. Lu, H. J. Ye, Z. Y. Yu, S. Y. Zhang, Y. Fu, W. L. Xu, S. C. Shen, and W. Giriat
*Local and Quasi-Local Modes of Fe and Zn in CdTe
(78.30; 63.20; S8.15)
(b) Vol. **147**, No. 2, p. 767—778 (1988).

M. P. dos Santos and C. Hirlimann
*Impurity Effects in the Raman and Luminescence Spectra of Nitrogen-Doped GaP
(78.30 and 78.55; S7.11)
(b) Vol. **147**, No. 2, p. 779—789 (1988).

W. S. Tse, C. S. Fang, K. F. Pai, and W. Lih
*A Raman Investigation of the Polymorphism of Potassium Nitrite
(78.30; 63.20; S11)
(b) Vol. **147**, No. 2, p. 791—795 (1988).

M. H. Kuok and S. H. Tang
*A Raman Study of $KCdBr_3$ Single Crystals
(78.30; 63.20; S9.16)
(b) Vol. **147**, No. 2, p. K195—K199 (1988).

A. N. Veis, D. D. Koditsa, and N. S. Popovich
*Optical Properties of $TlBiS_2$ Monocrystals
(78.30; S8.16)
(a) Vol. **107**, No. 2, p. K169—K173 (1988).

A. Lehmann
*Theory of Infrared Transmission Spectra of Thin Insulating Films
(78.30 and 78.65)
(b) Vol. **148**, No. 1, p. 401—405 (1988).

T. D. Ibragimov and N. T. Mamedov
*Two-Phonon Absorption in $TlInTe_2$ and $TlGaTe_2$
(63.20 and 78.30; S8.16)
(b) Vol. **148**, No. 1, p. K89—K92 (1988).

K. R. Allakhverdiev, N. D. Akhmedzade, M. M. Tagiev, M. M. Shirnov, and S. Häseler
*Long-Wavelegth IR-Active Phonons in the $TlInS_2$–$TlInSe_2$ System
(63.20 and 78.30; S8.16)
(b) Vol. **148**, No. 1, p. K93—K96 (1988).

S. Iselborn, B. Schröder, and J. Geiger
The Influence of Hydrogen Surface Reactions on the Growth of Evaporated a-Si:H Films
(68.55; 78.30; S5.11)
(a) Vol. **108**, No. 1, p. 275—284 (1988).

J. Otto
Stress-Induced Dichroism of the Vibrational Lines and Phonon Sidebands of SH^- Centres in CsBr
(63.20; 78.30; S9.11)
(b) Vol. **148**, No. 2, p. 489—499 (1988).

R. Riera, F. Comas, C. Trallero Giner, and S. T. Pavlov
*Electron Raman Scattering in Semiconductor Quantum Wells
(71.25 and 78.30; S7.12; S7.15)
(b) Vol. **148**, No. 2, p. 533—542 (1988).

T. S. Shi, Y. G. Zhao, P. X. Cai, H. F. Gu, and L. M. Xie
*IR Absorption Peaks and Oxygen Vacancies in $YBa_2Cu_3O_x$
(61.70 and 78.30; 78.50; S10)
(b) Vol. **148**, No. 2, p. 715—721 (1988).

V. V. Frolov and B. N. Mukashev
On the Origin of the Hydrogen-Related Complex Corresponding to the 2223 cm^{-1} (at 10 K) Infrared Band in c-Si:H)
(61.70; 78.30; S5.11)
(b) Vol. **148**, No. 2, p. K105—K107 (1988).

H. L. Park, H. K. Kim, C. H. Chung, Chang-Dae Kim, Tong-San Cho, and Wha-Tek Kim
*Raman Scattering of $CdGa_2Se_4$ and Its Space Group
(78.30; S8.16)
(b) Vol. **148**, No. 2, p. K173—K176 (1988).

H. D. HUSSEINOV, SH. N. YURAEV, S. H. HUSSEINOV, and S. S. ABDINBEKOV
*Some Features of IR Reflection Spectra in $Tl(InS_2)_{1-x}(FeSe_2)_x$ Single Crystals ($0 \leqq x \leqq 0.015$) near the Phase Transition Temperature
(78.30; 78.50; S8, S8.15)
(b) Vol. **148**, No. 2, p. K177—K182 (1988).

FANGQING ZHANG, GUOSHENG SUN, and GUANGHUA CHEN
*ESR and IR Studies on P-Doped a-Si:H/a-Si:H Junctions
(76.30 and 78.30; S5.11)
(a) Vol. **108**, No. 2, p. K139—K143 (1988).

M. PÜTZLER and F. FISCHER
*Magnetic Dipole Absorption of Colloidal Ag in Single Crystals of KCl
(78.30; S9.11)
(b) Vol. **149**, No. 1, p. 371—381 (1988).

A. V. SHERMAN
*A Recursion Method for Calculating Many-Particle Green's Functions. Resonant Secondary Radiation and Excitation Spectra in the Exciton Region
(71.35 and 78.30)
(b) Vol. **149**, No. 2, p. 725—737 (1988).

K. KUMAZAKI and K. IMAI
*Far-Infrared Reflection and Raman Scattering Spectra in γ-InSe
(63.20 and 78.30; S8)
(b) Vol. **149**, No. 2, p. K183—K186 (1988).

V. S. JAYAKUMAR, G. SEKAR, P. RAJAGOPAL, and G. ARULDHAS
*IR and Polarized Raman Spectra of $(NH_4)_2Mg(SO_4)_2 \cdot 6\,H_2O$
(78.30; S11)
(a) Vol. **109**, No. 2, p. 635—640 (1988).

A. DE ANDRÉS
Applications of a Point Polarizable Ion Model to Suzuki Phase Precipitates in Alkali Halides
(63.20; 61.70; 78.30; S9.11)
(b) Vol. **150**, No. 1, p. 31—37 (1988).

S. ZEKENG, B. PREVOT, and C. SCHWAB
*Raman Determination of the Faust-Henry Coefficient of GaAs in the 1.9 to 2.7 eV Range at Ordinary and Low Temperatures
(63.20 and 78.30; S7.12)
(b) Vol. **150**, No. 1, p. 65—72 (1988).

A. A. MAKISOMOV, I. I. TARTAKOVSKII, and T. A. KRIVENKO
Determination of Polariton–Phonon Interaction Parameters in Anthracene Crystals
(71.36; 78.30; 78.55; S12)
(b) Vol. **150**, No. 1, p. 125—132 (1988).

E. Z. Katsnelson, A. G. Karoza, and L. A. Meleshchenko
*Photoinduced Magnetization Effect on IR Spectra of Ferrites
(75.50 and 78.30; S11.2)
(a) Vol. **110**, No. 1, p. K51—K56 (1988).

W. J. Tomlinson
Nonlinear Phenomena in Single-Mode Optical Fibers
(78.20; 78.30; 78.35)
(b) Vol. **150**, No. 2, p. 851—862 (1988).

A. K. Garg
High-Pressure Raman Spectroscopic Study of the Ice Ih → Ice IX Phase Transition
(64.70; 62.50; 78.30; S10)
(a) Vol. **110**, No. 2, p. 467—480 (1988).

Er-wang Mao, Wen-qin Zhao, Hong-rui Zhang, Ai-zhen Li, Jian-min Chen, and Guo-ping Fang
The Influence of Strain and Dislocations on Transport Properties of GaAs/Si Strained-Layer Heterojunctions
(73.40; 78.30; S5.11; S7.12)
(a) Vol. **110**, No. 2, p. 515—520 (1988).

M. G. Kaplunov, N. D. Kushch, and E. B. Yagubskii
*Optical Properties of the New Organic Superconductor $(BEDT\text{-}TTF)_2$ $Cu(SCN)_2$
(78.30; 71.38; S12)
(a) Vol. **110**, No. 2, p. K111—K116 (1988).

78.35. Brillouin and Rayleigh scattering

W. J. Tomlinson
Nonlinear Phenomena in Single-Mode Optical Fibers
(78.20; 78.30; 78.35)
(b) Vol. **150**, No. 2, p. 851—862 (1988).

V. V. Gorbachev, V. M. Durasov, R. B. Zezin, E. V. Ivakin, A. S. Rubanov, and N. A. Tatyanina
*Light-Induced Thermal Gratings in Natural Diamond
(66.70 and 78.35; S6)
(b) Vol. **150**, No. 2, p. 901—905 (1988).

78.40. Visible and ultraviolet spectra

K. Takarabe
Optical Properties of InS under High Pressure
(71.25; 78.40; S8)
(b) Vol. **145**, No. 1, p. 219—225 (1988).

A. V. Sherman
*Temperature Dependence of Exciton Absorption Spectra. A Calculation Based on the Recursion Method
(71.35 and 78.40)
(b) Vol. **145**, No. 1, p. 319—332 (1988).

T. Tsuboi
*Energy Level Discreteness of CuCl Electrons Confined to a Three-Dimensional Potential Well
(71.35 and 78.40; S9; S9.12)
(b) Vol. **146**, No. 1, p. K11—K15 (1988).

O. Goede, W. Heimbrodt, M. Lamla, and V. Weinhold
*Energy Gap of MnS
(78.40 and 78.55; 71.35; S8.11)
(b) Vol. **146**, No. 1, p. K65—K69 (1988).

B. T. Deshmukh, S. V. Bodade, and S. V. Moharil
Thermoluminescence in Synthetic Langbeinite
(78.60; 78.40; S11)
(a) Vol. **106**, No. 1, p. K107—K111 (1988).

M. P. Lisitsa, I. D. Turyanitsa, and I. V. Fekeshgazi
*Non-Linear Light Absorption Spectroscopy in Sb–S–I Crystals and Glasses
(78.40; S11)
(b) Vol. **146**, No. 2, p. 723—726 (1988).

N. A. Moldovyan, S. I. Radautsan, V. F. Zhitar, E. D. Arama, and D. S. Remenko
*Direct Optical Transitions in $CdGaInS_4$
(78.40; S8.16)
(a) Vol. **106**, No. 2, p. K181—K184 (1988).

A. A. A. Hady
*Optical Study of $Hg_{1-x}Cd_xTe$ Graded Gap Material
(78.40; S8.15)
(a) Vol. **106**, No. 2, p. K191—K195 (1988).

K. R. Allakhverdiev, N. A. Bakhyshov, S. S. Guseinov, T. G. Mamedov, M. A. Nizametdinova, and I. K. Efendieva
*Band Gap Variation in Mixed Crystals with TlSe-Type Crystal Structure
(71.25 and 78.40; S8)
(b) Vol. **147**, No. 1, p. K99—K104 (1988).

J. Lipavičius, E. Čijauskas, and A. Audzijonis
*The Fundamental Absorption Edge Tail of $Sn_2P_2Se_6$
(78.40; S8.16)
(b) Vol. **148**, No. 1, p. K97—K100 (1988).

J. A. de Saja and K. Tanaka
Photoelectrochemical Cells with p-Type Poly(3-Methylthiophene)
(73.40; 72.40; 78.40; S12)
(a) Vol. **108**, No. 2, p. K109—K114 (1988).

V. V. Arutunyan, G. N. Eritsyan, R. K. Ezoyan, and V. A. Gevorkyan
*On the Mechanism of Radiation Induced Change in the Transparence of Corundum Single Crystals
(78.40; S10.1)
(b) Vol. **149**, No. 1, p. K77—K80 (1988).

A. A. Agasiev, V. E. Bagiev, A. M. Mamedov, and Ya. Yu. Guseinov
*VUV Reflection Spectra and Electron Structure of Bi_2O_3
(78.40; S10.1)
(b) Vol. **149**, No. 2, p. K191—K195 (1988).

N. A. Moldovyan, S. I. Radautsan, and V. Z. Chebotaru
*Photoconductivity, Cathodoluminescence, and Optical Absorption of $CdIn_2S_2Se_2$ Single Crystals
(72.40 and 78.40; 78.60; S8.16)
(a) Vol. **110**, No. 1, p. 277—282 (1988).

78.45. Stimulated emission

G. Beister, J. Maege, and G. Richter
Identification of Surface Influences on the Current–Voltage Characteristic of Narrow Oxide Stripe Laser Diodes
(73.25; 78.45; S7.15)
(a) Vol. **105**, No. 1, p. K45—K49 (1988).

A. A. Kaminskii, K. Kurbanov, K. L. Ovanesyan, and A. G. Petrosyan
*Stimulated Emission Spectroscopy of Pr^{3+} Ions in Orthorhombic $YAlO_3$ Single Crystals
(78.45; 78.55; S10)
(a) Vol. **105**, No. 2, p. K155—K159 (1988).

B. Rheinländer, R. Heilmann, and G. Oelgart
*Determination of Temperature-Dependent Carrier Losses in 1.3 μm InGaAsP/InP Double Heterostructures
(78.45 and 85; 73.40; S7.11; S7.16)
(a) Vol. **106**, No. 1, p. K113—K116 (1988).

T. Hönig and J. Gutowski
Stimulated Emission of the I_1–TA(P_M) Band In CdS
(71.35 and 71.55; 78.45; S8.11)
(b) Vol. **146**, No. 2, p. 589—595 (1988).

N. P. Bezhan, O. V. Kulikova, L. L. Kulyuk, S. I. Radautsan, A. V. Siminel, E. E. Strumban, L. I. Shulga, and V. A. Chumak
*On the Influence of a Buffer Layer on the Emissive Properties of Epitaxial Layers in InGaAsP (λ = 1.17 μm)
(78.45 and 78.65; S8.15)
(a) Vol. **109**, No. 2, p. K153—K156 (1988).

E. Schöll and M. Schell
*Theory of Ultrashort Pulse Generation and Amplification by Gain-Switched Semiconductor Laser
(78.45)
(b) Vol. **150**, No. 2, p. 575—579 (1988).

F. Šrobár
*Analysis of Absorptive Bistability in a Two-Section Semiconductor Injection Laser
(78.20 and 78.45)
(b) Vol. **150**, No. 2, p. 581—585 (1988).

V. L. Velichanskii, A. V. Gainer, V. I. Saptsov, V. A. Sautenkov, V. F. Seregin, V. F. Solodkov, and G. I. Surdutovich
Quasisoliton Mechanism of Generation of Bistable Injection Laser with External Cavity
(78.20; 78.45)
(b) Vol. **150**, No. 2, p. 605—609 (1988).

H. Kalt, A. L. Smirl, and T. F. Boggess
*Picosecond Optical Nonlinearities and Stimulated Emission in Indirect $Al_xGa_{1-x}As$
(78.20 and 78.45; 78.55; S7.15)
(b) Vol. **150**, No. 2, p. 895—900 (1988).

78.50. Impurity and defect absorption in solids

H. DONKER, W. M. A. SMIT, and G. BLASSE
On the Luminescence of CaO:Sn^{2+}
(78.55; 78.50; S10.1)
(b) Vol. **145**, No. 1, p. 333—342 (1988).

A. DE and K. V. RAO
*Optical Absorption and Thermoluminescence of CaF_2 Single Crystals Subjected to High Electric Fields and Later Irradiated with X-Rays
(78.50; 78.60; S9.12)
(a) Vol. **105**, No. 1, p. 297—302 (1988).

B. S. V. S. R. ACHARYULU and V. S. KISHAN KUMAR
*Optical and Thermoluminescence Studies on Thallium-Doped Suprapure Cesium Sulphate
(78.50 and 78.60; S11.1)
(a) Vol. **105**, No. 1, p. 303—307 (1988).

I. K. REBANE
*Theory of Two-Step Spectral Hole Burning by Pulses
(78.50)
(b) Vol. **145**, No. 2, p. 749—757 (1988).

C. VIJAYAN and Y. V. G. S. MURTI
Ionic Thermocurrents Due to Z-Centres in KCl:Gd^{2+}
(61.70; 77.30; 77.40; 78.50; S9.11)
(a) Vol. **105**, No. 2, p. 397—402 (1988).

R. FORNARI and L. DOZSA
Deep Level Characterization in Semi-Insulating LEC Gallium Arsenide
(71.55; 72.40; 78.50; S7.12)
(a) Vol. **105**, No. 2, p. 521—530 (1988).

F. R. KESSLER, U. BARKOW, R. NIES, and N. UNZNER
Optical Profile Analysis of Diffused and Ion-Implanted Silicon
(78.20; 61.70; 66.30; 78.50; S5.11)
(a) Vol. **105**, No. 2, p. 627—635 (1988).

D. NEOGY and T. PUROHIT
The Behaviour of Active Centers in a Laser Host. A Crystal Field Investigation on Dy^{3+} in LaF_3 Single Crystal
(71.70; 78.50; S9)
(b) Vol. **146**, No. 1, p. 181—187 (1988).

S. R. WILK and F. LÜTY
Spectroscopic Determination of the Hindered Rotational Behaviour of CN^- Defects in Alkali Halides
(78.30; 78.50; S9.11)
(b) Vol. **146**, No. 1, p. 303—310 (1988).

S. V. MELNICHUK, V. M. KRAMAR, and K. D. TOVSTYUK
*Vibronic Interaction in the Absorption Spectra of Vanadium-Doped $A^{II}B^{VI}$ Semiconductors
(71.70 and 78.50; S8.1)
(b) Vol. **146**, No. 2, p. 613—617 (1988).

A. A. Alybakov, V. A. Gubanova, K. Kudabaev, and K. Sharsheev
*Optical Absorption and EPR Spectra of Cu^{2+} Ions in $LiKSO_4$ Single Crystals
(76.30 and 78.50; S11.1)
(b) Vol. **146**, No. 2, p. K135—K139 (1988).

N. B. Manson and T. J. C. Hosea
*Anomalous Temperature Dependence of the Absorption Strength of $SrO:Ni^{2+}$ and $SrO:Co^{2+}$
(78.50; S10.1)
(b) Vol. **146**, No. 2, p. K153—K156 (1988).

S. G. Lyapin, M. I. Eremets, A. M. Shirokov, R. Herrmann, H.-U. Müller, and F. Ludwig
*Submillimetre Magnetospectroscopy of Tellurium under Hydrostatic Pressure
(71.55 and 78.50; 72.40; 76.40; S5)
(b) Vol. **147**, No. 1, p. 297—306 (1988).

G. Dlubek, A. Dlubek, R. Krause, and O. Brümmer
Point Defects in GaAs Studied by Correlated Positron Lifetime, Optical, and Electrical Measurements. II. Point Defects in GaAs Irradiated with Fast Neutrons
(61.80; 78.50; 78.55; 78.70; S7.12)
(a) Vol. **107**, No. 1, p. 111—121 (1988).

D. Wruck and A. Knauer
Doping Inhomogeneities and Compensation in n-Type LEC InP Wafers
(72.20; 78.50; S7.11)
(a) Vol. **107**, No. 1, p. 321—328 (1988).

B. M. Anghelov
*Spectroscopic Properties of $LiNdP_4S_{12}$
(71.70 and 78.50; S10)
(b) Vol. **148**, No. 1, p. K83—K88 (1988).

T. R. Volk and N. M. Rubinina
*X-Ray and UV Influence on the Optical Absorption Spectra of the Non-Photorefractive Lithium Niobate
(78.50; S11.1)
(a) Vol. **108**, No. 1, p. 437—442 (1988).

T. S. Shi, Y. G. Zhao, P. X. Cai, H. F. Gu, and L. M. Xie
IR Absorption Peaks and Oxygen Vacancies in $YBa_2Cu_3O_x$
(61.70 and 78.30; 78.50; S10)
(b) Vol. **148**, No. 2, p. 715—721 (1988).

J. L. Rao, R. M. Krishna, and S. V. J. Lakshman
Crystal Field Energy Levels, EPR Parameters, and Magnetic Susceptibility of the Co^{2+} Ion in Cs_3ZnCl_5
(71.70; 76.30; 78.50; S9.12)
(b) Vol. **148**, No. 2, p. K119—K123 (1988).

H. D. Husseinov, Sh. N. Yuraev, S. H. Husseinov, and S. S. Abdinbekov
Some Features of IR Reflection Spectra in $Tl(InS_2)_{1-x}(FeSe_2)_x$ Single Crystals ($0 \leqq x \leqq 0.015$) near the Phase Transition Temperature
(78.30; 78.50; S8; S8.15)
(b) Vol. **148**, No. 2, p. K177—K182 (1988).

X. L. YUAN and S. W. S. MCKEEVER
Impurity Clustering Effects in Magnesium-Doped Lithium Fluoride
(61.70; 78.50; 78.60; S9.11)
(a) Vol. **108**, No. 2, p. 545—551 (1988).

W. WESCH, E. WENDLER, N. P. KEKELIDSE, and V. V. TARBA
Investigation of the Amorphization Process in GaAs and GaP by Means of Optical Spectroscopy
(61.80; 78.50; S7.11; S7.12)
(a) Vol. **108**, No. 2, p. K93—K96 (1988).

B. NARASHIMHA, K. V. RAO, and R. N. P. CHAUDHURY
*Optical Absorption of Quenched and γ-Ray Irradiated BaF_2 Single Crystals Later Subjected to High Electric Fields of Laser Excitation
(61.70 and 78.50; S9.12)
(a) Vol. **108**, No. 2, p. K97—K100 (1988).

A. G. MAKHANEK, V. S. KOROLKOV, and L. A. YUGURYAN
Analysis of the Intensity and Polarization Dependence of Two-Photon Absorption in the Channel $E_{3/2}(4f)$–$E_{3/2}(5d)$ for Ce^{3+} in CaF_2
(71.70; 78.50; S9.12)
(b) Vol. **149**, No. 1, p. 231—240 (1988).

S. MURAMATSU
*Electron–Lattice Interaction Parameters Determined from Absorption and MCD Spectra of the F Centers in CsCl and CsBr
(71.55 and 78.50; 61.70; S9.11; S9.12)
(b) Vol. **149**, No. 1, p. K19—K23 (1988).

A. G. BADALYAN, T. I. MAKSIMOVA, and J. ROSA
EPR of MnO_4^{2-} Centres in Alkali Halide Crystals
(76.30; 78.50; S9.11)
(b) Vol. **149**, No. 1, p. K31—K34 (1988).

K. INABE, S. NAKAMURA, and N. TAKEUCHI
*Optical Bleaching of BaFCl Crystals X-Irradiated at Room Temperature
(61.70 and 78.50; S9.15)
(b) Vol. **149**, No. 1, p. K67—K71 (1988).

E. RADZHABOV
Reaction of OH^- and F Centers in Sodium Halides
(61.70 and 71.55; 78.50; S9.11)
(b) Vol. **149**, No. 2, p. 441—446 (1988).

J. P. JOUART and G. MARY
Up-Conversion from Er^{3+}–Er^{3+} Pairs in CdF_2 Crystals
(71.55; 78.50; 78.55; S9.12)
(b) Vol. **149**, No. 2, p. 633—639 (1988).

A. HAMAÏDIA and A. HACHIMI
*Colour Centres in Alkali Metal Doped CaF_2
(78.20 and 78.50; 61.70; 78.55; S9.12)
(b) Vol. **149**, No. 2, p. 711—724 (1988).

L. Bosi and M. Nimis
An Analysis of the Decay Properties of the F and M Centers in LiF Crystals
(61.70; 78.50; S9.11)
(b) Vol. **149**, No. 2, p. K89—K94 (1988).

K. Inabe, S. Nakamura, and N. Takeuchi
Optical Bleaching of F-Centers in BaFCl Crystals after X-Irradiation at Room Temperature
(61.70; 78.50; S9.15)
(b) Vol. **149**, No. 2, p. K101—K104 (1988).

A. M. Karguppikar and A. G. Vedeshwar
*Electrical and Optical Properties of Natural Iron Pyrite (FeS_2)
(72.80 and 78.50; S1.63)
(a) Vol. **109**, No. 2, p. 549—558 (1988).

W. Ulrici, K. Friedland, L. Eaves, D. P. Halliday, and C. A. Payling
Optical and Electrical Studies of GaP:Ti
(71.55; 72.20; 78.50; 78.55; S7.11)
(b) Vol. **150**, No. 1, p. 177—190 (1988).

A. Hamaïdia
*Colour Centres in Alkali Metal Doped SrF_2
(78.20 and 78.50; 61.70; 76.30; S9.12)
(b) Vol. **150**, No. 1, p. 323—336 (1988).

N. I. Tarbaev, J. Schreiber, and G. A. Shepelskii
Physical Properties of $A^{II}B^{VI}$ Semiconductor Crystals after Plastic Deformation at Low Temperature
(61.70 and 71.55; 72.40; 78.50; 78.55; S8.1)
(a) Vol. **110**, No. 1, p. 97—106 (1988).

A. M. Brodin, S. A. Boiko, M. Ya. Valakh, and G. G. Tarasov
Effect of Optical Feedback on Self-Induced Change of Light Polarization in Cubic Crystals with Directions Instable for Polarization
(78.20; 78.50; S9.11)
(b) Vol. **150**, No. 2, p. 557—560 (1988).

M. P. Lisitsa, M. Ya. Valakh, G. G. Tarasov, S. A. Boiko, A. M. Brodin, A. M. Shpak, R. Voska, and I. Feldvari
Nonlinear Polarization Spectroscopy of Impure Cubic Crystals
(78.20; 78.50; S9.11)
(b) Vol. **150**, No. 2, p. 907—911 (1988).

A. N. Guryanov, E. M. Dianov, V. M. Kim, V. M. Mashinskii, V. B. Neustruev, V. A. Tikhomirov, and V. F. Khopin
*UV and γ-Induced Paramagnetic Colour Centres in Germanium-Doped Silica Glass
(78.50; 76.30; S10.15)
(a) Vol. **110**, No. 2, p. K107—K110 (1988).

78.55. Photoluminescence

O. GOEDE and W. HEIMBRODT
Optical Properties of (Zn, Mn) and (Cd, Mn) Chalcogenide Mixed Crystals and Superlattices **(Review Article)**
(78.20; 71.35; 78.55; S8.1)
(b) Vol. **146**, No. 1, p. 11—62 (1988).

R. BOYN
*4f–4f Luminescence of Rare-Earth Centers in II–VI Compounds (**Review Article**)
(78.55 and 78.60; 71.55; S8.1)
(b) Vol. **148**, No. 1, p. 11—47 (1988).

H. DONKER, W. M. A. SMIT, and G. BLASSE
*On the Luminescence of $CaO:Sn^{2+}$
(78.55; 78.50; S10.1)
(b) Vol. **145**, No. 1, p. 333—342 (1988).

K. POLÁK, D. J. S. BIRCH, and M. NIKL
*Decay Kinetics of UV Liminescence from Undoped $PbCl_2$ Crystals
(78.55; 71.35; S9)
(b) Vol. **145**, No. 2, p. 741—747 (1988).

B. P. ANTONYUK, P. BUSSEMER, L. LEINE, and A. A. MUKHAMEDOV
*Bistability of Molecular Excitons
(71.35 and 78.55)
(b) Vol. **145**, No. 2, p. 759—766 (1988).

A. A. KAMINSKII, K. KURBANOV, K. L. OVANESYAN, and A. G. PETROSYAN
Stimulated Emission Spectroscopy of Pr^{3+} Ions in Orthorhombic $YAlO_3$ Single Crystals
(78.45; 78.55; S10)
(a) Vol. **105**, No. 2, p. K155—K159 (1988).

S. KISHIDA, K. MATSUURA, A. MATSUOKA, and I. TSURUMI
*The Transient Behaviour of the 2.5 eV Emission Band in Se-Treated ZnSe Crystals
(78.55; S8.12)
(a) Vol. **105**, No. 2, p. K165—K168 (1988).

A. P. ELISSEEV, A. M. YURKIN, and E. G. SAMOILOVA
*Cr^{3+} Energy Levels in Alexandrite
(78.55; 71.70; S10)
(a) Vol. **105**, No. 2, p. K169—K173 (1988).

L. E. NAGLI and S. V. DYACHENKO
*Influence of v_c^--Vacancy on Luminescence of Pb^+ Centres in Alkali Halides
(78.55; 61.70; S9.11)
(b) Vol. **146**, No. 1, p. 295—301 (1988).

O. Goede, W. Heimbrodt, M. Lamla, and V. Weinhold
*Energy Gap of MnS
(78.40 and 78.55; 71.35; S8.11)
(b) Vol. **146**, No. 1, p. K65—K69 (1988).

D. S. Domanevskii, S. V. Zhokhovets, and M. V. Prokopenya
*Peculiarities of Radiative Recombination in Gallium Arsenide Doped with Shallow Donors and Acceptors
(78.55; 71.55; S7.12)
(a) Vol. **106**, No. 1, p. 249—260 (1988).

S. Kishida, K. Matsuura, H. Mori, T. Yanagawa, I. Tsurumi, and C. Hamaguchi
*The 2.5 eV Emission Band in the Se-Treated ZnSe Crystals
(78.55; S8.12)
(a) Vol. **106**, No. 1, p. 283—289 (1988).

L. Werner and J. W. Tomm
*Photoluminescence in p-$Hg_{0.42}Cd_{0.58}Te$
(78.55; S8.15)
(a) Vol. **106**, No. 1, p. K83—K87 (1988).

T. Itho, Y. Iwabuchi, and T. Kirihara
Size-Quantized Excitons in Microcrystals of Cuprous Halides Embedded in Alkali-Halide Matrices
(71.35; 78.55; S9)
(b) Vol. **146**, No. 2, p. 531—543 (1988).

G. Dlubek, A. Dlubek, R. Krause, O. Brümmer, K. Friedland, and R. Rentzsch
Point Defects in GaAs Studied by Correlated Positron Lifetime, Optical, and Electrical Measurements. I. Native Point Defects and Their Complexes in As-Grown GaAs
(61.70; 72.80; 78.55; 78.70; S7.12)
(a) Vol. **106**, No. 2, p. 419—432 (1988).

J. W. Tomm, L. Werner, D. Genzow, K. Herrmann, D. Schikora, and J. Griesche
*Experimental Evidence of Strain-Induced Gap Shifts in PbTe Epitaxial Layers by Photoluminescence
(68.55 and 78.55; S8)
(a) Vol. **106**, No. 2, p. 509—514 (1988).

A. Chiari, M. Colocci, F. Fermi, Yuzhang Li, R. Querzoli, A. Vinattieri, and Weihua Zhuang
*Temperature Dependence of the Photoluminescence in GaAs–GaAlAs Multiple Quantum Well Structures
(78.55; 73.40; S7.15)
(b) Vol. **147**, No. 1, p. 421—429 (1988).

S. I. Pokutnyi
*Energy Transfer of Electron Excitation: Kinetics of Impurity Luminescence Decay in Thin Films
(73.60 and 78.55)
(b) Vol. **147**, No. 1, p. K21—K24 (1988).

G. Dlubek, A. Dlubek, R. Krause, and O. Brümmer
Point Defects in GaAs Studied by Correlated Positron Lifetime, Optical, and Electrical Measurements. II. Point Defects in GaAs Irradiated with Fast Neutrons
(61.80; 78.50; 78.55; 78.70; S7.12)
(a) Vol. **107**, No. 1, p. 111—121 (1988).

O. V. Kulikova, L. L. Kulyuk, S. I. Radautsan, S. A. Ratseev, E. E. Strumban, V. E. Tezlevan, and V. I. Tsitsanu
*Influence of Defect Generation Processes in $CdIn_2S_4$ Single Crystals on the Photoluminescence and Raman Scattering Spectra
(78.30 and 78.55; S8.16)
(a) Vol. **107**, No. 1, p. 373—377 (1988).

J. Misiewics
*The Low Temperature Photoluminescence in Zn_3P_2
(71.55 and 78.55; S7)
(a) Vol. **107**, No. 1, p. K65—K68 (1988).

I. I. Dobynde, A. I. Bobrysheva, I. M. Razdobreev, and Yu. G. Shekun
*Photoluminescence of GaSe
(78.55; S8)
(b) Vol. **147**, No. 2, p. 717—720 (1988).

A. A. Klochikhin, B. S. Razbirin, D. K. Nelson, T. Amand, M. Brousseau, J. Collet, A. Cornet, and M. Pugnet
*Picosecond Kinetics of Luminescence in Highly Excited Direct-Gap Semiconductors
(78.55; 71.35; S8.11; S8.12)
(b) Vol. **147**, No. 2, p. 727—732 (1988).

M. P. dos Santos and C. Hirlimann
*Impurity Effects in the Raman and Luminescence Spectra of Nitrogen-Doped GaP
(78.30 and 78.55; S7.11)
(b) Vol. **147**, No. 2, p. 779—789 (1988).

Th. Schmidt, J. Köhler, C. Kryschi, and D. Schmid
*Sensitized Luminescence of Pure and Doped $NaNO_2$ Single Crystals. III. Excitation Spectroscopy and Time Resolved Trap Fluorescence
(78.55; 71.35; 71.55; S11)
(b) Vol. **147**, No. 2, p. 797—809 (1988).

A. Overberg and F. Fischer
*Effect of Magnetic Field on $^3A_{1u} \rightarrow {}^1A_{1g}$ Luminescence of Pb^{2+} in CaF_2 and SrF_2 and of Au^- in NaCl
(78.55; 63.20; S9.11; S9.12)
(b) Vol. **147**, No. 2, p. 811—821 (1988).

T. X. Hoai, T. H. Nhung, D. X. Thanh, and N. D. Hung
*Experimental Study of the Photomagnetoelectric Effect near the Threshold of the Stimulated Emission in Semiconductors
(72.40 and 78.55; S8.12)
(b) Vol. **147**, No. 2, p. K165—K170 (1988).

S. Shimanuki
*Radiative and Non-Radiative Processes of Localized Impurity Centers in Alkali Halides: KBr: Ag^-
(71.55 and 78.55; S9.11)
(b) Vol. **148**, No. 1, p. 251—262 (1988).

K. Löbe and R. Boyn
Dye-Laser 4f-4f-Line Probing of Crystal-Field Fluctuations in $ZnS:Sm^{3+}$ Polycrystalline Films
(71.55; 68.60; 78.55; S8.11)
(b) Vol. **148**, No. 1, p. 263—272 (1988).

H. Donker, W. M. A. Smit, and G. Blasse
*Self-Trapped Exciton Emission of $SnSO_4$
(78.55; S11.1)
(b) Vol. **148**, No. 1, p. 413—419 (1988).

P. Schotanus, C. W. E. van Eijk, G. Blasse, and H. W. den Hartog
*Luminescence of the Divalent Lead Ion in Barium Fluoride Crystals
(78.55; S9.12)
(b) Vol. **148**, No. 1, p. K77—K81 (1988).

V. Thomas, J. Barrau, M. Brousseau, K. Kirouni, and G. Couloumiers
*The Luminescence of Palladium-Doped Silicon
(78.55; 71.55; S5.11)
(b) Vol. **148**, No. 2, p. 723—727 (1988).

K. Dunphy and W. W. Duley
*Multiphoton Excitation of the 520 nm Luminescence Band in MgO:Al
(78.55; 71.35; 71.55; S10.1)
(b) Vol. **148**, No. 2, p. 729—735 (1988).

A. Popov, A. Bahnev, and S. Dardjonov
*Photoluminescence and EPR Spectra of Fe-Doped $Ga_{1-x}In_xP$ Epitaxial Layers
(68.55 and 78.55; 76.30; S7.15)
(a) Vol. **108**, No. 2, p. 631—636 (1988).

J. M. Neto and T. Abritta
*Luminescence and Site Occupation of Cr^{3+} in Sodium-β″-Alumina
(78.55; 61.70; S10)
(a) Vol. **108**, No. 2, p. 795—802 (1988).

J. A. Sanz-García, E. Dieguez, and C. Zaldo
*Luminescence Characterization of $K_4Nb_6O_{17}$ Crystals
(78.55 and 78.60; S10)
(a) Vol. **108**, No. 2, p. K145—K149 (1988).

R. Baltramiejūnas and A. Žukauskas
*Manifestation of Hot Phonons in the Luminescence Spectra of Highly Photoexcited $A^{II}B^{VI}$ Crystals
(63.20 and 78.55; 71.45; S8.12)
(b) Vol. **149**, No. 1, p. 337—346 (1988).

A. J. de Vries, J. P. M. van Vliet, and G. Blasse
*Mechanism of Concentration Quenching of Gd^{3+} and Eu^{3+} Emission in Perovskite-Like Aluminates
(71.70 and 78.55; S11.1)
(b) Vol. **149**, No. 1, p. 391—401 (1988).

E. D. Arama, S. I. Radautsan, I. M. Tiginyanu, V. F. Zhitar, A Turković, B. Petrović, B. Etlinger, and N. B. Urli
*Photoconductivity and Luminescence Spectra of $ZnIn_2S_4$ Crystals Irradiated by γ-Quanta
(72.40 and 78.55; S8.16)
(a) Vol. **109**, No. 1, p. K55—K57 (1988).

J. P. Jouart and G. Mary
Up-Conversion from Er^{3+}–Er^{3+} Pairs in CdF_2 Crystals
(71.55; 78.50; 78.55; S9.12)
(b) Vol. **149**, No. 2, p. 633—639 (1988).

A. Hamaïdia and A. Hachimi
Colour Centres in Alkali Metal Doped CaF_2
(78.20 and 78.50; 61.70; 78.55; S9.12)
(b) Vol. **149**, No. 2, p. 711—724 (1988).

H. Scibior, I. Brylowska, P. Mazurek, and M. Subotowicz
*Influence of Ion Implantation on the Luminescence of GaAs
(78.55 and 78.60; 61.70; S7.12)
(a) Vol. **109**, No. 2, p. 597—601 (1988).

S. Kishida, K. Matsuura, H. Mori, Y. Mizuguchi, and I. Tsurumi
*Temperature Dependence of the 2.5 eV Emission Band in Se-Treated ZnSe Crystals
(78.55; 71.55; S8.12)
(a) Vol. **109**, No. 2, p. 617—623 (1988).

W. Hergert and S. Hildebrandt
*Unified Theoretical Description of EBIC, LBIC, CL, and PL Experiments. Transient Analysis
(78.55 and 78.60; 72.10)
(a) Vol. **109**, No. 2, p. 625—633 (1988).

A. A. Maksimov, I. I. Tartakovskii, and T. A. Krivenko
Determination of Polariton–Phonon Interaction Parameters in Anthracene Crystals
(71.36; 78.30; 78.55; S12)
(b) Vol. **150**, No. 1, p. 125—132 (1988).

R. Baltramiejūnas, V. Stepankevičius, G. Tamulaitis, and A. Žukauskas
Investigation of Hot Exciton Interaction in ZnTe Crystals
(71.35; 71.38; 78.55; S8.13)
(b) Vol. **150**, No. 1, p. 139—146 (1988).

W. Ulrici, K. Friedland, L. Eaves, D. P. Halliday, and C. A. Payling
Optical and Electrical Studies of GaP:Ti
(71.55; 72.20; 78.50; 78.55; S7.11)
(b) Vol. **150**, No. 1, p. 177—190 (1988).

K. Löbe and R. Boyn
Sm^{3+} in ZnS on Lattice Sites and Wurtzite-Type Interstitial Sites?
(61.70 and 71.55; 71.70; 78.55; S8.11)
(b) Vol. **150**, No. 1, p. 191—200 (1988).

N. I. Tarbaev, J. Schreiber, and G. A. Shepelskii
Physical Properties of $A^{II}B^{VI}$ Semiconductor Crystals after Plastic Deformation at Low Temperature
(61.70 and 71.55; 72.40; 78.50; 78.55; S8.1)
(a) Vol. **110**, No. 1, p. 97—106 (1988).

T. V. Torchinskaya, A. A. Shmatov, and M. K. Sheinkman
Recombination-Enhanced Transformation of Deep Centers in Red Light-Emitting AlGaAs Diodes
(71.55; 73.40; 78.55; S7.15)
(a) Vol. **110**, No. 1, p. 213—220 (1988).

D. S. Domanevskii, S. V. Zhokhovets, and M. V. Prokopenya
The Valence Band Tail State Filling in Heavily Doped and Compensated Crystals of Gallium Arsenide
(71.55; 78.55; S7.12)
(a) Vol. **110**, No. 1, p. 221—226 (1988).

A. G. J. Green, B. Ray, I. V. F. Viney, and J. W. Brightwell
*Luminescence Studies of CaS Phosphors Doped with Cerium and Copper
(71.55 and 78.55; 76.30; 78.60; S8)
(a) Vol. **110**, No. 1, p. 269—276 (1988).

H. Uchiki, T. Kobayashi, and E. Tokunaga
*Carrier Diffusion and Trapping by Quantum Wells
(73.40 and 78.55; S7.12; S7.15)
(b) Vol. **150**, No. 2, p. 667—672 (1988).

H. Weinert, V. Latinis, S. Juršenas, and R. Baltramiejunas
*Time-Resolved Photoluminescence of a GaAs/GaAlAs Superlattice
(71.35 and 78.55; S7.12; S7.15)
(b) Vol. **150**, No. 2, p. 699—704 (1988).

H.-E. Swoboda, F. A. Majumder, R. Renner, C. Weber, M. Sence, Lu Jie, G. Noll, E. O. Göbel, J. Vaitkus, and C. Klingshirn
Nonlinear Optical Properties of the System $CdS_{1-x}Se_x$
(78.20; 78.55; S8.15)
(b) Vol. **150**, No. 2, p. 749—759 (1988).

D. K. Nelson, A. A. Klochikhin, G. V. Mikhailov, A. G. Panfilov, and B. S. Razbirin
*Nonlinear Effects in the Emission Spectra of High Density Excitons in CdS Crystals
(71.35 and 78.55; S8.11)
(b) Vol. **150**, No. 2, p. 777—781 (1988).

H. Kalt, A. L. Smirl, and T. F. Boggess
Picosecond Optical Nonlinearities and Stimulated Emission in Indirect $Al_xGa_{1-x}As$
(78.20 and 78.45; 78.55; S7.15)
(b) Vol. **150**, No. 2, p. 895—900 (1988).

78.60. Other luminescence and radiative recombination

For photoconduction and photovoltaic effects, see 72.40

R. Boyn
*4f–4f Luminescence of Rare-Earth Centers in II–VI Compounds **(Review Article)**
(78.55 and 78.60; 71.55; S8.1)
(b) Vol. **148**, No. 1, p. 11—47 (1988).

A. De and K. V. Rao
Optical Absorption and Thermoluminescence of CaF_2 Single Crystals Subjected to High Electric Fields and Later Irradiated with X-Rays
(78.50; 78.60; S9.12)
(a) Vol. **105**, No. 1, p. 297—302 (1988).

B. S. V. S. R. Acharyulu and V. S. Kishan Kumar
*Optical and Thermoluminescence Studies on Thallium-Doped Suprapure Cesium Sulphate
(78.50 and 78.60; S11.1)
(a) Vol. **105**, No. 1, p. 303—307 (1988).

D. Köhler, G. Koschek, and E. Kubalek
Vacancy Diffusion Profiles at the Grain Boundaries of Electronic Ceramics
(66.30; 61.70; 78.60; S10.1)
(a) Vol. **105**, No. 2, p. 377—385 (1988).

J. Benoit, P. Benalloul, A. Geoffroy, and C. Barthou
*Decay of the Yellow Emission of Mn^{2+} in AC Thin Electroluminescent Devices
(78.60 and 85; 78.65; S8.12)
(a) Vol. **105**, No. 2, p. 637—647 (1988).

R. Stegmann, B. Jacobs, H. Kamleh, M. Albani, M. Heider, and F. Monzer
Carrier Injection at Graded Heterojunctions
(73.40 and 85; 78.60; S7.15)
(a) Vol. **105**, No. 2, p. 661—670 (1988).

G. Kitis and St. Charalambous
*Regenerated Thermoluminescence as a Function of Irradiation Temperature
(78.60; S9.11; S9.12)
(a) Vol. **105**, No. 2, p. K175—K179 (1988).

E. G. Bagduev, E. V. Korovkin, G. I. Peresada, E. G. Ponyatovskii, and M. Sh. Shikhsaidov
Dark Instability of F-Centres in Pressure-Treated KCl and KBr Crystals
(61.70; 78.60; S9.11)
(b) Vol. **146**, No. 1, p. 75—80 (1988).

K. Kawabata, T. Seiyama, S. Okuda, and Y. Shono
Thermally Stimulated Exo-Electron Emission and Thermally Stimulated Luminescence of Irradiated α-Al_2O_3 Powder
(71.55 and 79.75; 78.60; S10.1)
(a) Vol. **106**, No. 1, p. 181—190 (1988).

B. D. Bhasin, S. P. Kathuria, and S. V. Moharil
*Some Peculiarities of Photo-Transfer Thermoluminescence in LiF-TLD 100
(78.60; S9.11)
(a) Vol. **106**, No. 1, p. 271—276 (1988).

R. K. Gartia, S. J. Singh, and P. S. Mazumdar
*Symmetry Factor and Order of Kinetics in Thermally Stimulated Luminescence
(78.60)
(a) Vol. **106**, No. 1, p. 291—296 (1988).

S. Pakeva and R. Dafinova
*The Electroluminescence of ZnS–Pb, Cu, Br Phosphors
(78.60; S8.11)
(a) Vol. **106**, No. 1, p. K97—K100 (1988).

B. T. Deshmukh, S. V. Bodade, and S. V. Moharil
*Thermoluminescence in Synthetic Langbeinite
(78.60; 78.40; S11)
(a) Vol. **106**, No. 1, p. K107—K111 (1988).

S. B. S. Sastry and G. Muralidharan
*Thermoluminecence and Fluorescence Studies on RbBr:Gd^{2+}
(78.60; 61.70; S9.11)
(b) Vol. **146**, No. 2, p. 727—731 (1988).

S. A. Hakeem and R. K. Gartia
*On the Skewness of Thermally Stimulated Luminescence Peaks
(78.60)
(a) Vol. **106**, No. 2, p. 635—640 (1988).

G. Koschek, H. Lakner, and E. Kubalek
*On the Origin of Cathodoluminescence Contrast Phenomena in Semiinsulating GaAs
(78.60; 71.55; S7.12)
(a) Vol. **106**, No. 2, p. 651—658 (1988).

A. V. Egranov and A. I. Nepomnyachikh
Excitonic Processes and Thermoluminescence in NaCl Crystals Doped with Mg or Mn
(61.70 and 71.35; 78.60; S9.11)
(b) Vol. **147**, No. 1, p. 267—272 (1988).

P. Fernández, J. Llopis, and J. Piqueras
Annealing Effects on Mechanically Damaged ZnO Ceramics
(62.20; 78.60; S10.1)
(a) Vol. **107**, No. 1, p. 197—203 (1988).

B. Rheinländer, G. Grummt, J. Kováč, H. Haefner, and R. Heilmann
Minority-Carrier Generation and Recombination in 1.35 μm InGaAsP/InP Double Heterostructure Diodes
(73.40; 71.55; 72.40; 78.60; S7.11; S7.15)
(a) Vol. **107**, No. 1, p. 405—418 (1988).

G. Grummt, H. Haefner, G. Oelgart, W. Thielemann, R. Pickenhain, and B. Jacobs
*Properties of Red Light Emitting (Al, Ga)As Double Heterostructure Diodes
(73.40 and 78.60; 85; S7.12)
(a) Vol. **107**, No. 1, p. 419—428 (1988).

E. Zlateva and I. Gutzow
*Radiation-Induced Volume Changes in Thermoluminescent-Active Materials
(61.80 and 78.60; S9.12)
(a) Vol. **107**, No. 2, p. K99—K103 (1988).

V. B. Shipilo, E. M. Shishonok, A. M. Zaitsev, A. A. Melnikov, and A. I. Olekhnovich
*Influence of High Pressure on Cathodoluminescence of Cubic Boron Nitride
(78.60; S7.1)
(a) Vol. **108**, No. 1, p. 431—436 (1988).

Y. Nakagawa, S. Nakamura, and N. Takeuchi
*Decrease of the TL Glow Peak Temperature by Partial Cleaning
(78.60; S9.11)
(a) Vol. **108**, No. 1, p. 463—468 (1988).

A. Opanowicz
Comments on Interpretation of Thermally Stimulated Luminescence and Conductivity in Conductive ZnSe Crystals
(71.55; 78.60; S8.12)
(a) Vol. **108**, No. 1, p. K47—K51 (1988).

J. Piqueras, F. Domínguez-Adame, and B. Mendez
*SEM-CL of Reaction Bonded SiC
(61.70 and 78.60; S1.61)
(a) Vol. **108**, No. 1, p. K81—K84 (1988).

X. L. Yuan and S. W. S. McKeever
Impurity Clustering Effects in Magnesium-Doped Lithium Fluoride
(61.70; 78.50; 78.60; S9.11)
(a) Vol. **108**, No. 2, p. 545—551 (1988).

Ya. D. Lebedev, M. V. Fok, and M. D. Shagalov
*On the Mechanism of Polarized Electroluminescence from Light-Emitting Structures Based on GaN
(73.40 and 78.60; S7.1)
(a) Vol. **108**, No. 2, p. 657—667 (1988).

G. Koschek, H. Lakner, and E. Kubalek
On the Improvement of the Semiinsulating GaAs Substrate Homogeneity Due to Annealing
(71.55; 78.60; S7.12)
(a) Vol. **108**, No. 2, p. 683—688 (1988).

B. JASSEMNEJAD, R. J. ABBUNDI, M. D. BROWN, and S. W. S. MCKEEVER
*Thermoluminescence Processes in CaF_2:Ce, Mn
(78.60; S9.12)
(a) Vol. **108**, No. 2, p. 753—764 (1988).

D. LAPRAZ, P. IACCONI, Y. SAYADI, P. KELLER, J. BARTHE, and G. PORTAL
*Some Thermoluminescence Properties of an α-Al_2O_3 Sample. Sensitization Effects
(78.60; 61.80; S10)
(a) Vol. **108**, No. 2, p. 783—794 (1988).

J. A. SANZ-GARCÍA, E. DIEGUEZ, and C. ZALDO
*Luminescence Characterization of $K_4Nb_6O_{17}$ Crystals
(78.55 and 78.60; S10)
(a) Vol. **108**, No. 2, p. K145—K149 (1988).

I. FERNÁNDEZ and J. LLOPIS
*Reducing Effects on the CL Red Emission of MgO Doped Crystals
(78.60; S10.1)
(a) Vol. **108**, No. 2, p. K163—K167 (1988).

N. KRISTIANPOLLER, B. TRIEMAN, R. CHEN, and Y. KIRSH
The Application of Thermally Stimulated Processes to the Study of Defects in Perovskite Type Fluorides
(61.70; 72.20; 78.60; S9.16)
(b) Vol. **149**, No. 1, p. 45—54 (1988).

Z. DREGER, J. KALINOWSKI, I. DAVOLI, S. STIZZA, and S. FELIZIANI
*Thermoluminescence Study of Charge Traps in Polycrystalline Layers of Phenanthrene
(71.55 and 78.60; S12)
(b) Vol. **149**, No. 1, p. 363—370 (1988).

F. KOCH, W. HERGERT, G. OELGART, and N. PUHLMANN
Determination of Semiconductor Parameters by Electron Beam Induced Current and Cathodoluminescence Measurements
(73.40; 78.60; S7.12)
(a) Vol. **109**, No. 1, p. 261—272 (1988).

C. ANDRAUD, F. PELLE, O. PILLA, and B. BLANZAT
*Intrinsic Luminescence of $CsCdBr_3$
(78.60; S9.16)
(b) Vol. **149**, No. 2, p. 759—763 (1988).

V. D. RODRIGUEZ and C. ORTIZ
*Thermally Stimulated Relaxation in NaF X-Irradiated at Room Temperature
(61.70 and 78.60; S9.11)
(a) Vol. **109**, No. 2, p. 413—419 (1988).

H. SCIBIOR, I. BRYLOWSKA, P. MAZUREK, and M. SUBOTOWICZ
*Influence of Ion Implantation on the Luminescence of GaAs
(78.55 and 78.60; 61.70; S7.12)
(a) Vol. **109**, No. 2, p. 597—601 (1988).

W. Hergert and S. Hildebrandt
*Unified Theoretical Description of EBIC, LBIC, CL, and PL Experiments. Transient Analysis
(78.55 and 78.60; 72.10)
(a) Vol. **109**, No. 2, p. 625—633 (1988).

R. Zângaro and J. Gasiot
*A Surface Effect in $CaSO_4$:Mn/PVDF Mixtures
(68.45 and 78.60; S11.1)
(a) Vol. **109**, No. 2, p. K109—K113 (1988).

S. B. S. Sastry and G. Muralidharan
*Thermoluminescence and Other Optical Studies on γ-Irradiated RbCl:Gd^{2+}
(78.60; 61.70; S9.11)
(b) Vol. **150**, No. 1, p. 315—322 (1988).

F. Koch, R. Mitdank, G. Oelgart, and C. Chlupsa
*Excess Carrier Recombination in Indirect-Gap $GaAs_{1-x}P_x$:N
(73.60 and 78.60; S7.15)
(a) Vol. **110**, No. 1, p. 197—204 (1988).

A. G. J. Green, B. Ray, I. V. F. Viney, and J. W. Brightwell
Luminescence Studies of CaS Phosphors Doped with Cerium and Copper
(71.55 and 78.55; 76.30; 78.60; S8)
(a) Vol. **110**, No. 1, p. 269—276 (1988).

N. A. Moldovyan, S. I. Radaustan, and V. Z. Chebotaru
Photoconductivity, Cathodoluminescence, and Optical Absorption of $CdIn_2S_2Se_2$ Single Crystals
(72.40 and 78.40; 78.60; S8.16)
(a) Vol. **110**, No. 1, p. 277—288 (1988).

S. Hildebrandt, J. Schreiber, W. Hergert, and V. I. Petrov
*Determination of the Absorption Coefficient and the Internal Luminescence Spectrum of GaAs and $GaAs_{1-x}P_x$ ($x = 0.375, 0.78$) from Beam Voltage Dependent Measurements of Cathodoluminescence Spectra in the Scanning Electron Microscope
(78.60; S7.12; S7.15)
(a) Vol. **110**, No. 1, p. 283—291 (1988).

E. Krause
*Effects of Growth and Annealing Conditions on Deep Level Emissions in CVD Poly-ZnSe Layers
(71.55 and 78.60; S8.12)
(a) Vol. **110**, No. 1, p. K33—K37 (1988).

A. I. Ekimov and A. L. Efros
Nonlinear Optics of Semiconductor-Doped Glasses
(71.35 and 78.20; 78.60)
(b) Vol. **150**, No. 2, p. 627—633 (1988).

G. O. Müller, R. Mach, B. Selle, and G. Schulz
*Measuring on Thin Film Electroluminescent Devices
(73.40 and 78.60; 85; S8.11)
(a) Vol. **110**, No. 2, p. 657—669 (1988).

78.65. Optical properties of thin films

Z. T. Al-Dhhan, C. A. Hogarth, and N. Riddleston
*The Optical Absorption Edge in Thin Amorphous Oxide Films Based on Cerium Dioxide
(78.65; 68.55; S10.15)
(b) Vol. **145**, No. 1, p. 145—149 (1988).

A. L. Shabalov, M. S. Feldman, and M. Z. Bashirov
*Optical Properties of SiN_x Films of Variable Composition
(78.65; S7)
(b) Vol. **145**, No. 1, p. K71—K74 (1988).

S. Sen, S. Annamalai, and P. K. Acharya
*Study and Characterization of PECVD Oxides
(68.55 and 78.65; S10.1)
(a) Vol. **105**, No. 1, p. 171—176 (1988).

E. P. Pokatilov, S. I. Beril, V. M. Fomin, V. G. Litovchenko, D. V. Korbutyak, E. G. Lashkevich, and E. V. Mikhailovaskaya
*The Size-Quantized States of the Wannier-Mott Exciton in Structures with Superthin Films
(71.35 and 78.65; S9)
(b) Vol. **145**, No. 2, p. 535—544 (1988).

S. K. Biswas, S. Chaudhuri, and A. Choudhury
*Effects of Heat Treatment on the Optical and Structural Properties of InSe Thin Films
(78.65; 68.55; S8)
(a) Vol. **105**, No. 2, p. 467—475 (1988).

P. R. Vaya, K. R. Murali, and M. Sundaram
Some Studies on Zinc Phosphide Films Grown by Hot Wall Epitaxy
(68.55; 73.60; 78.65; S7)
(a) Vol. **105**, No. 2, p. 477—484 (1988).

P. Royer, J. P. Goudonnet, T. Inagaki, G. Chabrier, and E. T. Arakawa
Photoacoustic Study of the Optical Absorption of Oblate Silver Spheroids in Attenuated-Total-Reflection Geometry
(78.20; 78.65; S1.3)
(a) Vol. **105**, No. 2, p. 617—625 (1988).

J. Benoit, P. Benalloul, A. Geoffroy, and C. Barthou
Decay of the Yellow Emission of Mn^{2+} in AC Thin Electroluminescent Devices
(78.60 and 85; 78.65; S8.12)
(a) Vol. **105**, No. 2, p. 637—647 (1988).

D. Schalch, A. Scharmann, and R. Wolfrat
*Hydrogen-Ion-Implanted Silicon Nitride and Silicon Oxynitride Films
(61.80 and 78.65; S6)
(a) Vol. **105**, No. 2, p. K81—K86 (1988).

J. George, K. S. Joseph, B. Pradeep, and T. I. Palson
*Reactively Evaporated Films of Indium Sulphide
(68.55 and 78.65; S8)
(a) Vol. **106**, No. 1, p. 123—131 (1988).

F. Lukeš
Natural Surface Films on GeS
(68.55; 78.65; S8)
(a) Vol. **106**, No. 1, p. K27—K30 (1988).

M. T. Mohammad and W. A. Abdul-Ghafor
*Properties of Fluorine-Doped SnO_2 Films Prepared by the Spray Pyrolysis Technique
(78.65; 71.25; S10.1)
(a) Vol. **106**, No. 2, p. 479—484 (1988).

W. A. Murad, S. M. Al-Shamari, F. H. Al-Khateeb, and R. H. Misho
*Band Gap Measurements of Co_3O_4 Thin Films Prepared by the Method of Chemical Spray Pyrolysis
(68.55 and 78.65; S10.1)
(a) Vol. **106**, No. 2, p. K143—K146 (1988).

N. N. Dinh, V. T. Bich, N. H. Hoang, T. X. Hoai, V. N. Quynh, and N. V. Chanh
Vacuum Annealing Effects on Structural and Electrical Properties of Reactively Evaporated Indium Tin Oxide
(73.60; 78.65; S10.1)
(a) Vol. **106**, No. 2, p. K147—K151 (1988).

F. Lukeš
*Ellipsometric Strudies of Natural Films on GaAs
(78.65; S7.12)
(a) Vol. **107**, No. 1, p. 239—251 (1988).

K. Konan, G. Galibert, and J. Calas
*Silver Photodissolution and Optical Properties of $GeSe_3$ Vitreous Glass Thin Films
(66.30 and 78.65; 73.60; S8)
(a) Vol. **107**, No. 1, p. 273—279 (1988).

Zhigang He, Jiayou Feng, Yudian Fan, and Hengde Li
*In–Se Thin Films: Crystallization and Optical Properties
(68.55 and 78.65; S8)
(a) Vol. **107**, No. 2, p. K123—K128 (1988).

A. Lehmann
*Theory of Infrared Transmission Spectra of Thin Insulating Films
(78.30 and 78.65)
(b) Vol. **148**, No. 1, p. 401—405 (1988).

B. Schumann, R. Flagmeyer, and H. Weinert
Multi-Quantum-Well Structures Prepared by the Flash Evaporation Technique. Growth, Structural, and Optical Characterization
(73.40; 68.55; 78.65; S7.1; S9.15)
(a) Vol. **108**, No. 1, p. 257—264 (1988).

T. Drüsedau, M. Eckler, and R. Bindemann
Photoconductive a-Si:H with Dominant Monohydride Bonding Prepared by DC-Magnetron Sputtering
(68.55 and 73.60; 76.30; 78.65; S5.11)
(a) Vol. **108**, No. 1, p. 285—293 (1988).

S. Fries, P. Hertel, and H. P. Menzler
*Extraordinary versus Ordinary Refractive Index Change in Planar $LiNbO_3$:Ti Waveguides
(78.65; S11.1)
(a) Vol. **108**, No. 1, p. 449—455 (1988).

N. Q. Khan, M. Fried, G. Battistig, Z. Laczik, T. Lohner, E. Jároli, V. Schiller, and J. Gyulai
Ellipsometric and Ion Backscattering Measurements of the Properties of Silicon-On-Insulator Structure Formed by Thermally Activated Redistribution of High-Dose Ion Implanted Nitrogen
(61.70 and 68.48; 78.65; S5.11)
(a) Vol. **108**, No. 1, p. K35—K40 (1988).

S. Romanowski
*Study of the Polarized Light Absorption on Electrically Charged Thin Films of Copper Monocrystals by ECPA
(78.65; S1.2)
(b) Vol. **148**, No. 2, p. 515—524 (1988).

D. Dădârlat, R. M. Cândea, R. Turcu, L. P. Biró, I. I. Zasavitskii, M. V. Valeiko, and A. P. Shotov
*Size Effects in Polycrystalline PbSe Films Obtained by Chemical Deposition
(73.60 and 78.65; S8)
(a) Vol. **108**, No. 2, p. 637—641 (1988).

A. K. Abass, Z. A. Ahmed, and F. A. Ali
*Preparation and Interband Transitions of Chemically Deposited Fe_2O_3 Coatings in the Fundamental Absorption Region
(78.65; S10.1)
(a) Vol. **108**, No. 2, p. K151—K155 (1988).

N. N. Dinh, V. T. Bich, N. H. Hoang, and L. Q. Minh
*Photochromic Properties of Tungsten Trioxide Films Made by Electron Beam Deposition
(78.65; S10.1)
(a) Vol. **108**, No. 2, p. K157—K161 (1988).

Ž. T. Al-Dhhan, C. A. Hogarth, and A. Davidson
*The Optical Absorption Edge in Thin Films of CeO_2/Cd
(78.65; S.10.1)
(b) Vol. **149**, No. 1, p. K81—K84 (1988).

Fangqing Zhang, Guanghua Chen, Yafei Zhang, and Gong Yu
IR and ESR Studies of the Structural Properties of Hydrogenated Amorphous Carbon Films
(68.55; 76.30; 78.65; S5)
(a) Vol. **109**, No. 1, p. K39—K44 (1988).

G. K. Padam, G. L. Malhrota, and S. U. M. Rao
Electron Microscopy and Optical Studies of Chemically Deposited $CuInS_2$ Thin Films
(68.55; 78.65; S7.16)
(a) Vol. **109**, No. 1, p. K45—K49 (1988).

W. I. Khleif, C. A. Hogarth, and Z. T. Al-Dhhan
*Optical Absorption in Thin Films of Cerium Dioxide Containing Indium Oxide
(78.65; S10.1)
(b) Vol. **149**, No. 2, p. K187—K190 (1988).

A. Abdellaoui, L. Martin, and A. Donnadieu
*Structure and Optical Properties of MoO_3 Thin Films Prepared by Chemical Vapor Deposition
(68.55 and 78.65; S10.1)
(a) Vol. **109**, No. 2, p. 455—462 (1988).

M. Anwar and C. A. Hogarth
*Optical Properties of Amorphous Thin Films of MoO_3 Deposited by Vacuum Evaporation
(78.65; 68.55; S10.1)
(a) Vol. **109**, No. 2, p. 469—478 (1988).

R. H. Misho, W. A. Murad, G. H. Fatahalah, I. M. Abdul Aziz, and H. M. Al-Doori
*Preparation and Optical Properties of Thin Nickel Oxide Solid Films
(68.55 and 78.65; S10.1)
(a) Vol. **109**, No. 2, p. K101—K104 (1988).

N. P. Bezhan, O. V. Kulikova, L. L. Kulyuk, S. I. Radautsan, A. V. Siminel, E. E. Strumban, L. I. Shulga, and V. A. Chumak
*On the Influence of a Buffer Layer on the Emissive Properties of Epitaxial Layers in InGaAsP ($\lambda = 1.17\ \mu m$)
(78.45 and 78.65; S8.15)
(a) Vol. **109**, No. 2, p. K153—K156 (1988).

W. Andrä, G. Götz, H. Hobert, V. Misyuchenko, V. A. Samuilov, and V. Stelmakh
*IR and RBS Spectroscopy Investigation of Semi-Insulating Phosphorus-Doped Polycrystalline Silicon Layers
(68.55 and 78.65; S5.11)
(a) Vol. **110**, No. 1, p. 181—187 (1988).

R. Rousina and G. K. Shivakumar
*Effect of Heat Treatment on the Optical Absorption of Amorphous Indium Selenide Films
(78.65; 61.40; S8)
(a) Vol. **110**, No. 1, p. K57—K60 (1988).

G. V. Sinitsyn
Digital Circuits for All-Optical Computing Based on Thin-Film Semiconductor Interferometers
(78.20; 78.65; S8.1)
(b) Vol. **150**, No. 2, p. 455—464 (1988).

B. B. Boiko, N. S. Petrov, V. A. Shakin, and A. B. Zimin
*Optical Bistability under Conditions of Total Reflection
(78.65)
(b) Vol. **150**, No. 2, p. 527—531 (1988).

L. G. Zimin, S. V. Gaponenko, and V. Yu. Lebed
*Room-Temperature Optical Nonlinearity in Semiconductor-Doped Glasses
(78.65; S8.15)
(b) Vol. **150**, No. 2, p. 653—656 (1988).

G. Bongiovanni, J. L. Staehli, and D. Martin
*Many-Body Effects in the Electron–Hole Plasma Confined in GaAs–(GaAl)As Quantum Wells
(71.35 and 78.65; 71.45; S7.12; S.715)
(b) Vol. **150**, No. 2, p. 685—690 (1988).

I. Rückmann, J. Kolenda, F. Henneberger, and M. Petrauskas
ps-Laser Induced Transmission and Absorption in GaInAsP Epitaxial Layers
(78.20; 78.65; S7.15)
(b) Vol. **150**, No. 2, p. 735—742 (1988).

V. P. Gribkovskii, L. G. Zimin, S. V. Gaponenko, I. E. Malinovskii, P. I. Kuznetsov, and G. G. Yakushcheva
*Resonatorless Optical Bistability near the Excitonic Resonance of ZnSe
(71.35 and 78.65; S8.12)
(b) Vol. **150**, No. 2, p. 761—764 (1988).

I. Jánossy
Nonlinear Optical Effects in Amorphous Semiconductors
(78.20; 78.65; S8)
(b) Vol. **150**, No. 2, p. 783—789 (1988).

Lun-Biao Xu, Shou-Fu Cai, and Guo-Xin Xie
*Quantum-Statistical Investigation on the Nonlinear Optical Oscillation in Amorphous Semiconductor $GeSe_2$ Films
(78.20 and 78.65; 71.45; S8)
(b) Vol. **150**, No. 2, p. 797—803 (1988).

U. Trutschel, F. Lederer, and U. Langbein
*Optical Response of Nonlinear Absorbing and Saturable Multilayer Systems
(78.65)
(b) Vol. **150**, No. 2, p. 805—810 (1988).

M. Yudasaka, M. Tanaka, Y. Kuwae, K. Nakanishi, and S. Kurita
*Uniaxial Stress-Effect on Visible and Near-Infrared Absorption Spectra of Si(phthalocyaninato)$(OH)_2$ Epitaxial Films
(78.65; S12)
(a) Vol. **110**, No. 2, p. 645—649 (1988).

78.70. Other interactions of matter with particles and radiation

V. S. Malyshevskii
*Interaction of Channeled Electrons with Lattice Excitations **(Review Article)**
(61.80; 61.70; 63.20; 78.70)
(b) Vol. **147**, No. 1, p. 11—44 (1988).

S. V. Dyachenko, L. E. Nagli, and F. V. Pirogov
Radiation Defect Production in Activated KCl Crystals
(61.80; 78.70; S9.11)
(a) Vol. **105**, No. 1, p. 87—95 (1988).

G. Brauer, W. Hertle, and A. G. Balogh
*γ-Irradiation of Polyethylene Studied by Positron Annihilation
(61.40 and 78.70; 61.80; S12)
(a) Vol. **105**, No. 1, p. K7—K12 (1988).

P. P. Lottici, G. Antonioli, and C. Razzetti
*EXAFS in $Zn_xCd_{1-x}Ga_2S_4$ Defect Chalcopyrite Solid Solution
(61.60 and 78.70; S8.16)
(b) Vol. **145**, No. 2, p. 401—407 (1988).

A. P. Bezirganyan and P. A. Bezirganyan
*Solution of the Two-Dimensional Stationary Schrödinger Equation with Cosine-Like Coefficient (in View of X-Ray Diffraction)
(61.10 and 78.70; 68.20)
(a) Vol. **105**, No. 2, p. 345—355 (1988).

I. Ya. Dekhtyar, E. G. Madatova, A. Čížek, and M. Šob
Relation between the Strength Characteristics of Iron and Nickel and the Electronic Structure of Defects by Positron Annihilation Spectroscopy
(61.70; 78.70; S1.1; S1.2)
(a) Vol. **105**, No. 2, p. 357—363 (1988).

M. Dębowska, J. Ch. Abbé, and G. Duplâtre
*Positron Annihilation Studies in Zeolites
(61.80 and 78.70; S11)
(b) Vol. **146**, No. 1, p. 91—96 (1988).

G. Dräger, R. Frahm, G. Materlik, and O. Brümmer
*On the Multipole Character of the X-Ray Transitions in the Pre-Edge Structure of Fe K Absorption Spectra. An Experimental Study
(78.70; 71.50; S1.1; S10.1)
(b) Vol. **146**, No. 1, p. 287—294 (1988).

A. V. Soldatov, Yu. V. Sukhetskii, A. G. Khasabov, and A. N. Gusatinskii
*X-Ray Spectroscopic and Theoretical Investigation of the Energy Band Structure of SbSI
(71.25 and 78.70; S9)
(b) Vol. **146**, No. 1, p. K21—K24 (1988).

G. Dlubek, C. Ascheron, R. Krause, H. Erhard, and D. Klimm
Positron Study of Vacancy Defects in Proton and Neutron Irradiated GaP, InP, and Si
(61.80; 78.70; S5.11; S7.11)
(a) Vol. **106**, No. 1, p. 81—88 (1988).

G.-H. Wang, M.-K. Teng, D.-X. Shen, C.-Y. Yi, Y.-Y. Zhou, Y.-Y. Lu, H.-W. Wang, Y.-Z. Zhu, and L. Dou
*Positron Annihilation Lifetime and Doppler Broadening Studies of Electron-Irradiated Polypropylene
(61.70 and 78.70; S12)
(a) Vol. **106**, No. 1, p. K1—K6 (1988).

V. Švorčík, V. Rybka, and V. Myslík
*Photoelectrochemical Etching of n-GaAs and n-InP
(68.45 and 78.70; S7.11; S7.12)
(a) Vol. **106**, No. 1, p. K35—K39 (1988).

H. Solbrig
*The Coherent Field in Infinite Disordered System of Point-Like Scatterers. Refractive Index and Amplitude
(78.70)
(b) Vol. **146**, No. 2, p. 733—742 (1988).

G. Dlubek, A. Dlubek, R. Krause, O. Brümmer, K. Friedland, and R. Rentzsch
Point Defects in GaAs Studied by Correlated Positron Lifetime, Optical, and Electrical Measurements. I. Native Point Defects and Their Complexes in As-Grown GaAs
(61.70; 72.80; 78.55; 78.70; S7.12)
(a) Vol. **106**, No. 2, p. 419—432 (1988).

D. T. Britton and P. C. Rice-Evans
*Fast Fourier Transform Deconvolution of Positron Annihilation Doppler-Broadened Lineshapes
(61.70 and 78.70; S1.2; S5)
(b) Vol. **147**, No. 1, p. 89—96 (1988).

K. B. Garg, R. K. Singhal, K. V. R. Rao, U. Chandra, H. S. Chauhan, J. Singh, K. S. Jerath, and D. C. Jain
XANES Study of Copper Mixed Valence and Mechanism of Superconductivity in $YBa_2Cu_3O_{6.5+\delta}$ System
(74.10; 78.70; S11)
(b) Vol. **147**, No. 1, p. 343—348 (1988).

A. Urbaniak-Kucharczyk
*The Temperature Dependence of the Spin Autocorrelation Time in Thin Antiferromagnetic Films
(75.30 and 78.70)
(b) Vol. **147**, No. 1, p. K71—K74 (1988).

G. Dlubek, A. Dlubek, R. Krause, and O. Brümmer
Point Defects in GaAs Studied by Correlated Positron Lifetime, Optical, and Electrical Measurements. II. Point Defects in GaAs Irradiated with Fast Neutrons
(61.80; 78.50; 78.55; 78.70; S7.12)
(a) Vol. **107**, No. 1, p. 111—121 (1988).

A. F. Vavrek and K. K. Christova
*X-Ray Polarization and Depolarization Phenomena in $Bi_{12}SiO_{20}$
(77.30 and 78.70; 72.40; S10)
(a) Vol. **107**, No. 1, p. 379—386 (1988).

K. M. Podurets, S. Sh. Shilstein, and F. Eichhorn
*Refraction Shift of Bragg Reflection of Neutrons
(78.70; S 5.12)
(a) Vol. **107**, No. 1, p. 387—392 (1988).

J. Chojcan and M. Szuszkiewicz
*The Mössbauer Effect and Positron Annihilation Studies of Chromium–Iron Alloys
(76.80 and 78.70; S1.1)
(b) Vol. **147**, No. 2, p. K201—K204 (1988).

P. Rennert and Nguyen Van Hung
*Curved Wave and Multiple Scattering Corrections in Calculated EXAFS Spectra
(61.10 and 78.70; 71.10)
(b) Vol. **148**, No. 1, p. 49—61 (1988).

G. Das, B. K. Panda, D. P. Mahapatra, and H. C. Padhi
*Simple Numerical Method for the Estimation of Multiple Scattering in the Compton Effect
(78.70; S2; S4)
(b) Vol. **149**, No. 1, p. 355—362 (1988).

V. N. Baier, V. M. Katkov, and V. M. Strakhovenko
*Incoherent Radiation and Pair Creation in Crystals
(61.80 and 78.70)
(b) Vol. **149**, No. 1, p. 403—415 (1988).

A. P. Bezirganyan
*X-Ray Reflection from and Transmission through a Plane-Parallel Dielectric Plate with Cosine-Like Polarizability (Symmetrical Laue Case)
(61.10 and 78.70; 68.20)
(a) Vol. **109**, No. 1, p. 101—110 (1988).

Hong Zhang, Xiao-Gan Wang, and Yao-xian Fu
Study of the Movement of Oxygen Vacancies in the Orthorhombic Phase of $YBa_2Cu_3O_{7-x}$ by Positron Doppler Broadening Spectroscopy
(61.70; 74.70; 78.70; S10.15)
(a) Vol. **109**, No. 1, p. 135—139 (1988).

M. V. Pérez Diaz, J. Ch. Abbé, and G. Duplâtre
*Positron Annihilation Studies in Non-Metallic Solids: d-Camphor and Silver Iodide
(78.70; S9; S12)
(a) Vol. **109**, No. 1, p. 337—343 (1988).

A. Šimůnek and G. Wiech
*Angle-Resolved X-Ray Emission Spectra (ARXES) of Carbon in Potassium Graphite C_8K
(78.70; 71.20; S6)
(b) Vol. **149**, No. 2, p. 765—774 (1988).

A. V. Shishkin, S. Linderoth, and S. V. Svirida
*Positron Annihilation in BaF_2 and SrF_2
(78.70; S9.12)
(b) Vol. **149**, No. 2, p. 785—789 (1988).

M. KITAMURA, C. SUGIURA, and S. MURAMATSU
*CoK X-Ray Absorption Near-Edge Structure of Cobalt(III) Complex Compounds
(78.70; S9)
(b) Vol. **149**, No. 2, p. 791—797 (1988).

R. RAMÍREZ and M. C. BÖHM
*Analytic Expressions for the One-Center Elements of Compton Profiles. Slater-Type Orbitals
(71.10 and 78.70)
(b) Vol. **149**, No. 2, p. K135—K140 (1988).

XIANYI ZHOU, HUILIN JIANG, QIRUI ZHANG, GUOQING PAN, and QIWEI LONG (C. W. LUNG)
*Positron Trapping in $YBa_2Cu_3O_x$ Superconductors
(74.70 and 78.70; S10.15)
(a) Vol. **109**, No. 2, p. K129—K132 (1988).

HONG ZHANG, XIAO-GANG WANG, and YAO-XIAN FU
*Positron Annihilation in Thermally Quenched $YBa_2Cu_3O_{7-x}$
(74.70 and 78.70; S10.15)
(a) Vol. **109**, No. 2, p. K133—K138 (1988).

V. A. BUSHUEV, A. YU. KAZIMIROV, and M. V. KOVALCHUK
*Coherent Compton Effect under Conditions of X-Ray Dynamical Laue Diffraction
(61.10 and 78.70; S5.11)
(b) Vol. **150**, No. 1, p. 9—18 (1988).

R. V. VEDRINSKII, L. A. BUGAEV, and I. G. LEVIN
*Single and Multiple Electron Scattering Description in X-Ray Absorption Spectrum Theory and the Problem of Solid Structure Investigation
(78.70; 61.10; 71.10; S9.11; S10.1; S11.1)
(b) Vol. **150**, No. 1, p. 307—314 (1988).

J. DUBARD, A. L. SMIRL, A. G. CUI, G. C. VALLEY, and T. F. BOGGESS
*Beam Amplification by Transient Energy Transfer in GaAs and Si
(78.70; S5.11; S7.12)
(b) Vol. **150**, No. 2, p. 913—919 (1988).

A. SERRA and N. DE DIEGO
Characterization of Defects in Deformed Titanium
(61.70; 78.70; S1.2)
(a) Vol. **110**, No. 2, p. 409—414 (1988).

R. EWERTOWSKI, A. BARANOWSKI, and W. ŚWIATKOWSKI
*Positron Trapping at Ag/Na_3AlF_6 Interface; Life-Time Measurements
(78.70; S1.3; S9)
(a) Vol. **110**, No. 2, p. K101—K105 (1988).

79. Electron and ion emission by liquids and solids; impact phenomena

79.20. Impact phenomena, including electron spectra and sputtering

T. Bremer, W. Heiland, A. Klekamp, and E. Krätzig
*SIMS Investigations of Iron Profiles in $LiNbO_3$:Ti Waveguides
(66.30 and 79.20; 78.20; S11.1)
(a) Vol. **105**, No. 1, p. K17—K20 (1988).

C. Zwanziger, S. Petrowa, J. Reinhold, P. Streubel, and A. Meisel
Approximate Calculation of Inner-Shell Binding Energies and Auger Transition Energies of Solid Silicon Compounds
(71.50; 79.20; S5.11; S6)
(b) Vol. **145**, No. 2, p. 603—607 (1988).

K. Meinel, M. Klaua, and H. Bethge
Segregation and Sputter Effects on Perfectly Smooth (111) and (100) Surfaces of Au–Ag Alloys Studied by AES
(68.40 and 68.55; 64.75; 79.20; S1.3)
(a) Vol. **106**, No. 1, p. 133—144 (1988).

V. Drchal and J. Kudrnovský
*Impurity Auger Spectra. Host Crystal with Unfilled Valence Band
(71.55 and 79.20; S1)
(b) Vol. **146**, No. 2, p. 597—604 (1988).

E. Taute and K. Blum
*On the Nuclear Charge Dependence of Auger Neutralisation at Surfaces
(79.20; S4)
(b) Vol. **146**, No. 2, p. K87—K90 (1988).

V. Drchal and J. Kudrnovský
*Theory of Auger Spectra from Alloys with Partially Filled Narrow Bands
(71.10 and 79.20; S1.2)
(b) Vol. **147**, No. 1, p. 215—222 (1988).

M. F. Burkhard, H. Rothard, and K.-O. E. Groeneveld
Single-Electron Deexcitation of Volume Plasmons Induced by Heavy Ions in Thin Solid Foils
(71.45; 79.20; S1.2; S4; S5)
(b) Vol. **147**, No. 2, p. 589—592 (1988).

M. Rösler and W. Brauer
*Theory of Electron Emission from Solids by Proton and Electron Bombardment
(79.20; S4)
(b) Vol. **148**, No. 1, p. 213—226 (1988).

H.-J. Fitting and D. Hecht
*Secondary Electron Field Emission
(73.60 and 79.20; S10.1)
(a) Vol. **108**, No. 1, p. 265—273 (1988).

A. Subbarayan, C. Balasubramanian, and Sa. K. Narayandass
*Laser-Induced Damage Study of Vacuum Deposited Perylene Thin Films
(61.80 and 79.20; S12)
(a) Vol. **108**, No. 1, p. K31—K34 (1988).

L. Guoba, L. Pranevičius, V. Sargūnas, and G. Zubauskas
Protective Coatings on GaAs and InP. High-Energy Ion Scattering Analysis
(68.48; 79.20; S7.11; S7.12)
(a) Vol. **109**, No. 1, p. 239—244 (1988).

S. M. Javed Akhtar and D. Ristau
Mechanism of Voltage Generation in Bi Films Due to Laser Irradiation
(73.60; 79.20; S4)
(a) Vol. **109**, No. 1, p. 255—259 (1988).

M. Wautelet, E. D. Gehain, and P. Quenon
Modelling the Role of Electronically Excited States in Laser-Assisted Phenomena
(61.70; 79.20; S5.11)
(b) Vol. **149**, No. 2, p. 465—475 (1988).

P. Mrozek, M. Menyhárd, A. Jabłoński, and G. Tyuliev
Surface Composition of the Ordered Fe–Co Alloys
(68.20; 79.20; S1.1; S1.2)
(a) Vol. **110**, No. 2, p. 495—507 (1988).

79.40. Thermionic emission

K. Biedrzycki
*Thermally Stimulated Electron Emission from Triglycine Sulphate Crystals Excited by AC Electric Field
(77.80 and 79.40; S12)
(a) Vol. **109**, No. 1, p. K79—K83 (1988).

79.60. Photoemission and photoelectron spectra

E. Majková and P. Duhaj
*XPS Study of Co–Fe–B Metallic Glasses
(71.25 and 79.60; S1.2)
(b) Vol. **146**, No. 1, p. K17—K20 (1988).

V. Fritzsche
*Approximations for the Scattering of Excited Electrons in Solids
(79.60; 61.10; 68.20)
(b) Vol. **147**, No. 2, p. 485—494 (1988).

R. Hoogewijs, P. de Volder, L. Fiermans, and J. Vennik
Room Temperature Oxidation in Air of Polycrystalline CuZn
(68.20; 79.60; S1.2; S10)
(a) Vol. **107**, No. 2, p. 625—631 (1988).

N. J. Castellani and D. B. Leroy
Final State Core Level Shifts in Supported and Free Metal Particles
(68.40; 79.60; S4)
(b) Vol. **148**, No. 1, p. 121—128 (1988).

C. Majumdar, A. B. Maity, and A. N. Chakravarti
*Quantized Photoemission from Quantum Wells in Ultrathin Films of Wide-Gap Semiconductors in a Tilted Magnetic Field
(79.60; 73.60; S7.12)
(b) Vol. **148**, No. 2, p. 559—565 (1988).

A. B. Maity, C. Majumdar, and A. N. Chakravarti
*Effect of Magneto-Size Quantization on Electric-Field-Aided Photoemission from Quantum Wells in Ultrathin Films of Wide-Gap Semiconductors
(73.60 and 79.60; S7.12)
(b) Vol. **149**, No. 2, p. 565—570 (1988).

I. N. Shabanova
*System Terminal State Investigation by the Shape of Electron Spectra of Inner Levels of Three-Valence Rare-Earth Compounds under Photoionization
(79.60; S1.4)
(b) Vol. **149**, No. 2, p. K177—K181 (1988).

B. Mitra, A. Ghosal, and K. P. Ghatak
*On the Photoemission from Bi in the Presence of an Arbitrarily Oriented Quantizing Field
(79.60; S4)
(b) Vol. **150**, No. 1, p. K67—K72 (1988).

79.70. Field emission and field ionization

R. Z. Bakhtizin and V. G. Valeev
*Microroughness Field Emission Electron Energy Distribution and Electron–Phonon Interaction Spectroscopy near the Metal Surface
(79.70; 68.20; S1; S4)
(a) Vol. **108**, No. 1, p. 251—255 (1988).

R. Z. Bakhtizin and Yu. M. Yumaguzin
*High-Energy Tails in the Total Energy Distribution of Field Emitted Electrons from Tungsten
(68.20 and 79.70; S1)
(b) Vol. **150**, No. 1, p. 103—108 (1988).

79.75. Exoelectronic emission

K. Kawabata, T. Seiyama, S. Okuda, and Y. Shono
*Thermally Stimulated Exo-Electron Emission and Thermally Stimulated Luminescence of Irradiated α-Al_2O_3 Powder
(71.55 and 79.75; 78.60; S10.1)
(a) Vol. **106**, No. 1, p. 181—190 (1988).

H. Lanteri, P. Rostaing, and R. Bindi
*Monte-Carlo Simulation of Exoelectron Energy Spectra
(79.75; S10.1)
(b) Vol. **149**, No. 1, p. 203—214 (1988).

79.80. Resonance tunneling

V. N. Ermakov and E. A. Ponezha
*Resonance Tunneling with Dissipation Taken into Account
(79.80)
(b) Vol. **145**, No. 2, p. 545—554 (1988).

85. Devices

H. Pagnia and N. Sotnik
*Bistable Switching in Electroformed Metal–Insulator–Metal Devices **(Review Article)**
(73.40 and 85; S10)
(a) Vol. **108**, No. 1, p. 11—65 (1988).

J. Benoit, P. Benalloul, A. Geoffroy, and C. Barthou
*Decay of the Yellow Emission of Mn^{2+} in AC Thin Electroluminescent Devices
(78.60 and 85; 78.65; S8.12)
(a) Vol. **105**, No. 2, p. 637—647 (1988).

P. D. Pester and T. Wilson
*An Analysis of the Photocurrent Induced in a Semi-Infinite Schottky Barrier by a Time Varying Focussed Laser Beam
(73.30 and 85; 72.40)
(a) Vol. **105**, No. 2, p. 649—660 (1988).

R. Stegmann, B. Jacobs, H. Kamleh, M. Albani, M. Heider, and F. Monzer
*Carrier Injection at Graded Heterojunctions
(73.40 and 85; 78.60; S7.15)
(a) Vol. **105**, No. 2, p. 661—670 (1988).

D. Kindl and J. Toušková
*Charge Transport Study in Thin Film Au–CdTe Schottky Diodes
(73.30 and 85; 73.60; S8.13)
(a) Vol. **106**, No. 1, p. 297—304 (1988).

B. Rheinländer, R. Heilmann, and G. Oelgart
*Determination of Temperature-Depenedent Carrier Losses in 1.3 μm InGaAsP/InP Double-Heterostructures
(78.45 and 85; 73.40; S7.11; S7.16)
(a) Vol. **106**, No. 1, p. K113—K116 (1988).

E. Pinčík, B. Maťátko, J. Bartoš, I. Thurzo, M. Grendel, V. Nádaždy, M. Zubeková, and M. Morvik
*Some Properties of MIS Structures Prepared by Plasma Oxidation of Al Layers on GaAs
(73.40 and 85; S7.12)
(a) Vol. **106**, No. 2, p. 659—667 (1988).

A. P. Kovchavtsev, G. L. Kurishev, K. O. Postnikov, R. A. Sokolov, and I. M. Subbotin
*Inelastic Electron Tunneling Spectroscopy of Si MIS Structures with Ultrathin Thermal Silicon Nitride and Thermal Silica
(73.40 and 85; S5.11)
(a) Vol. **106**, No. 2, p. 669—684 (1988).

A. Opanowicz, B. Kościelniak-Mucha, and G. Boroś
*Electrical and Photovoltaic Properties of Au/n-$CuInSe_2$ Schottky Barrier Diodes
(73.30 and 85; S8.16)
(a) Vol. **106**, No. 2, p. K197—K201 (1988).

A. Srivastava
*Effect of Polysilicon–Silicon Interface Recombination Velocity on Current Transport Properties of Poly-Emitter Transistors
(73.40 and 85; S5.11)
(a) Vol. **106**, No. 2, p. K203—K207 (1988).

P. R. Vaya and J. Majhi
*Effect of Surface Treatment on Al—Si Schottky Structures
(73.30 and 85; S5.11)
(a) Vol. **106**, No. 2, p. K209—K213 (1988).

F. Januschewski, H.-J. Erzgräber, and W. Füssel
*Modelling the Local Damage of Short-Channel MOSFETs Due to Hot Electron Injection Using Results from Photoinjection Measurements on MOS Capacitors
(73.40 and 85; 72.40; S5.11)
(a) Vol. **106**, No. 2, p. K215—K220 (1988).

B. Cabon and G. Ghibaudo
*Degradation of Submicron MOSFETs after Aging
(73.40 and 85)
(a) Vol. **107**, No. 1, p. 393—404 (1988).

G. Grummt, H. Haefner, G. Oelgart, W. Thielemann, R. Pickenhain, and B. Jacobs
Properties of Red Light Emitting (Al, Ga)As Double Heterostructure Diodes
(73.40 and 78.60; 85; S7.12)
(a) Vol. **107**, No. 1, p. 419—428 (1988).

H. Ohyama and K. Nemoto
Effect of Lattice Defects in the Collector Region of npn Si Transistors on the Degradation of h_{FE}
(61.80 and 73.40; 85; S5.11)
(a) Vol. **107**, No. 1, p. 429—439 (1988).

V. K. Goyal, R. Pal, and K. Sen
*On the Photogenerated Current in Polycrystalline Silicon MIS Solar Cells
(73.40 and 85; 72.40; S5.11)
(a) Vol. **107**, No. 1, p. K83—K87 (1988).

R. E. Hummel
*On the Current Density Dependence of Electromigration in Thin Films
(66.30 and 85; S4)
(a) Vol. **107**, No. 2, p. K175—K179 (1988).

U. Henniger, B. Schmidt, R. Tempel, and H. Moeller
*Temperature Distribution in Square-Shaped Semiconductor Chips
(73.40 and 85)
(a) Vol. **108**, No. 1, p. 469—478 (1988).

R. Agrawal, R. Gopal, R. Dwivedi, and S. K. Srivastava
*Analytical Modelling of a MOSFET with Non-Uniform Impurity Profile
(73.40 and 85)
(a) Vol. **108**, No. 2, p. 803—811 (1988).

V. N. Katerinchuk, Z. D. Kovalyuk, and V. A. Manasson
On a Mechanism of Photo-E.M.F. Formation in SIS Structures
(73.40; 85; S8; S10.1)
(a) Vol. **108**, No. 2, p. K115—K118 (1988).

M. Igalson, K. Plankiewicz, and R. Trykozko
*On the Photoresponse of Cu-Treated Cu_2S/CdS Thin Film Solar Cells
(73.40 and 85; 72.40; S8.11)
(a) Vol. **108**, No. 2, p. K169—K171 (1988).

Sh. M. Efendiev, V. E. Bagiev, R. A. Aliev, and E. R. Mustafaev
*Nonlinear $I-U$ Characteristics and Photoelectret State in Sillenite-Type Crystals
(73.40 and 85; 73.30; S11)
(a) Vol. **109**, No. 1, p. 345—357 (1988).

J. P. Banerjee, S. P. Pati, and S. K. Roy
*Computer Simulation Experiment on the mm-Wave Properties of Indium Phosphide Double Drift Impatts
(73.40 and 85; S7.11)
(a) Vol. **109**, No. 1, p. 359—364 (1988).

P. Krispin, N. T. Nghi, N. T. Q. Hai, D. D. Khang, and D. A. Tuan
Defects in GaP Green Light-Emitting Diodes
(71.55; 73.40; 85; S7.11)
(a) Vol. **109**, No. 2, p. K115—K119 (1988).

H. Ohyama and K. Nemoto
Process of Defect Formation in the Collector Region of an Electron-Irradiated npn Si Transistor
(61.80 and 73.40; 85; S5.11)
(a) Vol. **110**, No. 1, p. 301—308 (1988).

D. M. Butusov, G. G. Gotsadze, V. R. Larionov, and B. S. Ryvkin
On the Possibility of the Optical Connection between SEED's within One Array
(78.20; 85; S7.12)
(b) Vol. **150**, No. 2, p. 599—603 (1988).

H. J. Eichler, T. Brand, M. Glotz, and B. Smandek
Optical Nonlinearity and Bistability in Silicon
(78.20; 85; S5.11)
(b) Vol. **150**, No. 2, p. 705—718 (1988).

S1. Transition metals and their alloys and compounds

(for oxides, see S10.)

XIAN-WU ZOU, ZHUN-ZHI JIN, and DE-CHENG TIAN
Structure Factors of Liquid Zn, Cd, and Hg
(61.25; S1; S1.2)
(b) Vol. **145**, No. 1, p. 73—76 (1988).

T. BHATTACHARYA, A. MANNA, and A. K. GHOSH
Construction of Electronic Charge Density in Solids. Calculation of Atomic Scattering Factors of Elements and Energy of Formation of Alloys
(71.10; 61.55; S1)
(b) Vol. **145**, No. 1, p. 181—186 (1988).

S. N. GUPTA and H. C. VERMA
Studies of $Zr_{0.8}Fe_{0.2}$ Alloys Using Mössbauer Spectroscopy
(76.80; S1; S1.1)
(a) Vol. **105**, No. 1, p. 275—280 (1988).

B. FELLMUTH and D. ELEFANT
Influence of Residual Resistance Measurements on the Superconducting Transition of Ultra-High-Purity Niobium
(74.10; S1)
(b) Vol. **146**, No. 1, p. 207—216 (1988).

V. DRCHAL and J. KUDRNOVSKÝ
Impurity Auger Spectra. Host Crystal with Unfilled Valence Band
(71.55 and 79.20; S1)
(b) Vol. **146**, No. 2, p. 597—604 (1988).

ZHUN-ZHI JIN, XIAN-WU ZOU, and YONG-QIAO YAO
Dependence of Structure Factors upon Temperature for Liquid Metals, Zn and Hg
(61.25; S1; S1.2)
(b) Vol. **147**, No. 1, p. 45—49 (1988).

P. A. MAKSIMIYUK, V. A. GLEY, M. HAMERSKÝ, P. LUKÁČ, and Z. TROJANOVÁ
Temperature Dependence of Young's Modulus of α-Zirconium Polycrystals
(62.20; S1)
(a) Vol. **107**, No. 1, p. K11—K13 (1988).

F. H. M. SPIT, P. I. LOEFF, and H. BAKKER
On the Use of the ^{95}Zr Radioisotope for Diffusion Measurements
(66.30; S1)
(b) Vol. **147**, No. 2, p. K105—K107 (1988).

H.-J. OU and J. M. COWLEY
The Surface Reaction of Pd/MgO Studied by Scanning Reflection Electron Microscopy
(68.20; 68.48; S1; S10)
(a) Vol. **107**, No. 2, p. 719—729 (1988).

R. Z. Bakhtizin and V. G. Valeev
Microroughness Field Emission Electron Energy Distribution and Electron–Phonon Interaction Spectroscopy near the Metal Surface
(79.70; 68.20; S1; S4)
(a) Vol. **108**, No. 1, p. 251—255 (1988).

M. Cinal and S. Olszewski
Quantum-Statistical Approach to Elastic Constants in Metals Influenced by Magnetization
(75.80; 62.20; S1; S1.1; S1.2)
(b) Vol. **148**, No. 2, p. 699—707 (1988).

G. Neumann, C. Tuijn, G. de Vries, and H. Bakker
Diffusion Mechanisms in B.C.C. Metals. Calculation of the Correlation Factor for Mixed Nearest and Next-Nearest Neighbour Vacancy Jumps
(66.30; S1)
(b) Vol. **149**, No. 2, p. 483—493 (1988).

S. K. Mishra and T. N. Singh
Lattice Dynamics of B.C.C. Transition Metals
(63.20; S1; S1.1)
(b) Vol. **149**, No. 2, p. 503—510 (1988).

R. Z. Bakhtizin and Yu. M. Yumaguzin
High-Energy Tails in the Total Energy Distribution of Field Emitted Electrons from Tungsten
(68.20 and 79.70; S1)
(b) Vol. **150**, No. 1, p. 103—108 (1988).

L. Nistor, A. De Veirman, and J. Van Landuyt
Cross Section Transmission Electron Microscopy Studies of Zr Samples Laser Irradiated in Air
(61.16; S1)
(a) Vol. **110**, No. 1, p. K1—K3 (1988).

S1.1. Fe

D. N. Jha
Calculation of Electrical Resistivity of Some Liquid Metals by the Mode-Coupling Method
(72.10; 72.15; S1.1; S1.2; S1.3; S1.4)
(b) Vol. **145**, No. 1, p. 277—282 (1988).

L. M. Clarebrough and C. T. Forwood
Burgers Vectors of Secondary Grain Boundary Dislocations in a $\Sigma 3$ Boundary in α-Iron
(61.70; S1.1)
(a) Vol. **105**, No. 1, p. 131—138 (1988).

E. Kokmeijer, E. Huizer, B. J. Thijsse, and A. van den Beukel
Resistivity Changes during Structural Relaxation of Amorphous $Fe_{40}Ni_{40}B_{20}$
(72.15; 61.40; S1.1)
(a) Vol. **105**, No. 1, p. 235—242 (1988).

J. L. Sánchez Llamazares, S. García García, O. Arés Muzio, and S. Díaz Castañón
Mössbauer and Thermomagnetic Analysis of Al-Containing Nd–Fe–B Permanent Magnet Alloys
(75.50; 76.80; S1.1)
(a) Vol. **105**, No. 1, p. 243—247 (1988).

A. A. Hirsch and T. Massalha
Magnetic Metastable States in the Critical Transitions of Nickel and Iron
(75.30; 76.80; S1.1; S1.2)
(a) Vol. **105**, No. 1, p. 249—262 (1988).

M. Takahashi, F. Sato, and T. Miyazaki
Magnetic Anisotropy and Magnetostriction in Fe–Nd Alloys
(75.50; 75.80; S1.1)
(a) Vol. **105**, No. 1, p. 263—273 (1988).

S. N. Gupta and H. C. Verma
Studies of $Zr_{0.8}Fe_{0.2}$ Alloys Using Mössbauer Spectroscopy
(76.80; S1; S1.1)
(a) Vol. **105**, No. 1, p. 275—280 (1988).

K. Sumiyama and Y. Nakamura
Nonequilibrium $Fe_{1-x}Co_x$ Alloys Produced by Vapor Quenching
(75.70; 61.10; 76.80; S1.1)
(a) Vol. **105**, No. 1, p. 291—295 (1988).

P. P. Pal-Val, V. D. Natsik, and S. Kadečková
Low-Temperature Dislocation Relaxations in High-Purity Iron Single Crystals Deformed at 295 and 78 K
(62.40; 61.70; S1.1)
(a) Vol. **105**, No. 1, p. K37—K40 (1988).

Tiesong Zhao, Yong Zhu, and Hanmin Jin
Estimation of the Crystal Field at R Sites in $R_2Fe_{14}B$ Compounds
(75.30; 71.70; S1.1; S1.4)
(b) Vol. **145**, No. 2, p. 679—686 (1988).

Ming Mao and Yizhen He
Two-Stage Structural Relaxation of the Metallic Glass $Fe_{46}Ni_{31}V_1Si_8B_{14}$
(61.40; 65.40; S1.1; S1.2)
(a) Vol. **105**, No. 2, p. 325—333 (1988).

I. Ya. Dekhtyar, E. G. Madatova, A. Čížek, and M. Šob
Relation between the Strength Characteristics of Iron and Nickel and the Electronic Structure of Defects by Positron Annihilation Spectroscopy
(61.70; 78.70; S1.1; S1.2)
(a) Vol. **105**, No. 2, p. 357—363 (1988).

C. T. Forwood and L. M. Clarebrough
Rigid Body Displacements at a Faceted $\Sigma 3$ Boundary in α-Iron
(61.70; S1.1)
(a) Vol. **105**, No. 2, p. 365—375 (1988).

J. P. Allemand, F. Fouquet, J. Perez, and B. de Guillebon
Evolution of the Mechanical and Micromechanical Properties as a Function of Quench Rate in a Metallic Fe–Cr–C–P–Si Glass
(62.20; 61.40; 75.80; S1.1)
(a) Vol. **105**, No. 2, p. 427—439 (1988).

Yan-Zhong Zhang
Magnetic Instability of Metallic Glass $(Fe_{0.1}Ni_{0.33}Co_{0.55}Cr_{0.02})_{78}Si_8B_{14}$. I. Reversible Disaccommodation Phenomenon with Respect to Demagnetization
(75.50; S1.1; S1.2)
(a) Vol. **105**, No. 2, p. 579—587 (1988).

H. K. Lachowicz, A. Siemko, N. Moser, A. Forkl, and H. Kronmüller
Direct Evidence for Stress-Dependent Saturation Magnetostriction in Metallic Glasses
(75.80; 78.20; S1.1)
(a) Vol. **105**, No. 2, p. 597—600 (1988).

V. Angelov, Ts. Bonchev, V. Rusanov, and I. Mandjukov
On the Probability of Recoilles Absorption f' and Reemission f'' in Some Fe-Containing Compounds
(76.80; S1.1)
(b) Vol. **146**, No. 1, p. 269—277 (1988).

G. Dräger, R. Frahm, G. Materlik, and O. Brümmer
On the Multipole Character of the X-Ray Transitions in the Pre-Edge Structure of Fe K Absorption Spectra. An Experimental Study
(78.70; 71.50; S1.1; S10.1)
(b) Vol. **146**, No. 1, p. 287—294 (1988).

M. Polcarová, K. Godwod, J. Bąk-Misiuk, S. Kadečkova, and J. Brádler
Lattice Parameters of Fe–Si Alloy Single Crystals
(61.55; 61.10; S1.1; S1.61)
(a) Vol. **106**, No. 1, p. 17—23 (1988).

P. Svoboda and P. Vašek
Low Temperature Thermal Conductivity of Metallic Glasses $Fe_xNi_{80-x}B_{20}$
(66.70; 72.15; S1.1)
(a) Vol. **106**, No. 1, p. 115—121 (1988).

YAN-ZHONG ZHANG
Magnetic Instability of Metallic Glass $(Fe_{0.1}Ni_{0.33}Co_{0.55}Cr_{0.02})_{78}Si_8B_{14}$. II. Irreversible Magnetic Permeability Decay with Respect to Demagnetization
(75.50; S1.1; S1.2)
(a) Vol. **106**, No. 1, p. 207—215 (1988).

J. W. MOROŃ and M. KŁOSEK
Remarks on the Nature of the High Internal Friction Peak near T_g in Glassy Alloys
(62.40; S1.1; S1.2)
(a) Vol. **106**, No. 1, p. K13—K17 (1988).

S. ROTH and G. STEPHANI
Influence of the Melt Temperature on the Formation of Soft Magnetic Properties of an Amorphous $Fe_{40}Ni_{40}P_{14}B_6$ Alloy
(75.50; 75.60; S1.1; S1.2)
(a) Vol. **106**, No. 1, p. K61—K63 (1988).

U. R. KEVORKYAN
The Simulation of Inactivation Reactions of Point Defect Interactions with Impurities and Impurity–Defect Complexes
(61.70; S1.1)
(a) Vol. **106**, No. 2, p. 379—386 (1988).

YU. S. NECHAEV and K. H. G. EGZIABHER
On Gas Permeability, Diffusivity, and Characteristics of Near-Dislocation Segregation Regions in Metals
(66.30; 61.70; 64.75; S1.1; S1.2)
(a) Vol. **106**, No. 2, p. 399—406 (1988).

A. WĘGRZYN, K. KUŁAKOWSKI, and Z. OBUSZKO
Magnetostriction and Magnetization of Fe-Rich $(Fe_{1-x}Ni_x)_{90}Zr_{10}$ Amorphous Alloys
(75.80; S1.1; S1.2)
(a) Vol. **106**, No. 2, p. K169—K172 (1988).

L. G. SHPINKOVA and A. A. SOROKIN
Magnetic Quenching of the Electric Quadrupole Interaction of ^{181}Ta in Ferromagnetic Laves Phases AFe_2 (A = Zr, Y, Lu)
(75.30 and 76.80; S1.1)
(b) Vol. **147**, No. 1, p. K75—K78 (1988).

JINGHUA TIAN, JINGKUI LIANG, DAYU YANG, HANJIE FAN, and FANGHUA LI
The Study of the One-Dimensional Composition-Incommensurate Structure of $Pr_{1+\varepsilon}Fe_4B_4$
(61.55; S1.1; S1.4)
(a) Vol. **107**, No. 1, p. 45—53 (1988).

P. KWAPULIŃSKI, J. RASEK, and Z. GIERAK
Scattering of Conductivity Electrons on Grain Boundaries in Metals
(72.15; S1.1; S1.2)
(a) Vol. **107**, No. 1, p. 299—304 (1988).

J. F. FUERTES and M. TEJEDOR
Wrinkle-Like Domain Structures on Imperfect Iron Whiskers
(75.60; S1.1)
(a) Vol. **107**, No. 1, p. 339—346 (1988).

F. SATO, S. ISHIO, and T. MIYAZAKI
Soft Magnetic Properties of Rapidly Quenched Fe–Al–M (M = Si, B, Ge) Alloys
(75.50; 75.60; S1.1)
(a) Vol. **107**, No. 1, p. 355—363 (1988).

GUOAN CHENG, HENGDE LI, and BAIXIN LIU
Enhanced Adhesion in Metal/Oxide Pairs by Ion Irradiation
(61.80 and 68.48; S1.1; S1.2; S10.1)
(a) Vol. **107**, No. 1, p. K29—K32 (1988).

A. KOWALCZYK, P. STEFAŃSKI, and A. WRZECIONO
Magnetic Properties of $Pr_2Fe_{14-x-y}Co_xCr_yB$ Compounds
(75.50; S1.1)
(a) Vol. **107**, No. 1, p. K61—64 (1988).

M. BUDZYŃSKI, M. SUBOTOWICZ, H. NIEZGODA, S. SPUSTEK, W. TAŃSKA-KRUPA, and R. WASIEWICZ
The Hyperfine Fields and Their Pressure Derivatives in $(Y_{0.9}Hf_{0.1})Fe_2$ and $(Zr_{0.9}Hf_{0.1})Fe_2$
(76.80; 71.70; S1.1)
(b) Vol. **147**, No. 2, p. 685—689 (1988).

J. CHOJCAN and M. SZUSZKIEWICZ
The Mössbauer Effect and Positron Annihilation Studies of Chromium–Iron Alloys
(76.80 and 78.70; S1.1)
(b) Vol. **147**, No. 2, p. K201—K204 (1988).

C. VAROTSOS, K. EFTAXIAS, and V. HADJICONTIS
Correlation of the Diffusion Coefficients of Various Elements Diffusing in Ferromagnetic and Paramagnetic α-Fe
(66.30; S1.1)
(a) Vol. **107**, No. 2, p. K109—K114 (1988).

A. H. WAFIK and G. C. HALLAM
Effect of Tensile Strain on Barkhausen Jumps in Amorphous Metal Ribbons of $Fe_{80}B_{20}$
(75.60; S1.1)
(a) Vol. **107**, No. 2, p. K157—K161 (1988).

M. HADA, K. SUMIYAMA, and Y. NAKAMURA
Structure and Magnetic Properties of Nonequilibrium Fe–Ru Alloys Produced by Vapor Quenching
(68.55 and 75.70; 76.80; S1.1)
(a) Vol. **108**, No. 1, p. 295—302 (1988).

A. RAOUFI and S. ARAJS
Stress Dependence of Magnetoresistance in Crystalline and Amorphous Ferromagnets
(72.15; 75.80; S1.1)
(a) Vol. **108**, No. 1, p. 369—374 (1988).

B. IDZIKOWSKI and A. WRZECIONO
Properties of Amorphous $Sm_xFe_{80-x}B_{20}$ Ribbons
(75.60; 75.80; S1.1; S1.4)
(a) Vol. **108**, No. 1, p. 375—380 (1988).

U. Hartmann
Observation of a New Domain Nucleation Process in Soft Magnetic Materials
(75.60; S1.1)
(a) Vol. **108**, No. 1, p. 387—401 (1988).

D. C. Jiles
Variation of the Magnetic Properties of AISI 4140 Steels with Plastic Strain
(75.60; 62.20; S1.1)
(a) Vol. **108**, No. 1, p. 417—429 (1988).

J. Gwiazda, E. Mariańska, and W. Zych
The Influence of Transition Metal Substitution and $B_{3/2}$ Values for Some Amorphous Alloys
(76.80; 75.50; S1.1)
(a) Vol. **108**, No. 1, p. K67—K72 (1988).

S. Yu. Korostelev, S. G. Psakhe, and V. E. Panin
A Molecular-Dynamic Study of the Changes in the Atomic Structure of a Material Induced by a Shock Wave with a Distorted Front
(61.50 and 62.50; S1.1)
(b) Vol. **148**, No. 2, p. 483—488 (1988).

M. Cinal and S. Olszewski
Quantum-Statistical Approach to Elastic Constants in Metals Influenced by Magnetization
(75.80; 62.20; S1; S1.1; S1.2)
(b) Vol. **148**, No. 2, p. 699—707 (1988).

K.-D. Durst, H. Kronmüller, and W. Ervens
Investigations of the Magnetic Properties and Demagnetization Processes of an Extremely High Coercive $Sm(Co, Cu, Fe, Zr)_{7.6}$ Permanent Magnet. II. The Coercivity Mechanism
(75.30; 75.50; 75.60; S1.1; S1.2)
(a) Vol. **108**, No. 2, p. 705—719 (1988).

I. Škorvánek, B. Idzikowski, A. Zentko, and E. Mosiniewicz-Szablewska
Influence of Neutron Irradiation on the Magnetic Properties of FeNiCrMoSiB Amorphous Alloys
(75.50; 61.80; S1.1; S1.2)
(a) Vol. **108**, No. 2, p. 747—751 (1988).

Yu. V. Zakharov, Yu. I. Mankov, and L. S. Titov
Electric Conductivity of a Metallic Ferromagnet and Magnetization Process
(72.15; 75.60; S1.1)
(b) Vol. **149**, No. 1, p. 241—248 (1988).

Ming Mao, Yuansheng Wang, and Yizhen He
Crystallization of Metallic Glass $Fe_{46}Ni_{31}VSi_{8}B_{14}$
(61.40 and 64.70; S1.1; S1.2)
(a) Vol. **109**, No. 1, p. 39—51 (1988).

R. Manaila, V. Florescu, A. Jianu, and A. Badescu
Icosahedral Phases in the Al–Fe–Cr Alloy System
(61.55; S1.1; S1.2)
(a) Vol. **109**, No. 1, p. 61—66 (1988).

A. E. Ligachev, O. S. Kuzmin, G. V. Potemkin, and V. P. Yanovskii
Structure and Properties of Implanted Steels
(61.70 and 62.20; S1.1)
(a) Vol. **109**, No. 1, p. K11—K14 (1988).

S. K. Mishra and T. N. Singh
Lattice Dynamics of B.C.C. Transition Metals
(63.20; S1; S1.1)
(b) Vol. **149**, No. 2, p. 503—510 (1988).

J. Świerczek and S. Szymura
Temperature Dependence of Magnetization in Amorphous Co–Fe–Si–B Alloys
(75.50; S1.1; S1.2)
(a) Vol. **109**, No. 2, p. 559—567 (1988).

J. L. Sánchez Llamazares, S. García García, S. Díaz Castañón, and N. Suarez Almodovar
On the Effect of Al Addition on the Phase Composition and Magnetic Properties of Nd–Fe–B Permanent Magnet Alloys
(75.30; 61.60; S1.1)
(a) Vol. **109**, No. 2, p. 569—573 (1988).

F. Yassin and V. Christoph
Hysteresis Loops of Fine Particle Magnets of the Type NdFeB
(75.60; S1.1)
(a) Vol. **109**, No. 2, p. K145—K148 (1988).

A. Schröder and M. Feller-Kniepmeier
The Ferromagnetic Diffusion Anomaly of the Diffusion System Iron–Antimony
(66.30; 75.50; S1.1)
(a) Vol. **110**, No. 1, p. 107—121 (1988).

A. Kowalczyk and A. Wrzeciono
Magnetic and Crystallographic Properties of $R_2Fe_{14-x}Cr_xB$ Compounds (R = Y, Nd, and Gd)
(75.30; 61.55; S1.1; S1.2)
(a) Vol. **110**, No. 1, p. 241—245 (1988).

U. Hartmann
Static Magnetization Irreversibilities in Soft Ferromagnetic Media of High Crystalline Perfection
(75.60; S1.1)
(a) Vol. **110**, No. 1, p. 247—257 (1988).

R. B. McLellan and M. L. Wasz
Carbon–Vacancy Interactions in B.C.C. Iron
(61.70; S1.1)
(a) Vol. **110**, No. 2, p. 421—427 (1988).

P. Mrozek, M. Menyhárd, A. Jabłoński, and G. Tyuliev
Surface Composition of the Ordered Fe–Co Alloys
(68.20; 79.20; S1.1; S1.2)
(a) Vol. **110**, No. 2, p. 495—507 (1988).

S.1.2. Other 3d metals

XIAN-WU ZOU, ZHUN-ZHI JIN, and DE-CHENG TIAN
Structure Factors of Liquid Zn, Cd, and Hg
(61.25; S1; S1.2)
(b) Vol. **145**, No. 1, p. 73—76 (1988).

A. P. DESHPANDE, V. B. SAPRE, and C. MANDE
Study of the Environment of Copper in Some Ternary Chalcogenides by EXAFS Spectroscopy
(61.10; S1.2; S8.16)
(b) Vol. **145**, No. 1, p. 77—88 (1988).

D. N. JHA
Calculation of Electrical Resistivity of Some Liquid Metals by the Mode-Coupling Method
(72.10; 72.15; S1.1; S1.2; S1.3; S1.4)
(b) Vol. **145**, No. 1, p. 277—282 (1988).

A. A. HIRSCH and T. MASSALHA
Magnetic Metastable States in the Critical Transitions of Nickel and Iron
(75.30; 76.80; S1.1; S1.2)
(a) Vol. **105**, No. 1, p. 249—262 (1988).

W. APPEL
Structure and Energy of 180° Domain Walls in H.C.P. Cobalt as Functions of the Temperature
(75.60 and 75.70; S1.2)
(a) Vol. **105**, No. 1, p. K67—K69 (1988).

S. ROMANOWSKI
An Application of the Extended CPA Method to Electron Work Function and Surface Charge Calculation for Cu Single Crystals
(73.30; 71.10; S1.2)
(b) Vol. **145**, No. 2, p. 467—482 (1988).

Z. KĄKOL, H. FIGIEL, J. CHMIST, and A. KOŁODZIEJCZYK
The Influence of Manganese on Magnetic Properties of $RE_2(Co_{1-x}Mn_x)_{17}$, RE = Y, Gd Compounds
(75.50; S1.2; S1.4)
(b) Vol. **145**, No. 2, p. 687—693 (1988).

MING MAO and YIZHEN HE
Two-Stage Structural Relaxation of the Metallic Glass $Fe_{46}Ni_{31}V_1Si_8B_{14}$
(61.40; 65.40; S1.1; S1.2)
(a) Vol. **105**, No. 2, p. 325—333 (1988).

I. YA. DEKHTYAR, E. G. MADATOVA, A. ČÍŽEK, and M. ŠOB
Relation between the Strength Characteristics of Iron and Nickel and the Electronic Structure of Defects by Positron Annihilation Spectroscopy
(61.70; 78.70; S1.1; S1.2)
(a) Vol. **105**, No. 2, p. 357—363 (1988).

N. Kapoor and A. N. Nigam
Influence of Hydrogen Diffusion on Thermal Conductivity of Nickel
(66.30 and 72.15; S1.2)
(a) Vol. **105**, No. 2, p. 403—409 (1988).

R. Tsumura, D. Ríos-Jara, M. Chávez, L. Rodríguez, T. Akachi, and R. Escudero
Specific Heat Measurements of the β_1 and β_1' Phases in a Cu–Zn–Al Alloy
(65.40; S1.2)
(a) Vol. **105**, No. 2, p. 411—418 (1988).

M. H. Xu, S. Patu, and Z. G. Wang
The Improvement in Fatigue Life of Pure Polycrystalline Nickel by Nitrogen Ion Implantation
(62.20; 61.70; S1.2)
(a) Vol. **105**, No. 2, p. 419—425 (1988).

Litian Wang and Tingsui Ge (T. S. Ke)
Nonlinear Internal Friction in Mn–Cu Martensite
(62.40; 64.70; S1.2)
(a) Vol. **105**, No. 2, p. 447—453 (1988).

Yan-Zhong Zhang
Magnetic Instability of Metallic Glass $(Fe_{0.1}Ni_{0.33}Co_{0.55}Cr_{0.02})_{78}Si_8B_{14}$. I. Reversible Disaccommodation Phenomenon with Respect to Demagnetization
(75.50; S1.1; S1.2)
(a) Vol. **105**, No. 2, p. 579—587 (1988).

D. M. Vanderwalker
The Formation of Hydrides in Titanium
(61.50; 61.70; S1.2)
(a) Vol. **105**, No. 2, p. K77—K80 (1988).

K. Rajanna and S. Mohan
Response of Thin Film Strain Gauge to Tensile and Compressive Strains by Cantilever Technique
(62.20 and 73.60; S1.2)
(a) Vol. **105**, No. 2, p. K181—K184 (1988).

E. Majková and P. Duhaj
XPS Study of Co–Fe–B Metallic Glasses
(71.25 and 79.60; S1.2)
(b) Vol. **146**, No. 1, p. K17–K20 (1988).

K. Z. Botros, S. M. Saleh, A. H. G. El-Dhaher, and S. Hajaig
On the Visibility of Edge Fringes in Strong Beam Images of Intrinsic and Extrinsic Stacking Faults in F.C.C. Structures
(61.14 and 61.70; S1.2)
(a) Vol. **106**, No. 1, p. 49—55 (1988).

T. T. Dung, N. P. Thuy, N. M. Hong, and T. D. Hien
Anomaly in the Magnetic Properries of the YCo_4B Compound
(75.30; S1.2; S1.4)
(a) Vol. **106**, No. 1, p. 201—206 (1988).

YAN-ZHONG ZHANG
Magnetic Instability of Metallic Glass $(Fe_{0.1}Ni_{0.33}Co_{0.55}Cr_{0.02})_{78}Si_8B_{14}$. II. Irreversible Magnetic Permeability Decay with Respect to Demagnetization
(75.50; S1.1; S1.2)
(a) Vol. **106**, No. 1, p. 207—215 (1988).

Y. F. LI, L. HEDMAN, and Ö. RAPP
Magnetic Susceptibility of Amorphous $Cu_{1-x}Y_x$ Alloys
(75.50; S1.2; S1.4)
(a) Vol. **106**, No. 1, p. 233—237 (1988).

J. W. MOROŃ and M. KŁOSEK
Remarks on the Nature of the High Internal Friction Peak near T_g in Glassy Alloys
(62.40; S1.1; S1.2)
(a) Vol. **106**, No. 1, p. K13—K17 (1988).

S. ROTH and G. STEPHANI
Influence of the Melt Temperature on the Formation of Soft Magnetic Properties of an Amorphous $Fe_{40}Ni_{40}P_{14}B_6$ Alloy
(75.50; 75.60; S1.1; S1.2)
(a) Vol. **106**, No. 1, p. K61—K63 (1988).

L. VATSKICHEV and M. VATSKICHEVA
Relations between Some Galvanomagnetic and Magnetic Properties of Thin Ferromagnetic Films
(72.15 and 75.70; S1.2; S1.4)
(a) Vol. **106**, No. 1, p. K75—K78 (1988).

V. B. GUPTA and H. C. GUPTA
Photon Dispersion Curves of the Superconducting A15 Compounds Nb_3Ge and Nb_3(Al–Ge)
(63.20; S1.2)
(b) Vol. **146**, No. 2, p. 481—486 (1988).

K. HAVANCSÁK, G. SZENES, V. A. SKURATOV, YUN DUN MAN, W. MALINOWSKI, and V. A. SHCHEGOLEV
233 MeV Ne Ion Irradiation Effects on the Mechanical Properties of Copper
(61.80; 62.20; S1.2)
(a) Vol. **106**, No. 2, p. 371—378 (1988).

YU. S. NECHAEV and K. H. G. EGZIABHER
On Gas Permeability, Diffusivity, and Characteristics of Near-Dislocation Segregation Regions in Metals
(66.30; 61.70; 64.75; S1.1; S1.2)
(a) Vol. **106**, No. 2, p. 399—406 (1988).

P. CZURRATIS, R. KROGGEL, V. SYNEČEK, M. SIMERSKA, and H. LÖFFLER
Dislocation Loops in Al–Zn and Al–Zn–Mg Alloys
(61.70; 61.55; S1.2; S4)
(a) Vol. **106**, No. 2, p. 407—418 (1988).

A. UNKROTH, W. ANDRÄ, and G. GÖTZ
Composition Dependence of Magnetization of Ion Implanted Cobalt Layers
(75.70; 61.80; S1.2; S1.61; S1.62)
(a) Vol. **106**, No. 2, p. 589—594 (1988).

R. E. Hummel, W. M. Goho, and V. Gerold
The Influence of Precipitation on the Optical Properties and Electronic Structure of Copper–Cobalt Alloys
(61.55 and 78.20; S1.2)
(a) Vol. **106**, No. 2, p. 641—649 (1988).

A. Węgrzyn, K. Kułakowski, and Z. Obuszko
Magnetostriction and Magnetization of Fe-Rich $(Fe_{1-x}Ni_x)_{90}Zr_{10}$ Amorphous Alloys
(75.80; S1.1; S1.2)
(a) Vol. **106**, No. 2, p. K169—K172 (1988).

Zhun-Zhi Jin, Xian-Wu Zou, and Yong-Qiao Yao
Dependence of Structure Factors upon Temperature for Liquid Metals, Zn and Hg
(61.25; S1; S1.2)
(b) Vol. **147**, No. 1, p. 45—49 (1988).

D. T. Britton and P. C. Rice-Evans
Fast Fourier Transform Deconvolution of Positron Annihilation Doppler-Broadened Lineshapes
(61.70 and 78.70; S1.2; S5)
(b) Vol. **147**, No. 1, p. 89—96 (1988).

V. Drchal and J. Kudrnovský
Theory of Auger Spectra from Alloys with Partially Filled Narrow Bands
(71.10 and 79.20; S1.2)
(b) Vol. **147**, No. 1, p. 215—222 (1988).

R. E. Hummel and W. M. Goho
Optical Properties and Electronic Structure of Copper–Cobalt Alloys
(71.50 and 78.20; S1.2)
(b) Vol. **147**, No. 1, p. 417—420 (1988).

R. Ramji Rao and A. Rajput
Phonon Dispersion Relations in Silver and Copper
(63.20; S1.2; S1.3)
(b) Vol. **147**, No. 1, p. K11—K14 (1988).

P. Czarnecki, B. Idzikowski, and A. Wrzeciono
Magnetoresistance of $Pr_xFe_{80-x}B_{20}$ and $Ho_xCo_{70-x}B_{30}$ Amorphous Alloys in Magnetic Fields up to 22 T
(72.20; S1.2)
(b) Vol. **147**, No. 1, p. K47—K50 (1988).

An Hu, Shu-yi Zhang, Xiang-yang Yuan, Qing Shen, Zhong-nan Lu, and Duan Feng
Lattice Expansion and Phonon Softening in Nb/Cu Superlattices
(63.20 and 68.30; 61.55; S1.2)
(a) Vol. **107**, No. 1, p. 153—159 (1988).

M. Z. Butt and I. M. Ghauri
Effect of Short-Range Order on the Temperature Dependence of Plastic Flow in α-Brasses
(62.20; S1.2)
(a) Vol. **107**, No. 1, p. 187—195 (1988).

A. J. Abu El-Haija, K. A. Al-Saleh, N. A. Halim, J. M. Khalifeh, and N. S. Saleh
Ion Beam Induced Mixing of Cu/Si System Using Electrical Resistivity and RBS Measurements
(61.80 and 73.60; S1.2; S5.11)
(a) Vol. **107**, No. 1, p. 253—260 (1988).

P. Kwapuliński, J. Rasek, and Z. Gierak
Scattering of Conductivity Electrons on Grain Boundaries in Metals
(72.15; S1.1; S1.2)
(a) Vol. **107**, No. 1, p. 299—304 (1988).

F. Sachslehner
The Influence of Precipitates on the Low-Field Hall Coefficient of Cu–Be 2
(72.15; S1.2)
(a) Vol. **107**, No. 1, p. 305—313 (1988).

J. M. Alameda, M. C. Contreras, F. Carmona, and F. López
Analysis of Magnetic Anisotropy and Magnetization Dispersion in Polycrystalline Films Evaporated at Low Incidence Angles ($\alpha < 60°$)
(75.70; S1.2)
(a) Vol. **107**, No. 1, p. 329—337 (1988).

Guoan Cheng, Hengde Li, and Baixin Liu
Enhanced Adhesion in Metal/Oxide Pairs by Ion Irradiation
(61.80 and 68.48; S1.1; S1.2; S10.1)
(a) Vol. **107**, No. 1, p. K29—K32 (1988).

Ningsheng Luo, Wenlan Xu, and Shuechu Shen
Application of the Embedded Atom Method to Phonon Dispersions and Local Surface Vibrations in Copper
(63.20; 68.30; S1.2)
(b) Vol. **147**, No. 2, p. 511—516 (1988).

S. B. Shrivastava, V. K. Gupta, and K. S. Bhadoriya
Positronium at Metal Surfaces
(71.60 and 73.20; S1.2; S1.3)
(b) Vol. **147**, No. 2, p. 553—557 (1988).

M. F. Burkhard, H. Rothard, and K.-O. E. Groeneveld
Single-Electron Deexcitation of Volume Plasmons Induced by Heavy Ions in Thin Solid Foils
(71.45; 79.20; S1.2; S4; S5)
(b) Vol. **147**, No. 2, p. 589—592 (1988).

A. W. Sheikh and M. El-Fazani
A Phenomenological Model for the Magnetic Resistivity of NM–TM Spin Glasses
(72.15; 75.30; S1.2; S1.3)
(b) Vol. **147**, No. 2, p. 601—612 (1988).

N. K. Flevaris, Th. Karakostas, and J. Stoemenos
Coherency Strains and Growth of Artificially Modulated Cu–Ni Thin Films
(68.55; S1.2)
(a) Vol. **107**, No. 2, p. 579—587 (1988).

M. de Graef, L. Delaey, and D. Broddin
High Resolution Electron Microscopic Study of the X-Phase in Cu–Al and Cu–Al–Zn Alloys
(61.55; 61.16; S1.2; S4)
(a) Vol. **107**, No. 2, p. 597—609 (1988).

Y. Hayashi
Ordering of Ni_3Mn and Diffusivity of Hydrogen in the Alloy
(61.55 and 66.30; S1.2)
(a) Vol. **107**, No. 2, p. 619—624 (1988).

R. Hoogewijs, P. de Volder, L. Fiermans, and J. Vennik
Room Temperature Oxidation in Air of Polycrystalline CuZn
(68.20; 79.60; S1.2; S10)
(a) Vol. **107**, No. 2, p. 625—631 (1988).

A. Loiseau and C. Vannuffel
$TiAl_2$: A Reentrant Phase in the Ti–Al System
(64.70; 61.16; S1.2; S4)
(a) Vol. **107**, No. 2, p. 655—671 (1988).

J. Menon and C. Suryanarayana
The Al–Co Decagonal Phase
(61.55; S1.2; S4)
(a) Vol. **107**, No. 2, p. 693—708 (1988).

E. Ruedl, G. Valdrè, P. Delavignette, and U. Valdrè
Structural and Analytical Electron Microscopy of Dislocation Loops and Precipitates Formed during Ageing of an α-Bombarded Mn–Cr Austenitic Steel
(61.55 and 61.70; 61.16; S1.2)
(a) Vol. **107**, No. 2, p. 745—758 (1988).

G. Torres-Villasenor and G. van Tendeloo
Guinier-Preston Zones in the α-Phase of Superplastic Zn–Al Alloy
(61.70; 62.20; S1.2; S4)
(a) Vol. **107**, No. 2, p. 785—790 (1988).

C. Sarrazin, R. J. Gaboriaud, and J. P. Rivière
TEM Investigations of Ni–Cu Thin Film Coatings Obtained by Multilayer Technique, Coevaporation, and Ion Beam Assisted Deposition
(68.55; 61.16; S1.2)
(a) Vol. **107**, No. 2, p. 867—871 (1988).

J. Mimkes
Thermal Expansion, Heat of Fusion, and Electron Density of Monovalent Metals
(65.70; 65.50; S1.2; S1.3; S2)
(b) Vol. **148**, No. 1, p. 91—100 (1988).

E. Dębowska and J. Rudzińska-Girulska
Electronic Structure of $Pd_{1-x}Ni_x$ Alloys Investigated by Positron Annihilation
(71.25; 71.60; S1.2)
(b) Vol. **148**, No. 1, p. 227—231 (1988).

V. FRITZSCHE
Calculation of Matrix Elements for Optical Transitions
(78.20; S1.2)
(b) Vol. **148**, No. 1, p. 395—399 (1988).

M. HASHIMOTO, H. SAKA, T. IMURA, S. JITSUKAWA, and K. SUZUKI
Effect of Alloying on the Void Swelling in Cu-Base and Ag-Base Alloys
(61.70; 64.75; S1.2; S1.3)
(a) Vol. **108**, No. 1, p. 163—170 (1988).

S. ARAJS and T. A. JENKINS
Electrical Resistivity of Some Directionally Solidified Nickel-Base Eutectic Composites
(72.15; S1.2)
(a) Vol. **108**, No. 1, p. 343—349 (1988).

M. A. FAHIM and A. R. ALI
On the Influence of Copper Atoms on Domain Wall Energy and Pinning Force in Ni–Cu Alloys
(75.60; S1.2)
(a) Vol. **108**, No. 1, p. 381—386 (1988).

K.-D. DURST, H, KRONMÜLLER, and W. ERVENS
Investigations of the Magnetic Properties and Demagnetization Processes of an Extremely High Coercive $Sm(Co, Cu, Fe, Zr)_{7.6}$ Permanent Magnet. I. Determination of Intrinsic Magnetic Material Parameters
(75.30; 75.50; 75.60; S1.2)
(a) Vol. **108**, No. 1, p. 403—416 (1988).

J. SZABLEWSKI and B. KUŹNICKA
Activation Energy for Precipitation in Undeformed and Deformed Cu–Cr Alloys
(64.75 and 66.30; S1.2)
(a) Vol. **108**, No. 1, p. K5—K8 (1988).

B. SCHÖNFELD, L. REINHARD, G. KOSTORZ, and W. BÜHRER
Short-Range Order and Atomic Displacements in Ni–20 at% Cr Single Crystals
(61.55 and 64.80; S1.2)
(b) Vol. **148**, No. 2, p. 457—471 (1988).

S. ROMANOWSKI
Study of the Polarized Light Absorption on Electrically Changed Thin Films of Copper Monocrystals by ECPA
(78.65; S1.2)
(b) Vol. **148**, No. 2, p. 515—524 (1988).

M. CINAL and S. OLSZEWSKI
Quantum-Statistical Approach to Elastic Constants in Metals Influenced by Magnetization
(75.80; 62.20; S1; S1.1; S1.2)
(b) Vol. **148**, No. 2, p. 699—707 (1988).

D. PRAKASH and J. C. UPADHYAYA
Three-Body Effect on the Lattice Dynamics of Pt and Co
(63.20; S1.2)
(b) Vol. **148**, No. 2, p. K109—K112 (1988).

Y. Kikuchi, K. Watanabe, S. Naitoh, K. Hiratsuka, and H. Yamaguchi
HREM Image of a Vacancy in Copper
(61.70; 61.16; S1.2)
(a) Vol. **108**, No. 2, p. 509—517 (1988).

P. J. Ratcliffe and R. A. Collins
Thermal Oxidation of Ion Implanted Copper
(61.70; 73.40; S1.2)
(a) Vol. **108**, No. 2, p. 537—543 (1988).

P. M. Ossi
Elastic Anomalies and Martensite Nucleation in β-Phase Alloys
(62.30 and 64.70; S1.2; S1.3)
(a) Vol. **108**, No. 2, p. 587—598 (1988).

Jin-Song Zhu, R. Schaller, and W. Benoit
Internal Friction Transistory Effects Associated with Martensitic Transformation in Ni–Ti Alloys
(62.40 and 64.70; S1.2)
(a) Vol. **108**, No. 2, p. 613—618 (1988).

K.-D. Durst, H. Kronmüller, and W. Ervens
Investigations of the Magnetic Properties and Demagnetization Processes of an Extremely High Coercive $Sm(Co, Cu, Fe, Zr)_{7.6}$ Permanent Magnet. II. The Coercivity Mechanism
(75.30; 75.50; 75.60; S1.1; S1.2)
(a) Vol. **108**, No. 2, p. 705—719 (1988).

I. Škorvánek, B. Idzikowski, A. Zentko, and E. Mosiniewicz-Szablewska
Influence of Neutron Irradiation on the Magnetic Properties of FeNiCrMoSiB Amorphous Alloys
(75.50; 61.80; S1.1; S1.2)
(a) Vol. **108**, No. 2, p. 747—751 (1988).

Ming Mao, Yuansheng Wang, and Yizhen He
Crystallization of Metallic Glass $Fe_{46}Ni_{31}VSi_8B_{14}$
(61.40 and 64.70; S1.1; S1.2)
(a) Vol. **109**, No. 1, p. 39—51 (1988).

R. Manaila, V. Florescu, A. Jianu, and A. Badescu
Icosahedral Phases in the Al–Fe–Cr Alloy System
(61.55; S1.1; S1.2)
(a) Vol. **109**, No. 1, p. 61—66 (1988).

J. Olfe and H. Neuhäuser
Dislocation Groups, Multipoles, and Friction Stresses in α-CuZn Alloys
(61.70; 62.20; S1.2)
(a) Vol. **109**, No. 1, p. 149—160 (1988).

A. R. Ali and M. A. Fahim
On the Influence of Internal Stress and Magnetizing Frequency on Domain Wall Motion in High-Purity Ni
(75.60; S1.2)
(a) Vol. **109**, No. 1, p. 313—318 (1988).

B. Decamps and M. Condat
$a/2\langle 110\rangle$ Dislocations at the γ/γ' Interface in a Ni-Based Superalloy
(61.70; S1.2)
(a) Vol. **109**, No. 1, p. K15—K18 (1988).

K. Schröder and H. Hejase
Magnetism in Thin Chromium Films
(75.70; S1.2)
(b) Vol. **149**, No. 2, p. 685—690 (1988).

J. Świerczek and S. Szymura
Temperature Dependence of Magnetization in Amorphous Co–Fe–Si–B Alloys
(75.50; S1.1; S1.2)
(a) Vol. **109**, No. 2, p. 559—567 (1988).

A. Unkroth and W. Andrä
Coercivity of Ion Implanted Cobalt Films
(75.60; S1.2)
(a) Vol. **109**, No. 2, p. K149—K152 (1988).

D. Kashchiev and Yu. O. Kanter
Oscillations of Specular Beam Intensity in Reflection Diffraction from the Surface of a Growing Epitaxial Film. A Theoretical Study
(61.14 and 68.55; S1.2)
(a) Vol. **110**, No. 1, p. 61—76 (1988).

F. J. Bremer, M. Beyss, and H. Wenzl
The Order–Disorder Transition of the Intermetallic Phase Ni_3Al
(61.50; S1.2)
(a) Vol. **110**, No. 1, p. 77—82 (1988).

J. M. Holender, J. Sołtys, G. Wendrock, and P. Czurratis
Non-Isothermal Studies of the Eutectoidal Decomposition of the Alloy Al–59 at% Zn
(64.70; 61.10; 61.55; S1.2; S4)
(a) Vol. **110**, No. 1, p. 123—132 (1988).

L. Smardz and J. Baszyński
Average Magnetization of Compositionally Modulated Cu/Ni Films
(75.70; S1.2)
(a) Vol. **110**, No. 1, p. 227—230 (1988).

K. Polański, I. Zasada, K. Dolecki, and E. Misiak
The Structural Phase Transition in Cobalt Monocrystals Investigated by Means of SEM
(64.70 and 75.60; S1.2)
(a) Vol. **110**, No. 1, p. 231—240 (1988).

A. Kowalczyk and A. Wrzeciono
Magnetic and Crystallographic Properties of $R_2Fe_{14-x}Cr_xB$ Compounds (R = Y, Nd, and Gd)
(75.30; 61.55; S1.1; S1.2)
(a) Vol. **110**, No. 1, p. 241—245 (1988).

A. Z. Mohamed, M. M. Mostafa, M. S. Sakr, and A. A. El-Daly
The Role of Grain Size on the Work-Hardening Characteristics of Cu–Al Solid Solutions
(62.20; S1.2)
(a) Vol. **110**, No. 1, p. K13—K17 (1988).

Guoan Wu, Yuanda Dong, Keqin Xiao, Yizhen He, Longshu Wang, and Jiaqing He
An Investigation of Fast Neutron Irradiation Effects on the Structure of $Cu_{50}Ti_{50}$ Metallic Glass
(61.40; 64.70; S1.2)
(a) Vol. **110**, No. 2, p. 317—321 (1988).

U. D. Kulkarni, S. Muralidhar, and S. Banerjee
Computer Simulation of the Early Stages of Ordering in Ni–Mo Alloys
(61.55; S1.2)
(a) Vol. **110**, No. 2, p. 331—345 (1988).

A. Serra and N. de Diego
Characterization of Defects in Deformed Titanium
(61.70; 78.70; S1.2)
(a) Vol. **110**, No. 2, p. 409—414 (1988).

H.-A. Kuhn and H. G. Sockel
Comparison between Experimental Determination and Calculation of Elastic Properties of Nickel-Base Superalloys between 25 and 1200 °C
(62.20; S1.2)
(a) Vol. **110**, No. 2, p. 449—458 (1988).

P. Mrozek, M. Menyhárd, A. Jabłoński, and G. Tyuliev
Surface Composition of the Ordered Fe–Co Alloys
(68.20; 79.20; S1.1; S1.2)
(a) Vol. **110**, No. 2, p. 495—507 (1988).

H. Krautz, Ch. Wenzel, K. Bornkessel, and G. Blasek
Barrier Behaviour of TiW between Copper and Aluminium
(68.48; S1.2; S4)
(a) Vol. **110**, No. 2, p. K77—K79 (1988).

C. H. Westphal and C. C. Becerra
Magnetic Transitions in $Mn_{1-x}Ni_x(HCOO)_2 \cdot 2\,H_2O$
(75.50; S1.2; S11)
(a) Vol. **110**, No. 2, p. K97—K100 (1988).

S1.3. Noble metals

D. N. Jha
Calculation of Electrical Resistivity of Some Liquid Metals by the Mode-Coupling Method
(72.10; 72.15; S1.1; S1.2; S1.3; S1.4)
(b) Vol. **145**, No. 1, p. 277—282 (1988).

I. J. Jabr, J. D. Meyer, K. A. Al-Saleh, and N. S. Saleh
Ion-Beam Mixing of Te/Au. Metastable Phase Formation
(61.80 and 64.75; 73.60; S1.3; S5)
(a) Vol. **105**, No. 1, p. 177—181 (1988).

A. Extremera
The Magnetic Behaviour of $NaHg_4$ Peritectic
(75.30; 61.55; S1.3)
(a) Vol. **105**, No. 1, p. 281—284 (1988).

P. Royer, J. P. Goudonnet, T. Inagaki, G. Chabrier, and E. T. Arakawa
Photoacoustic Study of the Optical Absorption of Oblate Silver Spheroids in Attenuated-Total-Reflection Geometry
(78.20; 78.65; S1.3)
(a) Vol. **105**, No. 2, p. 617—625 (1988).

D. B. Tran Thoai and E. Zeitler
Image Potential of an Electron in a Gap between Two Metals
(71.45 and 73.20; S1.3; S4)
(b) Vol. **146**, No. 1, p. 137—140 (1988).

K. Meinel, M. Klaua, and H. Bethge
Segregation and Sputter Effects on Perfectly Smooth (111) and (100) Surfaces of Au–Ag Alloys Studied by AES
(68.40 and 68.55; 64.75; 79.20; S1.3)
(a) Vol. **106**, No. 1, p. 133—144 (1988).

M. Pattabi, M. S. M. Sastry, and V. S. Sivaramakrishnan
Studies on the Stability of Discontinuous Silver Films with Overlayers of Al_2O_3 and SiO_2
(73.60; S1.3; S10.1)
(a) Vol. **106**, No. 1, p. 145—152 (1988).

E. M. Shpilevski and D. A. Gorbachevskii
Argon Ion Effect on Solid Phase Reaction in Copper–Tin, Copper–Antimony Films
(61.80 and 68.48; 66.30; S1.3; S4)
(a) Vol. **106**, No. 1, p. K23—K26 (1988).

K. Meinel, M. Klaua, Ch. Ammer, and H. Bethge
AES Investigations on Low-Temperature Interdiffusion in Thin Au Films
on Ag(111) Substrates
(66.30 and 68.48; 68.60; S1.3)
(a) Vol. **106**, No. 2, p. 493—507 (1988).

R. Ramji Rao and A. Rajput
Phonon Dispersion Relations in Silver and Copper
(63.20; S1.2; S1.3)
(b) Vol. **147**, No. 1, p. K11—K14 (1988).

E. Jároli, B. Pécz, R. Veresegyházy, F. Pászti, T. Lohner, M. Fried, I. Mojzes, and J. Gyulai
Effect of Ion Beam Mixing on the Evolution of Arsenic from the Au–GaAs System
(61.80 and 68.48; S1.3; S7.12)
(a) Vol. **107**, No. 1, p. K15—K17 (1988).

S. B. Shrivastava, V. K. Gupta, and K. S. Bhadoriya
Positronium at Metal Surfaces
(71.60 and 73.20; S1.2; S1.3)
(b) Vol. **147**, No. 2, p. 553—557 (1988).

J. Graefenstein, I. Mertig, E. Mrosan, V. N. Antonov, and Vl. N. Antonov
Calculated Fermi Surface Characteristics of the Noble Metals
(71.10; 71.25; 71.38; S1.3)
(b) Vol. **147**, No. 2, p. 575—582 (1988).

A. W. Sheikh and M. El-Fazani
A Phenomenological Model for the Magnetic Resisitivity of NM–TM Spin Glasses
(72.15; 75.30; S1.2; S1.3)
(b) Vol. **147**, No. 2, p. 601—612 (1988).

M. D. Tiwari, G. Thummes, and V. Kukermann
Deviations from Matthiessen's Rule in Ag–Au Alloys
(72.15; S1.3)
(b) Vol. **147**, No. 2, p. 661—675 (1988).

M. Nakagawa and K. Yasuda
Phase Changes and Structural Features of Cu_3Au II in a Gold–Copper–Silver Ternary Alloy during Isothermal Ageing
(61.55 and 64.70; S1.3)
(a) Vol. **107**, No. 2, p. 709—717 (1988).

J. Ledzion
Electrical Conductivity of Thin Metallic Films
(73.60; S1.3)
(a) Vol. **107**, No. 2, p. K115—K118 (1988).

J. Mimkes
Thermal Expansion, Heat of Fusion, and Electron Density of Monovalent Metals
(65.70; 65.50; S1.2; S1.3; S2)
(b) Vol. **148**, No. 1, p. 91—100 (1988).

Liang-Yao Chen and D. W. Lynch
The Optical Properties of $AuAl_2$ and $PtAl_2$
(78.20; S1.3; S4)
(b) Vol. **148**, No. 1, p. 387—394 (1988).

J. Kudrnovský and V. Drchal
The Electronic Structure of Palladium–Noble Metal Alloys
(71.20; 71.55; S1.3)
(b) Vol. **148**, No. 1, p. K23—K27 (1988).

M. Hashimoto, H. Saka, T. Imura, S. Jitsukawa, and K. Suzuki
Effect of Alloying on the Void Swelling in Cu-Base and Ag-Base Alloys
(61.70; 64.75; S1.2; S1.3)
(a) Vol. **108**, No. 1, p. 163—170 (1988).

F. Khater and M. El-Hiti
Strain Coefficients of Electrical Resistance of Double-Layer Thin Metallic Films
(73.60; 72.15; S1.3)
(a) Vol. **108**, No. 1, p. 241—249 (1988).

E. Heiner
Nonequilibrium Thermalized Stage of Noble Metals
(71.10; 65.40; 71.38; S1.3)
(b) Vol. **148**, No. 2, p. 599—609 (1988).

P. M. Ossi
Elastic Anomalies and Martensite Nucleation in β-Phase Alloys
(62.30 and 64.70; S1.2; S1.3)
(a) Vol. **108**, No. 2, p. 587—598 (1988).

S. Nara
A Note on the Icosahedral Morphology of Ultrafine Metal Particles
(68.55; 61.70; 73.20; S1.3)
(b) Vol. **149**, No. 2, p. 555—564 (1988).

F. Khater and M. El-Hiti
Temperature Coefficient of the Strain Coefficient of Electrical Resistivity of Double-Layer Thin Metallic Films
(72.10 and 73.60; S1.3)
(a) Vol. **109**, No. 2, p. 517—523 (1988).

K. Meinel and M. Klaua
On Pore Formation in Thin Film/Substrate Interdiffusion at Low Temperatures
(66.30 and 68.48; 68.60; S1.3)
(a) Vol. **109**, No. 2, p. K97—K100 (1988).

K. Meinel, M. Klaua, and H. Bethge
On Twin and Stacking Fault Formation during the Epitaxial Film Growth of F.C.C. Materials on (111) Substrates
(61.70 and 68.55; S1.3)
(a) Vol. **110**, No. 1, p. 189—196 (1988).

A. Heilmann, C. Hamann, and Do Ngoc Uan
Fractalic Growth in Metal-Cluster Polymer Composite Films
(68.55; S1.3)
(a) Vol. **110**, No. 1, p. K29—K31 (1988).

L. J. Huang, E. Ma, and B. X. Liu
The Band Structure and the Anomalous Alloying Behavior of Noble Metal Based Systems
(64.75; 71.25; S1.3)
(a) Vol. **110**, No. 2, p. 443—448 (1988).

Xingzhao Ding and Yizhen He
A Study of Structural Relaxation in Glassy $Pd_{77.5}Cu_{6.0}Si_{16.5}$ by Microcalorimetric Measurements
(64.70 and 65.50; S1.3)
(a) Vol. **110**, No. 2, p. K67—K71 (1988).

R. Ewertowski, A. Baranowski, and W. Świątkowski
Positron Trapping at Ag/Na_3AlF_6 Interface; Life-Time Measurements
(78.70; S1.3; S9)
(a) Vol. **110**, No. 2, p. K101—K105 (1988).

S1.4. Lanthanides

D. N. Jha
Calculation of Electrical Resistivity of Some Liquid Metals by the Mode-Coupling Method
(72.10; 72.15; S1.1; S1.2; S1.3; S1.4)
(b) Vol. **145**, No. 1, p. 277—282 (1988).

E. A. Goremychkin, E. Mühle, and I. Natkaniec
Refinement of the Crystal Field Parameters of $PrCu_5$ by Inelastic Neutron Scattering
(71.70; S1.4)
(b) Vol. **145**, No. 1, p. K15—K19 (1988).

A. A. Kazakov and S. M. Kleshchev
Parametric Green Function Method in the Heisenberg Model
(75.10; 75.30; 75.40; S1.4)
(b) Vol. **145**, No. 2, p. 659—667 (1988).

Tiesong Zhao, Yong Zhu, and Hanmin Jin
Estimation of the Crystal Field at R Sites in $R_2Fe_{14}B$ Compounds
(75.30; 71.70; S1.1; S1.4)
(b) Vol. **145**, No. 2, p. 679—686 (1988).

Z. Kąkol, H. Figiel, J. Chmist, and A. Kołodziejczyk
The Influence of Manganese on Magnetic Properties of $RE_2(Co_{1-x}Mn_x)_{17}$, RE = Y, Gd Compounds
(75.50; S1.2; S1.4)
(b) Vol. **145**, No. 2, p. 687—693 (1988).

A. Apostolov, M. Mikhov, N. Scheludko, P. Tcholakov, and V. Skumriev
The Effect of Zr-Substitution of Co on the Intersublattice Exchange in $Gd_2Co_{17-x}Zr_x$
(75.30; S1.4)
(a) Vol. **105**, No. 2, p. K145—K149 (1988).

B. N. Onwuagba
Phonon Spectrum in Metastable F.C.C. Lanthanum
(63.20; S1.4)
(b) Vol. **146**, No. 1, p. 131—136 (1988).

T. T. Dung, N. P. Thuy, N. M. Hong, and T. D. Hien
Anomaly in the Magnetic Properties of the YCo_4B Compound
(75.30; S1.2; S1.4)
(a) Vol. **106**, No. 1, p. 201—206 (1988).

Y. F. Li, L. Hedman, and Ö. Rapp
Magnetic Susceptibility of Amorphous $Cu_{1-x}Y_x$ Alloys
(75.50; S1.2; S1.4)
(a) Vol. **106**, No. 1, p. 233—237 (1988).

L. Vatskichev and M. Vatskicheva
Relations between Some Galvanomagnetic and Magnetic Properties of Thin Ferromagnetic Films
(72.15 and 75.70; S1.2; S1.4)
(a) Vol. **106**, No. 1, p. K75—K78 (1988).

A. E. Szukiel, P. J. Markowski, and K. Durczewski
Collective Excitation Effects on the Electrical Resistivity of Singlet Ground State Paramagnets of Metallic Conduction
(72.15; 71.70; S1.4)
(b) Vol. **147**, No. 1, p. 321—334 (1988).

Jinghua Tian, Jingkui Liang, Dayu Yang, Hanjie Fan, and Fanghua Li
The Study of the One-Dimensional Composition-Incommensurate Structure of $Pr_{1+\varepsilon}Fe_4B_4$
(61.55; S1.1; S1.4)
(a) Vol. **107**, No. 1, p. 45—53 (1988).

A. Czopnik, H. Mädge, and B. Staliński
Anomalous Strong Divergence of the Specific Heat of $ErIn_3$ at the Antiferromagnetic Transition
(75.40; S1.4)
(a) Vol. **107**, No. 2, p. K151—K155 (1988).

Z. Kletowski
Resistance of Some Lanthanum and Lutetium Compounds of the Type (La, Lu)Me_3
(72.10; 72.15; S1.4)
(a) Vol. **108**, No. 1, p. 363—367 (1988).

B. Idzikowski and A. Wrzeciono
Properties of Amorphous $Sm_xFe_{80-x}B_{20}$ Ribbons
(75.60; 75.80; S1.1; S1.4)
(a) Vol. **108**, No. 1, p. 375—380 (1988).

R. Sen Gupta and S. Chatterjee
Electronic Structure of SmH_2
(71.25; S1.4)
(b) Vol. **148**, No. 2, p. 579—584 (1988).

S. S. Kushwaha, M. P. Dikshit, and S. K. Srivastava
A Model for Lattice Dynamics of Lanthanides
(63.20; S1.4)
(b) Vol. **149**, No. 1, p. 69—79 (1988).

A. Urbaniak-Kucharczyk
Electrical Resistivity Due to Electron–Phonon Scattering in Thin Gadolinum Films
(73.60; S1.4)
(b) Vol. **149**, No. 2, p. 543—546 (1988).

I. N. Shabanova
System Terminal Investigation by the Shape of Electron Spectra of Inner Levels of Three-Valence Rare-Earth Compounds under Photoionization
(79.60; S1.4)
(b) Vol. **149**, No. 2, p. K177—K181 (1988).

A. Czopnik, H. Mädge, R. Pott, and B. Staliński
Thermal Properties of $NdIn_3$
(75.40; S1.4)
(a) Vol. **110**, No. 2, p. 601—614 (1988).

S1.5. Actinides

J. M. Wesselinowa
On the Theory of the Broad Spin Response in Uranium Compounds
(75.10; 63.10; S1.5)
(b) Vol. **145**, No. 1, p. K65—K69 (1988).

M. Zelený, F. Zounová, and A. V. Andreev
Magnetic Properties of UCuGe
(75.50; S1.5)
(a) Vol. **109**, No. 1, p. K67—K69 (1988).

F. López-Aguilar, S. Balle, and J. Costa-Quintana
Dynamic Potentials for Unstable Narrow Band Systems. Application to Intermediate Valence and Heavy Fermion Compounds
(71.20 and 71.25; S1.5; S1.63)
(b) Vol. **149**, No. 2, p. 585—593 (1988).

S1.61. Carbides, silicides, germanides

T. Luciński and J. Baszyński
The Influence of Two Amorphous Phases on Transport Properties of Amorphous $Fe_{1-x}Si_x$ Films
(73.60; 75.70; S1.61)
(a) Vol. **105**, No. 2, p. 501—509 (1988).

M. Polcarová, K. Godwod, J. Bąk-Misiuk, S. Kadečková, and J. Brádler
Lattice Parameters of Fe–Si Alloy Single Crystals
(61.55; 61.10; S1.1; S1.61)
(a) Vol. **106**, No. 1, p. 17—23 (1988).

S. U. Jen and C. J. Weng
Thermal Expansion of $(FeM)_{80}B_{14}Si_6$ Metallic Glasses (M = V, Cr, Mn)
(65.70; 75.80; S1.61)
(a) Vol. **106**, No. 1, p. 99—105 (1988).

P. Čapková, M. Merisalo, M. Laitinen, V. Valvoda, and L. Dobiášová
Thermal Vibrations and Thermal Expansion of $ZrC_{0.98}$ Studied by X-Ray Diffraction
(63.20 and 65.70; S1.61)
(a) Vol. **106**, No. 1, p. 107—114 (1988).

A. I. Gusev, A. A. Rempel, and V. N. Lipatnikov
Magnetic Susceptibility and Atomic Ordering in Tantalum Carbide
(64.60; 75.30; S1.61)
(a) Vol. **106**, No. 2, p. 459—466 (1988).

A. Unkroth, W. Andrä, and G. Götz
Composition Dependence of Magnetization of Ion Implanted Cobalt Layers
(75.70; 61.80; S1.2; S1.61; S1.62)
(a) Vol. **106**, No. 2, p. 589—594 (1988).

V. E. Borisenko, D. I. Zarovskii, and V. V. Tokarev
Influence of Argon Implantation on the Formation of Platinum Silicides
(61.70 and 68.55; S1.61)
(a) Vol. **107**, No. 1, p. K33—K35 (1988).

W. Cao, H. Q. Ye, and K. H. Kuo
A New Octagonal Quasicrystal and Related Crystalline Phases in Rapidly Solidified Mn_4Si
(61.55; 61.14; S1.61)
(a) Vol. **107**, No. 2, p. 511—519 (1988).

I. Ya. Nikiforov and A. B. Kolpachev
Electronic Struture of Nibium Nitrocarbides
(71.20; 74.70; S1.61; S1.62)
(b) Vol. **148**, No. 1, p. 205—211 (1988).

J. Piqueras, F. Domínguez-Adame, and B. Mendez
SEM-CL of Reaction Bonded SiC
(61.70 and 78.60; S1.61)
(a) Vol. **108**, No. 1, p. K81—K84 (1988).

V. P. ZHUKOV, N. I. MEDVEDEVA, D. L. NOVIKOV, and V. A. GUBANOV
The Study of Energy Band Structure and Chemical Bonding in Yttrium Carbide, Yttrium and Strontium Dicarbides by the LMTO and X_α DV Method
(71.25; S1.61)
(b) Vol. **149**, No. 1, p. 175—186 (1988).

R. MATTHEIS and D. HESSE
A Reinvestigation of Ni_2Si Thin Film Growth on Si(111) by TEM and RBS. Evidence of the Presence of an Interfacial NiSi Layer prior to Ni Consumption
(68.48 and 68.55; S1.61)
(a) Vol. **109**, No. 1, p. 217—229 (1988).

I. A. BACHILO, F. F. KOMAROV, A. P. NOVIKOV, and S. A. PETROV
Structural Analysis of Silicon Doped with High Doses of C^+ Ions
(61.80 and 68.55; S1.61; S5.11)
(a) Vol. **109**, No. 1, p. 231—238 (1988).

ZONGQUAN LI, HUI SHEN, LANPING YUE, and YIZHEN HE
A Special Metastable Phase of Metallic Glass $Pd_{77.5}Ni_6Si_{16.5}$
(61.40; S1.61)
(a) Vol. **109**, No. 1, p. K1—K2 (1988).

D. V. MORGAN, H. THOMAS, W. T. ANDERSON, P. THOMPSON, A. CHRISTOU, and D. J. DISKETT
High Temperature Metallisation for GaAs Device Processing
(61.80 and 73.40; S1.61; S4; S7.12)
(a) Vol. **110**, No. 2, p. 531—536 (1988).

S1.62. Nitrides, phosphides, arsenides

E. SOBESLAVSKY, H. U. JÄGER, U. KREISSIG, W. SKORUPA, and K. WOLLSCHÄGER
Nitrogen Profile Modification in High Dose Implantation Synthesis of Silicon Nitride
(61.80; S1.62; S5.11)
(a) Vol. **105**, No. 2, p. 387—396 (1988).

A. UNKROTH, W. ANDRÄ, and G. GÖTZ
Composition Dependence of Magnetization of Ion Implanted Cobalt Layers
(75.70; 61.80; S1.2; S1.61; S1.62)
(a) Vol. **106**, No. 2, p. 589—594 (1988).

K. BÄRNER
On the Electronic States of MnAs
(75.30; S1.62)
(b) Vol. **147**, No. 2, p. K189—K194 (1988).

I. YA. NIKIFOROV and A. B. KOLPACHEV
Electronic Structure of Niobium Nitrocarbides
(71.20; 74.70; S1.61; S1.62)
(b) Vol. **148**, No. 1, p. 205—211 (1988).

S1.63. Sulfides, selenides, tellurides

T. Satow, H. Ishikawa, and O. Uemura
Semiconductivity in the Liquid Cu–TlTe System
(72.10 and 72.80; 75.30; S1.63)
(b) Vol. **146**, No. 2, p. 661—669 (1988).

W. Sienicki
Investigation of Tungsten Disulphide and Diselenide Intercalated by Thallium
(72.80; 61.70; S1.63)
(a) Vol. **106**, No. 2, p. K157—K159 (1988).

O. S. Rajora and A. E. Curzon
A Study of the Diffusion of Indium into TaS_2 and $TaSe_2$ in a Scanning Electron Microscope
(66.30; 61.16; S1.63)
(a) Vol. **108**, No. 2, p. 529—536 (1988).

H. Ishikawa, Y. Kameda, O. Uemura, and T. Satow
Electronic Properties of Liquid Tl–$GeTl_2Se_4$ and Tl–$SnTl_2Se_4$ Systems
(72.20 and 75.30; S1.63; S8.16)
(b) Vol. **149**, No. 1, p. 257—266 (1988).

F. Lópeż-Aguilar, S. Balle, and J. Costa-Quintana
Dynamic Potentials for Unstable Narrow Band Systems. Application to Intermediate Valence and Heavy Fermion Compounds
(71.20 and 71.25; S1.5; S1.63)
(b) Vol. **149**, No. 2, p. 585—593 (1988).

A. M. Karguppikar and A. G. Vedeshwar
Electrical and Optical Properties of Natural Iron Pyrite (FeS_2)
(72.80 and 78.50; S1.63)
(a) Vol. **109**, No. 2, p. 549—558 (1988).

M. Stordeur, M. Stölzer, H. Sobotta, and V. Riede
Investigation of the Valence Band Structure of Thermoelectric $(Bi_{1-x}Sb_x)_2Te_3$ Single Crystals
(71.25; 72.20; 72.30; 78.20; S1.63)
(b) Vol. **150**, No. 1, p. 165—176 (1988).

T. Shiozawa and T. Kobayashi
High Resolution Electron Microscopic Study of Silver Sulfide Microcrystals Formed on Silver Bromide Emulsion Grains
(61.60; 61.16; S1.63; S9)
(a) Vol. **110**, No. 2, p. 375—382 (1988).

S2. Alkali metals and their alloys and compounds

(for alkaline halides, see S9.11)

L. C. R. Alfred
Ion Interaction in Simple Metals
(71.10; 71.25; S2)
(b) Vol. **145**, No. 1, p. 203—208 (1988).

I. T. Yakubov, A. G. Khrapak, V. V. Pogosov, and S. A. Trigger
Thermodynamical Characteristics of Liquid Metal Droplets
(68.10; 73.30; S2)
(b) Vol. **145**, No. 2, p. 455—466 (1988).

M. Kumari and N. Dass
The Melting Laws
(64.70; 62.20; S2; S5.2; S9)
(b) Vol. **146**, No. 1, p. 105—110 (1988).

R. Allub and S. M. Mujibur Rahman
On the Phase Stability of Rb under External Pressure
(64.70; S2)
(b) Vol. **147**, No. 1, p. K5—K8 (1988).

J. Mimkes
Thermal Expansion, Heat of Fusion, and Electron Density of Monovalent Metals
(65.70; 65.50; S1.2; S1.3; S2)
(b) Vol. **148**, No. 1, p. 91—100 (1988).

R. Ramji Rao and A. Padmaja
Fourth-Order Elastic Constants of Sodium
(62.20; S2)
(b) Vol. **148**, No. 1, p. K1—K3 (1988).

N. Singh, N. S. Banger, and S. P. Singh
Phonon Frequencies of B.C.C. Li: An Application of the Local Pseudopotential
(63.20; S2)
(b) Vol. **148**, No. 1, p. K5—K9 (1988).

L. K. Mishra and D. N. Tripathy
A Self-Consistent Method for Evaluating the Effective Mass of Electrons in Metals
(71.25; S2; S4)
(b) Vol. **148**, No. 2, p. 585—592 (1988).

N. SINGH, N. S. BANGER, and S. P. SINGH
Elastic Constants and Phonon Spectra of Alkali Metals. Dynamical Treatment with Temperature-Dependent Two-Body Interaction
(62.20; S2)
(b) Vol. **149**, No. 1, p. 63—68 (1988).

M. RANI, A. PRATAP, and N. S. SAXENA
Collective Excitations in Liquid Cesium Using a Theory of Phonons in Amorphous Solids
(63.50; S2)
(b) Vol. **149**, No. 1, p. 93—99 (1988).

G. DAS, B. K. PANDA, D. P. MAHAPATRA, and H. C. PADHI
Simple Numerical Method for the Estimation of Multiple Scattering in the Compton Effect
(78.70; S2; S4)
(b) Vol. **149**, No. 1, p. 355—362 (1988).

S3. Alkaline earth metals and their alloys and compounds
(for alkaline earth halides, see S9.12)

S. CHANDRA and H. C. VERMA
Charge Shift Model Calculation of Cohesive Energy of Magnesium
(61.50; S3)
(b) Vol. **145**, No. 1, p. K1—K3 (1988).

K. K. MUKHERJEE, S. M. RAFIQUE, M. MEHAR KHAN, and S. BAGCHI
Energetic Approach to the Electrical Resistivity of Liquid Magnesium
(72.15; S3)
(b) Vol. **147**, No. 1, p. K51—K55 (1988).

S4. Other metals and their alloys

N. M. Kabachnik, V. N. Kondratev, and O. V. Chumanova
The Impact-Parameter Dependence of Electron Stopping in Proton Collisions with Gas and Solid Atoms
(61.80; S4; S5.11)
(b) Vol. **145**, No. 1, p. 103—116 (1988).

R. Urban, P. Ehrhart, W. Schilling, H. R. Schober, and H. Lauter
Erratum to: Investigation of the Resonant Vibration Modes of Radiation Induced Interstitial Atoms in Aluminium by Diffuse Inelastic Neutron Scattering
(61.80 and 63.20; S4)
(b) Vol. **145**, No. 1, p. 351 (1988).

Tran Thoai Duy Bao
Plasmon Excitation in Metal Surface Layers by Fast Electrons
(71.45 and 73.20; 61.80; S4)
(b) Vol. **145**, No. 2, p. 509—515 (1988).

R. Gilde, H. J. van Daal, and H. H. Brongersma
Electromigration and Coagulation in Thin Films of Pure Indium and Pure Tin
(73.60; S4)
(a) Vol. **105**, No. 2, p. 493—499 (1988).

N. Clark and M. Segnini
The Electrical Conductance of Indium Thin Films Immediately after Being Deposited
(73.60; S4)
(a) Vol. **105**, No. 2, p. K125—K128 (1988).

D. B. Tran Thoai and E. Zeitler
Image Potential of an Electron in a Gap between Two Metals
(71.45 and 73.20; S1.3; S4)
(b) Vol. **146**, No. 1, p. 137—140 (1988).

E. M. Shpilevskii and D. A. Gorbachevskii
Argon Ion Effect on Solid Phase Reaction in Copper–Tin, Copper–Antimony Films
(61.80 and 68.48; 66.30; S1.3; S4)
(a) Vol. **106**, No. 1, p. K23—K26 (1988).

E. Taute and K. Blum
On the Nuclear Charge Dependence of Auger Neutralisation at Surfaces
(79.20; S4)
(b) Vol. **146**, No. 2, p. K87—K90 (1988).

M. Mondal and K. P. Ghatak
Theoretical Analysis of the Plasma Frequency in Bismuth in the Presence of a Quantizing Magnetic Field
(72.30; S4)
(b) Vol. **146**, No. 2, p. K97—K102 (1988).

P. Czurratis, R. Kroggel, V. Syneček, M. Simerska, and H. Löffler
Dislocation Loops in Al–Zn and Al–Zn–Mg Alloys
(61.70; 61.55; S1.2; S4)
(a) Vol. **106**, No. 2, p. 407—418 (1988).

F. Müller and H.-J. Köhler
Microwave-Induced Constant Voltage Steps at 0.1 Volt from a Series Array of Josephson Tunnel Junctions
(74.50; S4)
(a) Vol. **106**, No. 2, p. K165—K168 (1988).

A. P. Zhernov and T. N. Kulagina
The Microcontact Spectrum of Aluminium
(71.38; 72.10; S4)
(b) Vol. **147**, No. 1, p. 149—153 (1988).

S. Ohtaki and I. Hashimoto
Numerical Evaluation for Stage II_A Recovery in Cold-Worked Aluminium
(61.70; 62.20; 72.15; S4)
(a) Vol. **107**, No. 1, p. 123—130 (1988).

M. A. Kenawy, G. Saad, G. Graiss, and A. Fawzy
The Effect of Excitation-Induced Fatigue on the Recovery of Internal Friction of Al–1.0 wt% Mn–0.28 wt% Fe Alloy
(62.40; S4)
(a) Vol. **107**, No. 1, p. 171—176 (1988).

B. L. Cheng and T. S. Kê
Bamboo Boundary Internal-Fraction Peak in 99.9999% Aluminium and the Effect of Cold Work on the Peak
(62.40; S4)
(a) Vol. **107**, No. 1, p. 177—185 (1988).

M. S. M. Sastry
Post-Deposition Resistance Increase in Island Aluminium Films
(73.60; 68.55; S4)
(a) Vol. **107**, No. 1, p. 225—233 (1988).

M. F. Burkhard, H. Rothard, and K.-O. E. Groeneveld
Single-Electron Deexcitation of Volume Plasmons Induced by Heavy Ions in Thin Solid Foils
(71.45; 79.20; S1.2; S4; S5)
(b) Vol. **147**, No. 2, p. 589—592 (1988).

M. de Graef, L. Delaey, and D. Broddin
High Resolution Electron Microscopic Study of the X-Phase in Cu–Al and Cu–Al–Zn Alloys
(61.55; 61.16; S1.2; S4)
(a) Vol. **107**, No. 2, p. 597—609 (1988).

A. Loiseau and C. Vannuffel
$TiAl_2$: A Reentrant Phase in the Ti–Al System
(64.70; 61.16; S1.2; S4)
(a) Vol. **107**, No. 2, p. 655—671 (1988).

J. Menon and C. Suryanarayana
The Al–Co Decagonal Phase
(61.55; S1.2; S4)
(a) Vol. **107**, No. 2, p. 693—708 (1988).

G. TORRES-VILLASENOR and G. VAN TENDELOO
Guinier-Preston Zones in the α-Phase of Superplastic Zn–Al Alloy
(61.70; 62.20; S1.2; S4)
(a) Vol. **107**, No. 2, p. 785—790 (1988).

D. B. TRAN THOAI and E. ZEITLER
Inelastic Scattering of Fast Electrons by Small Metal Particles
(61.80 and 73.20; 71.45; S4)
(a) Vol. **107**, No. 2, p. 791—797 (1988).

R. E. HUMMEL
On the Current Density Dependence of Electromigration in Thin Films
(66.30 and 85; S4)
(a) Vol. **107**, No. 2, p. K175—179 (1988).

A. SACHS
Relations between Third Order Isothermal and Adiabatic Elastic Constants for an Isotropic Material
(62.20; S4)
(b) Vol. **148**, No. 1, p. 101—105 (1988).

N. J. CASTELLANI and D. B. LEROY
Final State Core Level Shifts in Supported and Free Metal Particles
(68.40; 79.60; S4)
(b) Vol. **148**, No. 1, p. 121—128 (1988).

M. RÖSLER and W. BRAUER
Theory of Electron Emission from Solids by Proton and Electron Bombardment
(79.20; S4)
(b) Vol. **148**, No. 1, p. 213—226 (1988).

LIANG-YAO CHEN and D. W. LYNCH
The Optical Properties of $AuAl_2$ and $PtAl_2$
(78.20; S1.3; S4)
(b) Vol. **148**, No. 1, p. 387—394 (1988).

F. M. MUNTYANU, V. G. KISTOL, and I. A. POPOV
Quantum Oscillations of Transport Phenomena of H-Holes in $Bi_{1-x}Sb_x$ Alloys ($x \geqq 0.25$)
(72.15; S4)
(b) Vol. **148**, No. 1, p. K37—K40 (1988).

M. HEYTING, B. J. THIJSSE, and A. VAN DEN BEUKEL
Change of Electrical Resistance during Crystallization of Icosahedral $Al_{86}Mn_{14}$
(61.50 and 72.15; S4)
(a) Vol. **108**, No. 1, p. 97—101 (1988).

CHU THIEN TRUONG, G. WENDROCKI and H. LÖFFLER
On the Existence Field of Guinier-Preston Zones in the Al–Zn System
(61.55; 64.70; S4)
(a) Vol. **108**, No. 1, p. 103—110 (1988).

P. Czurratis, E. Korngiebel, G. Jacobi, and H. Löffler
The Effect of Additions of Mn and Cr and the Applied Solidification Technique on the Structure of an Al–2.2 at% Zn–1.4 at% Mg Alloy
(61.70; 61.55; 62.20; S4)
(a) Vol. **108**, No. 1, p. 171—182 (1988).

M. Kobiyama, S. Takamura, and K. Nakata
Internal Friction in Single-Crystal Al Alloys after Electron Irradiation at Low Temperature
(62.40; S4)
(a) Vol. **108**, No. 1, p. 219—224 (1988).

R. Z. Bakhtizin and V. G. Valeev
Microroughness Field Emission Electron Energy Distribution and Electron–Phonon Interaction Spectroscopy near the Metal Surface
(79.70; 68.20; S1; S4)
(a) Vol. **108**, No. 1, p. 251—255 (1988).

V. P. Goltsev and S. I. Zhukova
The Loss of Strength of Al-Alloy after Sb^+ Implantation
(61.70 and 62.20; S4)
(a) Vol. **108**, No. 1, p. K1—K4 (1988).

I. Manna and S. K. Pabi
A Resistometric Study of Migrating Boundary Characteristics in Pb–Sn Alloys
(64.75 and 68.48; S4)
(a) Vol. **108**, No. 1, p. K19—K23 (1988).

L. K. Mishra and D. N. Tripathy
A Self-Consistent Method for Evaluating the Effective Mass of Electrons in Metals
(71.25; S2; S4)
(b) Vol. **148**, No. 2, p. 585—592 (1988).

T. S. Kê, L. D. Zhang, B. L. Cheng, and A. W. Zhu
On the Origin of the Macrocrystalline Internal-Friction Peak (the Bamboo Boundary Peak)
(62.40; S4)
(a) Vol. **108**, No. 2, p. 569—575 (1988).

S. E. Urreta de Pereyra, H. Bertorello, and A. A. Ghilarducci de Salva
Precipitation and Grain Boundary Internal Friction Peaks in Al–Mg–Si
(62.40; 61.70; S4)
(a) Vol. **108**, No. 2, p. 577—586 (1988).

G. Das, B. K. Panda, D. P. Mahapatra, and H. C. Padhi
Simple Numerical Method for the Estimation of Multiple Scattering in the Compton Effect
(78.70; S2; S4)
(b) Vol. **149**, No. 1, p. 355—362 (1988).

S. M. Javed Akhtar and D. Ristau
Mechanism of Voltage Generation in Bi Films Due to Laser Irradiation
(73.60; 79.20; S4)
(a) Vol. **109**, No. 1, p. 255—259 (1988).

J. Gastaldi, C. Jourdan, G. Grange, and C. L. Bauer
"In Situ" Determination of Grain Boundary Migration Rates by Synchrotron White Beam X-Ray Topography
(61.70; 61.10; S4)
(a) Vol. **109**, No. 2, p. 403—411 (1988).

A. A. Opalenko and A. A. Kornilova
Anisotropy of the Mössbauer Effect in β-Sn under Pressure and the Lattice Dynamics
(63.20 and 76.80; 62.50; S4)
(b) Vol. **150**, No. 1, p. 25—29 (1988).

B. Mitra, A. Ghosal, and K. P. Ghatak
On the Photoemission from Bi in the Presence of an Arbitrarily Oriented Quantizing Field
(79.60; S4)
(b) Vol. **150**, No. 1, p. K67—K72 (1988).

J. M. Holender, J. Sołtys, G. Wendrock, and P. Czurratis
Non-Isothermal Studies of the Eutectoidal Decomposition of the Alloy Al–59 at% Zn
(64.70; 61.10; 61.55; S1.2; S4)
(a) Vol. **110**, No. 1, p. 123—132 (1988).

T. Kanadani and A. Sakakibara
Short-Range Clustering in the Al–3 mass% Mg Alloy
(64.75; S4)
(a) Vol. **110**, No. 1, p. K9—K11 (1988).

G. Graiss, G. Saad, and A. Fawzy
Effect of Loading on the Recovery of Internal Friction of an Al–1.0 wt% Mn–0.28 wt%Fe Alloy
(62.40; S4)
(a) Vol. **110**, No. 2, p. 481—487 (1988).

E. Ma and M.-A. Nicolet
Pt_2Al_3 Formation on Evaporated and Large-Grained Al Substrates
(68.48 and 68.55; S4)
(a) Vol. **110**, No. 2, p. 509—514 (1988).

D. V. Morgan, H. Thomas, W. T. Anderson, P. Thompson, A. Christou, and D. J. Diskett
High Temperature Metallisation for GaAs Device Processing
(61.80 and 73.40; S1.61; S4; S7.12)
(a) Vol. **110**, No. 2, p. 531—536 (1988).

H. Krautz, Ch. Wenzel, K. Bornkessel, and G. Blasek
Barrier Behaviour of TiW between Copper and Aluminium
(68.48; S1.2; S4)
(a) Vol. **110**, No. 2, p. K77—K79 (1988).

S5. Nonmetallic elements

V. I. Gavrilenko, N. I. Klyui, V. G. Litovchenko, and V. E. Strelnitskii
Characteristic Features of the Electronic Structure of Carbon Films
(71.25; 78.20; S5)
(b) Vol. **145**, No. 1, p. 209—217 (1988).

I. J. Jabr, J. D. Meyer, K. A. Al-Saleh, and N. S. Saleh
Ion-Beam Mixing of Te/Au: Metastable Phase Formation
(61.80 and 64.75; 73.60; S1.3; S5)
(a) Vol. **105**, No. 1, p. 177—181 (1988).

B. I. Craig and P. V. Smith
The Periodic LUC-MINDO Method and Its Application to Diamond
(71.10; S5)
(b) Vol. **146**, No. 1, p. 149—160 (1988).

V. A. Shenderovskii, P. M. Tomchuk, P. N. Gorley, N. P. Gavaleshko, and I. I. Tarasyuk
The Effect of Structural Imperfections on Macroscopic Properties of Tellurium-Type Anisotropic Semiconductors
(72.10; 72.20; S5)
(b) Vol. **146**, No. 2, p. 671—681 (1988).

P. P. Lottici
Phonon and Fraction Dynamics in Amorphous Arsenic
(63.50; S5)
(b) Vol. **146**, No. 2, p. K81—K85 (1988).

D. T. Britton and P. C. Rice-Evans
Fast Fourier Transform Decenvolution of Positron Annihilation Doppler-Broadened Lineshapes
(61.70 and 78.70; S1.2; S5)
(b) Vol. **147**, No. 1, p. 89—96 (1988).

S. G. Lyapin, M. I. Eremets, A. M. Shirokov, R. Herrmann, H.-U. Müller, and F. Ludwig
Submillimetre Magnetospectroscopy of Tellurium under Hydrostatic Pressure
(71.55 and 78.50; 72.40; 76.40; S5)
(b) Vol. **147**, No. 1, p. 297—306 (1988).

D. K. Papademitriou
Scattering of X-Rays by Plasmons in Boron
(71.45 and 78.30; S5)
(b) Vol. **147**, No. 1, p. K95—K98 (1988).

M. F. Burkhard, H. Rothard, and K.-O. E. Groeneveld
Single-Electron Deexcitation of Volume Plasmons Induced by Heavy Ions in Thin Solid Foils
(71.45; 79.20; S1.2; S4; S5)
(b) Vol. **147**, No. 2, p. 589—592 (1988).

Zijing Lin and Kelin Wang
A Simple Scheme for Studies of Surface Phonons of Covalent Semiconductors. I. Calculation of Force Constants
(63.20 and 68.30; S5)
(b) Vol. **148**, No. 2, p. 525—532 (1988).

K. N. Madhusoodanan and J. Philip
Optical Energy Gap and Thermal Diffusivity of Ge–Se Semiconducting Glasses
(78.20; S5)
(a) Vol. **108**, No. 2, p. 775—782 (1988).

Fangqing Zhang, Guanghua Chen, Yafei Zhang, and Gong Yu
IR and ESR Studies of the Structural Properties of Hydrogenated Amorphous Carbon Films
(68.55; 76.30; 78.65; S5)
(a) Vol. **109**, No. 1, p. K39—K44 (1988).

L. N. Kostadinov and D. D. Dobrev
On the Structure of the C_8 Cubic Modification of Carbon
(61.55; S5)
(a) Vol. **109**, No. 2, p. K85—K88 (1988).

B. L. Evans and G. Y. Nasser
The DC Conductivity of Carbon Films
(73.60; S5)
(a) Vol. **110**, No. 1, p. 165—179 (1988).

S. S. Jiang, Y. Qiu, and G. S. Green
The Simulation of Modulated I_3 Fringes in an X-Ray Section Topograph of a Stacking Fault
(61.10; 61.70; S5)
(a) Vol. **110**, No. 2, p. 323—330 (1988).

J. C. Bernède, G. Safoula, A. Ameziane, and P. Burgaud
Experimental Study of Oxygen Effect on the Variation of Resistivity with Temperature for Polycrystalline Selenium Thin Films
(73.60; S5; S8)
(a) Vol. **110**, No. 2, p. 521—530 (1988).

S5.11. Silicon

J. GRIESCHE, R. ENDERLEIN, and D. SCHIKORA
Orientation and Strain in Heteroepitaxial Growth **(Review Article)**
(61.50 and 68.48; 68.20; S5.11; S7.12; S8.13)
(a) Vol. **109**, No. 1, p. 11—38 (1988).

V. M. KOROL
Sodium–Ion Implantation into Silicon **(Review Article)**
(61.80; 66.30; 72.80; 73.40; S5.11)
(a) Vol. **110**, No. 1, p. 9—34 (1988).

N. M. KABACHNIK, V. N. KONDRATEV, and O. V. CHUMANOVA
The Impact-Parameter Dependence of Electron Stopping in Proton Collisions with Gas and Solid Atoms
(61.80; S4; S5.11)
(b) Vol. **145**, No. 1, p. 103—116 (1988).

L. F. MAKARENKO and L. I. MURIN
Trapping of Minority Carriers in Thermal U^--Donors in n-Si
(71.55; 72.40; S5.11)
(b) Vol. **145**, No. 1, p. 241—253 (1988).

D. K. GAUTAM, W. S. KHOKLE, and K. B. GARG
Transition Probability of Impact Ionization by Holes in Silicon
(72.20; S5.11)
(b) Vol. **145**, No. 1, p. 269—275 (1988).

L. A. KAZAKEVICH, V. I. KUZNETSOV, and P. F. LUGAKOV
Clusters of Radiation Defects in Silicon with Dislocations
(61.80; S5.11)
(a) Vol. **105**, No. 1, p. 97—100 (1988).

J. JABŁOŃSKI, J. WOJCIECHOWSKI, and K. KUCHARSKI
The Nucleation Kinetics of Platelike Oxide Precipitates in Cz-Silicon
(61.70; S5.11)
(a) Vol. **105**, No. 1, p. 113—121 (1988).

GUANGHUA CHEN, GUOSHENG SUN, and FANGQING ZHANG
ESR in P/I Heterojunctions Based on a-Si:H
(73.40 and 76.30; S5.11)
(a) Vol. **105**, No. 1, p. K41—K43 (1988).

V. HOLÝ and J. HÄRTWIG
The Role of Diffuse Scattering on Microdefects in the Precise Lattice Parameter Measurement
(61.10 and 61.50; 61.70; S5.11)
(b) Vol. **145**, No. 2, p. 363—372 (1988).

S. SATPATHY and Z. PAWLOWSKA
Construction of Bond-Centered Wannier Functions for Silicon Valence Bands
(71.10; 71.25; S5.11)
(b) Vol. **145**, No. 2, p. 555—565 (1988).

C. Zwanziger, S. Petrowa, J. Reinhold, P. Streubel, and A. Meisel
Approximate Calculation of Inner-Shell Binding Energies and Auger Transition Energies of Solid Silicon Compounds
(71.50; 79.20; S5.11; S6)
(b) Vol. **145**, No. 2, p. 603—607 (1988).

J. Kreissl and W. Gehlhoff
Electron Paramagnetic Resonance of the Mn_4^0 Cluster in Silicon
(71.55 and 76.30; 61.70; S5.11)
(b) Vol. **145**, No. 2, p. 609—617 (1988).

N. Russo, M. Toscano, V. Barone, and C. Minichino
Model Clusters and Electronic Characteristics of Deep-Level Impurities in Silicon
(71.55; S5.11)
(b) Vol. **145**, No. 2, p. K125—K130 (1988).

E. Sobeslavsky, H. U. Jäger, U. Kreissig, W. Skorupa, and K. Wollschläger
Nitrogen Profile Modification in High Dose Implantation Synthesis of Silicon Nitride
(61.80; S1.62; S5.11)
(a) Vol. **105**, No. 2, p. 387—396 (1988).

W. Dąbrowski
Investigation of Deep Impurities in Schottky Diodes on High-Resistivity Silicon
(71.55; 73.40; S5.11)
(a) Vol. **105**, No. 2, p. 511—520 (1988).

F. R. Kessler, U. Barkow, R. Nies, and N. Unzner
Optical Profile Analysis of Diffused and Ion-Implanted Silicon
(78.20; 61.70; 66.30; 78.50; S5.11)
(a) Vol. **105**, No. 2, p. 627—635 (1988).

V. Hadjicontis, C. A. Londos, and K. Eftaxias
Correlation of Diffusivities of Various Elements in Silicon
(66.30; S5.11)
(a) Vol. **105**, No. 2, p. K87—K92 (1988).

G. Pastor, P. Tejedor, I. Jiménez, E. Domínguez, M. Torres, and J. V. García-Ramos
Low Pressure Chemical Vapour Deposition Amorphous Silicon Behaviour under Annealing
(61.40 and 68.55; 78.30; S5.11)
(a) Vol. **106**, No. 1, p. 11—16 (1988).

G. Dlubek, C. Ascheron, R. Krause, H. Erhard, and D. Klimm
Positron Study of Vacancy Defects in Proton and Neutron Irradiated GaP, InP, and Si
(61.80; 78.70; S5.11; S7.11)
(a) Vol. **106**, No. 1, p. 81—88 (1988).

I. R. Entin and I. A. Smirnova
The Effect of X-Ray Beam Intensity Oscillations as a Function of the Amplitude of Ultrasound Excited in a Crystal
(61.10; 62.60; S5.11)
(a) Vol. **106**, No. 2, p. 339—350 (1988).

A. V. ANDREEVA, A. V. BAZHENOV, N. A. BULENKOV, and A. A. FIRSOVA
Structural and Photoelectrical Properties of a Third-Order Twin Boundary ($\Sigma = 27$) in Silicon
(61.55; 61.70; 72.20; S5.11)
(a) Vol. **106**, No. 2, p. 351—362 (1988).

V. G. GORSHKOV, YU. K. DANILEIKO, V. V. OSIKO, A. V. SIDORIN, N. V. VESELOVSKAYA, YU. V. DANKOVSKII, and B. L. SHKLYAR
Mechanical Microstresses in Dislocation-Free Floating-Zone Silicon Monocrystals
(61.70; 62.20; S5.11)
(a) Vol. **106**, No. 2, p. 363—369 (1988).

G. PETERMANN
DC and AC Characterization of Grain Boundaries in Float Zone Silicon
(71.20 and 73.40; S5.11)
(a) Vol. **106**, No. 2, p. 535—549 (1988).

E. SUSI, L. PASSARI, M. MERLI, and M. C. CAROTTA
Determination of Minority Carrier Mobility in Silicon from Stationary and Transient Lifetime Measurements
(72.20; S5.11)
(a) Vol. **106**, No. 2, p. 583—587 (1988).

A. P. KOVCHAVTSEV, G. L. KURISHEV, K. O. POSTNIKOV, R. A. SOKOLOV, and I. M. SUBBOTIN
Inelastic Electron Tunneling Spectroscopy of Si MIS Structures with Ultrathin Thermal Silicon Nitride and Thermal Silica
(73.40 and 85; S5.11)
(a) Vol. **106**, No. 2, p. 669—674 (1988).

G. D. IVLEV, V. L. MALEVICH, and V. V. ZHIDKOV
Melting Temperature of Amorphized Silicon Heated by Nanosecond Laser Radiation
(64.70; S5.11)
(a) Vol. **106**, No. 2, p. K123—K127 (1988).

L. N. ALEKSANDROV
Mechanical Stress in the Crystallization Process of Amorphous Semiconductors by Pulse Action
(61.40 and 65.70; S5.11)
(a) Vol. **106**, No. 2, p. K135—K138 (1988).

L. F. MAKARENKO
On the Cross Section of Electron Capture into the Deep State of Thermal U^- Donors in Si
(71.55; 72.40; S5.11)
(a) Vol. **106**, No. 2, p. K153—K155 (1988).

A. SRIVASTAVA
Effect of Polysilicon–Silicon Interface Recombination Velocity on Current Transport Properties of Poly-Emitter Transistors
(73.40 and 85; S5.11)
(a) Vol. **106**, No. 2, p. K203—K207 (1988).

P. R. VAYA and J. MAJHI
Effect of Surface Treatment on Al–Si Schottky Structures
(73.30 and 85; S5.11)
(a) Vol. **106**, No. 2, p. K209—K213 (1988).

F. Januschewski, H.-J. Erzgräber, and W. Füssel
Modelling the Local Damage of Short-Channel MOSFETs Due to Hot Electron Injection Using Results from Photoinjection Measurements on MOS Capacitors
(73.40 and 85; 72.40; S5.11)
(a) Vol. **106**, No. 2, p. K215—K220 (1988).

G. Pompe and E. Hegenbarth
Thermal Conductivity of Amorphous Si at Low Temperatures
(66.70; S5.11)
(b) Vol. **147**, No. 1, p. 103—108 (1988).

J. Czerbniak, M. Moneta, B. Pawłowski, and M. Piwowarski
The Influence of Thermal Vibrations on RBS Proton Spectra Obtained from Subsurface Regions of Si Crystals
(61.80 and 68.30; S5.11)
(b) Vol. **147**, No. 1, p. 123—126 (1988).

B. I. Craig and P. V. Smith
Determination of the Bulk Properties of Silicon by the Periodic LUC-MINDO Method
(71.10; 62.20; S5.11)
(b) Vol. **147**, No. 1, p. K31—K35 (1988).

E. Förster, P. Glas, K. Goetz, St. Jokscн, P. V. Nickles, M. Schnürer, and I. Will
About the Sensitivity of in Situ Diffraction Measurements with X-Radiation of a Laser-Produced Plasma
(61.10; 65.70; 68.20; S5.11)
(a) Vol. **107**, No. 1, p. 85—93 (1988).

A. Henry, K. Saminadayar, J. L. Pautrat, and N. Magnea
The Kinetics of the Formation of the Thermal Donors in Silicon. Effect of Various Parameters
(61.70; 72.80; 78.30; S5.11)
(a) Vol. **107**, No. 1, p. 101—110 (1988).

L. N. Aleksandrov, S. V. Lozovskii, and S. Y. Knyazev
Silicon Zone Sublimation Regrowth
(68.55; 64.75; 71.55; S5.11)
(a) Vol. **107**, No. 1, p. 213—223 (1988).

A. J. Abu El-Haija, K. A. Al-Saleh, N. A. Halim, J. M. Khalifeh, and N. S. Saleh
Ion Beam Induced Mixing of Cu/Si System Using Electrical Resistivity and RBS Measurements
(61.80 and 73.60; S1.2; S5.11)
(a) Vol. **107**, No. 1, p. 253—260 (1988).

G. M. Gusakov, A. A. Komarnitskii, and A. S. Em
Dynamics of Optical Parameters of Silicon and Germanium Single Crytals during Two-Pulsed Nanosecond Laser Irradiation
(61.80 and 68.55; 64.70; 78.20; S5.11; S5.12)
(a) Vol. **107**, No. 1, p. 261—271 (1988).

B. Tillack, R. Banisch, and H. H. Richter
Generation and Characterization of Thick Silicon-on-Insulator Films
(68.55; S5.11)
(a) Vol. **107**, No. 1, p. 281—289 (1988).

H. Ohyama and K. Nemoto
Effect of Lattice Defects in the Collector Region of npn Si Transistors on the Degradation of h_{FE}
(61.80 and 73.40; 85; S5.11)
(a) Vol. **107**, No. 1, p. 429—439 (1988).

V. K. Goyal, R. Pal, and K. Sen
On the Photogenerated Current in Polycrystalline Silicon MIS Solar Cells
(73.40 and 85; 72.40; S5.11)
(a) Vol. **107**, No. 1, p. K83—K87 (1988).

H. Bender and J. Vanhellemont
Rod-Like Defects in Silicon: Coesite or Hexagonal Silicon?
(61.70; S5.11)
(a) Vol. **107**, No. 2, p. 455—467 (1988).

A. Bourret, J. L. Rouvière, and J. Spendeler
Direct Observation of Atomic Columns in Semiconductors by HREM at 400 kV
(61.55 and 61.60; 61.16; S5.11; S5.12; S7.1; S8.1)
(a) Vol. **107**, No. 2, p. 481—501 (1988).

M. Reiche, J. Reichel, and W. Nitzsche
Correlations between Thermal Donor Formation, Rod-Like Defect Formation, and Oxygen Reduction during Low-Temperature Annealing of CZ-Grown Silicon
(61.70; S5.11)
(a) Vol. **107**, No. 2, p. 851—865 (1988).

P. I. Gaiduk, F. F. Komarov, N. L. Prokhorenko, and V. S. Solovev
Structure of Silicon after B^+ and P^+ Implantation and Rapid Thermal Annealing
(61.80; S5.11)
(a) Vol. **107**, No. 2, p. K105—K107 (1988).

H. Übensee, G. Paasch, D. Bundfuss, and J.-P. Zöllner
Electrons in MIS Inversion and Accumulation Layers. Exchange and Correlation Effects within a Modified Thomas-Fermi Theory
(73.40; S5.11; S7.1)
(b) Vol. **148**, No. 1, p. 421—434 (1988).

Qing Gu and Xi-Mao Bao
Effect of the Structure of Laser Crystallized Silicon-on-Insulator on Impurity Diffusion
(61.50; 66.30; 72.80; S5.11)
(a) Vol. **108**, No. 1, p. 81—87 (1988).

G. Wallner, E. Burkel, H. Metzger, and J. Peisl
Grazing Incidence X-Ray Diffraction from Si with an Implantation Induced Amorphous Surface Layer
(61.10; 68.20; S5.11)
(a) Vol. **108**, No. 1, p. 129—133 (1988).

V. I. Kuznetsov, P. F. Lugakov, and V. V. Lukyanitsa
Influence of Irradiation Temperature on Radiation Defect Accumulation in Silicon
(61.80; 71.55; S5.11)
(a) Vol. **108**, No. 1, p. 143—148 (1988).

S. Iselborn, B. Schröder, and J. Geiger
The Influence of Hydrogen Surface Reactions on the Growth of Evaporated a-Si:H Films
(68.55; 78.30; S5.11)
(a) Vol. **108**, No. 1, p. 275—284 (1988).

T. Drüsedau, M. Eckler, and R. Bindemann
Photoconductive a-Si:H with Dominant Monohydride Bonding Prepared by DC-Magnetron Sputtering
(68.55 and 73.60; 76.30; 78.65; S5.11)
(a) Vol. **108**, No. 1, p. 285—293 (1988).

O. A. Golikova, M. M. Mezdrogina, N. A. Feoktistov, K. L. Sorokina, M. M. Kazanin, P. M. Karageorgy-Alkalaev, and A. Yu. Leiderman
Injection and Recombination Currents in a-Si:H Structures
(72.20 and 73.40; S5.11)
(a) Vol. **108**, No. 1, p. 323—329 (1988).

C. Manfredotti, F. Fizzotti, and G. Amato
Frenkel Effect in SIPOS Films
(73.25; S5.11)
(a) Vol. **108**, No. 1, p. K25—K30 (1988).

N. Q. Khanh, M. Fried, G. Battistig, Z. Laczik, T. Lohner, E. Jároli, V. Schiller, and J. Gyulai
Ellipsometric and Ion Backscattering Measurements of the Properties of Silicon-On-Insulator Structure Formed by Thermally Activated Redistribution of High-Dose Ion Implanted Nitrogen
(61.70 and 68.48; 78.65; S5.11)
(a) Vol. **108**, No. 1, p. K35—K40 (1988).

B. F. Minaev, Yu. A. Serebrennikov, and H. D. Rempel
Non-Equilibrium Spin Polarization of the Si-Sl Centre in Silicon Induced by Spin–Orbit Coupling
(71.70 and 76.30; S5.11)
(b) Vol. **148**, No. 2, p. 689—698 (1988).

V. Thomas, J. Barrau, M. Brousseau, K. Kirouni, and G. Couloumiers
The Luminescence of Palladium-Doped Silicon
(78.55; 71.55; S5.11)
(b) Vol. **148**, No. 2, p. 723—727 (1988).

V. V. Frolov and B. N. Mukashev
On the Origin of the Hydrogen-Related Complex Corresponding to the 2223 cm^{-1} (at 10 K) Infrared Band in c-Si:H
(61.70; 78.30; S5.11)
(b) Vol. **148**, No. 2, p. K105—K107 (1988).

E. Susi, A. Poggi, G. Lulli, C. Monteverdi, and M. Merli
Lifetime and Crystal Order in Annealed CZ Silicon
(61.70; 72.20; S5.11)
(a) Vol. **108**, No. 2, p. 503—508 (1988).

Ch. Schnittler and G. Teichmann
A Kinetic Investigation of Cluster Formation in Semiconductors at Elevated Temperatures
(61.70; 64.75; S5.11)
(a) Vol. **108**, No. 2, p. 519—528 (1988).

V. V. ARISTOV, U. WINTER, A. YU. NIKULIN, S. V. REDKIN, A. A. SNIGIREV, P. ZAUMSEIL, and V. A. YUNKIN
Interference Thickness Oscillations of an X-Ray Wave on Periodically Profiled Silicon
(61.10 and 68.20; S5.11)
(a) Vol. **108**, No. 2, p. 651—655 (1988).

V. B. SHIKIN and N. I. SHIKINA
Charged Dislocations in Semiconductors
(61.70 and 71.55; S5.11; S5.12)
(a) Vol. **108**, No. 2, p. 669—681 (1988).

A. S. BAKÀI and N. P. LAZAREV
Diffusion of Implanted Hydrogen in a Silicon Film under Stress of a Standing Sound Wave
(66.30; 62.30; S5.11)
(a) Vol. **108**, No. 2, p. K101—K103 (1988).

FANGQING ZHANG, GUOSHENG SUN, and GUANGHUA CHEN
ESR and IR Studies on P-Doped a-Si:H/a-Si:H Junctions
(76.30 and 78.30; S5.11)
(a) Vol. **108**, No. 2, p. K139—K143 (1988).

T. HALICIOGLU, H. O. PAMUK, and S. ERKOC
Interatomic Potentials with Multi-Body Interactions
(63.10; 61.50; 68.40; S5.11)
(b) Vol. **149**, No. 1, p. 81—92 (1988).

ZIJING LIN and KELIN WANG
A Simple Scheme for Studies of Surface Phonons of Covalent Semiconductors. II. Surface Phonons at Ideal and Relaxed Si(110) Surface
(63.20 and 68.30; S5.11)
(b) Vol. **149**, No. 1, p. 143—148 (1988).

A. ARMIGLIATO, A. BOURRET, S. FRABBONI, and A. PARISINI
Contrast of Small SiX Particles in Silicon by Computed HREM Images
(61.16; S5.11)
(a) Vol. **109**, No. 1, p. 53—60 (1988).

V. A. KHRAMTSOV, V. N. LOMASOV, YA. YA. PILKEVICH, M. P. VLASENKO, and L. S. VLASENKO
Accumulation and Transformation of Radiation Defects in Silicon under Different Doses and Intensities of Electron Irradiation
(61.80; 76.30; S5.11)
(a) Vol. **109**, No. 1, p. 127—134 (1988).

G. A. KACHURIN, I. E. TYSHCHENKO, E. WIESER, and CH. WEISE
Redistribution of Boron in Silicon by High-Temperature Irradiation with Heavy Ions
(61.80 and 66.30; S5.11)
(a) Vol. **109**, No. 1, p. 141—148 (1988).

V. M. KOLESHKO and I. V. KIRYUSHIN
Deformation Mechanism Maps and Gettering Diagrams for Single-Crystal Silicon
(62.20; S5.11)
(a) Vol. **109**, No. 1, p. 161—169 (1988).

D. A. Aronov, R. Kabulov, and Yu. M. Yuabov
The Current Characteristics of p–i–n and p^+–i–p^+ Structures Based on Hydrogenated Amorphous Silicon at Various Temperatures and Excitation Levels
(73.40; 73.60; S5.11)
(a) Vol. **109**, No. 1, p. 211—216 (1988).

I. A. Bachilo, F. F. Komarov, A. P. Novikov, and S. A. Petrov
Structural Analysis of Silicon with High Doses of C^+ Ions
(61.80 and 68.55; S1.61; S5.11)
(a) Vol. **109**, No. 1, p. 231—238 (1988).

H. Wieczorek and W. Fuhs
Transient Photocurrents in a-Si:H Diodes: Effects of Deep Trapping
(72.40 and 73.40; 61.40; 71.55; S5.11)
(a) Vol. **109**, No. 1, p. 245—253 (1988).

V. Pandian and V. Kumar
Silver Related Deep Levels in Silicon
(71.55; S5.11)
(a) Vol. **109**, No. 1, p. 273—278 (1988).

K. Schmalz, F.-G. Kirscht, and K. Tittelbach-Helmrich
DLTS Study of Deep Level Defects in Cz n-Si Due to Heat Treatment at 600 to 900 °C
(71.55; S5.11)
(a) Vol. **109**, No. 1, p. 279—294 (1988).

L. N. Aleksandrov, V. Yu. Balandin, A. V. Dvurechenskii, and O. A. Kulayasova
Melting of Multilayered Structures under Pulse Heating (Computational Experiment)
(64.70; S5.11; S10.1)
(a) Vol. **109**, No. 1, p. K27—K29 (1988).

M. Wautelet, E. D. Gehain, and P. Quenon
Modelling the Role of Electronically Excited States in Laser-Assisted Phenomena
(61.70; 79.20; S5.11)
(b) Vol. **149**, No. 2, p. 465—475 (1988).

T. Soma, N. Atarashi, and H.-Matsuo Kagaya
Stable Configuration of an Interstitial Atom with Relaxation Effect in Si and Ge
(61.70; S5.11; S5.12)
(b) Vol. **149**, No. 2, p. 477—482 (1988).

M. Fliyou, M. A. Khan, and J. Ringeissen
Dynamical Effects on the Binding Energies of Core Excitons in Covalent and Polar Semiconductors
(71.35; S5.11; S5.12; S7.1; S8.1)
(b) Vol. **149**, No. 2, p. 615—621 (1988).

V. G. Khlabutin, V. I. Gridnev, and E. I. Rozum
Quantum Modulation of Electron Dechanneling
(71.25; S5.11)
(b) Vol. **149**, No. 2, p. K129—K134 (1988).

V. E. Borisenko and S. G. Yudin
Implanted Impurity Behaviour in Silicon Subjected to Transient Heating
(61.70 and 66.30; S5.11)
(a) Vol. **109**, No. 2, p. 395—402 (1988).

C. A. Londos and J. Grammatikakis
Notes on the Carbon-Associated Deep Level Complex in Irradiated Silicon
(61.80; 71.55; S5.11)
(a) Vol. **109**, No. 2, p. 421—426 (1988).

J. Suñé, I. Placencia, E. Farrés, N. Barniol, and X. Aymerich
On the Si–SiO_2 Interface Roughness in VLSI–MOS Structures
(68.48 and 73.60; S5.11; S10.1)
(a) Vol. **109**, No. 2, p. 479—491 (1988).

J. Krynicki and H. Rzewuski
Deep Electronic Levels in Carbon-Implanted Silicon
(61.70 and 71.55; S5.11)
(a) Vol. **109**, No. 2, p. 531—536 (1988).

V. M. Babich, N. P. Baran, A. A. Bugai, A. A. Konchits, V. B. Kovalchuk, V. M. Maksimenko, and B. D. Shanina
Electrical and Paramagnetic Properties of Thermodonors-II in Silicon. Discussion of a Model
(71.55; 72.80; 76.30; S5.11)
(a) Vol. **109**, No. 2, p. 537—547 (1988).

V. A. Bushuev, A. Yu. Kazimirov, and M. V. Kovalchuk
Coherent Compton Effect under Conditions of X-Ray Dynamical Laue Diffraction
(61.10 and 78.70; S5.11)
(b) Vol. **150**, No. 1, p. 9—18 (1988).

N. T. Bagraev and I. S. Polovtsev
Reactions for Iron Centers in Silicon under Optical Pumping
(61.70 and 71.55; 72.40; 76.30; 76.60; S5.11)
(b) Vol. **150**, No. 1, p. 19—24 (1988).

V. Yu. Balandin, A. V. Dvurechenskii, and L. N. Aleksandrov
Intermediate Crystallization of Amorphous Layers of Silicon at Pulsed Laser Annealing
(64.60; 61.50; S5.11)
(a) Vol. **110**, No. 1, p. 133—139 (1988).

W. Andrä, G. Götz, H. Hobert, V. Misyuchenko, V. A. Samuilov, and V. Stelmakh
IR and RBS Spectroscopy Investigation of Semi-Insulating Phosphorus-Doped Polycrystalline Silicon Layers
(68.55 and 78.65; S5.11)
(a) Vol. **110**, No. 1, p. 181—187 (1988).

H. Ohyama and K. Nemoto
Process of Defect Formation in the Collector Region of an Electron-Irradiated npn Si Transistor
(61.80 and 73.40; 85; S5.11)
(a) Vol. **110**, No. 1, p. 301—308 (1988).

B. M. Ashkinadze, V. V. Belkov, and A. V. Subashiev
Optical and Electrical Bistability Induced by Exciton Ionization Processes
(78.20; 71.35; S5.11)
(b) Vol. **150**, No. 2, p. 533—537 (1988).

H. J. Eichler, T. Brand, M. Glotz, and B. Smandek
Optical Nonlinearity and Bistability in Silicon
(78.20; 85; S5.11)
(b) Vol. **150**, No. 2, p. 705—718 (1988).

A. Galeckas, V. Netiksis, M. Petrauskas, and J. Vaitkus
Reflectivity and Dynamic Gratings in Implanted Si Induced by Picosecond Laser Pulses
(78.20; 61.80; S5.11)
(b) Vol. **150**, No. 2, p. 743—748 (1988).

K. Jarašiūnas and J. Vaitkus
Transient Gratings in Metrology of Semiconductor Parameters and Optoelectronic Devices
(78.20; S5.11; S7.12; S7.13; S8.12)
(b) Vol. **150**, No. 2, p. 879—884 (1988).

A. A. Bugaev and B. P. Zakharchenya
Picosecond Holographic Diagnostic of Light Pulse Absorption in Semiconductors
(72.30 and 78.20; 71.38; S5.11; S7.12)
(b) Vol. **150**, No. 2, p. 891—894 (1988).

J. Dubard, A. L. Smirl, A. G. Cui, G. C. Valley, and T. F. Boggess
Beam Amplification by Transient Energy Transfer in GaAs and Si
(78.70; S5.11; S7.12)
(b) Vol. **150**, No. 2, p. 913—919 (1988).

P. F. Lugakov and T. A. Lukashevich
Peculiarities of Radiation Defect Accumulation under High-Flux γ-Ray and Electron Irradiation of Silicon
(61.80; 72.80; S5.11)
(a) Vol. **110**, No. 2, p. 403—408 (1988).

Er-wang Mao, Wen-qin Zhao, Hong-rui Zhang, Ai-zhen Li, Jian-min Chen, and Guo-ping Fang
The Influence of Strain and Dislocations on Transport Properties of GaAs/Si Strained-Layer Heterojunctions
(73.40; 78.30; S5.11; S7.12)
(a) Vol. **110**, No. 2, p. 515—520 (1988).

H. Ohyama and K. Nemoto
Recovery Mechanism of Lattice Defects Formed in the Collector Region for Electron-Irradiated npn Si Transistors
(61.70 and 73.40; 85; S5.11)
(a) Vol. **110**, No. 2, p. 677—686 (1988).

O. V. Aleksandrov, V. V. Kozlovskii, V. V. Popov, and B. E. Samorukov
Diffusion of Impurities from Implanted Silicon Layers by Rapid Thermal Annealing
(61.70 and 66.30; S5.11)
(a) Vol. **110**, No. 2, p. K61—K65 (1988).

S5.12. Germanium

H.-R. Höche, J. Nieber, M. Clausnitzer, and G. Materlik
Modification of Specularly Reflected X-Ray Intensity by Grazing Incidence Coplanar Bragg-Case Diffraction
(61.10; S5.12)
(a) Vol. **105**, No. 1, p. 53—60 (1988).

Yu. A. Firsov, E. P. Nakhmedov, V. N. Prigodin, and W. Weller
Low-Temperature Conductivity of a Weakly Disordered, Strongly Anisotropic Two-Dimensional System with Electron–Electron Interaction. Application to Germanium Bicrystals and to Surface Superlattices
(72.10; 73.40; S5.12)
(b) Vol. **145**, No. 2, p. 625—635 (1988).

V. M. Ratag
Influence of Bond-Angle and Dihedral-Angle Disorders on the Top of the Valence Band of Amorphous Germanium
(71.25; S5.12)
(b) Vol. **147**, No. 1, p. 243—251 (1988).

G. M. Gusakov, A. A. Komarnitskii, and A. S. Em
Dynamics of Optical Parameters of Silicon and Germanium Single Crystals during Two-Pulsed Nanosecond Laser Irradiation
(61.80 and 68.55; 64.70; 78.20; S5.11; S5.12)
(a) Vol. **107**, No. 1, p. 261—271 (1988).

K. M. Podurets, S. Sh. Shilstein, and F. Eichhorn
Refraction Shift of Bragg Reflection of Neutrons
(78.70; S5.12)
(a) Vol. **107**, No. 1, p. 387—392 (1988).

U. Pietsch
The Adventage of Internal Force Calculations in the X-Ray Crystal Structure Analysis
(61.10 and 62.20; S5.12; S7.12)
(b) Vol. **147**, No. 2, p. 517—527 (1988).

D. Kremp, M. Schlanges, and T. Bornath
Kinetic Theory of Chemically Reacting Systems. Nonideality Effects in Rate Equations
(71.35; S5.12)
(b) Vol. **147**, No. 2, p. 747—756 (1988).

F. R. Kessler, J. Nieder, R. Nies, and E. Gaulke
Indirect Exciton Excitation and the Oscillatory Magnetophotoconductivity of Germanium
(71.35 and 72.40; 78.20; S5.12)
(b) Vol. **147**, No. 2, p. K143—K148 (1988).

A. Bourret, J. L. Rouvière, and J. Spendeler
Direct Observation of Atomic Columns in Semiconductors by HREM at 400 kV
(61.55 and 61.60; 61.16; S5.11; S5.12; S7.1; S8.1)
(a) Vol. **107**, No. 2, p. 481—501 (1988).

V. V. Kochelab, V. B. Molodkin, S. I. Olikhovskii, and M. E. Osinovskii
The Thickness Dependence of the X-Ray Diffuse Scattering Intensity for Crystals with Microdefects at Laue-Case Diffraction
(61.10; 61.70; S5.12)
(a) Vol. **108**, No. 1, p. 67—79 (1988).

V. V. Litvinov, G. V. Palchik, and V. I. Urenev
Extrinsic Self-Trapping and Negative *U* of Oxygen-Related Thermal Donors in Germanium
(71.55; 61.70; S5.12)
(a) Vol. **108**, No. 1, p. 311—321 (1988).

A. Kumar and I. Singh
Electronic Structure of the Ge–ZnSe, Ge–GaAs, and Ge–AlAs(110) Interfaces and Superlattices
(71.25 and 73.40; S5.12; S7.12; S8.12)
(b) Vol. **148**, No. 2, p. 549—558 (1988).

A. Dargys
Addendum to "Tunneling in ***k***-Space"
(72.10; 71.25; S5.12)
(b) Vol. **148**, No. 2, p. K125—K127 (1988).

U. Jendrich and P. Haasen
Internal Friction Due to Kink Motion and Kink Pair Formation in Intrinsic and n-Doped Germanium
(61.70 and 62.40; S5.12)
(a) Vol. **108**, No. 2, p. 553—568 (1988).

V. B. Shikin and N. I. Shikina
Charged Dislocations in Semiconductors
(61.70 and 71.55; S5.11; S5.12)
(a) Vol. **108**, No. 2, p. 669—681 (1988).

T. Soma, N. Atarashi, and H.-Matsuo Kagaya
Stable Configuration of an Interstitial Atom with Relaxation Effect in Si and Ge
(61.70; S5.11; S5.12)
(b) Vol. **149**, No. 2, p. 477—482 (1988).

M. Fliyou, M. A. Khan, and J. Ringeissen
Dynamical Effects on the Binding Energies of Core Excitons in Covalent and Polar Semiconductors
(71.35; S5.11; S5.12; S7.1; S8.1)
(b) Vol. **149**, No. 2, p. 615—621 (1988).

O. Pacherová and R. Bubáková
A Many-Beam Diffraction Effect in the $CoK\alpha_1$ Spectral Lineshape
(61.10; S5.12)
(a) Vol. **110**, No. 1, p. 35—42 (1988).

J. K. Požela, Z. N. Tamaševičienė, A. V. Tamaševičius, J. K. Ulbikas, and G. V. Bandurkina
Self-Generated Chaos in n-Ge After Electron Irradiation
(72.20; 71.55; S5.12)
(a) Vol. **110**, No. 2, p. 555—563 (1988).

S5.2. Solidified Gases

M. KUMARI and N. DASS
The Melting Laws
(64.70; 62.20; S2; S5.2; S9)
(b) Vol. **146**, No. 1, p. 105—110 (1988).

R. BAHADUR and C. M. KACHHAVA
Green Function Formalism and Superfluid Properties of ^{4}He
(67.40; S5.2)
(b) Vol. **147**, No. 2, p. 583—587 (1988).

I. YA. FUGOL, A. M. RATNER, L. V. KHASHCHINA, and I. M. PRITULA
Infrared Absorption and Anharmonicity of Crystalline Nitrogen
(78.30; 63.20; S5.2)
(b) Vol. **147**, No. 2, p. 757—766 (1988).

W. VOGELSBERGER, H.-G. FRITSCHE, and E. MÜLLER
Size Dependence of Surface Thermodynamic Parameters of Microclusters and the Location of the Gibbs' "Surface of Tension"
(68.40; 64.10; S5.2)
(b) Vol. **148**, No. 1, p. 155—164 (1988).

H.-G. FRITSCHE, D. BONCHEV, and O. MEKENYAN
Are Small Clusters of Inert-Gas Atoms Polyhedra of Minimum Surfaces?
(61.55; S5.2)
(b) Vol. **148**, No. 2, p. K101—K104 (1988).

P. MOHAZZABI and F. BEHROOZI
Vibrational Energy as a Function of Interatomic Distance in Rare-Gas Solids. A Universal Relationship
(61.50 and 65.40; S5.2)
(b) Vol. **149**, No. 2, p. 495—501 (1988).

S6. Group IV compounds
(for transition metal compounds, see S1.61)

C. Zwanziger, S. Petrowa, J. Reinhold, P. Streubel, and A. Meisel
Approximate Calculation of Inner-Shell Binding Energies and Auger Transition Energies of Solid Silicon Compounds
(71.50; 79.20; S5.11; S6)
(b) Vol. **145**, No. 2, p. 603—607 (1988).

D. Schalch, A. Scharmann, and R. Wolfrat
Hydrogen-Ion-Implanted Silicon Nitride and Silicon Oxynitride Films
(61.80 and 78.65; S6)
(a) Vol. **105**, No. 2, p. K81—K86 (1988).

I. I. Burshta, E. F. Venger, Yu. A. Pasechnik, and O. V. Snitko
Simplest Properties of Guided Wave Polaritons in Three-Layer Structures
(71.36 and 73.20; S6)
(b) Vol. **146**, No. 2, p. 517—524 (1988).

P. E. Van Camp, V. E. Van Doren, and J. T. Devreese
Ground State and Electronic Properties of Silicon Carbide and Boron Nitride
(71.25; S6; S7.1)
(b) Vol. **146**, No. 2, p. 573—587 (1988).

B. K. Ghosh and Bal K. Agrawal
Vibrational Structure of Hydrogenated Amorphous Silicon Carbide Alloys
(63.50; S6)
(b) Vol. **147**, No. 1, p. 97—102 (1988).

T. Soma, K. Kamada, and H.-Matsuo Kagaya
The Pressure Effect on the Phase Diagrams of the Ge–Sn and Si–Sn Systems
(64.75; 62.50; S6)
(b) Vol. **147**, No. 1, p. 109—114 (1988).

A. Šimůnek and G. Wiech
Angle-Resolved X-Ray Emission Spectra (ARXES) of Carbon in Potassium Graphite C_8K
(78.70; 71.20; S6)
(b) Vol. **149**, No. 2, p. 765—774 (1988).

Yong Yan, Fuquan Ji, Jun Chen, and Duang Feng
Imaging of the SOI Interface by High Resolution Electron Microscopy
(68.48; S6; S10.1)
(a) Vol. **110**, No. 1, p. K25—K28 (1988).

V. V. Gorbachev, V. M. Durasov, R. B. Zezin, E. V. Ivakin, A. S. Rubanov, and N. A. Tatyanina
Light-Induced Thermal Gratings in Natural Diamond
(66.70 and 78.35; S6)
(b) Vol. **150**, No. 2, p. 901—905 (1988).

S7. Group V compounds
(for transition metal compounds, see S1.62)

A. L. SHABALOV, M. S. FELDMAN, and M. Z. BASHIROV
Optical Properties of SiN_x Films of Variable Composition
(78.65; S7)
(b) Vol. **145**, No. 1, p. K71—K74 (1988).

P. R. VAYA, K. R. MURALI, and M. SUNDARAM
Some Studies on Zinc Phosphide Films Grown by Hot Wall Epitaxy
(68.55; 73.60; 78.65; S7)
(a) Vol. **105**, No. 2, p. 477—484 (1988).

J. SZATKOWSKI and K. SIERAŃSKI
Influence of Interface States on the Electrical Properties of Mg–Zn_3P_2 Junctions
(72.40 and 73.30; S7)
(a) Vol. **106**, No. 1, p. K31—K34 (1988).

M. A. ANGADI and R. ANGADI
Electrical Properties of MnSb Films
(73.60; 75.70; S7)
(a) Vol. **106**, No. 2, p. 485—491 (1988).

J. SZATKOWSKI, K. SIERAŃSKI, and J. MISIEWICZ
Characteristics of Mg–Zn_3P_2:Cd Contacts
(72.40 and 73.30; S7)
(a) Vol. **106**, No. 2, p. 551—555 (1988).

J. MISIEWICZ
The Low Temperature Photoluminescence in Zn_3P_2
(71.55 and 78.55; S7)
(a) Vol. **107**, No. 1, p. K65—K68 (1988).

V. P. DYMONT, G. I. MAKOVETSKII, and V. M. RYZHKOVSKII
Neutron Diffraction Determination of Mn and Zn Structure Positions in MnZnSb
(61.60; S7)
(a) Vol. **107**, No. 2, p. K89—K91 (1988).

S7.1. III–V compounds

K. S. Srivastava, A. Sinha, R. Srivastava, and A. Tandon
Interaction between Two Surface Excitations in Degenerate Polar Semiconductors
(68.30 and 71.45; S7.1)
(b) Vol. **146**, No. 1, p. 141—147 (1988).

P. E. Van Camp, V. E. Van Doren, and J. T. Devreese
Ground State and Electronic Properties of Silicon Carbide and Boron Nitride
(71.25; S6; S7.1)
(b) Vol. **146**, No. 2, p. 573—587 (1988).

J. Szatkowski and K. Sierański
Electrical Properties of Mg–Zn_3P_2 Contacts
(73.30; 72.40; S7.1)
(a) Vol. **106**, No. 2, p. 473—477 (1988).

H. Übensee, G. Paasch, J.-P. Zöllner, and S. Handschack
Modified Thomas-Fermi Approximation for Accumulation Layers in MIS Structures
(73.40; S7.1)
(b) Vol. **147**, No. 2, p. 823—837 (1988).

A. Bourret, J. L. Rouvière, and J. Spendeler
Direct Observation of Atomic Columns in Semiconductors by HREM at 400 kV
(61.55 and 61.60; 61.16; S5.11; S5.12; S7.1; S8.1)
(a) Vol. **107**, No. 2, p. 481—501 (1988).

H. Übensee, G. Paasch, D. Bundfuss, and J.-P. Zöllner
Electrons in MIS Inversion and Accumulation Layers. Exchange and Correlation Effects within a Modified Thomas-Fermi Theory
(73.40; S5.11; S7.1)
(b) Vol. **148**, No. 1, p. 421—434 (1988).

B. Schumann, R. Flagmeyer, and H. Weinert
Multi-Quantum-Well Structures Prepared by the Flash Evaporation Technique. Growth, Sructural, and Optical Characterization
(73.40; 68.55; 78.65; S7.1; S9.15)
(a) Vol. **108**, No. 1, p. 257—264 (1988).

V. B. Shipilo, E. M. Shishonok, A. M. Zaitsev, A. A. Melnikov, and A. I. Olekhnovich
Influence of High Pressure on Cathodoluminescence of Cubic Boron Nitride
(78.60; S7.1)
(a) Vol. **108**, No. 1, p. 431—436 (1988).

Ya. D. Lebedev, M. V. Fok, and M. D. Shagalov
On the Mechanism of Polarized Electroluminescence from Light-Emitting Structures Based on GaN
(73.40 and 78.60; S7.1)
(a) Vol. **108**, No. 2, p. 657—667 (1988).

P. ENDERS and U. SCHADE
Atomic Diffusion in Tetrahedral Semiconductors. Approximative Treatment of Diffusion Theories
(66.30; S7.1)
(b) Vol. **149**, No. 2, p. 447—456 (1988).

M. FLIYOU, M. A. KHAN, and J. RINGEISSEN
Dynamical Effects on the Binding Energies of Core Excitons in Covalent and Polar Semiconductors
(71.35; S5.11; S5.12; S7.1; S8.1)
(b) Vol. **149**, No. 2, p. 615—621 (1988).

S7.11. Phosphides

G. IRMER and W. SIEGEL
Carrier Concentration Profiles across GaP Grain Boundaries Observed by Raman Measurements
(61.70 and 72.20; 78.30; S7.11)
(a) Vol. **105**, No. 2, p. 549—553 (1988).

C. ASCHERON, H. A. KLOSE, W. FRENTRUP, and M. GRIEPENTROG
Gettering of Copper in Proton- and Helium-Bombarded Buried Regions of Gallium Phosphide
(61.80; S7.11)
(a) Vol. **106**, No. 1, p. 73—79 (1988).

G. DLUBEK, C. ASCHERON, R. KRAUSE, H. ERHARD, and D. KLIMM
Positron Study of Vacancy Defects in Proton and Neutron Irradiated GaP, InP, and Si
(61.80; 78.70; S5.11; S7.11)
(a) Vol. **106**, No. 1, p. 81—88 (1988).

I. I. RESHINA
Raman Study of Disorder in Laser-Annealed III–V Semiconductors
(78.30; 63.50; S7.11; S7.12)
(a) Vol. **106**, No. 1, p. 261—270 (1988).

V. ŠVORČÍK, V. RYBKA, and V. MYSLÍK
Photoelectrochemical Etching of n-GaAs and n-InP
(68.45 and 78.70; S7.11; S7.12)
(a) Vol. **106**, No. 1, p. K35—K39 (1988).

B. RHEINLÄNDNER, R. HEILMANN, and G. OELGART
Determination of Temperature-Dependent Carrier Losses in 1.3 μm InGaAsP/InP Double-Heterostructures
(78.45 and 85; 73.40; S7.11; S7.16)
(a) Vol. **106**, No. 1, p. K113—K116 (1988).

J. BĄK-MISIUK, H. G. BRÜHL, W. PASZKOWICZ, and U. PIETSCH
Temperature Dependence of the Lattice Constant in Doped and Nonstoichiometric GaAs, $GaAs_{1-x}P_x$, and GaP
(65.40; 61.70; S7.11; S7.12)
(a) Vol. **106**, No. 2, p. 451—457 (1988).

A. R. HASSAN
Phonon-Assisted Two-Photon Exciton Transitions in Semiconductors
(71.25 and 71.35; S7.11; S9)
(b) Vol. **147**, No. 1, p. 211—214 (1988).

D. WRUCK and A. KNAUER
Doping Inhomogeneities and Compensation in n-Type LEC InP Wafers
(72.20; 78.50; S7.11)
(a) Vol. **107**, No. 1, p. 321—328 (1988).

B. RHEINLÄNDER, G. GRUMMT, J. KOVÁČ, H. HAEFNER, and R. HEILMANN
Minority-Carrier Generation and Recombination in 1.35 μm InGaAsP/InP Double Heterostructure Diodes
(73.40; 71.55; 72.40; 78.60; S7.11; S7.15)
(a) Vol. **107**, No. 1, p. 405—418 (1988).

O. V. KULIKOVA, L. L. KULYUK, N. M. NARTYA, S. I. RADAUTSAN, E. V. RUSSU, A. V. SIMINEL, and E. E. STRUMBAN
On the Possibility of Formation of Associated Impurity Centers in Semi-Insulating Indium Phosphide Doped by Oxygen
(71.55; 72.40; S7.11)
(a) Vol. **107**, No. 1, p. K57—K60 (1988).

M. P. DOS SANTOS and C. HIRLIMANN
Impurity Effects in the Raman and Luminescence Spectra of Nitrogen-Doped GaP
(78.30 and 78.55; S7.11)
(b) Vol. **147**, No. 2, p. 779—789 (1988).

B. LEPLEY, A. BATH, C. CARABATOS-NEDELEC, G. J. PAPAIOANNOU, P. C. EUTHYMIOU, and S. RAVELET
Determination by Optical DLTS of the Distribution of States near the Valence Band of Plasma Oxidized n-Type InP
(73.40; S7.11)
(a) Vol. **108**, No. 1, p. K41—K46 (1988).

S. I. RADAUTSAN, I. M. TIGINYANU, and N. B. PYSHNAYA
Electrical Properties of InP: Fe Single Crystals Implanted by Phosphorus Ions
(71.55 and 72.80; S7.11)
(a) Vol. **108**, No. 1, p. K59—K61 (1988).

W. WESCH, E. WENDLER, N. P. KEKELIDSE, and V. V. TARBA
Investigation of the Amorphization Process in GaAs and GaP by Means of Optical Spectroscopy
(61.80; 78.50; S7.11; S7.12)
(a) Vol. **108**, No. 2, p. K93—K96 (1988).

L. GUOBA, L. PRANEVIČIUS, V. SARGŪNAS, and G. ZUBAUSKAS
Protective Coatings on GaAs and InP. High-Energy Ion Scattering Analysis
(68.48; 79.20; S7.11; S7.12)
(a) Vol. **109**, No. 1, p. 239—244 (1988).

J. P. BANERJEE, S. P. PATI, and S. K. ROY
Computer Simulation Experiment on the mm-Wave Properties of Indium Phosphide Double Drift Impatts
(73.40 and 85; S7.11)
(a) Vol. **109**, No. 1, p. 359—364 (1988).

P. KRISPIN, N. T. NGHI, N. T. Q. HAI, D. D. KHANG, and D. A. TUAN
Defects in GaP Green Light-Emitting Diodes
(71.55; 73.40; 85; S7.11)
(a) Vol. **109**, No. 2, p. K115—K119 (1988).

W. ULRICI, K. FRIEDLAND, L. EAVES, D. P. HALLIDAY, and C. A. PAYLING
Optical and Electrical Studies of GaP:Ti
(71.55; 72.20; 78.50; 78.55; S7.11)
(b) Vol. **150**, No. 1, p. 177—190 (1988).

J. KREISSL, W. ULRICI, and W. GEHLHOFF
Electron Paramagnetic Resonance of Ti^{3+} in GaP
(76.30; S7.11)
(b) Vol. **150**, No. 1, p. K57—K61 (1988).

S7.12. Arsenides

J. GRIESCHE, R. ENDERLEIN, and D. SCHIKORA
Orientation and Strain in Heteroepitaxial Growth **(Review Article)**
(61.50 and 68.48; 68.20; S5.11; S7.12; S8.13)
(a) Vol. **109**, No. 1, p. 11—38 (1988).

R. ENDERLEIN, DESHENG JIANG, and YINSHENG TANG
On the Mechanisms of Photoreflectance in Multiple Quantum Wells
(78.20; 71.25; 71.35; S7.12; S7.15)
(b) Vol. **145**, No. 1, p. 167—180 (1988).

T. BAUMBACH, H.-G. BRÜHL, U. PIETSCH, and H. TERAUCHI
Characterization of $Ga_{1-x}Al_xAs$/GaAs Superlattices and Thin Single Layers by X-Ray Diffraction
(61.10 and 68.55; S7.12; S7.15)
(a) Vol. **105**, No. 1, p. 197—205 (1988).

V. N. BRUDNYI and V. V. PESCHEV
Electron Traps in n-GaAs Irradiated with High Electron Beam Fluxes at High Temperatures
(61.80 and 71.55; S7.12)
(a) Vol. **105**, No. 1, p. K57—K60 (1988).

W. B. COSTA and E. L. ALBUQUERQUE
Impurity Binding Energies in Heterostructures
(71.20; 71.55; S7.12; S7.15)
(b) Vol. **145**, No. 2, p. 597—602 (1988).

P. H. JIANG, Y. T. ZHU, D. Z. SUN, and Y. P. ZENG
Modulation-Doped $Al_xGa_{1-x}As$/GaAs Heterostructures with Parallel Conducting Layer in $Al_xGa_{1-x}As$
(73.40; S7.12; S7.15)
(b) Vol. **145**, No. 2, p. K111—K114 (1988).

R. Fornari and L. Dozsa
Deep Level Characterization in Semi-Insulating LEC Gallium Arsenide
(71.55; 72.40; 78.50; S7.12)
(a) Vol. **105**, No. 2, p. 521—530 (1988).

R. Rentzsch, K. J. Friedland, and A. N. Ionov
Negative Magnetoresistance of Neutron-Transmutation-Doped Gallium Arsenide at Variable-Range Hopping
(72.20; 72.60; S7.12)
(b) Vol. **146**, No. 1, p. 199—206 (1988).

D. S. Domanevskii, S. V. Zhokhovets, and M. V. Prokopenya
Peculiarities of Radiative Recombination in Gallium Arsenide Doped with Shallow Donors and Acceptors
(78.55; 71.55; S7.12)
(a) Vol. **106**, No. 1, p. 249—260 (1988).

I. I. Reshina
Raman Study of Disorder in Laser-Annealed III–V Semiconductors
(78.30; 63.50; S7.11; S7.12)
(a) Vol. **106**, No. 1, p. 261—270 (1988).

V. Švorčík, V. Rybka, and V. Myslík
Photoelectrochemical Etching of n-GaAs and n-InP
(68.45 and 78.70; S7.11; S7.12)
(a) Vol. **106**, No. 1, p. K35—K39 (1988).

M. K. Bose, C. Majumdar, A. B. Maity, and A. N. Chakravarti
On Electron States of Quantum Wells in Ultrathin Films of Wide-Gap Semiconductors in an Oriented Magnetic Field
(71.25 and 73.60; S7.12)
(b) Vol. **146**, No. 2, p. 525—530 (1988).

G. Dlubek, A. Dlubek, R. Krause, O. Brümmer, K. Friedland, and R. Rentzsch
Point Defects in GaAs Studied by Correlated Positron Lifetime, Optical, and Electrical Measurements. I. Native Point Defects and Their Complexes in As-Grown GaAs
(61.70; 72.80; 78.55; 78.70; S7.12)
(a) Vol. **106**, No. 2, p. 419—432 (1988).

J. Bąk-Misiuk, H. G. Brühl, W. Paszkowicz, and U. Pietsch
Temperature Dependence of the Lattice Constant in Doped and Nonstoichiometric GaAs, $GaAs_{1-x}P_x$, and GaP
(65.40; 61.70; S7.11; S7.12)
(a) Vol. **106**, No. 2, p. 451—457 (1988).

D. Bose and B. R. Nag
High-Field Electron Transport in GaAs and Ga(0.47)In(0.53)As Rectangular Quantum Wells
(72.20; S7.12; S7.15)
(a) Vol. **106**, No. 2, p. 569—575 (1988).

G. Koschek, H. Lakner, and E. Kubalek
On the Origin of Cathodoluminescence Contrast Phenomena in Semiinsulating GaAs
(78.60; 71.55; S7.12)
(a) Vol. **106**, No. 2, p. 651—658 (1988).

E. Pinčík, B. Maťátko, J. Bartoš, I. Thurzo, M. Grendel, V. Nádaždy, M. Zubeková, and M. Morvic
Some Properties of MIS Structures Prepared by Plasma Oxidation of Al Layers on GaAs
(73.40 and 85; S7.12)
(a) Vol. **106**, No. 2, p. 659—667 (1988).

R. Pérez-Alvarez, J. L. Parra-Santiago, and P. Pajón-Suarez
Cylindrical Quantum Well with Position-Dependent Mass
(71.25 and 73.40; S7.12; S7.15)
(b) Vol. **147**, No. 1, p. 127—133 (1988).

C. K. Sarkar, S. Kundu, and P. K. Basu
Cyclotron Resonance Linewidth in a Two-Dimensional Electron Gas
(73.40 and 76.40; S7.12)
(b) Vol. **147**, No. 1, p. 135—140 (1988).

H. van Cong, S. Charar, and S. Brunet
Band-Gap Narrowing Due to Many-Body Effects in n-Type Degenerate GaAs Crystals
(71.25; S7.12)
(b) Vol. **147**, No. 1, p. 253—260 (1988).

G. Dlubek, A. Dlubek, R. Krause, and O. Brümmer
Point Defects in GaAs Studied by Correlated Positron Lifetime, Optical, and Electrical Measurements. II. Point Defects in GaAs Irradiated with Fast Neutrons
(61.80; 78.50; 78.55; 78.70; S7.12)
(a) Vol. **107**, No. 1, p. 111—121 (1988).

B. E. Mdivanyan and M. Sh. Shikhsaidov
Photostimulated Enhancement of Dislocation Glide in Gallium Arsenide Crystals
(62.20; 61.80; S7.12)
(a) Vol. **107**, No. 1, p. 131—140 (1988).

F. Lukeš
Ellipsometric Studies of Natural Films on GaAs
(78.65; S7.12)
(a) Vol. **107**, No. 1, p. 239—251 (1988).

G. Grummt, H. Haefner, G. Oelgart, W. Thielemann, R. Pickenhain, and B. Jacobs
Properties of Red Light Emitting (Al, Ga)As Double Heterostructure Diodes
(73.40 and 78.60; 85; S7.12)
(a) Vol. **107**, No. 1, p. 419—428 (1988).

E. Jároli, B. Pécz, R. Veresegyházy, F. Pászti, T. Lohner, M. Fried, I. Mojzes, and J. Gyulai
Effect of Ion Beam Mixing on the Evolution of Arsenic from the Au–GaAs System
(61.80 and 68.48; S1.3; S7.12)
(a) Vol. **107**, No. 1, p. K15—K17 (1988).

R. Flagmeyer, K. Lenkeit, T. Baumbach, Yu. O. Kanter, and A. A. Fedorov
Characterization of $In_xGa_{1-x}As/GaAs$ Strained Layer Superlattices by Ion Backscattering-Channeling and X-Ray Diffraction
(68.48; 62.20; S7.12; S7.15)
(a) Vol. **107**, No. 1, p. K19—K24 (1988).

U. Pietsch
The Advantage of Internal Force Calculations in the X-Ray Crystal Structure Analysis
(61.10 and 62.20; S5.12; S7.12)
(b) Vol. **147**, No. 2, p. 517—527 (1988).

R. Pérez-Alvarez and P. Pajón-Suárez
Optical Absorption Coefficient in Cylindrical Quantum Wells
(71.25 and 73.40; S7.12)
(b) Vol. **147**, No. 2, p. 547—552 (1988).

B. N. Narahari Achar and J. R. Ferguson
Quantized Damping of Helicons in a GaAs/AlGaAs Superlattice
(72.20 and 73.60; S7.12; S7.15)
(b) Vol. **147**, No. 2, p. K137—K141 (1988).

A. Lefebvre and G. Vanderschaeve
The Origin of Microtwinning in Room Temperature Indented Undoped and n-Doped GaAs Crystals
(61.70 and 62.20; S7.12)
(a) Vol. **107**, No. 2, p. 647—653 (1988).

D. Chattopadhyay, P. C. Rakshit, and C. K. Sarkar
Small-Signal AC Mobility of Two-Dimensional Hot Electrons in Semiconductor Quantum Wells at Low Temperatures
(72.20 and 73.40; S7.12)
(b) Vol. **148**, No. 1, p. K17—K22 (1988).

R. Riera, F. Comas, C. Trallero Giner, and S. T. Pavlov
Electron Raman Scattering in Semiconductor Quantum Wells
(71.25 and 78.30; S7.12; S7.15)
(b) Vol. **148**, No. 2, p. 533—542 (1988).

E. L. Albuquerque and W. B. Costa
Impurity Binding Energies in Hetrostructures. Two-Band Model
(71.20 and 73.40; S7.12; S7.15)
(b) Vol. **148**, No. 2, p. 543—548 (1988).

A. Kumar and I. Singh
Electronic Structure of the Ge–ZnSe, Ge–GaAs, and Ge–AlAs(110) Interfaces and Superlattices
(71.25 and 73.40; S5.12; S7.12; S8.12)
(b) Vol. **148**, No. 2, p. 549—558 (1988).

C. MAJUMDAR, A. B. MAITY, and A. N. CHAKRAVARTI
Quantized Photoemission from Quantum Wells in Ultrathin Films of Wide-Gap Semiconductors in a Tilted Magnetic Field
(79.60; 73.60; S7.12)
(b) Vol. **148**, No. 2, p. 559—565 (1988).

K. KRÁL, Z. KHÁS, and B. HEJDA
Hot Phonon Generation Rate with Electron Energy Collision Broadening in Polar Semiconductors
(72.10; S7.12)
(b) Vol. **148**, No. 2, p. K129—K131 (1988).

G. KOSCHEK, H. LAKNER, and E. KUBALEK
On the Improvement of the Semiinsulating GaAs Substrate Homogeneity Due to Annealing
(71.55; 78.60; S7.12)
(a) Vol. **108**, No. 2, p. 683—688 (1988).

W. SIEGEL, H. WITTE, G. KÜHNEL, H. A. SCHNEIDER, and T. FLADE
Thermal Activation Energies of Undoped Semi-Insulating LEC GaAs. Correlation of TDH and TSC Results
(71.55 and 72.20; S7.12)
(a) Vol. **108**, No. 2, p. 689—697 (1988).

W. WESCH, E. WENDLER, N. P. KEKELIDSE, and V. V. TARBA
Investigation of the Amorphization Process in GaAs and GaP by Means of Optical Spectroscopy
(61.80; 78.50; S7.11; S7.12)
(a) Vol. **108**, No. 2, p. K93—K96 (1988).

P. S. GLADKOV and K. B. OZANYAN
Photoconductivity in Bulk and Epitaxial GaAs:V
(71.55 and 72.40; 61.70; S7.12)
(a) Vol. **108**, No. 2, p. K125—K130 (1988).

W. W. RÜHLE and K. LEO
Carrier Heating in GaAs by Nonradiative Recombination
(71.38; S7.12)
(b) Vol. **149**, No. 1, p. 215—220 (1988).

P. FRANCOIS, A. LEFEBVRE, and G. VANDERSCHAEVE
Low Temperature Plasticity of Brittle Materials. A New Device for Compressive Testing under Confining Pressure
(62.20; S7.12)
(a) Vol. **109**, No. 1, p. 187—192 (1988).

L. GUOBA, L. PRANEVIČIUS, V. SARGŪNAS, and G. ZUBAUSKAS
Protective Coatings on GaAs and InP. High-Energy Ion Scattering Analysis
(68.48; 79.20; S7.11; S7.12)
(a) Vol. **109**, No. 1, p. 239—244 (1988).

F. Koch, W. Hergert, G. Oelgart, and N. Puhlmann
Determination of Semiconductor Parameters by Electron Beam Induced Current and Cathodoluminescence Measurements
(73.40; 78.60; S7.12)
(a) Vol. **109**, No. 1, p. 261—272 (1988).

G. T. Baumbach, H. Rhan, and U. Pietsch
Detection of the Active Layer of $A^{III}B^{V}$ Semiconductor Quantum-Well Structures by High Resolution X-Ray Diffractometry
(61.10 and 68.55; S7.12; S7.15)
(a) Vol. **109**, No. 1, p. K7—K10 (1988).

B. Sandow, E. Nebauer, and H. Strusny
Contact Resistance Reduction by Implanted n^+-Layers at the Ni/AuGe–GaAs System
(73.40; S7.12)
(a) Vol. **109**, No. 1, p. K35—K37 (1988).

H. Leon and F. Comas
On the Scattering of Electrons by Confined LO Phonons in a Semiconductor Quantum Well
(71.38 and 73.40; S7.12; S7.15)
(b) Vol. **149**, No. 2, p. 533—542 (1988).

Chr. Pratsch and M. Suhrke
Low Temperature Electron Transport in Quasi-Two-Dimensional Systems
(73.40; S7.12; S7.15)
(b) Vol. **149**, No. 2, p. 547—553 (1988).

A. Maity, C. Majumdar, and A. N. Chakravarti
Effect of Magneto-Size Quantization on Electric-Field-Aided Photoemission from Quantum Wells in Ultrathin Films of Wide-Gap Semiconductors
(73.60 and 79.60; S7.12)
(b) Vol. **149**, No. 2, p. 565—570 (1988).

T. Yu. Bilgildeeva and T. A. Polyanskaya
The Quantum Corrections to the Conductivity in Strong Spin–Orbit Coupling Semiconductors
(72.20; 71.30; S7.12)
(b) Vol. **149**, No. 2, p. 649—655 (1988).

B. Mitra and K. P. Ghatak
On the Einstein Relation in Semiconductor Superlattices under Magnetic Quantization
(72.20 and 73.40; S7.12; S7.15)
(b) Vol. **149**, No. 2, p. K117—K122 (1988).

S. Fujita, K. Maeda, and S. Hyodo
Radiation Enhanced Dislocation Glide Effect on Indentation-Induced Fracture of GaAs Single Crystals
(61.70 and 62.20; S7.12)
(a) Vol. **109**, No. 2, p. 383—393 (1988).

V. S. Postnikov, B. I. Sysoev, A. V. Budanov, N. N. Bezryadin, Yu. K. Shlyk, and B. L. Agapov
Heterostructures on the Basis of Indium Arsenide with Semi-Insulating $A_2^{III}B_3^{VI}$ Compound Layers
(73.40; S7.12; S8)
(a) Vol. **109**, No. 2, p. 463—467 (1988).

H. Scibior, I. Brylowska, P. Mazurek, and M. Subotowicz
Influence of Ion Implantation on the Luminescence of GaAs
(78.55 and 78.60; 61.70; S7.12)
(a) Vol. **109**, No. 2, p. 597—601 (1988).

S. Zekeng, B. Prevot, and C. Schwab
Raman Determination of the Faust-Henry Coefficient of GaAs in the 1.9 to 2.7 eV Range at Ordinary and Low Temperatures
(63.20 and 78.30; S7.12)
(b) Vol. **150**, No. 1, p. 65—72 (1988).

D. S. Domanevskii, S. V. Zhokhovets, and M. V. Prokopenya
The Valence Band Tail State Filling in Heavily Doped and Compensated Crystals of Gallium Arsenide
(71.55; 78.55; S7.12)
(a) Vol. **110**, No. 1, p. 221—226 (1988).

S. Hildebrandt, J. Schreiber, W. Hergert, and V. I. Petrov
Determination of the Absorption Coefficient and the Internal Luminescence Spectrum of GaAs and $GaAs_{1-x}P_x$ ($x = 0.375, 0.78$) from Beam Voltage Dependent Measurements of Cathodoluminescence Spectra in the Scanning Electron Microscope
(78.60; S7.12; S7.15)
(a) Vol. **110**, No. 1, p. 283—291 (1988).

S. Schmitt-Rink
Coherent Nonlinear Optical Processes in Semiconductors. Physics and Applications (*invited*)
(71.35 and 78.20; S7.12; S7.15; S12)
(b) Vol. **150**, No. 2, p. 349—356 (1988).

I. Balslev and A. Stahl
Coherent Nonlinear Response of Excitons in Semiconductors
(71.35 and 78.20; S7.12; S7.15)
(b) Vol. **150**, No. 2, p. 413—418 (1988).

R. Mickevičius and A. Reklaitis
Subpicosecond Dynamics of the Nonequilibrium Electron–Phonon System in GaAs
(78.20; S7.12)
(b) Vol. **150**, No. 2, p. 437—441 (1988).

R. Kuszelewicz, J. L. Oudar, R. Azoulay, J. C. Michel, J. Brandon, and O. Emile
All-Epitaxial GaAs/AlAs Nonlinear Etalons. Towards Continuous and Parallel Operation
(68.55 and 78.20; S7.12; S7.15)
(b) Vol. **150**, No. 2, p. 465—470 (1988).

B. S. Ryvkin
Semiconductor Cells with N-Type Photocurrent–Voltage Characteristic: A New Class of Nonlinear Optical Elements
(78.20; 72.40; S7.12)
(b) Vol. **150**, No. 2, p. 567—574 (1988).

D. M. Butusov, G. G. Gotsadze, V. R. Larionov, and B. S. Ryvkin
On the Possibility of the Optical Connection between SEED's within One Array
(78.20; 85; S7.12)
(b) Vol. **150**, No. 2, p. 599—603 (1988).

H. Uchiki, T. Kobayashi, and E. Tokunaga
Carrier Diffusion and Trapping by Quantum Wells
(73.40 and 78.55; S7.12; S7.15)
(b) Vol. **150**, No. 2, p. 667—672 (1988).

I. N. Uraltsev, E. L. Ivchenko, P. S. Kopev, V. P. Kochereshko, and D. R. Yakovlev
Exciton Parameters and Electron Miniband Structure of GaAs/AlGaAs Superlattices
(71.35 and 78.20; 71.25; S7.12; S7.15)
(b) Vol. **150**, No. 2, p. 673—678 (1988).

E. Lach, M. Walther, G. Tränkle, A. Forchel, and G. Weimann
Optical Investigation of Exciton Screening in GaAs/GaAlAs MQW Structures
(71.35 and 78.20; S7.12; S7.15)
(b) Vol. **150**, No. 2, p. 679—683 (1988).

G. Bongiovanni, J. L. Staehli, and D. Martin
Many-Body Effects in the Electron–Hole Plasma Confined in GaAs–(GaAl)As Quantum Wells
(71.35 and 78.65; 71.45; S7.12; S7.15)
(b) Vol. **150**, No. 2, p. 685—690 (1988).

K. Henneberger and G. Manzke
Dynamics of Incoherent Excitonic Optical Nonlinearities in GaAs–GaAlAs MQWS's
(71.35 and 78.20; S7.12; S7.15)
(b) Vol. **150**, No. 2, p. 691—693 (1988).

V. K. Kononenko
Nonlinear Absorption in Quantum-Size Heterostructures
(78.20; S7.12)
(b) Vol. **150**, No. 2, p. 695—698 (1988).

H. Weinert, V. Latinis, S. Juršenas, and R. Baltramiejunas
Time-Resolved Photoluminescence of a GaAs/GaAlAs Superlattice
(71.35 and 78.55; S7.12; S7.15)
(b) Vol. **150**, No. 2, p. 699—704 (1988).

F. V. Karpushko
Fast and Strong Nonlinearity of Thin-Film Semiconductor Interferometers Due to Intrinsic Photo- and Electrooptical Effects of the Semiconductor Interlayer
(73.40 and 78.20; S7.12; S8.12)
(b) Vol. **150**, No. 2, p. 791—796 (1988).

R. P. Seisyan, A A. Toropov, V. P. Khvostikov, and T. V. Shubina
Excitonic Electroabsorption in GaAs–GaAlAs Single Quantum Well Waveguide Structures
(71.35 and 78.20; S7.12; S7.15)
(b) Vol. **150**, No. 2, p. 863—866 (1988).

K. Jarašiūnas and J. Vaitkus
Transient Gratings in Metrology of Semiconductor Parameters and Optoelectronic Devices
(78.20; S5.11; S7.12; S7.13; S8.12)
(b) Vol. **150**, No. 2, p. 879—884 (1988).

H. Bergner, V. Brückner, M. Lenzner, and R. Strobel
Investigation of the Field-Dependent Carrier Mobility in GaAs by Picosecond Photoconductivity Measurements
(72.20; 72.40; S7.12)
(b) Vol. **150**, No. 2, p. 885—889 (1988).

A. A. Bugaev and B. P. Zakharchenya
Picosecond Holographic Diagnostic of Light Pulse Absorption in Semiconductors
(72.30 and 78.20; 71.38; S5.11; S7.12)
(b) Vol. **150**, No. 2, p. 891—894 (1988).

J. Dubard, A. L. Smirl, A. G. Cui, G. C. Valley, and T. F. Boggess
Beam Amplification by Transient Energy Transfer in GaAs and Si
(78.70; S5.11; S7.12)
(b) Vol. **150**, No. 2, p. 913—919 (1988).

Yu. B. Bolkhovityanov, R. I. Bolkhovityanova, T. A. Gavrilova, and A. E. Dolbak
Relaxation Kinetics of a Non-Equilibrium Interface of a Multicomponent Liquid Phase–Binary Substrate on the Example of In–Ga–As/InAs System
(68.45; 68.55; S7.12)
(a) Vol. **110**, No. 2, p. 489—494 (1988).

Er-wang Mao, Wen-Qin Zhao, Hong-rui Zhang, Ai-zhen Li, Jian-min Chen, and Guo-ping Fang
The Influence of Strain and Dislocations on Transport Properties of GaAs/Si Strained-Layer Heterojunctions
(73.40; 78.30; S5.11; S7.12)
(a) Vol. **110**, No. 2, p. 515—520 (1988).

D. V. Morgan, H. Thomas, W. T. Anderson, P. Thompson, A. Christou, and D. J. Diskett
High Temperature Metallisation for GaAs Device Processing
(61.80 and 73.40; S1.61; S4; S7.12)
(a) Vol. **110**, No. 2, p. 531—536 (1988).

S7.13. Antimonides

R. Herrmann, W. Kraak, S. Handschack, Th. Schurig, D. Kusnick, and B. Schnackenburg
Magnetotransport Properties and Subband Structure of the Two-Dimensional Electron Gas in the Inversion Layer of InSb Bicrystals under Hydrostatic Pressure
(73.40; S7.13)
(b) Vol. **145**, No. 1, p. 157—166 (1988).

P. Aghamkar, P. Sen, and P. K. Sen
Effect of Doping on Parametric Amplificaton in Piezoelectric Semiconductors
(78.20; S7.13)
(b) Vol. **145**, No. 1, p. 343—349 (1988).

M. M. Aleksandrova, V. D. Blank, A. E. Golobokov, Yu. S. Konyaev, A. Yu. Zerr, and E. I. Estrin
Amorphisation of Gallium Antimonide under the Conditions of Shear Deformation under Pressure
(61.50 and 62.20; S7.13)
(a) Vol. **105**, No. 1, p. K29—K32 (1988).

M. de Dios Leyva, J. López Gondar, and J. Sabín del Valle
Erratum to: Nonparabolicity Parameter in the Conduction Band of GaAs–$Al_xGa_{1-x}As$ Quantum Wells
(71.25; S7.13; S7.15)
(b) Vol. **145**, No. 2, p. K155 (1988).

I. Gyúrù, B. Pödör, V. V. Popov, and R. V. Parfenev
Hall Effect Studies on p-Type GaSb Grown under Microgravity Conditions
(72.80; S7.13)
(a) Vol. **105**, No. 2, p. K129—K133 (1988).

V. N. Brudnyi and I. V. Kamenskaya
The Electrical Properties and Fermi Level Pinning in Proton-Irradiated GaSb
(61.80 and 72.80; S7.13)
(a) Vol. **105**, No. 2, p. K141—K144 (1988).

S. Ghosh and A. N. Chakravarti
Effect of Broadening of Tail States on the Einstein Relation in Heavily Doped Compensated Semiconductors
(72.20; S7.13)
(b) Vol. **147**, No. 1, p. 355—360 (1988).

V. V. Kosarev, V. V. Popov, V. S. Vekshina, and N. I. Pepik
Intrinsic and Deep Centre Charge Carriers in Indium Antimonide
(71.55; S7.13)
(a) Vol. **107**, No. 1, p. K43—K48 (1988).

S. Handschack, G. Paasch, and G. Gobsch
Self-Consistent Calculation of the Electronic Structure of an Inversion Layer in a Perpendicular Magnetic Field
(71.25 and 73.40; S7.13)
(b) Vol. **147**, No. 2, p. K125—K129 (1988).

F. Ludwig, H.-U. Müller, and R. Herrmann
Magneto-Optical Investigations on p-InSb Bicrystals in the Submillimeter Wavelength Range
(76.40 and 78.20; 73.40; S7.13)
(b) Vol. **148**, No. 1, p. 143—153 (1988).

M. K. BOSE, C. MAJUMDAR, A. B. MAITY, and A. N. CHAKRAVARTI
Effect of a Tilted Magnetic Field on the Size-Quantized Levels of Quantum Wells in Ultrathin Films of Narrow-Gap Semiconductors
(71.25 and 73.60; S7.13)
(b) Vol. **148**, No. 1, p. 165—172 (1988).

W. KRAAK, U. TROPPENZ, R. HERRMANN, S. M. CHUDINOV, and V. A. KULBACHINSKII
Magnetic Freeze-Out and Magnetic-Field-Induced Semiconductor–Semimetal Transition in $Bi_{1-x}Sb_x$ Alloys under High Hydrostatic Pressure
(71.30 and 72.20; S7.13)
(b) Vol. **148**, No. 1, p. 333—347 (1988).

G. NACHTWEI, D. SCHULZE, G. GOBSCH, G. PAASCH, W. KRAAK, H. KRÜGER, and R. HERRMANN
Calculations of Magnetotransport Properties of InSb Bicrystals with Several Populated Q2D Subbands
(73.25 and 73.40; S7.13)
(b) Vol. **148**, No. 1, p. 349—359 (1988).

V. A. BEREZOVETS, W. BRAUNE, N. KUBICKI, and A. O. SMIRNOV
Investigation of Interface Peculiarities in Anodized p-InSb MOS Structures
(73.40; 72.40; 73.25; S7.13)
(a) Vol. **108**, No. 1, p. 303—309 (1988).

W. KRAAK, G. NACHTWEI, R. HERRMANN, and M. GLIŃSKI
Properties of the Quantum Hall Effect of the Two-Dimensional Electron Gas in the n-Inversion Layer of InSb Grain Boundaries under High Hydrostatic Pressure
(73.40; S7.13)
(b) Vol. **148**, No. 2, p. 567—578 (1988).

B. WOLF and A. ZEHE
A SIMS Analysis of Hydrogen Involved Reactions on InSb Surfaces
(68.20; 68.30; S7.13)
(b) Vol. **149**, No. 1, p. 149—156 (1988).

G. R. ALLAN, H. A. MACKENZIE, J. J. HUNTER, and B. S. WHERRETT
The Identification of Urbach, Acceptor Impurity, and Phonon-Assisted Components of the Optical Absorption Band Tail of n-Type InSb
(78.20; 71.55; S7.13)
(b) Vol. **149**, No. 1, p. 383—389 (1988).

F. LUDWIG, H.-U. MÜLLER, and R. HERRMANN
Effective Scattering Times of Inversion Electrons in InSb Grain Boundaries Deduced from Cyclotron Resonance Measurements
(73.40 and 76.40; 72.10; S7.13)
(b) Vol. **150**, No. 1, p. 119—123 (1988).

G. R. ALLAN, P. L. CHUA, J. J. HUNTER, H. A. MACKENZIE, C. R. PIDGEON, and A. C. WALKER
Dynamical Nonlinearity and Bistability of Narrow Gap Semiconductors
(78.20; S7.13)
(b) Vol. **150**, No. 2, p. 719—727 (1988).

U. TRUTSCHEL, M. GOLZ, and M. ABRAHAM
Optical Bistability in Two-Dimensional Waveguide Structures
(78.20; S7.13)
(b) Vol. **150**, No. 2, p. 867—871 (1988).

K. JARAŠIŪNAS and J. VAITKUS
Transient Gratings in Metrology of Semiconductor Parameters and Optoelectronic Devices
(78.20; S5.11; S7.12; S7.13; S8.12)
(b) Vol. **150**, No. 2, p. 879—884 (1988).

S7.15. Mixed crystals

R. ENDERLEIN, DESHENG JIANG, and YINSHENG TANG
On the Mechanisms of Photoreflectance in Multiple Quantum Wells
(78.20; 71.25; 71.35; S7.12; S7.15)
(b) Vol. **145**, No. 1, p. 167—180 (1988).

G. IRMER, M. HERMS, J. MONECKE, and H. BOCK
Raman Scattering from LO-Phonon–Plasmon Modes in $GaAs_{1-x}P_x$ Crystals
(63.20 and 78.30; 71.45; S7.15)
(b) Vol. **145**, No. 1, p. K79—K82 (1988).

T. BAUMBACH, H.-G. BRÜHL, U. PIETSCH, and H. TERAUCHI
Characterization of $Ga_{1-x}Al_xAs$/GaAs Superlattices and Thin Single Layers by X-Ray Diffraction
(61.10 and 68.55; S7.12; S7.15)
(a) Vol. **105**, No. 1, p. 197—205 (1988).

G. BEISTER, J. MAEGE, and G. RICHTER
Identification of Surface Influences on the Current–Voltage Characteristic of Narrow Oxide Stripe Laser Diodes
(73.25; 78.45; S7.15)
(a) Vol. **105**, No. 1, p. K45—K49 (1988).

K. ŽĎÁNSKÝ and LE THANH BINH
Characterization of DX Centers in GaAlAs Grown by Liquid-Phase Epitaxy
(71.55; S7.15)
(a) Vol. **105**, No. 1, p. K51—K55 (1988).

W. B. COSTA and E. L. ALBUQUERQUE
Impurity Binding Energies in Heterostructures
(71.20; 71.55; S7.12; S7.15)
(b) Vol. **145**, No. 2, p. 597—602 (1988).

P. H. JIANG, Y. T. ZHU, D. Z. SUN, and Y. P. ZENG
Modulation-Doped $Al_xGa_{1-x}As$/GaAs Heterostructures with Parallel Conducting Layer in $Al_xGa_{1-x}As$
(73.40; S7.12; S7.15)
(b) Vol. **145**, No. 2, p. K111—K114 (1988).

M. DE DIOS LEYVA, J. LÓPEZ GONDAR, and J. SABÍN DEL VALLE
Erratum to: Nonparabolicity Parameter in the Conduction Band of GaAs–$Al_xGa_{1-x}As$ Quantum Wells
(71.25; S7.13; S7.15)
(b) Vol. **145**, No. 2, p. K155 (1988).

R. STEGMANN, B. JACOBS, H. KAMLEH, M. ALBANI, M. HEIDER, and F. MONZER
Carrier Injection at Graded Heterojunctions
(73.40 and 85; 78.60; S7.15)
(a) Vol. **105**, No. 2, p. 661—670 (1988).

A. I. Bobrysheva, S. S. Russu, and V. A. Zaloi
Light Absorption and Hyper-Raman Scattering by Many-Electron Complexes in Quantum Well Structures
(78.20 and 78.30; 71.35; S7.15)
(b) Vol. **146**, No. 1, p. 329—334 (1988).

N. A. Gushchina, S. M. Dunaevskii, and V. K. Nikulin
Energy Band Structure of $(GaAs)_m$ $(AlAs)_m$ Semiconductor Superlattices with Ultrathin Layers
(73.40; 71.25; S7.15)
(b) Vol. **146**, No. 2, p. 511—516 (1988).

D. Bose and B. R. Nag
High-Field Electron Transport in GaAs and Ga(0.47)In(0.53)As Rectangular Quantum Wells
(72.20; S7.12; S7.15)
(a) Vol. **106**, No. 2, p. 569—575 (1988).

R. Pérez-Alvarez, J. L. Parra-Santiago, and P. Pajón-Suarez
Cylindrical Quantum Well with Position-Dependent Mass
(71.25 and 73.40; S7.12; S7.15)
(b) Vol. **147**, No. 1, p. 127—133 (1988).

A. Chiari, M. Colocci, F. Fermi, Yuzhang Li, R. Querzoli, A. Vinattieri, and Weihua Zhuang
Temperature Dependence of the Photoluminescence in GaAs–GaAlAs Multiple Quantum Well Structures
(78.55; 73.40; S7.15)
(b) Vol. **147**, No. 1, p. 421—429 (1988).

B. Rheinländer, G. Grummt, J. Kováč, H. Haefner, and R. Heilmann
Miniority-Carrier Generation and Recombination in 1.35 μm InGaAsP/InP Double Heterostructure Diodes
(73.40; 71.55; 72.40; 78.60; S7.11; S7.15)
(a) Vol. **107**, No. 1, p. 405—418 (1988).

R. Flagmeyer, K. Lenkeit, T. Baumbach, Yu. O. Kanter, and A. A. Fedorov
Characterization of $In_xGa_{1-x}As$/GaAs Strained Layer Superlattices by Ion Backscattering-Channeling and X-Ray Diffraction
(68.48; 62.20; S7.12; S7.15)
(a) Vol. **107**, No. 1, p. K19—K24 (1988).

D. K. Ghosh, U. Chatterjee, and L. K. Samanta
On the Band Gap Dependence of Refractive Indices of Some Quaternary III–V and II–VI Compounds of Device Interest
(78.20; S7.15; S8.15)
(a) Vol. **107**, No. 1, p. K79—K81 (1988).

G. Manzke and K. Henneberger
Nonlinear Propagation of Laser Pulses in GaAs–GaAlAs MQWS
(71.35 and 78.20; 73.40; S7.15)
(b) Vol. **147**, No. 2, p. 733—739 (1988).

B. N. NARAHARI ACHAR and J. R. FERGUSON
Quantized Damping of Helicons in a GaAs/AlGaAs Superlattice
(72.20 and 73.60; S7.12; S7.15)
(b) Vol. **147**, No. 2, p. K137—K141 (1988).

H. RHAN and U. PIETSCH
Investigation of Nanometer Layer Heterostructures by X-Ray Grazing Incidence Diffraction
(68.55; 61.10; S7.15)
(a) Vol. **107**, No. 2, p. K93—K98 (1988).

R. RIERA, F. COMAS, C. TRALLERO GINER, and S. T. PAVLOV
Electron Raman Scattering in Semiconductor Quantum Wells
(71.25 and 78.30; S7.12; S7.15)
(b) Vol. **148**, No. 2, p. 533—542 (1988).

E. L. ALBUQUERQUE and W. B. COSTA
Impurity Binding Energies in Heterostructures. Two-Band Model
(71.20 and 73.40; S7.12; S7.15)
(b) Vol. **148**, No. 2, p. 543—548 (1988).

A. POPOV, A. BAHNEV, and S. DARDJONOV
Photoluminescence and EPR Spectra of Fe-Doped $Ga_{1-x}In_xP$ Epitaxial Layers
(68.55 and 78.55; 76.30; S7.15)
(a) Vol. **108**, No. 2, p. 631—636 (1988).

G. T. BAUMBACH, H. RHAN, and U. PIETSCH
Detection of the Active Layer of $A^{III}B^{V}$ Semiconductor Quantum-Well Structures by High Resolution X-Ray Diffractometry
(61.10 and 68.55; S7.12; S7.15)
(a) Vol. **109**, No. 1, p. K7—K10 (1988).

H. LEON and F. COMAS
On the Scattering of Electrons by Confined LO Phonons in a Semiconductor Quantum Well
(71.38 and 73.40; S7.12; S7.15)
(b) Vol. **149**, No. 2, p. 533—542 (1988).

CHR. PRATSCH and M. SUHRKE
Low-Temperature Electron Transport in Quasi-Two-Dimensional Systems
(73.40; S7.12; S7.15)
(b) Vol. **149**, No. 2, p. 547—553 (1988).

B. MITRA and K. P. GHATAK
On the Einstein Relation in Semiconductor Superlattices under Magnetic Quantization
(72.20 and 73.40; S7.12; S7.15)
(b) Vol. **149**, No. 2, p. K117—K122 (1988).

V. A. KULBACHINSKII and D. YU. RODICHEV
Peculiarities of the Localization of Electrons in GaAs–$Ga_{1-x}Al_xAs$ Heterostructures
(73.40; S7.15)
(b) Vol. **150**, No. 1, p. K25—K29 (1988).

J.-W. SCHÜNEMANN, K. BÄRNER, H. FJELLVAG, A. KJEKSHUS, J. EIGEMANN, E. GMELIN, and U. SONDERMANN
Classification of Martensitic-Like Transformation in $Cr_xMn_{1-x}As_{1-y}P_y$ Crystals
(64.70; 75.40; S7.15)
(a) Vol. **110**, No. 1, p. 141—154 (1988).

F. KOCH, R. MITDANK, G. OELGART, and C. CHLUPSA
Excess Carrier Recombination in Indirect-Gap $GaAs_{1-x}P_x$:N
(73.60 and 78.60; S7.15)
(a) Vol. **110**, No. 1, p. 197—204 (1988).

T. V. TORCHINSKAYA, A. A. SHMATOV, and M. K. SHEINKMAN
Recombination-Enhanced Transformation of Deep Centers in Red Light-Emitting AlGaAs Diodes
(71.55; 73.40; 78.55; S7.15)
(a) Vol. **110**, No. 1, p. 213—220 (1988).

S. HILDEBRANDT, J. SCHREIBER, W. HERGERT, and V. I. PETROV
Determination of the Absorption Coefficient and the Internal Luminecence Spectrum of GaAs and $GaAs_{1-x}P_x$ ($x = 0.375, 0.78$) from Beam Voltage Dependent Measurements of Cathodoluminescence Spectra in the Scanning Electron Microscope
(78.60; S7.12; S7.15)
(a) Vol. **110**, No. 1, p. 283—291 (1988).

S. SCHMITT-RINK
Coherent Nonlinear Optical Processes in Semiconductors. Physics and Applications
(71.35 and 78.20; S7.12; S7.15; S12)
(b) Vol. **150**, No. 2, p. 349—356 (1988).

I. BALSLEV and A. STAHL
Coherent Nonlinear Response of Excitons in Semiconductors
(71.35 and 78.20; S7.12; S7.15)
(b) Vol. **150**, No. 2, p. 413—418 (1988).

R. KUSZELEWICZ, J. L. OUDAR, R. AZOULAY, J. C. MICHEL, J. BRANDON, and O. EMILE
All-Epitaxial GaAs/AlAs Nonlinear Etalons. Towards Continuous and Parallel Operation
(68.55 and 78.20; S7.12; S7.15)
(b) Vol. **150**, No. 2, p. 465—470 (1988).

H. UCHIKI, T. KOBAYASHI, and E. TOKUNAGA
Carrier Diffusion and Trapping by Quantum Wells
(73.40 and 78.55; S7.12; S7.15)
(b) Vol. **150**, No. 2, p. 667—672 (1988).

I. N. Uraltsev, E. L. Ivchenko, P. S. Kopev, V. P. Kochereshko, and D. R. Yakovlev
Exciton Parameters and Electron Miniband Structure of GaAs/AlGaAs Superlattices
(71.35 and 78.20; 71.25; S7.12; S7.15)
(b) Vol. **150**, No. 2, p. 673—678 (1988).

E. Lach, M. Walther, G. Tränkle, A. Forchel, and G. Weimann
Optical Investigation of Exciton Screening in GaAs/GaAlAs MQW Structures
(71.35 and 78.20; S7.12; S7.15)
(b) Vol. **150**, No. 2, p. 679—683 (1988).

G. Bongiovanni, J. L. Staehli, and D. Martin
Many-Body Effects in the Electron–Hole Plasma Confined in GaAs–(GaAl)As Quantum Wells
(71.35 and 78.65; 71.45; S7.12; S7.15)
(b) Vol. **150**, No. 2, p. 685—690 (1988).

K. Henneberger and G. Manzke
Dynamics of Incoherent Excitonic Optical Nonlinearities in GaAs–GaAlAs MQWS's
(71.35 and 78.20; S7.12; S7.15)
(b) Vol. **150**, No. 2, p. 691—693 (1988).

H. Weinert, V. Latinis, S. Juršenas, and R. Baltramiejunas
Time-Resolved Photoluminescence of a GaAs/GaAlAs Superlattice
(71.35 and 78.55; S7.12; S7.15)
(b) Vol. **150**, No. 2, p. 699—704 (1988).

I. Rückmann, J. Kolenda, F. Henneberger, and M. Petrauskas
ps-Laser Induced Transmission and Absorption in GaInAsP Epitaxial Layers
(78.20; 78.65; S7.15)
(b) Vol. **150**, No. 2, p. 735—742 (1988).

R. P. Seisyan, A. A. Toropov, V. P. Khvostikov, and T. V. Shubina
Excitonic Electroabsorption in GaAs–GaAlAs Single Quantum Well Waveguide Structures
(71.35 and 78.20; S7.12; S7.15)
(b) Vol. **150**, No. 2, p. 863—866 (1988).

H. Kalt, A. L. Smirl, and T. F. Boggess
Picosecond Optical Nonlinearities and Stimulated Emission in Indirect $Al_xGa_{1-x}As$
(78.20 and 78.45; 78.55; S7.15)
(b) Vol. **150**, No. 2, p. 895—900 (1988).

S7.16. Higher compounds

G. C. Bhar and N. P. Ghosh
Broadband Tuning in an Infrared Parametric Oscillator
(78.20; S7.16; S8.12; S8.16)
(a) Vol. **106**, No. 1, p. K93—K96 (1988).

B. Rheinländer, R. Heilmann, and G. Oelgart
Determination of Temperature-Dependent Carrier Losses in 1.3 μm InGaAsP/InP Double-Heterostructures
(78.45 and 85; 73.40; S7.11; S7.16)
(a) Vol. **106**, No. 1, p. K113—K116 (1988).

R. Flagmeyer, V. Gottschalch, and H. Frey
Internal Distortion in GaInAsP Quaternary Alloys Studied by Ion Channeling
(61.60; S7.16)
(a) Vol. **108**, No. 1, p. 123—128 (1988).

P. Deus, U. Voland, and H. Neumann
Low-Temperature Thermal Expansion of $ZnSiAs_2$
(65.70; S7.16)
(a) Vol. **108**, No. 1, p. 225—231 (1988).

M. Levalois and G. Allais
X-Ray Diffraction Study of Bonding in the Ternary Semiconductors $ZnSiAs_2$, $ZnGeAs_2$, and $ZnSnAs_2$
(61.60; S7.16)
(a) Vol. **109**, No. 1, p. 111—118 (1988).

G. K. Padam, G. L. Malhotra, and S. U. M. Rao
Electron Microscopy and Optical Studies of Chemically Deposited $CuInS_2$ Thin Films
(68.55; 78.65; S7.16)
(a) Vol. **109**, No. 1, p. K45—K49 (1988).

S8. Chalcogenides

(for transition metal compounds, see S1.63)

K. K. MAMEDOV, A. YU. YANGIROV, A. G. GUSEINOV, and A. M. ABDULLAEV
Heat Capacity and Phase Transitions in Highly Anisotropic $A^{III}B^{VI}$-Type Semiconductors and Their Analogs at Low Temperatures ((**Review Article**)
(65.40; S8)
(a) Vol. **106**, No. 2, p. 315—331 (1988).

T. N. MAMONTOVA, A. S. KOCHEMIROVSKII, and L. V. PIVOVAROVA
Electronic Processes on the Surface of As–Se Chalcogenide Glassy Semiconductors (**Review Article**)
(73.25; 68.45; 72.40; 73.20; S8)
(a) Vol. **107**, No. 1, p. 11—43 (1988).

N. KRISTOFFEL and P. KONSIN
Vibronic Theory of Structural Phase Transitions and Displacive Ferroelectrics (**Review Article**)
(63.75 and 77.80; 64.70; S8; S11)
(b) Vol. **149**, No. 1, p. 11—40 (1988).

K. TAKARABE
Optical Properties of InS under High Pressure
(71.25; 78.40; S8)
(b) Vol. **145**, No. 1, p. 219—225 (1988).

F. POMER, X. BONET, A. SEGURA, and A. CHEVY
Electrical Conductivity Anisotropy in Tin-Doped n-Type Indium Selenide
(72.20; S8)
(b) Vol. **145**, No. 1, p. 261—268 (1988).

E. K. MARTYNCHUK and F. F. SIZOV
The Farraday Effect Approach to a Study of $A^{IV}B^{VI}$ Narrow-Gap Semiconductor Band Structure
(78.20; S8; S8.15)
(b) Vol. **145**, No. 1, p. 309—318 (1988).

A. RIZZO, C. DE BLASI, M. CATALANO, and P. CAVALIERE
Dislocations in $A^{III}B^{VI}$ Single Crystals
(61.70; S8)
(a) Vol. **105**, No. 1, p. 101—112 (1988).

B. B. NAYAK, S. K. SINGH, and B. S. ACHARYA
Characterization of As-Grown and Iodinated Bi_2S_3 Thin Films on Silicon and Quartz Substrates by XRD
(68.55; S8)
(a) Vol. **105**, No. 2, p. 455—459 (1988).

N. GANESAN and V. SIVARAMAKRISHNAN
Variation in Electrical Properties of SnTe Thin Films with Substrate Temperature
(73.60; 68.55; S8)
(a) Vol. **105**, No. 2, p. 461—466 (1988).

S. K. BISWAS, S. CHAUDHURI, and A. CHOUDHURY
Effect of Heat Treatment on the Optical and Structural Properties of InSe Thin Films
(78.65; 68.55; S8)
(a) Vol. **105**, No. 2, p. 467—475 (1988).

A. V. Golubkov, V. M. Egorov, T. S. Orlova, V. M. Sergeeva, B. I. Smirnov, and I. A. Smirnov
The Semiconductor–Metal Phase Transition under Uniaxial Compression $Sm_{1-x}Gd_xS$ Single Crystals
(62.20 and 64.70; S8)
(a) Vol. **105**, No. 2, p. K93—K96 (1988).

S. P. Nikanorov, Yu. A. Burenkov, A. B. Lebedev, A. V. Golubkov, T. B. Zhukova, and I. A. Smirnov
Temperature Dependence of Elastic Properties of $Sm_{1-x}Gd_xS$ Single Crystals in the Vicinity of the "Black-Gold" Phase Transition
(62.20; S8)
(a) Vol. **105**, No. 2, p. K103—K106 (1988).

I. M. Askerov, A. O. Makhrabov, G. K. Aslanov, B. G. Tagiev, and S. M. Nakhmetov
Transition-Metal Electron States in Imperfect Ga_2Se_3 Crystals
(76.30 and 76.80; S8)
(a) Vol. **105**, No. 2, p. K151—K154 (1988).

A. M. Elkorashy
Temperature Dependence of the Two-Dimensional Optical Energy Gap for Germanium Selenide Single Crystals
(78.20; S8)
(b) Vol. **146**, No. 1, p. 279—285 (1988).

V. S. Dneprovskii, A. I. Furtichev, V. I. Klimov, E. V. Nazvanova, D. K. Okorokov, and U. V. Vandishev
Excitons at High Density in CdS and GaSe, and Optical Bistability
(71.35 and 78.20; S8; S8.11)
(b) Vol. **146**, No. 1, p. 341—350 (1988).

R. Klabes, A. Thomas, G. Kluge, W. Beyer, R. Grötzschel, and P. Süptitz
Ion-Beam Induced Silver Doping in the $Ag_2Se/GeSe_2$-Resist System
(61.80 and 68.55; 66.30; S8)
(a) Vol. **106**, No. 1, p. 57—65 (1988).

J. George, K. S. Joseph, B. Pradeep, and T. I. Palson
Reactively Evaporated Films of Indium Sulphide
(68.55 and 78.65; S8)
(a) Vol. **106**, No. 1, p. 123—131 (1988).

Yu. Dmitriev, V. Koshkin, and U. Ulmanis
Defocusing of Atomic Knock-Ons in Crystals with Low Local Symmetry of Structural Elements
(61.80; 72.80; S8)
(a) Vol. **106**, No. 1, p. K7—K11 (1988).

F. Lukeš
Natural Surface Films on GeS
(68.55; 78.65; S8)
(a) Vol. **106**, No. 1, p. K27—K30 (1988).

A. I. Popov, I. A. Domoryad, and N. I. Michalev
Structural Modification of the Arsenic Chalcogenide Glasses under γ-Radiation
(61.40; 62.20; S8)
(a) Vol. **106**, No. 2, p. 333—337 (1988).

J. W. Tomm, L. Werner, D. Genzow, K. Herrmann, D. Schikora, and J. Griesche
Experimental Evidence of Strain-Induced Gap Shifts in PbTe Epitaxial Layers by Photoluminescence
(68.55 and 78.55; S8)
(a) Vol. **106**, No. 2, p. 509—514 (1988).

P. Lošťák, R. Novotný, J. Navrátil, and J. Šrámková
Optical Properties of Gallium-Doped Bi_2Te_3 Single Crystals
(78.20; 78.30; S8)
(a) Vol. **106**, No. 2, p. 619—625 (1988).

L. E. Oliveira
On the Static Dielectric Constant and Plasmon Energy of TiS_2 and $TiSe_2$
(71.25; 71.45; S8)
(b) Vol. **147**, No. 1, p. 223—228 (1988).

K. R. Allekhverdiev, N. A. Bakhyshov, S. S. Guseinov, T. G. Mamedov, M. A. Nizametdinova, and I. K. Efendieva
Band Gap Variation in Mixed Crystals with TlSe-Type Crystal Structure
(71.25 and 78.40; S8)
(b) Vol. **147**, No. 1, p. K99—K104 (1988).

K. Konan, G. Galibert, and J. Calas
Silver Photodissolution and Optical Properties of $GeSe_3$ Vitreous Glass Thin Films
(66.30 and 78.65; 73.60; S8)
(a) Vol. **107**, No. 1, p. 273—279 (1988).

F. S. Nasredinov, V. F. Masterov, Ch. S. Saidov, P. P. Seregin, N. N. Troitskaya, and H. U. Tschirner
Nature of the Electrical Inactivity of Tin and Iron Impurity Atoms in In_2Te_3 and Ga_2Te_3
(71.55; 76.80; S8)
(a) Vol. **107**, No. 1, p. 291—298 (1988).

H. Abe and Y. Nakamura
The Effect of Cu Doping on the DC Conductivity and Thermoelectric Power of As_2Se_3-Based Glasses
(72.20; S8)
(a) Vol. **107**, No. 1, p. 315—319 (1988).

J. Horak, Z. Starý, and J. Klikorka
Relations between Structure, Bonding, and Nature of Point Defects in Layered Crystals of Tetradymite Structure
(61.50 and 61.70; 72.80; S8; S8.15)
(b) Vol. **147**, No. 2, p. 501—510 (1988).

J. Calas, S. M. Abdel-Maksoud, and M. Averous
Photocarrier Drift Mobility in Bulk Vitreous $GeSe_3$ in the Temperature Range 60 to 300 K
(72.40; S8)
(b) Vol. **147**, No. 2, p. 629—634 (1988).

I. I. Dobynde, A. I. Bobrysheva, I. M. Razdobreev, and Yu. G. Shekun
Photoluminescence of GaSe
(78.55; S8)
(b) Vol. **147**, No. 2, p. 717—720 (1988).

L. Tichý and P. Nagels
Non-Isothermal Crystallization of Bi_2S_3 from Glassy $(GeS_2)_{0.5}(Bi_2S_3)_{0.5}$
(61.40; 64.60; S8)
(a) Vol. **107**, No. 2, p. 769—774 (1988).

A. Thomas, M. Guymont, M. P. Pardo, M. Guittard, and J. Flahaut
X-Ray Diffraction and Electron Microscopy Studies of α- and β-Ga_2S_3
(61.60; 61.10; 61.16; S8)
(a) Vol. **107**, No. 2, p. 775—784 (1988).

G. A. Wiegers, R. J. Haange, and F. van Bolhuis
The Crystal Structure of Stage-2 4H–$Ag_{0.22}NbS_2$
(61.60; S8)
(a) Vol. **107**, No. 2, p. 817—824 (1988).

J. Mahy, G. A. Wiegers, F. van Bolhuis, A. Diedering, and R. J. Haange
X-Ray and Electron Diffraction Study of Intercalates Ag_xTaS_2
(61.60; S8)
(a) Vol. **107**, No. 2, p. 873—887 (1988).

Zhigang He, Jiayou Feng, Yudian Fan, and Hengde Li
In–Se Thin Films: Crystallization and Optical Properties
(68.55 and 78.65; S8)
(a) Vol. **107**, No. 2, p. K123—K128 (1988).

G. I. Gashimov, A. G. Rustamov, A. A. Mustafaev, and A. D. Namazov
Kinetic Properties of the $Fe_{1-x}Zn_xCr_2S_4$ $(0 \leqq x < 1)$ System
(72.80; S8)
(a) Vol. **107**, No. 2, p. K129—K133 (1988).

G. B. Demishev, S. S. Kabalkina, and T. N. Kolobyanina
X-Ray Studies of Thallium Chalcogenides TlS and TlSe up to 37 GPa
(61.10; 64.70; S8)
(a) Vol. **108**, No. 1, p. 89—95 (1988).

R. A. Yakshibaev, N. N. Mukhamadeeva, and R. F. Almukhametov
Phase Transformations and Ionic Transport in $Cu_{2-\delta}Te$ Superionic Conductor
(64.70 and 66.30; S8)
(a) Vol. **108**, No. 1, p. 135—141 (1988).

H. Kahnt, F. Schirrmeister, and A. Feltz
Model of Photostimulated Diffusion of Ag in Amorphous Chalcogenides
(66.30; 72.20; S8)
(a) Vol. **108**, No. 1, p. 149—153 (1988).

R. M. Cândea, L. P. Biró, N. Dădârlat, G. Borodi, Al. Darabont, P. Fitori, and R. Turcu
The Influence of the Film History on Some Electrophysical Properties of VE, CAD, and COD PbSe Films
(68.55 and 73.60; S8)
(a) Vol. **108**, No. 1, p. 233—240 (1988).

S. Shigetomi, T. Ikari, Y. Koga, and S. Shigetomi
Electrical Properties of Cd-Doped p-InSe
(71.55 and 72.80; S8)
(a) Vol. **108**, No. 1, p. K53—K57 (1988).

W. KULLMANN, H. RAUH, R. GEICK, U. STEIGENBERGER, and W. F. KUHS
Thermal Parameters and Probability Density Functions of the Narrow Gap Semiconductor Bi_2Te_3 under Hydrostatic Pressure
(61.12 and 61.60; 62.50; 65.90; S8)
(b) Vol. **148**, No. 2, p. 445—456 (1988).

J. CALAS, S. M. ABDEL MAKSOUD, and M. AVEROUS
Energy Distribution of Trapping Centers in Vitreous $GeSe_3$ as Obtained from Transit Time Measurements in the Temperature Range 60 to 300 K
(71.55; 72.40; S8)
(b) Vol. **148**, No. 2, p. 629—634 (1988).

S. MOTIEJŪNAS, D. K. BALTRŪNAS, V. I. GARASIM, and P. M. STARIK
A Correlation between Ligand Electronegativity and the Value of the Mössbauer Isomer Shift of Stannous Chalcogenides
(76.80; S8)
(b) Vol. **148**, No. 2, p. K161—K164 (1988).

H. D. HUSSEINOV, SH. N. YURAEV, S. H. HUSSEINOV, and S. S. ABDINBEKOV
Some Features of IR Reflection Spectra in $Tl(InS_2)_{1-x}(FeSe_2)_x$ Single Crytals ($0 \leqq x \leqq 0.015$) near the Phase Transition Temperature
(78.30; 78.50; S8; S8.15)
(b) Vol. **148**, No. 2, p. K177—K182 (1988).

A. G. KYAZYMZADE, A. A. AKHMEDOV, S. KH. KHALILOV, and YA. A. BAIRAMOV
Simulation of Residual Photoconductivity at a Metal–Semiconductor Contact
(72.40 and 73.40; S8)
(a) Vol. **108**, No. 2, p. 619—623 (1988).

D. DĂDÂRLAT, R. M. CÂNDEA, R. TURCU, L. P. BIRÓ, I. I. ZASAVITSKII, M. V. VALEIKO, and A. P. SHOTOV
Size Effects in Polycrystalline PbSe Films Obtained by Chemical Deposition
(73.60 and 78.65; S8)
(a) Vol. **108**, No. 2, p. 637—641 (1988).

M. MOLDOVANOVA, R. ASSENOV, and L. PARTHIER
Carrier Mobility in PbTe(I)
(72.20; S8)
(a) Vol. **108**, No. 2, p. 699—704 (1988).

S. KOKAI, I. KIKUCHI, O. UEMURA, and T. SATOW
Short-Range Order in Amorphous $GeSe_2$
(61.40; S8)
(a) Vol. **108**, No. 2, p. K89—K92 (1988).

V. N. KATERINCHUK, Z. D. KOVALYUK, and V. A. MANASSON
On a Mechanism of Photo-E.M.F. Formation in SIS Structures
(73.40; 85; S8; S10.1)
(a) Vol. **108**, No. 2, p. K115—K118 (1988).

R. K. SINGH, D. C. GUPTA, and S. P. SANYAL
Anharmonic Properties of IV–VI Compound Semiconductors
(62.20; S8)
(b) Vol. **149**, No. 1, p. 121—125 (1988).

Z. Starý, L. Beneš, and J. Horák
Interaction of Anti-Site Defects with Ge Impurity Atoms in Sb_2Te_3 Single Crystals
(61.60; 61.70; S8)
(a) Vol. **109**, No. 1, p. 93—99 (1988).

G. A. Bordovskii, N. A. Savinova, and L. N. Seregina
Electrical Activity Peculiarities of Tin and Iron Impurities in the Semiconducting Glass $Ge_{28.5}Pb_{15}S_{56.5}$
(72.80 and 76.80; S8)
(a) Vol. **109**, No. 1, p. K51—K53 (1988).

A. M. Elkorashy
Optical Constants of Germanium Selenide Single Crystals in the Transparency Region
(78.20; S8)
(b) Vol. **149**, No. 2, p. 747—758 (1988).

K. Kumazaki and K. Imai
Far-Infrared Reflection and Raman Scattering Spectra in γ-InSe
(63.20 and 78.30; S8)
(b) Vol. **149**, No. 2, p. K183—K186 (1988).

J. R. Gavarri, R. Mokrani, G. Vacquier, and C. Boulesteix
Relations between Anisotropic Defects, Structural Evolution and van der Waals Bonding in 2H–$NbSe_2$
(61.70 and 65.70; 62.20; S8)
(a) Vol. **109**, No. 2, p. 445—454 (1988).

V. S. Postnikov, B. I. Sysoev, A. V. B. Udanov, N. N. Bezryadin, Yu. K. Shlyk, and B. L. Agapov
Heterostructures on the Basis of Indium Arsenide with Semi-Insulating $A_2^{III}B_3^{V}$ Compound Layers
(73.40; S7.12; S8)
(a) Vol. **109**, No. 2, p. 463—467 (1988).

S. Chaudhuri, S. K. Biswas, and A. Choudhury
The Effects of Heat Treatment on the Structural Properties of Amorphous $In_{0.3}Se_{0.7}$ Films
(68.55; S8)
(a) Vol. **109**, No. 2, p. 511—516 (1988).

K.-P. Möllmann
Additional Deep Level in Strongly Indium-Doped Lead Tin Telluride
(71.55; 72.80; S8)
(a) Vol. **109**, No. 2, p. K121—K124 (1988).

B. R. K. Gupta
Acoustic Mode Vibrational Anharmonicity in the Intermediate Valent Compound Samarium Sulfide
(62.20; S8)
(b) Vol. **150**, No. 1, p. 59—63 (1988).

K. H. Aharonian, H. L. Erknapetian, and D. R. Tilley
Collective Plasmon Modes in Thin Semiconductor Films Including the Effect of the Image Charges
(71.45; 73.60; S8)
(b) Vol. **150**, No. 1, p. 133—138 (1988).

V. A. Kulbachinskii, N. B. Brandt, P. A. Cheremnykh, S. A. Azou, J. Horak, and P. Lošťák
Magnetoresistance and Hall Effect in Bi_2Te_3 ⟨Sn⟩ in Ultrahigh Magnetic Fields and under Pressure
(72.20; S8)
(b) Vol. **150**, No. 1, p. 237—243 (1988).

M. Iwamatsu
On the Temperature Dependence of the Average Gap of SnTe and PbTe
(71.25; S8)
(b) Vol. **150**, No. 1, p. K39—K44 (1988).

B. I. Smirnov and I. A. Smirnov
Defect Structure and Physico-Mechanical Properties of SmS Single Crystals
(61.70; 62.20; 72.20; S8)
(a) Vol. **110**, No. 1, p. 89—96 (1988).

A. G. J. Green, B. Ray, I. V. F. Viney, and J. W. Brightwell
Luninescence Studies of CaS Phosphors Doped with Cerium and Copper
(71.55 and 78.55; 76.30; 78.60; S8)
(a) Vol. **110**, No. 1, p. 269—276 (1988).

R. Srivastava, V. M. Pathak, and V. V. Rao
Flat Band Potential of As-Grown n-$MoSe_2$ Single Crystal Electrodes
(72.40 and 73.40; S8)
(a) Vol. **110**, No. 1, p. 293—299 (1988).

E. A. Eivazov, A. F. Safarov, and A. O. Ismailov
Thermal Properties of $Co_{1-x}Cu_xCr_2S_4$ in the Magnetic Phase Transition Region
(65.70; 75.50; S8)
(a) Vol. **110**, No. 1, p. K45—K49 (1988).

R. Rousina and G. K. Shivakumar
Effect of Heat Treatment on the Optical Absorption of Amorphous Indium Selenide Films
(78.65; 61.40; S8)
(a) Vol. **110**, No. 1, p. K57—K60 (1988).

I. Jánossy
Nonlinear Optical Effects in Amorphous Semiconductors
(78.20; 78.65; S8)
(b) Vol. **150**, No. 2, p. 783—789 (1988).

Lun-Biao Xu, Shou-Fu Cai, and Guo-Xin Xie
Quantum-Statistical Investigation on the Nonlinear Optical Oscillation in Amorphous Semiconductor $GeSe_2$ Films
(78.20 and 78.65; 71.45; S8)
(b) Vol. **150**, No. 2, p. 797—803 (1988).

J. C. Bernède, G. Safoula, A. Ameziane, and P. Burgaud
Experimental Study of Oxygen Effect on the Variation of Resistivity with Temperature for Polycrystalline Selenium Thin Films
(73.60; S5; S8)
(a) Vol. **110**, No. 2, p. 521—530 (1988).

R. Klabes, A. Thomas, and G. Kluge
Utilization of Ag:$GeSe_2$ Films as Inorganic Positive Resist
(68.60; S8)
(a) Vol. **110**, No. 2, p. K81—K84 (1988).

S8.1. II–VI compounds

O. Goede and W. Heimbrodt
Optical Properties of (Zn, Mn) and (Cd, Mn) Chalcogenide Mixed Crystals and Superlattices **(Review Article)**
(78.20; 71.35; 78.55; S8.1)
(b) Vol. **146**, No. 1, p. 11—62 (1988).

R. Boyn
4f–4f Luminescence of Rare-Earth Centers in II–VI Compounds **(Review Article)**
(78.55 and 78.60; 71.55; S8.1)
(b) Vol. **148**, No. 1, p. 11—47 (1988).

T. J. Bastow and S. N. Stuart
NMR Study of the Zinc Chalcogenides (ZnX, X = O, S, Se, Te)
(76.60; S8.1)
(b) Vol. **145**, No. 2, p. 719—728 (1988).

S. V. Melnichuk, V. M. Kramar, and K. D. Tovstyuk
Vibronic Interaction in the Absorption Spectra of Vanadium-Doped $A^{II}B^{VI}$ Semiconductors
(71.70 and 78.50; S8.1)
(b) Vol. **146**, No. 2, p. 613—617 (1988).

A. Bourret, J. L. Rouvière, and J. Spendeler
Direct Observation of Atomic Columns in Semiconductors by HREM at 400 kV
(61.55 and 61.60; 61.16; S5.11; S5.12; S7.1; S8.1)
(a) Vol. **107**, No. 2, p. 481—501 (1988).

M. I. Auslender, E. M. Kogan, and S. V. Tretyakov
Anderson Localization in Ferromagnetic Semiconductors Due to Spin Disorder. II. Wide Conduction Band
(72.20; 72.60; S8.1; S10)
(b) Vol. **148**, No. 1, p. 289—295 (1988).

M. Fliyou, M. A. Khan, and J. Ringeissen
Dynamical Effects on the Binding Energies of Core Excitons in Covalent and Polar Semiconductors
(71.35; S5.11; S5.12; S7.1; S8.1)
(b) Vol. **149**, No. 2, p. 615—621 (1988).

N. I. Tarbaev, J. Schreiber, and G. A. Shepelskii
Physical Properties of $A^{II}B^{VI}$ Semiconductor Crystals after Plastic Deformation at Low Temperature
(61.70 and 71.55; 72.40; 78.50; 78.55; S8.1)
(a) Vol. **110**, No. 1, p. 97—106 (1988).

G. V. Sinitsyn
Digital Circuits for All-Optical Computing Based on Thin-Film Semiconductor Interferometers
(78.20; 78.65; S8.1)
(b) Vol. **150**, No. 2, p. 455—466 (1988).

U. Woggon, F. Henneberger, and M. Müller
Optical Nonlinearities of Semiconductor Microcrystallites
(78.20; 71.35; S8.1; S9.11)
(b) Vol. **150**, No. 2, p. 641—645 (1988).

Th. Peschel
Phenomenological Model for the Optical Properties of Semiconductor-Doped Glasses
(78.20; 71.35; S8.1)
(b) Vol. **150**, No. 2, p. 657—661 (1988).

S8.11. Sulfides

V. S. Dneprovskii, A. I. Furtichev, V. I. Klimov, E. V. Nazvanova, D. K. Okorokov, and U. V. Vandishev
Excitons at High Density in CdS and GaSe, and Optical Bistability
(71.35 and 78.20; S8; S8.11)
(b) Vol. **146**, No. 1, p. 341—350 (1988).

V. D. Egorov, P. Flögel, Hoang Xuan Nguyen, and M. Kaschke
Picosecond Transient Optical Nonlinearities in CdS and CdSSe
(71.35 and 78.20; S8.11; S8.15)
(b) Vol. **146**, No. 1, p. 351—356 (1988).

O. Goede, W. Heimbrodt, M. Lamla, and V. Weinhold
Energy Gap of MnS
(78.40 and 78.55; 71.35; S8.11)
(b) Vol. **146**, No. 1, p. K65—K69 (1988).

S. Pakeva and R. Dafinova
The Electroluminescence of ZnS–Pb, Cu, Br Phosphors
(78.60; S8.11)
(a) Vol. **106**, No. 1, p. K97—K100 (1988).

T. Hönig and J. Gutowski
Stimulated Emission of the I_1—TA(P_M) Band in CdS
(71.35 and 71.55; 78.45; S8.11)
(b) Vol. **146**, No. 2, p. 589—595 (1988).

E. Jahne, O. Goede, and V. Weinhold
Far Infrared Reflectivity of Tetrahedrally and Octahedrally Coordinated MnS
(63.20 and 78.30; S8.11)
(b) Vol. **146**, No. 2, p. K157—K160 (1988).

A. A. Klochikhin, B. S. Razbirin, D. K. Nelson, T. Amand, M. Brousseau, J. Collet, A. Cornet, and M. Pugnet
Picosecond Kinetics of Luminescence in Highly Excited Direct-Gap Semiconductors
(78.55; 71.35; S8.11; S8.12)
(b) Vol. **147**, No. 2, p. 727—732 (1988).

K. Löbe and R. Boyn
Dye-Laser 4f–4f-Line Probing of Crystal-Field Fluctuations in ZnS:Sm^{3+} Polycrystalline Films
(71.55; 68.60; 78.55; S8.11)
(b) Vol. **148**, No. 1, p. 263—272 (1988).

H. X. Nguyen, V. D. Egorov, A. Harendt, and E. V. Nazvanova
Transverse Spatial Resolution of Thermo-Optical Hysteresis in CdS
(78.20; S8.11)
(b) Vol. **148**, No. 1, p. 407—412 (1988).

M. Igalson, K. Plankiewicz, and R. Trykozko
On the Photoresponse of Cu-Treated Cu_2S/CdS Thin Film Solar Cells
(73.40 and 85; 72.40; S8.11)
(a) Vol. **108**, No. 2, p. K169—K171 (1988).

J. M. Galligan and P. Haasen
Two Trapping Centers in the Photoplastic Effect in CdS
(62.20; S8.11)
(a) Vol. **109**, No. 1, p. K19—K22 (1988).

K. P. Tchakpele and J. P. Albert
Excitons Bound to Te Impurities in CdS, ZnS, and Their Mixed Compounds with Wurtzite Structure
(71.35 and 71.55; S8.11; S8.15)
(b) Vol. **149**, No. 2, p. 641—648 (1988).

K. Löbe and R. Boyn
Sm^{3+} in ZnS on Lattice Sites and Wurtzite-Type Interstitial Sites?
(61.70 and 71.55; 71.70; 78.55; S8.11)
(b) Vol. **150**, No. 1, p. 191—200 (1988).

B. D. Flügel, J. P. Sokoloff, A. Jarka, S. W. Koch, M. Lindberg, N. Peyghambarian, M. Joffre, D. Hulin, A. Migus, A. Antonetti, C. Ell, L. Banyai, and H. Haug
Measurements of Ultrafast Optical Nonlinearities in Semiconductors
(71.35 and 78.20; S8.11)
(b) Vol. **150**, No. 2, p. 357—363 (1988).

M. Lindberg and S. W. Koch
Theory of the Optical Stark Effect in Semiconductors under Ultrashort-Pulse Excitation
(71.35 and 78.20; S811)
(b) Vol. **150**, No. 2, p. 379—385 (1988).

J. Puls, W. Rudolph, F. Henneberger, and D. Lab
Femtosecond Studies of Room Temperature Optical Nonlinearities in Wide-Gap II–VI Compounds
(71.35 and 78.20; S8.11; S8.12; S8.15)
(b) Vol. **150**, No. 2, p. 419—423 (1988).

K. Henneberger and F. Jahnke
Laser Excited Semiconductors: Light-Induced Gaps and Saturation Kinetics
(71.35 and 78.20; S8.11)
(b) Vol. **150**, No. 2, p. 425—430 (1988).

A. L. Ivanov and G. S. Vygovskii
The Exciton Reflection Shifting and Modification in Presence of Intensive Polariton Wave Interacting with LO-Phonons in Polar Semiconductors
(78.20; 71.35; 71.36; S8.11)
(b) Vol. **150**, No. 2, p. 443—446 (1988).

A. A. Borshch and V. I. Volkov
Cavity-Less Optical Bistability in Semiconductors
(78.20; S8.11)
(b) Vol. **150**, No. 2, p. 471—475 (1988).

A. Daunois, J.-Y. Bigot, J. Oberle, N. Cherkaoui Eddeqaqi, M. Wegener, A. Witt, and C. Klingshirn
Spectral Investigation of the Switching Dynamics in CdS Absorptive Bistable Devices
(78.20; S8.11)
(b) Vol. **150**, No. 2, p. 477—481 (1988).

V. A. Kochelap, A. V. Kuznetsov, and V. N. Sokolov
Resonatorless Dissipative Optical Bistability of Electronic Origin in Direct-Gap Semiconductors
(78.20; S8.11)
(b) Vol. **150**, No. 2, p. 489—493 (1988).

H. X. Nguyen and V. D. Egorov
Spatial Resolution of Dynamic Thermo-Optical Hysteresis in CdS
(78.20; S8.11)
(b) Vol. **150**, No. 2, p. 519—525 (1988).

S. V. Bogdanov and V. G. Lyssenko
Optoelectronic Bistability with Internal Feedback in Cadmium Sulfide
(78.20; 71.35; S8.11)
(b) Vol. **150**, No. 2, p. 593—598 (1988).

F. Yang, G. Xiong, and X. Xu
Mechanism of Optical Nonlinearity in CdS Due to Increasing Absorption at Room Temperature
(78.20; S8.11)
(b) Vol. **150**, No. 2, p. 765—768 (1988).

Ch. Spiegelberg, M. Kretzschmar, J. Puls, and F. Henneberger
Strong Dispersive Nonlinearity in the Excitonic Region of Wide-Gap II–VI Compounds at Room Temperature
(71.35 and 78.20; S8.11)
(b) Vol. **150**, No. 2, p. 769—775 (1988).

D. K. Nelson, A. A. Klochikhin, G. V. Mikhailov, A. G. Panfilov, and B. S. Razbirin
Nonlinear Effects in the Emission Spectra of High Density Excitons in CdS Crystals
(71.35 and 78.55; S8.11)
(b) Vol. **150**, No. 2, p. 777—781 (1988).

T. Hönig and J. Gutowski
Optical Nonlinearity and Bistability in the Bound Exciton Energy Range of CdS
(78.20; S8.11)
(b) Vol. **150**, No. 2, p. 833—838 (1988).

V. S. Dneprovskii, A. I. Furtichev, V. I. Klimov, Shen Li, E. V. Nazvanova, D. K. Okorokov, and U. V. Vandishev
Nonlinear Optical Properties of CdS and Optical Bistability
(78.20; S8.11)
(b) Vol. **150**, No. 2, p. 839—844 (1988).

G. O. Müller, R. Mach, B. Selle, and G. Schulz
Measuring on Thin Film Electroluminescent Devices
(73.40 and 78.60; 85; S8.11)
(a) Vol. **110**, No. 2, p. 657—669 (1988).

S8.12. Selenides

F. Henneberger, H. Rossmann, and A. Schülzgen
Mirrorless Dispersive Optical Bistability at the Brewster Angle Using a Nonlinear Prism
(78.20; S8.12)
(b) Vol. **145**, No. 1, p. K83—K88 (1988).

M. Grinberg
The Substitutional Shallow Donor Size Effect
(71.55; S8.12)
(b) Vol. **145**, No. 2, p. 619—624 (1988).

J. Benoit, P. Benalloul, A. Geoffroy, and C. Barthou
Decay of the Yellow Emission of Mn^{2+} in AC Thin Electroluminescent Devices
(78.60 and 85; 78.65; S8.12)
(a) Vol. **105**, No. 2, p. 637—647 (1988).

S. Kishida, K. Matsuura, A. Matsuoka, and I. Tsurumi
The Transient Behaviour of the 2.5 eV Emission Band in Se-Treated ZnSe Crystals
(78.55; S8.12)
(a) Vol. **105**, No. 2, p. K165—K168 (1988).

V. A. Kochelap, N. R. Kulish, M. P. Lisitsa, N. I. Malysh, and V. N. Sokolov
Band Edge Absorption Saturation Dynamics of Semiconductors
(78.20; S8.12)
(b) Vol. **146**, No. 1, p. 319—327 (1988).

J. H. Collet and M. Pugnet
Picosecond Plasma Dynamics and All-Optical Data Processing
(78.20; 71.35; S8.12)
(b) Vol. **146**, No. 1, p. 393—401 (1988).

S. Kishida, K. Matsuura, H. Mori, T. Yanagawa, I. Tsurumi, and C. Hamaguchi
The 2.5 eV Emission Band in the Se-Treated ZnSe Crystals
(78.55; S8.12)
(a) Vol. **106**, No. 1, p. 283—289 (1988).

G. C. Bhar and N. P. Ghosh
Broadband Tuning in an Infrared Parametric Oscillator
(78.20; S7.16; S8.12; S8.16)
(a) Vol. **106**, No. 1, p. K93—K96 (1988).

A. A. Klochikhin, B. S. Razbirin, D. K. Nelson, T. Amand, M. Brousseau, J. Collet, A. Cornet, and M. Pugnet
Picosecond Kinetics of Luminescence in Highly Excited Direct-Gap Semiconductors
(78.55; 71.35; S8.11; S8.12)
(b) Vol. **147**, No. 2, p. 727—732 (1988).

T. X. Hoai, T. H. Nhung, D. X. Thanh, and N. D. Hung
Experimental Study of the Photomagnetoelectric Effect near the Threshold of the Stimulated Emission in Semiconductors
(72.40 and 78.55; S8.12)
(b) Vol. **147**, No. 2, p. K165—K170 (1988).

V. P. Gribkovskii, L. G. Zimin, S. V. Gaponenko, I. E. Malinovskii, P. I. Kuznetsov, and G. G. Yakushcheva
Optical Hysteresis near Excitonic Resonance in Zinc Selenide Monocrystals
(71.35 and 78.20; S8.12)
(b) Vol. **148**, No. 1, p. K65—K69 (1988).

A. Opanowicz
Comments on the Interpretation of Thermally Stimulated Luminescence and Conductivity in Conductive ZnSe Crystals
(71.55; 78.60; S8.12)
(a) Vol. **108**, No. 1, p. K47—K51 (1988).

A. Kumar and I. Singh
Electronic Structure of the Ge–ZnSe, Ge–GaAs, and Ge–AlAs(110) Interfaces and Superlattices
(71.25 and 73.40; S5.12; S7.12; S8.12)
(b) Vol. **148**, No. 2, p. 549—558 (1988).

T. K. Bandyopadhyay, S. B. Chakraborty, and S. R. Chaudhuri
Effect of Surface Treatment on the Photoanodic Behaviour of Thin Film n-CdSe
(73.60; 72.40; S8.12)
(a) Vol. **108**, No. 2, p. K119—K123 (1988).

R. Baltramiejūnas and A. Žukauskas
Manifestation of Hot Phonons in the Luminescence Spectra of Highly Photoexcited $A^{II}B^{VI}$ Crystals
(63.20 and 78.55; 71.45; S8.12)
(b) Vol. **149**, No. 1, p. 337—346 (1988).

S. Kishida, K. Matsuura, H. Mori, Y. Mizuguchi, and I. Tsurumi
Temperature Dependence of the 2.5 eV Emission Band in Se-Treated ZnSe Crystals
(78.55; 71.55; S8.12)
(a) Vol. **109**, No. 2, p. 617—623 (1988).

E. Krause
Effects of Growth and Annealing Conditions on Deep Level Emissions in CVD Poly-ZnSe Layers
(71.55 and 78.60; S8.12)
(a) Vol. **110**, No. 1, p. K33—K37 (1988).

J. M. Hvam, C. Dörnfeld, and H. Schwab
Optical Nonlinearity and Phase Coherence in CdSe and $CdSe_xS_{1-x}$
(71.35 and 78.20; S8.12; S8.15)
(b) Vol. **150**, No. 2, p. 387—391 (1988).

J. Puls, W. Rudolph, F. Henneberger, and D. Lab
Femtosecond Studies of Room Temperature Optical Nonlinearities in Wide-Gap II–VI Compounds
(71.35 and 78.20; S8.11; S8.12; S8.15)
(b) Vol. **150**, No. 2, p. 419—423 (1988).

A. Schülzgen, F. Henneberger, and H. Rossmann
Cavityless Dispersive Bistability and Optical Addressing Using a Nonlinear Prism
(78.20; S8.12)
(b) Vol. **150**, No. 2, p. 495—499 (1988).

S. E. Esipov and V. A. Stadnik
Optical Bistability and Formation of Localized and Moving High Absorption Domains in ZnSe and ZnCdS
(78.20; S8.12; S8.15)
(b) Vol. **150**, No. 2, p. 501—505 (1988).

V. P. Gribkovskii, L. G. Zimin, S. V. Gaponenko, I. E. Malinovskii, P. I. Kuznetsov, and G. G. Yakushcheva
Resonatorless Optical Bistability near the Excitonic Resonance of ZnSe
(71.35 and 78.65; S8.12)
(b) Vol. **150**, No. 2, p. 761—764 (1988).

F. V. Karpushko
Fast and Strong Nonlinearity of Thin-Film Semiconductor Interferometers Due to Intrinsic Photo- and Electrooptical Effects of the Semiconductor Interlayer
(73.40 and 78.20; S7.12; S8.12)
(b) Vol. **150**, No. 2, p. 791—796 (1988).

K. Jarašiūnas and J. Vaitkus
Transient Gratings in Metrology of Semiconductor Parameters and Optoelectronic Devices
(78.20; S5.11; S7.12; S7.13; S8.12)
(b) Vol. **150**, No. 2, p. 879—884 (1988).

S8.13. Tellurides

J. Griesche, R. Enderlein, and D. Schikora
Orientation and Strain in Heteroepitaxial Growth **(Review Article)**
(61.50 and 68.48; 68.20; S5.11; S7.12; S8.13)
(a) Vol. **109**, No. 1, p. 11—38 (1988).

Y. Madhavan, K. Ramachandran, and T. M. Haridasan
Mean Square Displacements of Neighbours and Phonon Resonance Due to a Vacancy in HgTe and CdTe
(63.20; 61.70; S8.13)
(b) Vol. **145**, No. 1, p. 139—144 (1988).

J. Griesche and D. Schikora
Thermodynamic Model of the Gas Transport in a HWE System
(68.55; S8.13)
(a) Vol. **105**, No. 1, p. 189—195 (1988).

J. Tatarkiewicz and A. M. Witowski
Hydrogen Vibrations in ZnTe
(61.70 and 63.20; 78.30; S8.13)
(b) Vol. **145**, No. 2, p. K95—K98 (1988).

D. Kindl and J. Toušková
Charge Transport Study in Thin Film Au–CdTe Schottky Diodes
(73.30 and 85; 73.60; S8.13)
(a) Vol. **106**, No. 1, p. 297—304 (1988).

B. H. Bairamov, V. V. Toporov, N. V. Agrinskaya, E. A. Samedov, G. Irmer, and J. Monecke
Raman Scattering by Coupled LO-Phonon–Plasmon Modes in CdTe
(63.20 and 78.30; S8.13)
(b) Vol. **146**, No. 2, p. K161—K165 (1988).

D. M. Zayachuk, P. M. Starik, and V. A. Shenderovskii
The Effect of Nonstoichiometric Tellurium on the Properties of Lead and Tin Tellurides
(61.70; 72.20; 75.20; S8.13)
(a) Vol. **107**, No. 1, p. 95—99 (1988).

Ping Lu and David J. Smith
The Observation of In Situ and Ex Situ Oxidation Processes for ZnTe Surfaces by High-Resolution Electron Microscopy
(68.20; 61.16; S8.13)
(a) Vol. **107**, No. 2, p. 681—691 (1988).

O. Goede, W. Heimbrodt, and Th. Köpp
Photoacoustic Spectroscopy of CdTe and (Hg, Cd)Te
(78.20; S8.13)
(a) Vol. **108**, No. 1, p. 443—447 (1988).

H. Blumtritt, G. N. Panin, E. B. Yakimov, and J. Heydenreich
SEM–EBIC Investigations of Grain Boundaries in Cadmium Telluride
(61.14; 61.70; S8.13)
(a) Vol. **109**, No. 1, p. K3—K6 (1988).

R. Baltramiejūnas, V. Stepankevičius, G. Tamulaitis, and A. Žukauskas
Investigation of Hot Exciton Interaction in ZnTe Crystals
(71.35; 71.38; 78.55; S8.13)
(b) Vol. **150**, No. 1, p. 139—146 (1988).

S8.15. Mixed crystals

E. K. Martynchuk and F. F. Sizov
The Faraday Effect Approach to a Study of $A^{IV}B^{VI}$ Narrow-Gap Semiconductor Band Structure
(78.20; S8; S8.15)
(b) Vol. **145**, No. 1, p. 309—318 (1988).

J. G. Fleming and D. A. Stevenson
Interdiffusion in Mercury Cadmium Telluride Evaluated from Vapor Phase Growth Kinetics
(66.30; 68.55; S8.15)
(a) Vol. **105**, No. 1, p. 77—85 (1988).

K. Ponnuraju, P. R. Vaya, and J. Sobhanadri
Structural and Electrical Properties of $Pb_{1-x}Sn_xTe$ Films Grown by Molecular Beam Epitaxy
(68.55 and 73.60; S8.15)
(a) Vol. **105**, No. 1, p. 161—169 (1988).

P. Höschl, P. Moravec, V. Prosser, V. Szöcs, and R. Grill
Galvanomagnetic Properties of p-$Hg_{1-x}Cd_xTe$
(72.20; S8.15)
(b) Vol. **145**, No. 2, p. 637—648 (1988).

A. E. Curzon and O. S. Rajora
A Study of NbS_xSe_{2-x} Using Electron Channeling Patterns
(61.60; 61.80; S8.15)
(a) Vol. **105**, No. 2, p. 335—343 (1988).

S. Pfeffer and B. Schubert
The Dependence of Layer Thickness on Final Cell Voltage for Anodic Oxide Films Grown Galvanostatically on $Hg_{1-x}Cd_xTe$ ($x = 0.02, 0.16, 0.20, 0.22, 0.34, 1.00$)
(68.55; S8.15)
(a) Vol. **105**, No. 2, p. K115—K119 (1988).

R. K. Singh and N. V. K. Prabhakar
Phase Transition and Anharmonic Properties of Semimagnetic Semiconductors
(62.20 and 64.30; S8.15)
(b) Vol. **146**, No. 1, p. 111—116 (1988).

V. D. Egorov, P. Flögel, Hoang Xuan Nguyen, and M. Kaschke
Picosecond Transient Optical Nonlinearities in CdS and CdSSe
(71.35 and 78.20; S8.11; S8.15)
(b) Vol. **146**, No. 1, p. 351—356 (1988).

L. Werner and J. W. Tomm
Photoluminescence on p-$Hg_{0.42}Cd_{0.58}Te$
(78.55; S8.15)
(a) Vol. **106**, No. 1, p. K83—K87 (1988).

D. Baltrūnas, D. M. Zayachuk, S. Motiejūnas, and P. M. Starik
Direct Observation of Giant Deviations from Stoichiometry in Single Crystalline $Pb_{1-x}Sn_xTe$ by Means of the Mössbauer Effect on the ^{119}Sn Nuclei
(76.80; S8.15)
(a) Vol. **106**, No. 2, p. K173—K175 (1988).

A. A. A. Hady
Optical Study of $Hg_{1-x}Cd_xTe$ Graded Gap Material
(78.40; S8.15)
(a) Vol. **106**, No. 2, p. K191—K195 (1988).

P. Grima, M. Quintero, G. S. Pérez, R. Tovar, and J. C. Woolley
Phase Diagram and Lattice Parameter Values for the $Hg_{2x}(AgIn)_{1-x}Te_2$ Alloys
(61.55 and 64.70; S8.15)
(a) Vol. **107**, No. 1, p. 165—169 (1988).

M. Quintero, P. Grima, R. Tovar, G. S. Pérez, and J. C. Woolley
Phase Relations and the Effects of Ordering in $(AgIn)_{1-z}Mn_{2z}Te_2$ and $(CuIn)_{1-z}Mn_{2z}Te_2$ Alloys
(64.70; S8.15)
(a) Vol. **107**, No. 1, p. 205—211 (1988).

A. A. A. Hady
Annealing Effects on the Electrical Properties of $Hg_{1-x}Cd_xTe$ Grown by Two Different Techniques
(73.60; 68.55; S8.15)
(a) Vol. **107**, No. 1, p. K25—K28 (1988).

G. Eidžiūnas, J. Kavaliauskas, G. Krivaitė, and A. Šileika
Plasma Reflectivity Spectra of Inhomogeneous $Cd_{0.2}Hg_{0.8}Te$ Crystals
(78.30; S8.15)
(a) Vol. **107**, No. 1, p. K75—K78 (1988).

D. K. Ghosh, U. Chatterjee, and L. K. Samanta
On the Band Gap Dependence of Refractive Indices of Some Quaternary III–V and II–VI Compounds of Device Interest
(78.20; S7.15; S8.15)
(a) Vol. **107**, No. 1, p. K79—K81 (1988).

J. Horák, Z. Starý, and J. Klikorka
Relations between Structure, Bonding, and Nature of Point Defects in Layered Crystals of Tetradymite Structure
(61.50 and 61.70; 72.80; S8; S8.15)
(b) Vol. **147**, No. 2, p. 501—510 (1988).

P. Höschl, P. Moravec, E. Belas, J. Franc, and R. Grill
Thermoelectric Properties of p-$Hg_{1-x}Cd_xTe$
(72.20; S8.15)
(b) Vol. **147**, No. 2, p. 621—628 (1988).

W. Lu, H. J. Ye, Z. Y. Yu, S. Y. Zhang, Y. Fu, W. L. Xu, S. C. Shen, and W. Giriat
Local and Quasi-Local Modes of Fe and Zn in CdTe
(78.30; 63.20; S8.15)
(b) Vol. **147**, No. 2, p. 767—778 (1988).

I. M. Tsidilkovskii, Yu. G. Arapov, M. L. Zvereva, N. B. Brandt, V. A. Kulbachinskii, I. N. Gorbatyuk, and A. I. Elizarov
The Influence of Compensation on the Hall Effect and Magnetoresistance of n-$Hg_{0.8}Cd_{0.2}Te$
(71.30; 72.20; S8.15)
(b) Vol. **148**, No. 1, p. 197—204 (1988).

P. Hörschl, P. Moravec, R. Grill, and V. Čápek
Weak Impurity Metallic Behaviour of p-$Hg_{1-x}Cd_xTe$
(72.20; S8.15)
(b) Vol. **148**, No. 1, p. 319—324 (1988).

D. Baltrūnas, B. Vengalis, S. Motiejūnas, and N. Shiktorov
Effect of a Deviation from Stoichiometry on the Mössbauer Parameters of $Sn_{0.8}Ge_{0.2}Te$ during Phase Transition
(64.70; 76.80; S8.15)
(a) Vol. **108**, No. 1, p. 197—204 (1988).

H. W. Jeon, K. M. Kim, H. K. Kim, H. L. Park, and C. H. Chung
Determination of x in $Hg_{1-x}Zn_xTe$ Single Crystals by the Temperature Coefficient of the Energy Gap
(78.20; S8.15)
(a) Vol. **108**, No. 1, p. K77—K79 (1988).

J.-P. Zöllner, G. Paasch, G. Gobsch, and H. Übensee
Theory for n-Surface Charge Layers in $Hg_{1-x}Cd_xTe$ MIS Structures
(71.25 and 73.40; S8.15)
(b) Vol. **148**, No. 2, p. 611—618 (1988).

K. P. Ghatak and M. Mondal
On the Einstein Relation in n-Channel Inversion Layers on Ternary Semiconductors under Magnetic Quantization
(72.20; S8.15)
(b) Vol. **148**, No. 2, p. 645—652 (1988).

H. D. Husseinov, Sh. N. Yuraev, S. H. Husseinov, and S. S. Abdinbekov
Some Features of IR Reflection Spectra in $Tl(InS_2)_{1-x}(FeSe_2)_x$ Single Crystals ($0 \leqq x \leqq 0.015$) near the Phase Transition Temperature
(78.30; 78.50; S8; S8.15)
(b) Vol. **148**, No. 2, p. K177—K182 (1988).

K. P. Tchakpele and J. P. Albert
Excitons Bound to Te Impurities in CdS, ZnS, and Their Mixed Compounds with Wurtzite Structure
(71.35 and 71.55; S8.11; S8.15)
(b) Vol. **149**, No. 2, p. 641—648 (1988).

P. P. Paskov, L. I. Pavlov, and P. A. Atanasov
Nonlinear Refraction Due to Dynamical Burstein-Moss Effect in Lead Chalcogenides
(78.20; S8.15)
(b) Vol. **149**, No. 2, p. 739—746 (1988).

N. Romeo, A. Bosio, and V. Canevari
$Zn_{0.15}Cd_{0.85}S$ Thin Films by RF Sputtering in an Ar–H_2 Atmosphere
(68.55 and 73.60; S8.15)
(a) Vol. **109**, No. 2, p. K105—K108 (1988).

N. P. Bezhan, O. V. Kulikova, L. L. Kulyuk, S. I. Radautsan, A. V. Siminel, E. E. Strumban, L. I. Shulga, and V. A. Chumak
On the Influence of a Buffer Layer on the Emissive Properties of Epitaxial Layers in InGaAsP ($\lambda = 1.17\,\mu m$)
(78.45 and 78.65; S8.15)
(a) Vol. **109**, No. 2, p. K153—K156 (1988).

J. M. Hvam, C. Dörnfeld, and H. Schwab
Optical Nonlinearity and Phase Coherene in CdSe and $CdSe_xS_{1-x}$
(71.35 and 78.20; S8.12; S8.15)
(b) Vol. **150**, No. 2, p. 387—391 (1988).

J. Puls, W. Rudolph, F. Henneberger, and D. Lab
Femtosecond Studies of Room Temperature Optical Nonlinearities in Wide-Gap II–VI Compounds
(71.35 and 78.20; S8.11; S8.12; S8.15)
(b) Vol. **150**, No. 2, p. 419—423 (1988).

S. E. Esipov and V. A. Stadnik
Optical Bistability and Formation of Localized and Moving High Absorption Domains in ZnSe and ZnCdS
(78.20; S8.12; S8.15)
(b) Vol. **150**, No. 2, p. 501—505 (1988).

V. P. Gribkovskii, V. A. Zyulkov, A. E. Kazachenko, and S. A. Tikhomirov
Optical Nonlinearity of Semiconductor Microcrystals CdS_xSe_{1-x} under the Action of Picosecond and Nanosecond Laser Pulses
(78.20; S8.15)
(b) Vol. **150**, No. 2, p. 647—652 (1988).

L. G. Zimin, S. V. Gaponenko, and V. Yu. Lebed
Room-Temperature Optical Nonlinearity in Semiconductor-Doped Glasses
(78.65; S8.15)
(b) Vol. **150**, No. 2, p. 653—656 (1988).

Zhao Feng, Han Li, Zhang Yun, and Fei Hao Sheng
Laser-Induced Transient Grating in Semiconductor-Doped Glasses
(78.20; S8.15)
(b) Vol. **150**, No. 2, p. 663—666 (1988).

P. P. Paskov, L. I. Pavlov, P. A. Atanasov, D. B. Kushev, and N. N. Zheleva
Optical Nonlinearity and Fast Switching in PbSnTe Fabry-Perot Etalon
(78.20; S8.15)
(b) Vol. **150**, No. 2, p. 729—733 (1988).

H.-E. Swoboda, F. A. Majumder, R. Renner, C. Weber, M. Sence, Lu Jie, G. Noll, E. O. Göbel, J. Vaitkus, and C. Klingshirn
Nonlinear Optical Properties of the System $CdS_{1-x}Se_x$
(78.20; 78.55; S8.15)
(b) Vol. **150**, No. 2, p. 749—759 (1988).

K. Ferencz and R. Szipöcs
Nonlinear Optical Behaviour of Absorbing CdS_xSe_{1-x} Interference Filters
(78.20; S8.15)
(b) Vol. **150**, No. 2, p. 811—815 (1988).

J. Pellegrino and J. M. Galligan
The Photoplastic Effect in II–VI Compounds. Mercury Cadmium Telluride
(72.20; 61.70; S8.15)
(a) Vol. **110**, No. 2, p. 565—573 (1988).

S8.16. Higher compounds

A. P. Deshpande, V. B. Sapre, and C. Mande
Study of the Environment of Copper in Some Ternary Chalcogenides by EXAFS Spectroscopy
(61.10; S1.2; S8.16)
(b) Vol. **145**, No. 1, p. 77—88 (1988).

I. V. Bodnar, G. F. Smirnova, T. V. Smirnova, Yu. A. Aleshchenko, and L. K. Vodopyanov
Lattice Vibrations of $CuGaSe_2$, $CuInSe_2$ Ternary Compounds and $CuGa_xIn_{1-x}Se_2$ Solid Solutions
(63.20; 78.30; S8.16)
(b) Vol. **145**, No. 1, p. 117—122 (1988).

P. P. Lottici, G. Antonioli, and C. Razzetti
EXAFS in $Zn_xCd_{1-x}Ga_2S_4$ Defect Chalcopyrite Solid Solution
(61.60 and 78.70; S8.16)
(b) Vol. **145**, No. 2, p. 401—407 (1988).

T. D. Ibragimov and N. T. Mamedov
Two-Phonon Absorption Spectra and Phase Transition in $TlGaSe_2$
(63.20 and 64.70; 78.30; S8.16)
(b) Vol. **145**, No. 2, p. K103—K105 (1988).

S. A. Eshraghi, S. Kianian, B. Ostrom, O. M. Stafsudd, and A. L. Gentile
The Electrical and Optical Properties of n-Type $HgIn_2Te_4$
(72.80 and 78.20; S8.16)
(a) Vol. **105**, No. 2, p. 563—566 (1988).

I. A. Aksenov, L. A. Makovetskaya, G. P. Popelnyuk, and I. P. Shilovich
Phase Diagram of $CuAl_xIn_{1-x}S_2$ Solid Solutions
(64.70; S8.16)
(a) Vol. **105**, No. 2, p. K97—K102 (1988).

A. Opanowicz and B. Kościelniak-Mucha
Determination of the Electron Lifetime in p-$CuInS_2$ from a Photovoltaic Study of In/p-$CuInS_2$ Schotty Barrier Diodes
(72.40 and 73.30; S8.16)
(a) Vol. **105**, No. 2, p. K135—K139 (1988).

L. Jansa, R. Novotný, P. Lošťák, and J. Klikorka
Optical Properties of n-Type $Bi_{2-x}In_xTe_3$ Single Crystals
(78.30; S8.16)
(b) Vol. **146**, No. 1, p. K75—K80 (1988).

G. Kluge and J. Schmal
On the Application of the Ordinary and Photoconductivity Meyer-Neldel Rule to Amorphous $Ge_{30}Se_{60}Te_{10}$ Films
(72.40; S8.16)
(a) Vol. **106**, No. 1, p. K47—K52 (1988).

G. C. Bhar and N. P. Ghosh
Broadband Tuning in an Infrared Parametric Oscillator
(78.20; S7.16; S8.12; S8.16)
(a) Vol. **106**, No. 1, p. K93—K96 (1988).

P. P. Lottici
A MREI Model for the Zincblende-Like Phonons in the Mixed Defect Chalcopyrite $ZnGa_2(S_xSe_{1-x})_4$
(63.20; 78.30; S8.16)
(b) Vol. **146**, No. 2, p. 503—510 (1988).

N. A. Molodvyan, S. I. Radautsan, V. F. Zhitar, E. D. Arama, and D. S. Remenko
Direct Optical Transitions in $CdGaInS_4$
(78.40; S8.16)
(a) Vol. **106**, No. 2, p. K181—K184 (1988).

A. Opanowicz, B. Kościelniak-Mucha, and G. Boroś
Electrical and Photovoltaic Properties of Au/n-$CuInSe_2$ Schottky Barrier Diodes
(73.30 and 85; S8.16)
(a) Vol **106**, No. 2, p. K197—K201 (1988).

O. V. Kulikova, L. L. Kulyuk, S. I. Radautsan, S. A. Ratseev, E. E. Strumban, V. E. Tezlevan, and V. I. Tsitsanu
Influence of Defect Generation Processes in $CdIn_2S_4$ Single Crystals on the Photoluminescence and Raman Scattering Spectra
(78.30 and 78.55; S8.16)
(a) Vol. **107**, No. 1, p. 373—377 (1988).

M. Bredol and V. Leute
The Crystal Structure of Ga_2Se_2Te
(61.60; S8.16)
(a) Vol. **107**, No. 1, p. K7—K10 (1988).

T. K. Kasumov, F. I. Mamedov, and I. K. Gasymov
Negative Photoconductivity in $AgGaSe_2$ Single Crystals Due to the Negative Photogalvanic Effect
(72.40; S8.16)
(a) Vol. **107**, No. 1, p. K49—K52 (1988).

A. Galdikas, S. Grebinskii, and S. Mickevičius
Piezoresistance Anisotropy of the Ferromagnetic Semiconductor $HgCr_2Se_4$
(72.20; S8.16)
(a) Vol. **107**. No. 1, p. K53—K55 (1988).

E. M. Kogan and M. I. Auslender
Anderson Localization in Ferromagnetic Semiconductors Due to Spin Disorder. I. Narrow Conduction Band
(72.20; 72.60; S8.16)
(b) Vol. **147**, No. 2, p. 613—620 (1988).

N. Frangis and C. Manolikas
Composition Faults in $Zn_mIn_2S_{3+m}$ Crystals
(61.70; S8.16)
(a) Vol. **107**, No. 2, p. 589—595 (1988).

J. A. Kalomiros and J. Spyridelis
Determination of the Optical Constants *n*, *k* of $Zn_3In_2S_6$ and $Zn_5In_2S_8$ from Transmission Measurements
(78.20; S8.16)
(a) Vol. **107**, No. 2, p. 633—637 (1988).

A. N. Veis, D. D. Koditsa, and N. S. Popovich
Optical Properties of $TlBiS_2$ Monocrystals
(78.30; S8.16)
(a) Vol. **107**, No. 2, p. K169—K173 (1988).

T. D. Ibragimov and N. T. Mamedov
Two-Phonon Absorption in $TlInTe_2$ and $TlGaTe_2$
(63.20 and 78.30; S8.16)
(b) Vol. **148**, No. 1, p. K89—K92 (1988).

K. R. Allakhverdiev, N. D. Akhmedzade, M. M. Tagiev, M. M. Shirinov, and S. Häseler
Long-Wavelength IR-Active Phonons in the $TlInS_2$–$TlInSe_2$ System
(63.20 and 78.30; S8.16)
(b) Vol. **148**, No. 1, p. K93—K96 (1988).

J. Lipavičius, E. Čijauskas, and A. Audzijonis
The Fundamental Absorption Edge Tail of $Sn_2P_2Se_6$
(78.40; S8.16)
(b) Vol. **148**, No. 1, p. K97—K100 (1988).

A. Serpi and J. P. Zielinger
On the Photoconductivity Relaxation in $ZnIn_2S_4$
(72.40; 71.55; S8.16)
(a) Vol. **108**, No. 1, p. 351—362 (1988).

I. A. Aksenov, L. A. Makovetskaya, and V. A. Savchuk
Electrical Properties of Cadmium and Zinc Doped $CuAlS_2$
(72.80; S8.16)
(a) Vol. **108**, No. 1, p. K63—K65 (1988).

H. Nakanishi, M. Isomae, S. Endo, and T. Irie
Piezoresistance in $CdIn_2S_4$
(72.20; 71.25; S8.16)
(b) Vol. **148**, No. 2, p. 673—682 (1988).

J. Lipavičius and A. Audzijonis
Optical Birefringence in $Sn_2P_2S_6$
(78.20; S8.16)
(b) Vol. **148**, No. 2, p. K169—K171 (1988).

H. L. Park, H. K. Kim, C. H. Chung, Chang-Dae Kim, Tong-San Cho, and Wha-Tek Kim
Raman Scattering of $CdGa_2Se_4$ and Its Space Group
(78.30; S8.16)
(b) Vol. **148**, No. 2, p. K173—K176 (1988).

M. Quintero, P. Grima, J. E. Avon, G. Lamarche, and J. C. Woolley
Phase Diagram, Optical Energy Gap, and Magnetic Susceptibility of $(CuIn)_{1-z}Mn_{2z}Se_2$ Alloys
(64.75; 75.30; S8.16)
(a) Vol. **108**, No. 2, p. 599—606 (1988).

V. Parizek and T. A. Chuong
Influence of In on Transport Properties and Phase Transition in $Pb_{1-x}Sn_xTe$ Thin Films
(73.60; 64.70; S8.16)
(a) Vol. **108**, No. 2, p. 643—650 (1988).

H. Ishikawa, Y. Kameda, O. Uemura, and T. Satow
Electronic Properties of Liquid $Tl-GeTl_2Se_4$ and $Tl-SnTl_2Se_4$ Systems
(72.20 and 75.30; S1.63; S8.16)
(b) Vol. **149**, No. 1, p. 257—266 (1988).

M. Kranjčec, B. Čelustka, B. Etlinger, and D. Desnica
The Indirect Allowed Optical Transitions in $(Ga_{0.3}In_{0.7})_2Se_3$
(78.20; S8.16)
(a) Vol. **109**, No. 1, p. 329—336 (1988).

E. D. Arama, S. I. Radautsan, I. M. Tiginyanu, V. F. Zhitar, A. Turković, B. Petrović, B. Etlinger, and N. B. Urli
Photoconductivity and Luminescence Spectra of $ZnIn_2S_4$ Crystals Irradiated by γ-Quanta
(72.40 and 78.55; S8.16)
(a) Vol. **109**, No. 1, p. K55—K57 (1988).

E. A. Zhilinskaya, V. N. Lazukin, E. A. Bychkov, and S. B. Rosenkov
Correlation between the Structures of Ge–S Glasses and that of Impurity Mn^{2+} Complexes
(61.60 and 71.55; 76.30; S8.16)
(b) Vol. **149**, No. 2, p. 427—433 (1988).

K.-P. Möllmann
Light Quenching of Photoconductivity in Indium-Doped Lead Tin Telluride
(72.40; S8.16)
(b) Vol. **149**, No. 2, p. 663—667 (1988).

N. A. Moldovyan, S. I. Radautsan, and V. Z. Chebotaru
Photoconductivity, Cathodoluminescence, and Optical Absorption of $CdIn_2S_2Se_2$ Single Crystals
(72.40 and 78.40; 78.60; S8.16)
(a) Vol. **110**, No. 1, p. 277—282 (1988).

V. Bhaskara Sastry, K. Satyanarayana Murthy, N. T. Mamedov, and Prabhavathy
Powder Data on $TlInSe_2$, $TlInTe_2$, and $TlGaTe_2$
(61.60; S8.16)
(a) Vol. **110**, No. 1, p. K5—K7 (1988).

S. M. Wasim and J. G. Albornóz
Electrical and Optical Properties of n- and p-Type $CuInTe_2$
(72.20 and 78.20; 71.55; S8.16)
(a) Vol. **110**, No. 2, p. 575—583 (1988).

G. A. Medvedkin, Yu. V. Rud, and M. A. Tairov
Fundamental Optical Absorption Edge in $MnGa_2Te_4$ Single Crystals
(78.20; 72.40; S8.16)
(a) Vol. **110**, No. 2, p. 631—643 (1988).

S9. Halides

S. S. Sastry, G. Satyanandam, A. Subrahmanyam, and V. R. K. Murthy
Dielectric Properties of $HgCl_2$:2 KCl:H_2O Single Crystals in the Radio Frequency Region
(77.20 and 77.40; S9)
(a) Vol. **105**, No. 1, p. K71—K75 (1988).

E. P. Pokatilov, S. I. Beril, V. M. Fomin, V. G. Litovchenko, D. V. Korbutyak, E. G. Lashkevich, and E. V. Mikhailovskaya
The Size-Quantized States of the Wannier-Mott Exciton in Structures with Superthin Films
(71.35 and 78.65; S9)
(b) Vol. **145**, No. 2, p. 535—544 (1988).

T. Itoh, Y. Iwabuchi, and M. Kataoka
Study on the Size and Shape of CuCl Microcrystals Embedded in Alkali-Chloride Matrices and Their Correlation with Exciton Confinement
(61.10 and 71.35; 78.20; S9)
(b) Vol. **145**, No. 2, p. 567—577 (1988).

J. Röseler, R. I. Shekhmametev, and J. Neugebauer
Diamagnetic Shift of Bielectrons in BiI_3
(71.35; 78.20; S9)
(b) Vol. **145**, No. 2, p. 579—584 (1988).

K. Polák, D. J. S. Birch, and M. Nikl
Decay Kinetics of UV Luminescence from Undoped $PbCl_2$ Crystals
(78.55; 71.35; S9)
(b) Vol. **145**, No. 2, p. 741—747 (1988).

D. Skrzypek, K. Bielas-Borgiel, and Z. Celinski
Investigations on Magnetically Diluted $KNi_xMg_{1-x}F_3$ Perovskite Solid Solution
(75.30 and 76.30; S9)
(a) Vol. **105**, No. 2, p. 567—572 (1988).

M. Kumari and N. Dass
The Melting Laws
(64.70; 62.20; S2; S5.2; S9)
(b) Vol. **146**, No. 1, p. 105—110 (1988).

D. Neogy and T. Purohit
The Behavior of Active Centers in a Laser Host. A Crystal Field Investigation on Dy^{3+} in LaF_3 Single Crystal
(71.70; 78.50; S9)
(b) Vol. **146**, No. 1, p. 181—187 (1988).

T. Tsuboi
Energy Level Discreteness of CuCl Electrons Confined to a Three-Dimensional Potential Well
(71.35 and 78.40; S9; S9.12)
(b) Vol. **146**, No. 1, p. K11—K15 (1988).

A. V. Soldatov, Yu. V. Sukhetskii, A. G. Khasabov, and A. N. Gusatinskii
X-Ray Spectroscopic and Theoretical Investigation of the Energy Band Structure of SbSI
(71.25 and 78.70; S9)
(b) Vol. **146**, No. 1, p. K21—K24 (1988).

M. Pawłowska, W. Bronowska, and W. Wojciechowski
Anomalies in the Magnetic Susceptibility of $RbMnCl_3$ and $RbMnCl_3:Sm^{3+}$
(75.50; S9)
(b) Vol. **146**, No. 1, p. K51—K54 (1988).

T. Itoh, Y. Iwabuchi, and T. Kirihara
Size-Quantized Excitons in Microcrystals of Cuprous Halides Embedded in Alkali-Halide Matrices
(71.35; 78.55; S9)
(b) Vol. **146**, No. 2, p. 531—543 (1988).

S. D. Chaturvedi, R. C. Dixit, and A. A. S. Sangachin
Analysis of Crystal Binding and Thermoelastic Properties of Cuprous Halides, Silver Halides, and Thallous Halides
(61.50 and 63.90; 62.20; 65.70; S9)
(b) Vol. **147**, No. 1, p. 115—121 (1988).

A. R. Hassan
Phonon-Assisted Two-Photon Exciton Transitions in Semiconductors
(71.25 and 71.35; S7.11; S9)
(b) Vol. **147**, No. 1, p. 211—214 (1988).

P. Sekar and T. M. Haridasan
Diffusion Coefficient of F^- Ions in Perfect β-PbF_2 Crystals
(66.30; S9)
(b) Vol. **147**, No. 2, p. 495—500 (1988).

J. Giordano and J. L. Sauvajol
Study of the Temperature Dependence of the Raman Active Modes of ZnF_2
(63.20 and 78.30; S9)
(b) Vol. **147**, No. 2, p. 537—545 (1988).

I. Ch. Akopyan, D. N. Gromov, A. A. Klochikhin, B. V. Novikov, and S. G. Ogloblin
Effect of Lattice Disorder on the Excitonic States in Superionic Crystals
(71.35; 78.20; S9)
(b) Vol. **148**, No. 1, p. 185—195 (1988).

T. Hattori, T. Idogaki, and N. Uryû
Magnetic Susceptibility of the Tetrameric Complex Compound of Ruthenium Pentafluoride RuF_5
(75.30; S9)
(b) Vol. **148**, No. 1, p. K47—K51 (1988).

I. Kosacki and K. Zalibowski
Polarisation Phenomena in PbF_2 Superionic Crystals
(77.20 and 77.40; S9)
(a) Vol. **108**, No. 2, p. 765—770 (1988).

S. N. Martynov and S. B. Petrov
Single-Ion Anisotropy of Exchange-Coupled Mn^{2+} Ions in $(CH_3)_4NCdCl_3$ and $CsCdCl_3$
(71.55 and 76.30; S9)
(b) Vol. **149**, No. 1, p. K41—K44 (1988).

M. V. Pérez Diaz, J. Ch. Abbé, and G. Duplâtre
Positron Annihilation Studies in Non-Metallic Solids: d-Camphor and Silver Iodide
(78.70; S9; S12)
(a) Vol. **109**, No. 1, p. 337—343 (1988).

T. Takeuchi
Electronic Structure of $CuCl_4^{-3}$, $CuCl_4^{-4}$, and $CuCl_4^{-2}$ Clusters
(71.25: S9)
(b) Vol. **149**, No. 2, p. 579—584 (1988).

B. Hönerlage, M. Frindi, and J. Miletic
Time Dependence of Wave-Mixing Processes under Pulsed Resonant Excitation in Three-Level Systems
(78.20; S9)
(b) Vol. **149**, No. 2, p. 775—784 (1988).

M. Kitamura, C. Sugiura, and S. Muramatsu
CoK X-Ray Absorption Near-Edge Structure of Cobalt(III) Complex Compounds
(78.70; S9)
(b) Vol. **149**, No. 2, p. 791—797 (1988).

J. Batarūnas, A. Audzijonis, N. Mykolaitinė, and K. Žičkus
The Electronic Potential of Paraelectric SbSI
(71.25; S9)
(b) Vol. **150**, No. 1, p. K31—K34 (1988).

I. Akai, T. Karasawa, T. Komatsu, and Y. Kaifu
Large Optical Nonlinearity of the Two-Dimensional Exciton States at a Stacking Fault Interface in BiI_3
(71.35 and 78.20; 71.70; 73.40; S9)
(b) Vol. **150**, No. 2, p. 635—639 (1988).

N. Nagasawa
Optical Nonlinearity Due to the Excitonic System in CuCl
(71.35 and 78.20; S9)
(b) Vol. **150**, No. 2, p. 817—823 (1988).

R. Lévy, B. Hönerlage, and J. B. Grun
Time-Resolved Light-Induced Grating Spectroscopy in Semiconductors
(78.20; S9)
(b) Vol. **150**, No. 2, p. 825—832 (1988).

T. Shiozawa and T. Kobayashi
High Resolution Electron Microscopic Study of Silver Sulfide Microcrystals Formed on Silver Bromide Emulsion Grains
(61.60; 61.16; S1.63; S9)
(a) Vol. **110**, No. 2, p. 375—382 (1988).

R. Ewertowski, A. Baranowski, and W. Świątkowski
Positron Trapping at Ag/Na_3AlF_6 Interface; Life-Time Measurements
(78.70; S1.3; S9)
(a) Vol. **110**, No. 2, p. K101—K105 (1988).

S9.11. Alkali halides

F. Beerwerth, D. Fröhlich, and V. Leinweber
Temperature Dependence of Exciton Linewidth in CsI and CsBr
(71.35; 78.20; S9.11)
(b) Vol. **145**, No. 1, p. 195—201 (1988).

S. V. Dyachenko, L. E. Nagli, and F. V. Pirogov
Radiation Defect Production Processes in Activated KCl Crystals
(61.80; 78.70; S9.11)
(a) Vol. **105**, No. 1, p. 87—95 (1988).

T. Ohgaku and T. Takeuchi
The Relation of the Blaha Effect with Internal Friction for Alkali Halide Crystals
(62.20; 62.40; S9.11)
(a) Vol. **105**, No. 1, p. 153—159 (1988).

S. V. Nistor, A. Bouwen, and D. Schoemaker
ESR of Pb^{3+} Centers in Alkali Chlorides
(61.70 and 76.30; S9.11)
(b) Vol. **145**, No. 2, p. K135—K139 (1988).

C. Vijayan and Y. V. G. S. Murti
Ionic Thermocurrents Due to Z-Centres in $KCl:Gd^{2+}$
(61.70; 77.30; 77.40; 78.50; S9.11)
(a) Vol. **105**, No. 2, p. 397—402 (1988).

G. Kitis and St. Charalambous
Regenerated Thermoluminescence as a Function of Irradiation Temperature
(78.60; S9.11; S9.12)
(a) Vol. **105**, No. 2, p. K175—K179 (1988).

E. G. Bagduev, E. V. Korovkin, G. I. Peresada, E. G. Ponyatovskii, and M. Sh. Shikhsaidov
Dark Instability of F-Centres in Pressure-Treated KCl and KBr Crystals
(61.70; 78.60; S9.11)
(b) Vol. **146**, No. 1, p. 75—80 (1988).

T. R. Middya, D. Roy, and A. N. Basu
Multiple-Scattering-Theory Calculation of Elastic and Dielectric Properties of Alkali-Halide Solid Solutions
(62.20 and 77.20; S9.11)
(b) Vol. **146**, No. 1, p. 117—124 (1988).

M. Kumar, A. K. Pachauri, S. D. Chaturvedi, and A. K. Sharma
Equation of State and Pressure Dependence of the Grüneisen Parameter for NaCl Crystal
(64.30; 63.90; S9.11)
(b) Vol. **146**, No. 1, p. 125—130 (1988).

L. E. Nagli and S. V. Dyachenko
Influence of v_c^--Vacancy on Luminescence of Pb^+ Centres in Alkali Halides
(78.55; 61.70; S9.11)
(b) Vol. **146**, No. 1, p. 295—301 (1988).

S. R. Wilk and F. Lüty
Spectroscopic Determination of the Hindered Rotational Behavior of CN^- Defects in Alkali Halides
(78.30; 78.50; S9.11)
(b) Vol. **146**, No. 1, p. 303—310 (1988).

J. Prakash and F. Fischer
TSD in Pressed Pellets of $NaCl:HCO_2^-$
(77.30; 61.70; S9.11)
(b) Vol. **146**, No. 1, p. K71—K73 (1988).

B. D. Bhasin, S. P. Kathuria, and S. V. Moharil
Some Peculiarities of Photo-Transfer Thermoluminescence in LiF-TLD 100
(78.60; S9.11)
(a) Vol. **106**, No. 1, p. 271—276 (1988).

T. Ohgaku and N. Takeuchi
Temperature Dependence of the Effective Stress for KCl Single Crystals
(62.20; S9.11)
(a) Vol. **106**, No. 1, p. K19—K21 (1988).

S. B. S. Sastry and G. Muralidharan
Thermoluminescence and Fluorescence Studies on $RbBr:Gd^{2+}$
(78.60; 61.70; S9.11)
(b) Vol. **146**, No. 2, p. 727—731 (1988).

J. Prakash and A. K. Nishad
On the Analysis of the Ionic Thermocurrent Spectrum
(77.30; S9.11)
(a) Vol. **106**, No. 2, p. K177—K179 (1988).

K. Eftaxias, C. Varotsos, and V. Hadjicontis
Cation Vacancy Migration Entropy in Alkali Halides
(61.70; S9.11)
(b) Vol. **147**, No. 1, p. 83—88 (1988).

A. V. Egranov and A. I. Nepomnyachikh
Excitonic Processes and Thermoluminescence in NaCl Crystals Doped with Mg or Mn
(61.70 and 71.35; 78.60; S9.11)
(b) Vol. **147**, No. 1, p. 267—272 (1988).

C. Zaldo and E. Orozco
Raman Spectroscopy of Sr Phases in Potassium Halides
(78.30; S9.11)
(b) Vol. **147**, No. 1, p. 431—440 (1988).

D. Hans, A. A. S. Sangachin, and M. Kumar
Elastic, Dielectric, and Photoelastic Constants of Alkali Halide Crystals
(62.20; 63.90; 77.20; S9.11)
(b) Vol. **147**, No. 2, p. 529—536 (1988).

C. RUIZ-MEJÍA
About the F Center in Cesium Halides
(71.55; 61.70; S9.11)
(b) Vol. **147**, No. 2, p. 593—600 (1988).

A. OVERBERG and F. FISCHER
Effect of Magnetic Field on $^3A_{1u} \rightarrow {}^1A_{1g}$ Luminescence of Pb^{2+} in CaF_2 and SrF_2 and of Au^- in NaCl
(78.55; 63.20; S9.11; S9.12)
(b) Vol. **147**, No. 2, p. 811—821 (1988).

YU. M. ANNENKOV, V. F. STOLYARENKO, and T. S. FRANGULIAN
Influence of Li Ions on the Accumulation of Radiation Defects in KBr under Pulsed Electron Irradiation
(61.80; S9.11)
(b) Vol. **147**, No. 2, p. K109—K113 (1988).

N. G. ROMANOV, V. V. DYAKONOV, V. A. VETROV, and P. G. BARANOV
EPR Detected via Magnetic Circular Dichroism of Optical Absorption in Irradiated Alkali Halide Crystals Doped with Cadmium
(71.55 and 76.30; S9.11)
(b) Vol. **147**, No. 2, p. K171—K174 (1988).

S. SHIMANUKI
Radiative and Non-Radiative Processes of Localized Impurity Centers in Alkali Halides: KBr: Ag^-
(71.55 and 78.55; S9.11)
(b) Vol. **148**, No. 1, p. 251—262 (1988).

Y. NAKAGAWA, S. NAKAMURA, and N. TAKEUCHI
Decrease of the TL Glow Peak Temperature by Partial Cleaning
(78.60; S9.11)
(a) Vol. **108**, No. 1, p. 463—468 (1988).

J. OTTO
Stress-Induced Dichroism of the Vibrational Lines and Phonon Sidebands of SH^- Centres in CsBr
(63.20; 78.30; S9.11)
(b) Vol. **148**, No. 2, p. 489—499 (1988).

R. CALLENS, F. CALLENS, P. MATTHYS, and E. BOESMAN
EPR of a New O_2^- Centre in KCl
(71.55 and 76.30; S9.11)
(b) Vol. **148**, No. 2, p. 683—688 (1988).

X. L. YUAN and S. W. S. MCKEEVER
Impurity Clustering Effects in Magnesium-Doped Lithium Fluoride
(61.70; 78.50; 78.60; S9.11)
(a) Vol. **108**, No. 2, p. 545—551 (1988).

S. Lakshmi Raghavan, N. Ramesh, and K. Srinivasan
Electro-Optical Effect in KCl:Cd^{2+} and KCl:Sr^{2+}
(78.20; S9.11)
(a) Vol. **108**, No. 2, p. 771—774 (1988).

M. Pützler and F. Fischer
Magnetic Dipole Absorption of Colloidal Ag in Single Crystals of KCl
(78.30; S9.11)
(b) Vol. **149**, No. 1, p. 371—381 (1988).

S. Muramatsu
Electron–Lattice Interaction Parameters Determined from Absorption and MCD Spectra of the F Centers in CsCl and CsBr
(71.55 and 78.50; 61.70; S9.11; S9.12)
(b) Vol. **149**, No. 1, p. K19—K23 (1988).

P. Matthys, F. Callens, G. Vekeman, and E. Boesman
g-Tensor Analysis of Sulfur Centres in Alkali Halides
(71.70 and 76.30; S9.11)
(b) Vol. **149**, No. 1, p. K25—K30 (1988).

A. G. Badalyan, T. I. Maksimova, and J. Rosa
EPR of MnO_4^{2-} Centres in Alkali Halide Crystals
(76.30; 78.50; S9.11)
(b) Vol. **149**, No. 1, p. K31—K34 (1988).

M. P. Madan and M. N. Sharma
On the Change in Energy of Free Ions in Alkali Halides
(77.30; S9.11)
(b) Vol. **149**, No. 1, p. K55—K59 (1988).

E. Radzhabov
Reaction of OH^- and F Centers in Sodium Halides
(61.70 and 71.55; 78.5; S9.11)
(b) Vol. **149**, No. 2, p. 441—446 (1988).

L. Bosi and M. Nimis
An Analysis of the Decay Properties of the F and M Centers in LiF Crystals
(61.70; 78.50; S9.11)
(b) Vol. **149**, No. 2, p. K89—K94 (1988).

V. D. Rodriguez and C. Ortiz
Thermally Stimulated Relaxation in NaF X-Irradiated at Room Temperature
(61.70 and 78.60; S9.11)
(a) Vol. **109**, No. 2, p. 413—419 (1988).

R. Capelletti, R. Cywiński, M. Manfredi, and H. Opyrchał
Influence of Temperature on the Aggregation Processes of the NaCl:Eu^{2+} System
(78.20; S9.11)
(a) Vol. **109**, No. 2, p. 603—610 (1988).

A. DE ANDRÉS
Applications of a Point Polarizable Ion Model to Suzuki Phase Precipitates in Alkali Halides
(63.20; 61.70; 78.30; S9.11)
(b) Vol. **150**, No. 1, p. 31—37 (1988).

R. V. GOPALA RAO and A. SATPATHY
On the Evaluation of Entropy and Thermodynamic Properties of 1:1 Type of Molten Salts through the Generalized Debye Inverse Screening Length
(65.50; S9.11)
(b) Vol. **150**, No. 1, p. 53—58 (1988).

R. V. VEDRINSKII, L. A. BUGAEV, and I. G. LEVIN
Single and Multiple Electron Scattering Description in X-Ray Absorption Spectrum Theory and the Problem of Solid Structure Investigation
(78.70; 61.10; 71.10; S9.11; S10.1; S11.1)
(b) Vol. **150**, No. 1, p. 307—314 (1988).

S. B. S. SASTRY and G. MURALIDHARAN
Thermoluminescence and Other Optical Studies on γ-Irradiated $RbCl:Gd^{2+}$
(78.60; 61.70; S9.11)
(b) Vol. **150**, No. 1, p. 315—322 (1988).

M. KUMAR, I. S. ARORA, and J. SHANKER
On the Narayan-Ramaseshan Interionic Potentials
(61.50; S9.11)
(b) Vol. **150**, No. 1, p. K7—K10 (1988).

D. B. SIRDESHMUKH and K. G. SUBHADRA
The Pressure Dependence of Debye Temperatures of Alkali Halides
(63.70; S9.11)
(b) Vol. **150**, No. 1, p. K11—K14 (1988).

A. M. BRODIN, S. A. BOIKO, M. YA. VALAKH, and G. G. TARASOV
Effect of Optical Feedback on Self-Induced Change of Light Polarization in Cubic Crystals with Directions Instable for Polarization
(78.20; 78.50; S9.11)
(b) Vol. **150**, No. 2, p. 557—560 (1988).

U. WOGGON, F. HENNEBERGER, and M. MÜLLER
Optical Nonlinearities of Semiconductor Microcrystallites
(78.20; 71.35; S8.1; S9.11)
(b) Vol. **150**, No. 2, p. 641—645 (1988).

M. P. LISITSA, M. YA. VALAKH, G. G. TARASOV, S. A. BOIKO, A. M. BRODIN, A. M. SHPAK, R. VOSKA, and I. FELDVARI
Nonlinear Polarization Spectroscopy of Impure Cubic Crystals
(78.20; 78.50; S9.11)
(b) Vol. **150**, No. 2, p. 907—911 (1988).

R. V. GOPALA RAO and RATNA DAS
Structure Factor Computation of RbCl and RbBr
(61.25; S9.11)
(a) Vol. **110**, No. 2, p. 347—357 (1988).

S9.12. Alkaline earth halides

Y. Q. SHENG, E. RECKNAGEL, A. WEIDINGER, Z. G. GU, Z. Y. LAI, and Z. C. ZHUANG
Raman Scattering at the CaF_2 Crystal Surface Excited by an Evanescent Optical Wave Field
(71.36 and 73.20; 78.30; S9.12)
(b) Vol. **145**, No. 1, p. 151—156 (1988).

A. DE and K. V. RAO
Optical Absorption and Thermoluminescence of CaF_2 Single Crystals Subjected to High Electric Fields and Later Irradiated with X-Rays
(78.50; 78.60; S9.12)
(a) Vol. **105**, No. 1, p. 297—302 (1988).

M. LAZARIDOU and C. VAROTSOS
Defect Formation and Migration Entropies in Alkaline Earth Fluorides
(61.70; S9.12)
(a) Vol. **105**, No. 1, p. K13—K15 (1988).

S. C. GOYAL and A. K. SHARMA
Fourth-Order Elastic Constants of Fluorite Structure Solids
(62.20; S9.12)
(b) Vol. **145**, No. 2, p. 435—443 (1988).

G. KITIS and ST. CHARALAMBOUS
Regenerated Thermoluminescence as a Function of Irradiation Temperature
(78.60; S9.11; S9.12)
(a) Vol. **105**, No. 2, p. K175—K179 (1988).

T. TSUBOI
Energy Level Discreteness of CuCl Electrons Confined to a Three-Dimensional Potential Well
(71.35 and 78.40; S9; S9.12)
(b) Vol. **146**, No. 1, p. K11—K15 (1988).

R. P. GOYAL, O. P. GUPTA, and B. R. K. GUPTA
Theory of Compression in Some Fluorite Structure Crystals
(64.30; 62.20; 65.50; S9.12; S11)
(b) Vol. **146**, No. 2, p. 475—480 (1988).

M. KIKNADZE, R. MIRIANASHVILI, and T. SANADZE
Nature of Optically Induced Trigonal Centers of Tm^{2+} in BaF_2
(76.70; S9.12)
(b) Vol. **146**, No. 2, p. K141—K143 (1988).

G. P. TRIBERIS, N. GUSKOS, and G. KIRIAKIDIS
New Aspects on the Electronic Conduction in Semi-Insulating $CdF_2:Gd^{3+}$ Single Crystals
(72.20; S9.12)
(b) Vol. **147**, No. 1, p. 349—353 (1988).

A. OVERBERG and F. FISCHER
Effect of Magnetic Field on $^3A_{1u} \rightarrow {}^1A_{1g}$ Lumonescence of Pb^{2+} in CaF_2 and SrF_2 and of Au^- in NaCl
(78.55; 63.20; S9.11; S9.12)
(b) Vol. **147**, No. 2, p. 811—821 (1988).

E. ZLATEVA and I. GUTZOW
Radiation-Induced Volume Changes in Thermoluminescent-Active Materials
(61.80 and 78.60; S9.12)
(a) Vol. **107**, No. 2, p. K99—K103 (1988).

A. ZEHE, A. ADÁN, and M. MÄDER
A Regenerative Switching Device in a Silicon–Earthalkalifluoride Heterostructure Grown by Molecular Beam Epitaxy
(68.55; 73.60; S9.12)
(a) Vol. **107**, No. 2, p. K119—K122 (1988).

P. SCHOTANUS, C. W. E. VAN EIJK, G. BLASSE, and H. W. DEN HARTOG
Luminescence of the Divalent Lead Ion in Barium Fluoride Crystals
(78.55; S9.12)
(b) Vol. **148**, No. 1, p. K77—K80 (1988).

G. JEANDEL, A. WARIN, and G. MORLOT
Dielectric Constant and ITC Measurements of $BaF_2:Dy^{3+}$
(66.30 and 77.50; S9.12)
(a) Vol. **108**, No. 1, p. 183—187 (1988).

A. DE and K. V. RAO
Dielectric Properties of CaF_2 Single Crystals Irradiated with X-Rays under a High Electric Field or γ-Rays
(77.20 and 77.40; S9.12)
(a) Vol. **108**, No. 1, p. 457—462 (1988).

J. L. RAO, R. M. KRISHNA, and S. V. J. LAKSHMAN
Crystal Field Energy Levels, EPR Parameters, and Magnetic Susceptibility of the Co^{2+} Ion in Cs_3ZnCl_5
(71.70; 76.30; 78.50; S9.12)
(b) Vol. **148**, No. 2, p. K119—K123 (1988).

B. JASSEMNEJAD, R. J. ABBUNDI, M. D. BROWN, and S. W. S. MCKEEVER
Thermoluminescence Processes in CaF_2:Ce, Mn
(78.60; S9.12)
(a) Vol. **108**, No. 2, p. 753—764 (1988).

B. NARASHIMHA, K. V. RAO, and R. N. P. CHAUDHURY
Optical Absorption of Quenched and γ-Ray Irradiated BaF_2 Single Crystals Later Subjected to High Electric Fields or Laser Excitation
(61.70 and 78.50; S9.12)
(a) Vol. **108**, No. 2, p. K97—K100 (1988).

A. G. Makhanek, V. S. Korolkov, and L. A. Yuguryan
Analysis of the Intensity and Polarization Dependence of Two-Photon Absorption in the Channel $E_{3/2}(4f)$–$E_{3/2}(5d)$ for Ce^{3+} in CaF_2
(71.70; 78.50; S9.12)
(b) Vol. **149**, No. 1, p. 231—240 (1988).

S. Muramatsu
Electron–Lattice Interaction Parameters Determined from Absorption and MCD Spectra of the F Centers in CsCl and CsBr
(71.55 and 78.50; 61.70; S9.11; 9.12)
(b) Vol. **149**, No. 1, p. K19—K23 (1988).

J. P. Jouart and G. Mary
Up-Conversion from Er^{3+}–Er^{3+} Pairs in CdF_2 Crystals
(71.55; 78.50; 78.55; S9.12)
(b) Vol. **149**, No. 2, p. 633—639 (1988).

A. Hamaïdia and A. Hachimi
Colour Centres in Alkali Metal Doped CaF_2
(78.20 and 78.50; 61.70; 78.55; S9.12)
(b) Vol. **149**, No. 2, p. 711—724 (1988).

A. V. Shishkin, S. Linderoth, and S. V. Svirida
Positron Annihilation in BaF_2 and SrF_2
(78.70; S9.12)
(b) Vol. **149**, No. 2, p. 785—789 (1988).

A. Tempel, M. Mäder, A. Zehe, and R. Grötzschel
Thermal Mismatch Biased Rhombohedral Structure of Strained Epitaxial CaF_2 Films on Si (111)
(68.55; 65.70; S9.12)
(a) Vol. **109**, No. 2, p. 493—501 (1988).

D. Hans
Third-Order Elastic Constants for Alkaline Earth Oxides and Fluorides
(62.20; S9.12; S10.1)
(b) Vol. **150**, No. 1, p. 47—51 (1988).

A. Hamaïdia
Colour Centres in Alkali Metal Doped SrF_2
(78.20 and 78.50; 61.70; 76.30; S9.12)
(b) Vol. **150**, No. 1, p. 323—336 (1988).

S9.15. Mixed crystals

A. Anedda, G. Bongiovanni, and E. Fortin
Raman Scattering from Vibrational Modes in Layered $Cd_xHg_{1-x}I_2$ Solid Solution
(78.30; 63.20; S9.15)
(b) Vol. **146**, No. 2, p. 757—765 (1988).

B. Schumann, R. Flagmeyer, and H. Weinert
Multi-Quantum-Well Structures Prepared by the Flash Evaporation Technique. Growth, Structural, and Optical Characterization
(73.40; 68.55; 78.65; S7.1; S9.15)
(a) Vol. **108**, No. 1, p. 257—264 (1988).

M. Tanaka
EPR Linewidth of Two-Dimensional Diluted Magnetic Systems
(76.30; S9.15)
(b) Vol. **149**, No. 1, p. K45—K48 (1988).

K. Inabe, S. Nakamura, and N. Takeuchi
Optical Bleaching of BaFCl Crystals X-Irradiated at Room Temperature
(61.70 and 78.50; S9.15)
(b) Vol. **149**, No. 1, p. K67—K71 (1988).

K. Inabe, S. Nakamura, and N. Takeuchi
Optical Bleaching of F-Centers in BaFCl Crystals after X-Irradiation at Room Temperature
(61.70; 78.50; S9.15)
(b) Vol. **149**, No. 2, p. K101—K104 (1988).

S9.16. Higher compounds

Wan-Lun Yu and J. L. Rao
Pressure Dependence of the EPR Parameters of Cs_3CoCl_5 Crystal
(71.70; 76.30; S9.16)
(b) Vol. **145**, No. 1, p. 255—259 (1988).

Huang Ming
High-Order Perturbation Formula for the EPR Trigonal Symmetric $g_{\parallel}$-Value of the d^8 Ion
(71.70; S9.16)
(b) Vol. **145**, No. 1, p. K21—K25 (1988).

M. V. Chernitskii, A. E. Nikiforov, and A. I. Krotkii
Microscopic Mechanisms of Paramagnetic Probe–Lattice Interaction for the System $RbCdF_3:Cr^{3+}$ near the Structural Phase Transition
(64.60 and 76.30; S9.16)
(b) Vol. **145**, No. 2, p. 425—434 (1988).

T. Tsuboi
Light Scattering by Magnetic Polaritons in the Ferromagnet K_2CuF_4
(75.25 and 76.50; S9.16)
(b) Vol. **145**, No. 2, p. K145—K148 (1988).

K. S. Aleksandrov, S. V. Melnikova, I. N. Flerov, A. D. Vasilev, A. I. Kruglik, and I. T. Kokov
Phase Transitions in Cs_2CdI_4 Single Crystals
(64.70; 61.60; 65.40; 77.20; 78.20; S9.16)
(a) Vol. **105**, No. 2, p. 441—446 (1988).

T. Takeuchi
Note on MD Calculation of $RbAg_4I_5$
(63.90; S9.16)
(b) Vol. **147**, No. 1, p. K9—K10 (1988).

M. H. Kuok and S. H. Tang
A Raman Study of $KCdBr_3$ Single Crystals
(78.30; 63.20; S9.16)
(b) Vol. **147**, No. 2, p. K195—K199 (1988).

N. Kristianpoller, B. Trieman, R. Chen, and Y. Kirsh
The Application of Thermally Stimulated Processes to the Study of Defects in Perovskite Type Fluorides
(61.70; 72.20; 78.60; S9.16)
(b) Vol. **149**, No. 1, p. 45—54 (1988).

C. Andraud, F. Pelle, O. Pilla, and B. Blanzat
Intrinsic Luminescence of $CsCdBr_3$
(78.60; S9.16)
(b) Vol. **149**, No. 2, p. 759—763 (1988).

S10. Oxides

H. Pagnia and N. Sotnik
Bistable Switching in Electroformed Metal–Insulator–Metal Devices **(Review Article)**
(73.40 and 85; S10)
(a) Vol. **108**, No. 1, p. 11—65 (1988).

C. A. C. Sequeira and J. M. B. Fernandez Diniz
Conduction Processes in Oxide Solid Electrolytes
(66.30 and 72.20; 61.70; S10)
(a) Vol. **105**, No. 1, p. 123—130 (1988).

H. W. Zandbergen, R. Gronsky, and G. Thomas
Surface Decomposition of Superconducting $YBa_2Cu_3O_7$
(68.20; 61.16; S10)
(a) Vol. **105**, No. 1, p. 207—218 (1988).

A. A. Kaminskii, K. Kurbanov, K. L. Ovanesyan, and A. G. Petrosyan
Stimulated Emission Spectroscopy of Pr^{3+} Ions in Orthorhombic $YAlO_3$ Single Crystals
(78.45; 78.55; S10)
(a) Vol. **105**, No. 2, p. K155—K159 (1988).

A. P. Elisseev, A. M. Yurkin, and E. G. Samoilova
Cr^{3+} Energy Levels in Alexandrite
(78.55; 71.70; S10)
(a) Vol. **105**, No. 2, p. K169—K173 (1988).

V. L. Aksenov, S. Flach, and N. M. Plakida
A Microscopical Model of Structural Phase Transition in La_2CuO_4
(63.20 and 63.75; S10)
(b) Vol. **146**, No. 1, p. K1—K5 (1988).

F. Yassin, V. Christoph, and L. Jahn
On the Determination of the Curie Temperature of Anisotropic Ferromagnets
(75.50; S10)
(a) Vol. **106**, No. 1, p. K79—K82 (1988).

V. Dvořák
Erratum to: On the Structural Phase Transition in La_2CuO_4
(63.20 and 64.70; S10)
(b) Vol. **146**, No. 2, p. K167 (1988).

A. K. Abass, N. A. Al-Liabi, and W. A. Taha
Optical Properties of Bromine-Doped SnO_2 Coatings for Solar Applications
(78.20; S10)
(a) Vol. **106**, No. 2, p. 613—618 (1988).

A. F. Vavrek and K. K. Christova
X-Ray Polarization and Depolarization Phenomena in $Bi_{12}SiO_{20}$
(77.30 and 78.70; 72.40; S10)
(a) Vol. **107**, No. 1, p. 379—386 (1988).

M. Roy, R. N. P. Chowdhury, and H. N. Acharya
Structural and Thermal Properties of $(Sm_{1-x}Nd_x)_2(MoO_4)_3$
(61.60 and 65.90; S10)
(a) Vol. **107**, No. 1, p. K1—K5 (1988).

M. Koyano, S. Ōhara, H. Negishi, M. Sasaki, M. Inoue, M. Nomura, and H. Fujiwara
Effect of Pressure on Electrical Resistivity and Crystal Structure of Quasi-Two-Dimensional Mo_4O_{11}
(61.60 and 72.20; S10)
(b) Vol. **147**, No. 2, p. 559—565 (1988).

C. Boulesteix, M. Ben Salem, B. Yangui, Z. Kang, and L. Eyring
Thickness of Interfaces between Twins, Glide Domains, and Grain Boundaries in Oxides from HREM Studies (Sm_2O_3, Nd_2O_3, Pr_7O_{12}, $Tb_{12}O_{22}$, and $YBa_2Cu_3O_{7-x}$)
(68.48; 61.16; S10)
(a) Vol. **107**, No. 2, p. 469—480 (1988).

L. A. Bursill and XuDong Fan
Surface Structure of the Ceramic Superconductor $YBa_2Cu_3O_7$
(68.20; 61.16; 74.70; S10)
(a) Vol. **107**, No. 2, p. 503—510 (1988).

S. C. Farmer and A. H. Heuer
Surface Phase Transformation in Monoclinic ZrO_2
(68.20; 64.60; S10)
(a) Vol. **107**, No. 2, p. 571—578 (1988).

R. Hoogewijs, P. de Volder, L. Fiermans, and J. Vennik
Room Temperature Oxidation in Air of Polycrystalline CuZn
(68.20; 79.60; S1.2; S10)
(a) Vol. **107**, No. 2, p. 625—631 (1988).

H.-J. Ou and J. M. Cowley
The Surface Reaction of Pd/MgO Studied by Scanning Reflection Electron Microscopy
(68.20; 68.48; S1; S10)
(a) Vol. **107**, No. 2, p. 719—729 (1988).

H. R. Wenk, S. J. Shaffer, and G. van Tendeloo
Planar Defects in Low Temperature Quartz
(61.70 and 64.80; S10)
(a) Vol. **107**, No. 2, p. 799—805 (1988).

M. I. Auslender, E. M. Kogan, and S. V. Tretyakov
Anderson Localization in Ferromagnetic Semiconductors Due to Spin Disorder. II. Wide Conduction Band
(72.20; 72.60; S8.1; S10)
(b) Vol. **148**, No. 1, p. 289—295 (1988).

B. M. Anghelov
Spectroscopic Properties of $LiNdP_4O_{12}$
(71.70; 78.50; S10)
(b) Vol. **148**, No. 1, p. K83—K88 (1988).

C. A. C. Sequeira and J. M. B. Fernandes Diniz
Transport Equations in Common Solid State Electrochemical Situations
(66.30 and 72.10; S10)
(a) Vol. **108**, No. 1, p. 155—162 (1988).

M. Inoue, S. Ōhara, S. Horisaka, M. Koyano, and H. Negishi
Transport Properties of Quasi-Two-Dimensional Mo_4O_{11} Crystals
(72.20; S10)
(b) Vol. **148**, No. 2, p. 659—671 (1988).

T. S. Shi, Y. G. Zhao, P. X. Cai, H. F. Gu, and L. M. Xie
IR Absorption Peaks and Oxygen Vacancies in $YBa_2Cu_3O_x$
(61.70 and 78.30; 78.50; S10)
(b) Vol. **148**, No. 2, p. 715—721 (1988).

D. Lapraz, P. Iacconi, Y. Sayadi, P. Keller, J. Barthe, and G. Portal
Some Thermoluminescence Properties of an α-Al_2O_3 Sample. Sensitization Effects
(78.60; 61.80; S10)
(a) Vol. **108**, No. 2, p. 783—794 (1988).

J. M. Neto and T. Abritta
Luminescence and Site Occupation of Cr^{3+} in Sodium–β''-Alumina
(78.55; 61.70; S10)
(a) Vol. **108**, No. 2, p. 795—802 (1988).

J. A. Sanz-García, E. Dieguez, and C. Zaldo
Luminescence Characterization of $K_4Nb_6O_{17}$ Crystals
(78.55 and 78.60; S10)
(a) Vol. **108**, No. 2, p. K145—K149 (1988).

V. R. Belosludov, M. Yu. Lavrentiev, and S. A. Syskin
Dynamical Properties of the Molecular Crystals with Electrostatic Interaction Taken into Account. Low Pressue Ice Phases (I_h and I_c)
(63.20; S10)
(b) Vol. **149**, No. 1, p. 133—142 (1988).

V. L. AKSENOV, S. FLACH, and N. M. PLAKIDA
A Theoretical Study of Ultrasonic Anomalies in $La_{2-x}Sr_xCuO_4$
(62.80 and 63.75; S10)
(b) Vol. **149**, No. 1, p. K7—K12 (1988).

JUN CHEN and DUAN FENG
TEM Study of Phases and Domains in $NaNbO_3$ at Room Temperature
(64.70; 77.80; S10)
(a) Vol. **109**, No. 1, p. 171—185 (1988).

A. V. ZARETSKII, V. F. PETRENKO, and V. A. CHESNAKOV
The Protonic Conductivity of Heavily KOH-Doped Ice
(66.90 and 77.20; 64.70; S10)
(a) Vol. **109**, No. 2, p. 373—381 (1988).

A. BONNET, A. CONAN, M. MORSLI, M. GANNE, and M. TOURNOUX
Electronic and Structural Properties of Tungsten Bronzes
(72.20; 71.25; S10)
(b) Vol. **150**, No. 1, p. 225—235 (1988).

V. N. ASTRATOV, A. V. ILINSKII, and A. S. FURMAN
Photoinduced Charge Dynamics in Photorefractive Semiconductors
(78.20; 72.40; S10)
(b) Vol. **150**, No. 2, p. 611—615 (1988).

X. M. XIE, T. G. CHEN, and J. HUANG
Diffusitivity of Oxygen in the Orthorhombic $YBa_2Cu_3O_y$ Phase
(66.30; S10)
(a) Vol. **110**, No. 2, p. 415—419 (1988).

A. K. GARG
High-Pressure Raman Spectroscopic Study of the Ice Ih → Ice IX Phase Transition
(64.70; 62.50; 78.30; S10)
(a) Vol. **110**, No. 2, p. 467—480 (1988).

S10.1. Simple oxides

H. Donker, W. M. A. Smit, and G. Blasse
On the Luminescence of CaO:Sn^{2+}
(78.55; 78.50; S10.1)
(b) Vol. **145**, No. 1, p. 333—342 (1988).

S. Mochizuki
Infrared Absorption in Nickel Oxide Microcrystals
(78.30; S10.1)
(b) Vol. **145**, No. 1, p. K75—K78 (1988).

J. Härtwig, V. Holý, R. Kittner, J. Kuběna, and V. Lerche
Investigation of Quartz Crystals with Non-Homogeneous Distributions of Impurity Atoms by X-Ray Methods
(61.10 and 61.70; S10.1)
(a) Vol. **105**, No. 1, p. 61—75 (1988).

S. Sen, S. Annamalai, and P. K. Acharya
Study and Characterization of PECVD Oxides
(68.55; 78.65; S10.1)
(a) Vol. **105**, No. 1, p. 171—176 (1988).

A. E. Rakhshani and J. Varghese
The Effect of Temperature on Electrodeposition of Cuprous Oxide
(68.55; S10.1)
(a) Vol. **105**, No. 1, p. 183—188 (1988).

Wen-Chen Zheng
Determination of the Local Thermal Expansion Coefficient in the Vicinity of Cr^{3+} Ions Associated with Vacancies in MgO
(61.70 and 65.70; 71.70; S10.1)
(b) Vol. **145**, No. 2, p. K99—K102 (1988).

D. Köhler, G. Koschek, and E. Kubalek
Vacancy Diffusion Profiles at the Grain Boundaries of Electronic Ceramics
(66.30; 61.70; 78.60; S10.1)
(a) Vol. **105**, No. 2, p. 377—385 (1988).

G. Dräger, R. Frahm, G. Materlik, and O. Brümmer
On the Multipole Character of the X-Ray Transitions in the Pre-Edge Structure of Fe K Absorption Spectra. An Experimental Study
(78.70; 71.50; S1.1; S10.1)
(b) Vol. **146**, No. 1, p. 287—294 (1988).

M. Pattabi, M. S. M. Sastry, and V. Sivaramakrishnan
Studies on the Stability of Discontinuous Silver Films with Overlayers of Al_2O_3 and SiO_2
(73.60; S1.3; S10.1)
(a) Vol. **106**, No. 1, p. 145—152 (1988).

K. Kawabata, T. Seiyama, S. Okuda, and Y. Shono
Thermally Stimulated Exo-Electron Emission and Thermally Stimulated Luminescence of Irradiated α-Al_2O_3 Powder
(71.55 and 79.75; 78.60; S10.1)
(a) Vol. **106**, No. 1, p. 181—190 (1988).

N. Sasi, C. Balasubramanian, and Sa. K. Narayandass
Electrical Conduction in Germanium Dioxide Thin Films
(73.60; S10.1)
(a) Vol. **106**, No. 1, p. K41—K45 (1988).

A. De and K. V. Rao
Dielectric Properties of Synthetic Quartz Crystals Irradiated with γ-Rays of X-Rays under a High Electric Field
(77.20; 77.40; S10.1)
(a) Vol. **106**, No. 1, p. K101—K105 (1988).

N. B. Manson and T. J. C. Hosea
Anomalous Temperature Dependence of the Absorption Strength of SrO:Ni^{2+} and SrO:Co^{2+}
(78.50; S10.1)
(b) Vol. **146**, No. 2, p. K153—K156 (1988).

J. D. Stephenson
Quasi-Realtime Observation of Partial Dislocation Growth in ($0\bar{1}.1$) Synthetic Quartz near the α–β Phase Transition Temperature (T_c) Using Electric Potential Switching
(61.70; 64.60; S10.1)
(a) Vol. **106**, No. 2, p. 441—449 (1988).

M. T. Mohammad and W. A. Abdul-Ghafor
Properties of Fluorine-Doped SnO_2 Films Prepared by the Spray Pyrolysis Technique
(78.65; 71.25; S10.1)
(a) Vol. **106**, No. 2, p. 479—484 (1988).

C. B. Steele and J. Beynon
Thermopower and DC Electrical Conductivity in Au/SiO Thin Cermet Films
(73.60; 72.20; S10.1)
(a) Vol. **106**, No. 2, p. 515—523 (1988).

A. Singh
Oxidise–Etch–Oxidise and Anneal Method for Growing High Quality Ultrathin SiO_2
(68.55 and 73.40; S10.1)
(a) Vol. **106**, No. 2, p. K139—K142 (1988).

W. A. Murad, S. M. Al-Shamari, F. H. Al-Khateeb, and R. H. Misho
Band Gap Measurements of Co_3O_4 Thin Films Prepared by the Method of Chemical Spray Pyrolysis
(68.55 and 78.65; S10.1)
(a) Vol. **106**, No. 2, p. K143—K146 (1988).

N. N. Dinh, V. T. Bich, N. H. Hoang, T. X. Hoai, V. N. Quynh, and N. V. Chanh
Vacuum Annealing Effects on Structural and Electrical Properties of Reactively Evaporated Indium Tin Oxide
(73.60; 78.65; S10.1)
(a) Vol. **106**, No. 2, p. K147—K151 (1988).

E. M. Dianov, V. O. Sokolov, and V. B. Sulimov
Theory of H(I) Centers in Vitreous Silicon Dioxide
(71.55; S10.1)
(b) Vol. **147**, No. 1, p. 261—266 (1988).

M. Kumeda, M. Kiwaki, and T. Shimizu
Temperature Dependence of the E′ Center Creation in Silica Glasses
(61.70; 76.30; S10.1)
(b) Vol. **147**, No. 1, p. K1—K4 (1988).

P. Fernández, J. Llopis, and J. Piqueras
Annealing Effects on Mechanically Damaged ZnO Ceramics
(62.20; 78.60; S10.1)
(a) Vol. **107**, No. 1, p. 197—203 (1988).

Guoan Cheng, Hengde Li, and Baixin Liu
Enhanced Adhesion in Metal/Oxide Pairs by Ion Irradiation
(61.80 and 68.48; S1.1; S1.2; S10.1)
(a) Vol. **107**, No. 1, p. K29—K32 (1988).

G. Dolino, F. Mogeon, and P. Bastie
Birefringence and γ-Ray Study of the Irreversible Behaviour of the Incommensurate Phase of Quartz
(64.60; 78.20; S10.1)
(a) Vol. **107**, No. 2, p. 559—570 (1988).

H. Poelman, J. Vennik, and G. Dalmai
Temperature Effects on the V_2O_5 Surface Phonon Spectrum
(63.20 and 68.30; S10.1)
(a) Vol. **107**, No. 2, p. 731—737 (1988).

A. A. Samokhvalov, N. A. Viglin, B. A. Gizhevskii, T. I. Arbuzova, and N. M. Chebotaev
Magnetic Polarons in EuO
(72.20 and 75.30; 76.30; S10.1)
(b) Vol. **148**, No. 1, p. 361—367 (1988).

V. A. Gubanov, A. F. Zatsepin, V. S. Kortov, S. P. Freidman, and G. B. Cherlov
Electronic Structure of Intrinsic Defects in Amorphous GeO_2
(71.55; S10.1)
(b) Vol. **148**, No. 1, p. K33—K36 (1988).

H.-J. Fitting and D. Hecht
Secondary Electron Field Emission
(73.60 and 79.20; S10.1)
(a) Vol. **108**, No. 1, p. 265—273 (1988).

K. Dunphy and W. W. Duley
Multiphoton Excitation of the 520 nm Luminescence Band in MgO:Al
(78.55; 71.35; 71.55; S10.1)
(b) Vol. **148**, No. 2, p. 729—735 (1988).

E. G. Salman and V. N. Vertoprakhov
Thermally Stimulated Depolarization Current Controlled by Surface Charge Change
(73.40 and 77.30; 77.55; S10.1)
(a) Vol. **108**, No. 2, p. 625—630 (1988).

V. N. Katerinchuk, Z. D. Kovalyuk, and V. A. Manasson
On a Mechanism of Photo-E.M.F. Formation in SIS Structures
(73.40; 85; S8; S10.1)
(a) Vol. **108**, No. 2, p. K115—K118 (1988).

H. Krause, H.-P. Bär, and H. Schulze
Correlations between Current–Voltage Characteristics and Visible Defect Channels in Thin SiO_2 Films after Field Stress
(61.70 and 72.20; S10.1)
(a) Vol. **108**, No. 2, p. K131—K133 (1988).

H.-P. Bär, H. Krause, and H. Schulze
On the Nature of Structural Defects in SiO_2 Films Visible with the Aid of the Liquid Crystal Technique
(61.70 and 72.20; S10.1)
(a) Vol. **108**, No. 2, p. K135—K137 (1988).

A. K. Abass, Z. A. Ahmed, and F. A. Ali
Preparation and Interband Transitions of Chemically Deposited Fe_2O_3 Coatings in the Fundamental Absorption Region
(78.65; S10.1)
(a) Vol. **108**, No. 2, p. K151—K155 (1988).

N. N. Dinh, V. T. Bich, N. H. Hoang, and L. Q. Minh
Photochromic Properties of Tungsten Trioxide Films Made by Electron Beam Deposition
(78.65; S10.1)
(a) Vol. **108**, No. 2, p. K157—K161 (1988).

I. Fernández and J. Llopis
Reducing Effects on the CL Red Emission of MgO Doped Crystals
(78.60; S10.1)
(a) Vol. **108**, No. 2, p. K163—K167 (1988).

H. Lanteri, P. Rostaing, and R. Bindi
Monte-Carlo Simulation of Exoelectron Energy Spectra
(79.75; S10.1)
(b) Vol. **149**, No. 1, p. 203—214 (1988).

V. V. Arutunyan, G. N. Eritsyan, R. K. Ezoyan, and V. A. Gevorkyan
On the Mechanism of Radiation Induced Change in the Transparence of Corundum Single Crystals
(78.40; S10.1)
(b) Vol. **149**, No. 1, p. K77—K80 (1988).

Z. T. Al-Dhhan, C. A. Hogarth, and A. Davidson
The Optical Absorption Edge in Thin Films of CeO_2/Cd
(78.65; S10.1)
(b) Vol. **149**, No. 1, p. K81—K84 (1988).

J. Härtwig and V. Lerche
Anisotropic Deformation of a Crystal Plate and Its Analysis with X-Ray Diffraction Methods
(61.10; 61.70; 62.20; S10.1)
(a) Vol. **109**, No. 1, p. 79—91 (1988).

G. Völkel, A. Pöppl, and B. Voigtsberger
Investigation of the Oxygen Vacancy Balance in ZnO Ceramics by Means of EPR
(71.55; 76.30; S10.1)
(a) Vol. **109**, No. 1, p. 295—299 (1988).

M. Sacchi, M. Antonini, and M. Rrudenziati
Ru-Metal Segregation in Borosilicate Glasses
(64.75; S10.1)
(a) Vol. **109**, No. 1, p. K23—K26 (1988).

L. N. Aleksandrov, V. Yu. Balandin, A. V. Dvurechenskii, and O. A. Kulyasova
Melting of Multilayered Structures under Pulse Heating (Computational Experiment)
(64.70; S5.11; S10.1)
(a) Vol. **109**, No. 1, p. K27—K29 (1988).

S. A. Marshall, C. Yu, and Y. N. Zhang
Comments on the EPR Spectrum of Tetravalent Chromium in Ruby
(76.30; S10.1)
(b) Vol. **149**, No. 2, p. 691—695 (1988).

W. I. Khleif, C. A. Hogarth, and Z. T. Al-Dhhan
Optical Absorption in Thin Films of Cerium Dioxide Containing Indium Oxide
(78.65; S10.1)
(b) Vol. **149**, No. 2, p. K187—K190 (1988).

A. A. Agasiev, V. E. Bagiev, A. M. Mamedov, and Ya. Yŭ. Guseinov
VUV Reflection Spectra and Electron Structure of Bi_2O_3
(78.40; S10.1)
(b) Vol. **149**, No. 2, p. K191—K195 (1988).

A. Abdellaoui, L. Martin, and A. Donnadieu
Structure and Optical Properties of MoO_3 Thin Films Prepared by Chemical Vapor Deposition
(68.55 and 78.65; S10.1)
(a) Vol. **109**, No. 2, p. 455—462 (1988).

M. Anwar and C. A. Hogarth
Optical Properties of Amorphous Thin Films of MoO_3 Deposited by Vacuum Evaporation
(78.65; 68.55; S10.1)
(a) Vol. **109**, No. 2, p. 469—478 (1988).

J. Suñé, I. Placencia, E. Farrés, N. Barniol, and X. Aymerich
On the Si–SiO_2 Interface Roughness in VLSI–MOS Structures
(68.48 and 73.60; S5.11; S10.1)
(a) Vol. **109**, No. 2, p. 479—491 (1988).

R. H. Misho, W. A. Murad, G. H. Fatahalah, I. M. Abdul Aziz, and H. M. Al-Doori
Preparation and Optical Properties of Thin Nickel Oxide Solid Films
(68.55 and 78.65; S10.1)
(a) Vol. **109**, No. 2, p. K101—K104 (1988).

D. Hans
Third-Order Elastic Constants for Alkaline Earth Oxides and Fluorides
(62.20; S9.12; S10.1)
(b) Vol. **150**, No. 1, p. 47—51 (1988).

R. V. Vedrinskii, L. A. Bugaev, and I. G. Levin
Single and Multiple Electron Scattering Description in X-Ray Absorption Spectrum Theory and the Problem of Solid Structure Investigation
(78.70; 61.10; 71.10; S9.11; S10.1; S11.1)
(b) Vol. **150**, No. 1, p. 307—314 (1988).

R. Dorbez, B. Yangui, M. Ben Salem, M. H. Pischedda and G. Nihoul
Three-Dimensional Effects in the Electronic Images of Hexagonal Neodymium Oxide
(61.14; S10.1)
(a) Vol. **110**, No. 1, p. 43—54 (1988).

B. A. Budkevich, I. A. Ges, S. P. Zhvavyi, G. D. Ivlev, and V. A. Pilipovich
Modification of Electrochromic WO_3 Films by Pulsed Ultraviolet Laser Irradiation
(61.80 and 73.60; 68.60; S10.1)
(a) Vol. **110**, No. 1, p. 155—164 (1988).

Yong Yan, Fuquan Ji, Jun Chen, and Duan Feng
Imaging of the SOI Interface by High Resolution Electron Microscopy
(68.48; S6; S10.1)
(a) Vol. **110**, No. 1, p. K25—K28 (1988).

Z. T. Al-Dhhan and C. A. Hogarth
DC Conduction in Thin Films of SiO/In_2O_3 before and after Electroforming
(73.60; S10.1)
(a) Vol. **110**, No. 1, p. K39—K44 (1988).

H. Krause and H.-P. Bär
Charge Injection into SiO_2 Films at Fields between 1 and 3 MV cm^{-1} after Electrical Stress
(73.40; 73.60; S10.1)
(a) Vol. **110**, No. 2, p. 537—547 (1988).

A. Stesmans and G. de Vos
MgO Powder Containing Low Concentrations of Isotopically Pure $^{52}Cr^{3+}$ Ions. Its Application as ESR Marker
(76.30; 61.70; S10.1)
(a) Vol. **110**, No. 2, p. 615—629 (1988).

S10.15. Mixed crystals

Z. T. Al-Dhhan, C. A. Hogarth, and N. Riddleston
The Optical Absorption Edge in Thin Amorphous Oxide Films Based on Cerium Dioxide
(78.65; 68.55; S10.15)
(b) Vol. **145**, No. 1, p. 145—149 (1988).

M. Hidaka, H. Fujii, Y. Horie, and S. Mase
Superconductor–Semiconductor Transition in $BaPb_{1-x}Bi_xO_3$
(72.60 and 74.10; S10.15)
(b) Vol. **145**, No. 2, p. 649—658 (1988).

M. A. Hassan, M. N. Khan, and C. A. Hogarth
Electrical and Optical Properties of V_2O_5–GeO_2–NiO Glasses
(72.20 and 78.20; 78.30; S10.15)
(a) Vol. **105**, No. 2, p. 609—616 (1988).

R. Herrmann, N. Kubicki, H.-U. Müller, H. Dwelk, N. Pruss, A. Krapf, and L. Rothkirch
Effect of Fluorine on the Superconductivity of $YBa_2Cu_3O_{7-\delta}$
(74.70; S10.15)
(b) Vol. **146**, No. 1, p. K25—K28 (1988).

G. Krabbes, M. Ritschel, H. Vinzelberg, E. Wolf, H. Eschrig, W. Holzhäuser, D. Elefant, and U. Wiesner
Relationship between Oxygen Chemical Potential, Annealing Conditions, and Superconducting Properties of $YBa_2Cu_3O_x$
(74.70; S10.15)
(b) Vol. **146**, No. 1, p. K29—K33 (1988).

M. N. Khan, M. A. Hassan, and C. A. Hogarth
The Electronic and Optical Properties of Germanium Tellurite Glasses Containing Various Transition Metal Oxides
(72.80; 76.30; 78.30; S10.15)
(a) Vol. **106**, No. 1, p. 191—200 (1988).

B. R. Singh, H. S. Kothari, O. P. Daga, J. K. Singh, G. S. T. Rao, J. D. Jain, and W. S. Khokle
X-Ray Diffraction and EDS Analysis of $Y_1Bs_2Cu_3O_{9-y}$ Superconducting Compound
(61.50 and 74.70; S10.15)
(a) Vol. **106**, No. 1, p. K53—K56 (1988).

K. Bente and D. Siebert
EPR Investigations on High Temperature Superconducting $YBa_2Cu_3O_{7-x}$
(74.70 and 76.30; S10.15)
(a) Vol. **106**, No. 1, p. K57—K59 (1988).

J. Petzelt, S. Kamba, S. Pačesová, E. Pollert, J. Šrámek, O. Smrčková, and D. Sýkorová
Infrared Reflectivity of High-T_c $YBa_2Cu_3O_{7-x}$ and Related Ceramics
(63.20 and 78.30; 74.70; S10.15)
(b) Vol. **146**, No. 2, p. 743—755 (1988).

K. ROGACKI, C. SUŁKOWSKI, A. J. ZALESKI, Z. BUKOWSKI, B. GREŃ, R. HORYŃ, E. TROJNAR, and J. KLAMUT
Upper Critical Fields of High-T_c Superconductor $YBa_2Cu_3O_{7-\delta}$: Possibility of 350 T
(74.70; S10.15)
(b) Vol. **146**, No. 2, p. K103—K107 (1988).

Z. FRAIT, D. FRAITOVÁ, E. POLLERT, and L. PŮST
Observation of the Meissner-Ochsenfeld Field of the High-Temperature Superconductor Y–Ba–Cu–O by Menas of FREPR
(74.30 and 76.30; S10.15)
(b) Vol. **146**, No. 2, p. K119—K123 (1988).

Z. HENKIE, P. J. MARKOWSKI, R. HORYŃ, Z. BUKOWSKI, and J. KLAMUT
The Thermopower Sign Inversion for the High T_c Superconductor $YBa_2Cu_3O_{7-\delta}$
(72.15 and 74.70; S10.15)
(b) Vol. **146**, No. 2, p. K131—K134 (1988).

OM PARKASH, CH. DURGA PRASAD, and DEVENDRA KUMAR
Dielectric Properties of Lanthanum Lead Cobalt Titanium Oxide $(La_{1-x}Pb_x)(Co_{1-x}Ti_x)O_3$ $(0.50 \leqq x \leqq 0.90)$
(77.20 and 77.40; 77.80; S10.15)
(a) Vol. **106**, No. 2, p. 627—633 (1988).

H. PFEIFFER
On Oxygen Ordering in the High Temperature Superconductor Y–Ba–Cu–O
(64.70 and 74.70; S10.15)
(a) Vol. **106**, No. 2, p. K161—K164 (1988).

T. K. KOPEĆ
Charge Density Waves in Systems with Condensed Local Pairs and Superconductivity in $La_{2-x}M_xCuO_4$ Compounds
(74.20; S10.15)
(b) Vol. **147**, No. 1, p. K37—K42 (1988).

B. LIPPOLD, H. SCHMIEDEL, and H. BÖRNER
Evidence of Frozen Magnetic Flux in $Y_1Ba_2Cu_3O_{7-x}$ Ceramics
(74.30; 74.70; S10.15)
(b) Vol. **147**, No. 1, p. K43—K45 (1988).

Y. H. LIANG, J. CHEN, L. C. WANG, Y. YIAN, X. Q. PAN, M. WEI, G. Q. WANG, X. JING, S. Y. DING, D. FENG, and Q. R. ZHANG
A HREM Study on the Defects in Y–Ba–Cu–O Superconductors
(61.60 and 74.70; S10.15)
(a) Vol. **107**, No. 1, p. 63—72 (1988).

B. DOMENGÈS, M. HERVIEU, C. MICHEL, A. MAIGNAN, and B. RAVEAU
$La_3Ba_3Cu_6O_{14+x}$: A Single Phase or a Mixture?
(61.60; 64.80; S10.15)
(a) Vol. **107**, No. 1, p. 73—84 (1988).

M. HROBÁR and I. TRAVĚNEC
Time-Domain Spectroscopy. Method for Investigation of YSZ
(73.40; 77.50; S10.15)
(a) Vol. **107**, No. 1, p. 235—238 (1988).

S. X. Dou, A. J. Bourdillon, X. Y. Sun, H. K. Liu, J. P. Zhou, N. Savvides, C. C. Sorrell, and K. E. Easterling
Dependence of the Superconducting Transition Temperature on Radii of Alkali and Alkaline Earth Dopants in $Y_1Ba_2Cu_3O_{7-x}$
(74.70; S10.15)
(b) Vol. **147**, No. 2, p. K153—K157 (1988).

R. Herrmann, G. Nachtwei, H. Krüger, P. Svoboda, C. Breitlow, R. Holzberger, H. Dwelk, and A. Jacobi
Measurements of the Upper Critical Fields of Single Phase $YBa_2Cu_3O_{7-\delta}$
(74.70; S10.15)
(b) Vol. **147**, No. 2, p. K159—K163 (1988).

H. W. Zandbergen and G. Thomas
HREM Investigations of Defects in Sintered $YBa_2Cu_3O_{7-\delta}$
(61.16 and 61.70; 74.70; S10.15)
(a) Vol. **107**, No. 2, p. 825—843 (1988).

O. Smrčková, D. Sýkorová, J. Dominec, L. Smrčka, and P. Vašek
Resistivity and Magnetic Properties of Bi–Sr–Ca–Cu–O Superconductor
(74.70; S10.15)
(a) Vol. **107**, No. 2, p. K135—K138 (1988).

B. B. Boiko, F. P. Korshunov, G. V. Gatalskii, A. I. Akimov, V. I. Gatalskaya, S. E. Demyanov, and E. K. Stribuk
Radiation Effect on the Superconductivity in the Y–Sm–Ba–Cu–O Ceramic System
(74.70; S10.15)
(a) Vol. **107**, No. 2, p. K139—K144 (1988).

K. V. Rao, D.-X. Chen, H. Szymczak, M. Baran, P. Przysłupski, and R. Puźniak
On Magnetic Properties of the Superconductor $Y_1Ba_2Cu_3O_x$
(74.70 and 75.60; S10.15)
(a) Vol. **107**, No. 2, p. K145—K150 (1988).

D. Mercurio, B. Frit, G. Harburn, B. H. Parry, R. P. Williams, and R. J. D. Tilley
Diffraction Patterns from Ordered Intergrowth Phases in the Bi_2O_3–TeO_2 System
(61.14 and 64.80; S10.15)
(a) Vol. **108**, No. 1, p. 111—122 (1988).

Changyong Sun, Wenhui Su, Hongjian Liu, Jianshi Zhou, Fatang Wang, and Yupu Zhang
Magnetic and Superconducting Properties of $REBa_2Cu_3O_{9-y}$ (RE = Sm, Eu, Gd, Dy, Ho, Er, Tm)
(74.70; 75.20; S10.15)
(a) Vol. **108**, No. 1, p. 337—342 (1988).

M. Chybicki, J. Liwo, and K. Trzebiatowski
Modification of the Electrical Surface Conductivity of Lead–Silicate Glasses by H_2^+ Ion Bombardment
(73.25; S10.15)
(a) Vol. **108**, No. 2, p. K105—K107 (1988).

Hong Zhang, Xiao-gan Wang, and Yao-xian Fu
Study of the Movement of Oxygen Vacancies in the Orthorhombic Phase of $YBa_2Cu_3O_{7-x}$ by Positron Doppler Broadening Spectroscopy
(61.70; 74.70; 78.70; S10.15)
(a) Vol. **109**, No. 1, p. 135—139 (1988).

P. LOBOTKA and F. GÖMÖRY
The Complex AC Susceptibility of Superconducting Y–Ba–CuO Thin Film and Bulk Samples
(73.60 and 75.30; 74.70; S10.15)
(a) Vol. **109**, No. 1, p. 205—210 (1988).

YUHUAN XU, ZHONGRONG LI, CHAORUI LI, XIANDUAN LIN, WU LI, WAMMING LAO, HONG WANG, and GUOWEI CHEN
The International Strain Effect on T_c in the Y–Ba–Al–Cu Oxide Superconductor
(74.70; S10.15)
(a) Vol. **109**, No. 1, p. 301—306 (1988).

I. O. TROYANCHUK, L. A. BASHKIROV, A. A. SHEMYAKOV, and V. K. PROKOPENKO
NMR-Spectroscopy Study of $Ca(Cu_{3-x}Mn_x)Mn_4O_{12}$ Perovskites
(76.60; S10.15)
(a) Vol. **109**, No. 1, p. K59—K61 (1988).

E. M. GOLOLOBOV, A. S. GUREVICH, E. L. MAGER, Z. V. MEZHEVICH, N. A. PRYTKOVA, ZH. M. TOMILO, D. M. TURTSEVICH, and N. M. SHIMANSKAYA
Orthorhombic Lattice Distortion of $YBa_2Cu_3O_{7-y}$ and the Superconducting Transition Temperature
(61.70 and 74.70; S10.15)
(b) Vol. **149**, No. 2, p. K145—K148 (1988).

N. GUSKOS, TH. LEVENTOURI, CH. TRIKALINOS, and M. CALAMIOTOU
EPR Spectra of Cu^{2+} Ions in Orthorhombic Complexes of the $YBa_2Cu_3O_x$ Compound
(74.70 and 76.30; S10.15)
(b) Vol. **149**, No. 2, p. K157—K162 (1988).

W. GAWALEK, W. MICHALKE, H. BRUCHLOS, T. EICK, R. HERGT, and G. SCHMIDT
Preparation of High-T_c Y–Ba–Cu–O and Bi–Sr–Ca–Cu–O Thin Films by RF-Magnetron Sputtering
(68.55; 74.70; S10.15)
(a) Vol. **109**, No. 2, p. 503—509 (1988).

XIANYI ZHOU, HUILIN JIANG, QIRUI ZHANG, GUOQING PAN, and QIWEI LONG (C. W. LUNG)
Positron Trapping in $YBa_2Cu_3O_x$ Superconductors
(74.70 and 78.70; S10.15)
(a) Vol. **109**, No. 2, p. K129—K132 (1988).

HONG ZHANG, XIAO-GANG WANG, and YAO-XIAN FU
Positron Annihilation in Thermally Quenched $YBa_2Cu_3O_{7-x}$
(74.70 and 78.70; S10.15)
(a) Vol. **109**, No. 2, p. K133—K138 (1988).

A. GHOSH
Fundamental Absorption Edge in Bismuth–Vanadate Glasses
(78.20; S10.15)
(a) Vol. **110**, No. 2, p. 651—656 (1988).

C. S. SUNANDANA

On the Normal-State Electrical Resistivity of $YBa_2Cu_3O_7$ and Other Oxide Superconductors
(74.70; S10.15)
(a) Vol. **110**, No. 2, p. K85—K88 (1988).

HUIMIN SHAO, HUAQIN WANG, YUANFU HSIA, RONGCHUAN LIU, and XIN JIN

Tetragonal-to-Orthohombic Phase Transition of Fe-Substituted $YBa_2Cu_3O_7$
(64.70 and 74.70; S10.15)
(a) Vol. **110**, No. 2, p. K89—K91 (1988).

A. K. PRADHAN, P. C. JANA, and B. K. ROUL

Bismuth Strontium Calcium Copper Oxide High-T_c Superconductors from Nitrate Solutions
(74.70; S10.15)
(a) Vol. **110**, No. 2, p. K93—K96 (1988).

A. N. GURYANOV, E. M. DIANOV, V. M. KIM, V. M. MASHINSKII, V. B. NEUSTRUEV, V. A. TIKHOMIROV, and V. F. KHOPIN

UV and γ-Induced Paramagnetic Colour Centres in Germanium-Doped Silica Glass
(78.50; 76.30; S10.15)
(a) Vol. **110**, No. 2, p. K107—K110 (1988).

S11. Other higher compounds

N. Kristoffel and P. Konsin
Vibronic Theory of Structural Phase Transitions and Displacive Ferroelectrics **(Review Article)**
(63.75 and 77.80; 64.60; S8; S11)
(b) Vol. **149**, No. 1, p. 11—40 (1988).

Yi-Yang Zhou
Theoretical Study of the Optical Absorption, Mössbauer, and EPR Spectra of Fe^{2+} Ion in Orthopyroxene
(71.70; 76.30; 76.80; 78.30; S11)
(b) Vol. **145**, No. 1, p. 227—232 (1988).

H. Pykacz, Z. Czapla, and J. Mróz
Dilatometric Study of the Phase Transitions in NH_4HSeO_4
(64.70; 65.70; S11)
(a) Vol. **105**, No. 1, p. K33—K35 (1988).

M. Salagram
Infrared Spectrum of VO^{2+} Entity in $K_2C_2O_4 \cdot H_2O$ Crystals
(78.30; S11)
(a) Vol. **105**, No. 2, p. K161—K164 (1988).

M. Dębowska, J. Ch. Abbé, and G. Duplâtre
Positron Annihilation Studies in Zeolites
(61.80 and 78.70; S11)
b) Vol. **146**, No. 1, p. 91—96 (1988).

S. K. S. Chauhan, R. Agrawal, P. C. Pathak, B. R. K. Gupta, and R. P. Goyal
Thermal Effects on Thermophysical Properties of Alkali Cyanide Crystals
(65.50; 62.20; 65.70; S11)
(b) Vol. **146**, No. 1, p. 97—103 (1988).

A. N. Timoshevskii
The Peculiarities of the Electronic Structure of the $La_{2-x}Sr_xCuO_4$ Compound
(71.20 and 71.25; S11)
(b) Vol. **146**, No. 1, p. 161—171 (1988).

A. K. Mukherjee and D. Neogy
Magnetism and Crystal Field Analysis of Gadolinium Bromate Enneahydrate Crystal
(71.70 and 75.50; S11)
(b) Vol. **146**, No. 1, p. K45—K50 (1988).

B. T. Deshmukh, S. V. Bodade, and S. V. Moharil
Thermoluminescence in Synthetic Langbeinite
(78.60; 78.40; S11)
(a) Vol. **106**, No. 1, p. K107—K111 (1988).

R. P. Goyal, O. P. Gupta, and B. R. K. Gupta
Theory of Compression in Some Fluorite Structure Crystals
(64.30; 62.20; 65.50; S9.12; S11)
(b) Vol. **146**, No. 2, p. 475—480 (1988).

M. P. Lisitsa, I. D. Turyanitsa, and I. V. Fekeshgazi
Non-Linear Light Absorption Spectroscopy in Sb–S–I Crystals and Glasses
(78.40; S11)
(b) Vol. **146**, No. 2, p. p. 723—726 (1988).

Yi-Yang Zhou
A Theoretical Study of EPR Spectra in Cu^{2+}- and Mn^{2+}-Doped $CaCd \cdot (CH_3COO)_4 \cdot 6\,H_2O$
(71.70 and 76.30; S11)
(b) Vol. **147**, No. 1, p. 273—280 (1988).

K. B. Garg, R. K. Singhal, K. V. R. Rao, U. Chandra, H. S. Chauhan, J. Singh, K. S. Jerath, and D. C. Jain
XANES Study of Copper Mixed Valence and Mechanism of Superconductivity in $YBa_2Cu_3O_{6.5+\delta}$ System
(74.10; 78.70; S11)
(b) Vol. **147**, No. 1, p. 343—348 (1988).

S. F. Sagnowski, S. A. Hodorowicz, and B. Borzecka-Prokop
^1H and ^{19}F Magnetic Resonance Study on Molecular Dynamics in Nickel(II) Hexaammino-Tetrafluoroborate
(76.60; S11)
(a) Vol. **107**, No. 1, p. 347—353 (1988).

S. K. Misra and C. Rudowicz
Effect of Monoclinic Symmetry on the EPR Spectra of Gd^{3+}-Doped Hydrated Single Crystals of Rare-Earth Trichlorides
(76.30; S11)
(b) Vol. **147**, No. 2, p. 677—684 (1988).

A. B. Vassilikou-Dova and G. Lehmann
EPR of V^{4+} in Apophyllite
(61.70 and 76.30; S11)
(b) Vol. **147**, No. 2, p. 691—697 (1988).

W. S. Tse, C. S. Fang, K. F. Pai, and W. Lih
A Raman Investigation of the Polymorphism of Potassium Nitrite
(78.30; 63.20; S11)
(b) Vol. **147**, No. 2, p. 791—795 (1988).

Th. Schmidt, J. Köhler, C. Kryschi, and D. Schmid
Sensitized Luminescence of Pure and Doped $NaNO_2$ Single Crystals. III. Excitation Spectroscopy and Time Resolved Trap Fluorescence
(78.55; 71.35; 71.55; S11)
(b) Vol. **147**, No. 2, p. 797—809 (1988).

P. Werner and J. Heydenreich
Experimental Study of Structure Modification of $2M_1$-Muscovite by High-Resolution Electron Microscopy
(61.60; 61.16; S11)
(a) Vol. **107**, No. 2, p. 807—816 (1988).

G. I. Bersuker
Dynamics in the System of Dipolar Impurity Centers in Crystals
(63.20 and 71.55; S11)
(b) Vol. **148**, No. 1, p. 243—249 (1988).

N. M. Pladika and W. Salejda
The Improper Ferroelastic Phase Transition in Superionic $Rb_3H(SeO_4)_2$ Crystals
(64.70; 61.50; S11)
(b) Vol. **148**, No. 2, p. 473—481 (1988).

M. Mačkowiak and P. Kozioł
Effect of Pressure on the Symmetric Hydrogen Bond in $(CCl_3COO)_2HK$
(76.60; 62.50; S11)
(a) Vol. **108**, No. 2, p. 739—745 (1988).

J. Nandi and D. Neogy
Magnetic Behavior and Crystal Field of Pr^{3+} in Praseodymium Selenate Octahydrate
(71.70 and 75.30; 75.40; S11)
(b) Vol. **149**, No. 1, p. 275—281 (1988).

S. K. Asadov, E. A. Zavadskii, Yu. D. Zavorotnev, and B. M. Todris
Low-Temperature Magnetic Transformations in Ferrous Fluorsilicate Induced by Pressure
(75.30; S11)
(a) Vol. **109**, No. 1, p. 307—312 (1988).

Sh. M. Efendiev, V. E. Bagiev, R. A. Aliev, and E. R. Mustafaev
Nonlinear *I–U* Characteristics and Photoelectret State in Sillenite-Type Crystals
(73.40 and 85; 73.30; S11)
(a) Vol. **109**, No. 1, p. 345—357 (1988).

Wen-Cheng Zheng
On the Equations of State of $ZnSiF_6 \cdot 6\ H_2O:Ni^{2+}$ Crystal Determined from the Pressure Dependence of EPR Spectra
(64.30; 71.70; 76.30; S11)
(b) Vol. **149**, No. 2, p. K105—K109 (1988).

Ye Tao, Mei-shen Hu, and Duan Feng
Direct Observation of α–β Phase Transition in $KNbW_2O_9$ by Transmission Electron Microscopy
(64.70; 61.16; S11)
(a) Vol. **109**, No. 2, p. 435—444 (1988).

V. S. Jayakumar, G. Sekar, P. Rajagopal, and G. Aruldhas
IR and Polarized Raman Spectra of $(NH_4)_2Mg(SO_4)_2 \cdot 6\ H_2O$
(78.30; S11)
(a) Vol. **109**, No. 2, p. 635—640 (1988).

J. Kuriata, L. Sadłowski, B. Bojanowski, J. Walczak, M. Kurzawa, and J. Pichet
The Magnetic Resonance of $Fe_2V_{4-x}Mo_xO_{13}$ Type Phase
(76.30; 75.50; S11)
(a) Vol. **109**, No. 2, p. K139—K143 (1988).

P. Stefányi, A. Orendáčová, and A. Feher
The Influence of a Magnetic Field on the Heat Capacity of $CsDy(MoO_4)_2$
(75.30 and 75.40; S11)
(b) Vol. **150**, No. 1, p. K63—K66 (1988).

J. K. Liang, S. S. Xie, Y. L. Zhang, G. C. Che, G. H. Rao, J. Q. Huang, and X. R. Cheng
Crystal Structure and Superconductivity of the $Tl_{2+x}Ba_2Ca_{2-x}Cu_3O_{10}$ Compound
(61.60 and 74.70; S11)
(a) Vol. **110**, No. 1, p. 55—60 (1988).

H. Hempel, H. Maack, and G. Sorge
Ferroelastic Phase Transition in $LiRb_4H(SO_4)_3 \cdot H_2SO_4$. Landau-Theory and Experimental Results
(64.70 and 77.80; S11)
(a) Vol. **110**, No. 2, p. 459—465 (1988).

C. H. Westphal and C. C. Becerra
Magnetic Transitions in $Mn_{1-x}Ni_x(HCOO)_2 \cdot 2\,H_2O$
(75.50; S1.2; S11)
(a) Vol. **110**, No. 2, p. K97—K100 (1988).

S11.1. Salts of oxy-acids

E. D. Yakushkin and V. N. Anisimova
Phase Transition in Monoclinic KD_2PO_4 Crystal
(64.70; 77.20; S11.1)
(a) Vol. **105**, No. 1, p. 139—143 (1988).

A. A. Sukhovskii, I. P. Aleksandrova, Yu. N. Moskvich, and O. V. Rozanov
Phase Diagram of the Monoclinic and Orthorhombic Ammonium Hydrogen Selenate
(64.70; 76.60; S11.1)
(a) Vol. **105**, No. 1, p. 145—152 (1988).

H. Engelmann, W. Gatzweiler, I. Dézsi, and U. Gonser
A Comparative Study of Impurity Defects in Crystalline and Amorphous $LiNbO_3$
(71.55; 76.80; S11.1)
(a) Vol. **105**, No. 1, p. 219—230 (1988).

B. S. V. S. R. Acharyulu and V. S. Kishan Kumar
Optical and Thermoluminescence Studies on Thallium-Doped Suprapure Cesium Sulphate
(78.50 and 78.60; S11.1)
(a) Vol. **105**, No. 1, p. 303—307 (1988).

T. Bremer, W. Heiland, A. Klekamp, and E. Krätzig
SIMS Investigations of Iron Profiles in $LiNbO_3$:Ti Waveguides
(66.30 and 79.20; 78.20; S11.1)
(a) Vol. **105**, No. 1, p. K17—K20 (1988).

B. PILAWA, G. HESS, H. G. KAHLE, and A. KASTEN
Linear Birefringence Investigations of the Jahn-Teller Distorted Phase in the Mixed Crystal Systems (Tb_x, Tm_{1-x})VO_4 and (Tb_x, Yb_{1-x})VO_4
(78.20; 64.60; S11.1)
(b) Vol. **145**, No. 2, p. 729—739 (1988).

S. L. CHAPLOT
Potential Barriers for Molecular Rotations in Solids
(61.50; S11.1)
(b) Vol. **145**, No. 2, p. K89—K93 (1988).

R. SOMMERFELDT, L. HOLTMANN, E. KRÄTZIG, and B. C. GRABMAIER
Influence of Mg Doping and Composition on the Light-Induced Charge Transport in $LiNbO_3$
(61.70; 72.40; 78.20; S11.1)
(a) Vol. **106**, No. 1, p. 89—98 (1988).

A. A. ALYBAKOV, V. A. GUBANOVA, K. KUDABAEV, and K. SHARSHEEV
Optical Absorption and EPR Spectra of Cu^{2+} Ions in $LiKSO_4$ Single Crystals
(76.30 and 78.50; S11.1)
(b) Vol. **146**, No. 2, p. K135—K139 (1988).

XIAOQING PAN and DUAN FENG
Direct Observation of Nucleation of Discommensurations in Barium Sodium Niobate
(61.50; S11.1)
(a) Vol. **106**, No. 2, p. K117—K121 (1988).

M. SALAGRAM, K. SEETHARAM, P. SIDDAMBARY, and T. S. N. MURTHY
Infrared Characterisation of the Oxyanion in Kainite
(78.30; S11.1)
(a) Vol. **106**, No. 2, p. K185—190 (1988).

A. MAYR, W. PAUL, and H. G. KAHLE
$Er_2O_2SO_4$ — An Ising System with Six Magnetic Phases
(75.50; 75.25; 75.30; S11.1)
(b) Vol. **147**, No. 1, p. 383—390 (1988).

M. D. FONTANA, A. RIDAH, and G. E. KUGEL
Frequency Dependence of the Soft Mode in $KNbO_3$
(77.80 and 78.30; 63.20; S11.1)
(b) Vol. **147**, No. 1, p. 441—447 (1988).

YU. V. PISAREVSKII, I. M. SILVESTROVA, R. VOSZKA, Á. PÉTER, I. FÖLDVÁRI, and J. JANSZKY
Elastic and Acoustic Properties of $ZnWO_4$ Single Crystal
(62.20 and 62.65; S11.1)
(a) Vol. **107**, No. 1, p. 161—164 (1988).

V. D. SAJFERT, LJ. D. MAŠKOVIĆ, R. P. DJAJIĆ-JOVANOVIĆ, and U. F. KOZMIDIS-LUBURIĆ
Phase Transition in 1D KDP-Type Ferroelectrics
(64.60; 77.80; S11.1)
(b) Vol. **148**, No. 1, p. 81—90 (1988).

H. Donker, W. M. A. Smit, and G. Blasse
Self-Trapped Exciton Emission of $SnSO_4$
(78.55; S11.1)
(b) Vol. **148**, No. 1, p. 413—419 (1988).

K. Bielecki
A High Pressure EPR Study of Cr^{3+} Doped Deuterated Potassium Aluminium Alum
(76.30; 62.50; S11.1)
(b) Vol. **148**, No. 1, p. K53—K56 (1988).

V. A. Ivanshin, I. N. Kurkin, and G. Völkel
The Influence of Optical Modes on the Spin–Lattice Relaxation of Paramagnetic Centres in Irradiated Rochelle Salt Single Crystals
(76.30 and 77.80; S11.1)
(b) Vol. **148**, No. 1, p. K61—K63 (1988).

A. E. Aliev, A. Sh. Akramov, L. N. Fershtat, and P. K. Khabibullaev
Mechanism of a Superion Phase Transition in α-$LiIO_3$
(64.60; 77.40; S11.1)
(a) Vol. **108**, No. 1, p. 189—196 (1988).

T. R. Volk and N. M. Rubinina
X-Ray and UV Influence on the Optical Absorption Spectra of the Non-Photorefractive Lithium Niobate
(78.50; S11.1)
(a) Vol. **108**, No. 1, p. 437—442 (1988).

S. Fries, P. Hertel, and H. P. Menzler
Extraordinary versus Ordinary Refractive Index Change in Planar $LiNbO_3$:Ti Waveguides
(78.65; S11.1)
(a) Vol. **108**, No. 1, p. 449—455 (1988).

R. Kawashima and T. Matsuda
Temperature Dependence of the Dielectric Constant in Samarium Nitrate Crystal
(77.80; S11.1)
(a) Vol. **108**, No. 1, p. K73—K75 (1988).

O. Eibl, P. Pongratz, P. Skalicky, and H. Schmelz
Dislocations in $BaTiO_3$ Ceramics
(61.70; S11.1)
(a) Vol. **108**, No. 2, p. 495—502 (1988).

A. J. de Vries, J. P. M. van Vliet, and G. Blasse
Mechanism of Concentration Quenching of Gd^{3+} and Eu^{3+} Emission in Perovskite-Like Aluminates
(71.70 and 78.55; S11.1)
(b) Vol. **149**, No. 1, p. 391—401 (1988).

M. Verwerft, G. van Tendeloo, J. van Landuyt, W. Coene, and S. Amelinckx
Trial Model for the Tilting Scheme in $AgNbO_3$ Derived by Electron Diffraction and Imaging
(61.14; S11.1)
(a) Vol. **109**, No. 1, p. 67—78 (1988).

J. Minge and T. Krajewski
EPR Study of SO_4^- Radicals in $K_4LiH_3(SO_4)_4$ Crystals
(64.70; 76.30; S11.1)
(a) Vol. **109**, No. 1, p. 193—198 (1988).

Jun Chen and Duan Feng
In Situ TEM Studies of Para–Ferro Phase Transitions in $NaNbO_3$
(64.70; 77.80; S11.1)
(a) Vol. **109**, No. 2, p. 427—434 (1988).

W. Lottermoser and H. Fuess
Magnetic Structure of the Orthosilicates Mn_2SiO_4 and Co_2SiO_4
(75.25; S11.1)
(a) Vol. **109**, No. 2, p. 589—595 (1988).

R. Zângaro and J. Gasiot
A Surface Effect in $CaSO_4$:Mn/PVDF Mixtures
(68.45 and 78.60; S11.1)
(a) Vol. **109**, No. 2, p. K109—K113 (1988).

R. V. Vedrinskii, L. A. Bugaev, and I. G. Levin
Single and Multiple Electron Scattering Description in X-Ray Absorption Spectrum Theory and the Problem of Solid Structure Investigation
(78.70; 61.10; 71.10; S9.11; S10.1; S11.1)
(b) Vol. **150**, No. 1, p. 307—314 (1988).

V. A. Ganshin, Yu. N. Korkishko, and T. V. Morozova
Properties of Proton Exchanged Optical Waveguiding Layers in $LiNbO_3$ and $LiTaO_3$
(61.70 and 78.20; 61.10; S11.1)
(a) Vol. **110**, No. 2, p. 397—402 (1988).

R. Mizeris, J. Grigas, V. Samulionis, V. Skritski, A. I. Baranov, and L. A. Shuvalov
Microwave and Ultrasonic Investigations of Superionic Phase Transitions in $CsDSO_4$ and $CsDSeO_4$
(64.70; 62.65; 77.20; S11.1)
(a) Vol. **110**, No. 2, p. 429—436 (1988).

M. Czajkowski, M. Drozdowski, and M. Kozielski
Propagation of Plane Elastic Waves in $LiCsSO_4$ Single Crystals
(62.30 and 63.20; S11.1)
(a) Vol. **110**, No. 2, p. 437—442 (1988).

R. Poprawski
Comments on the Results of Dilatometric Studies of NH_4HSeO_4 Crystals
(64.70 and 77.80; S11.1)
(a) Vol. **110**, No. 2, p. K73—K76 (1988).

S11.2. Spinels, garnets, ferrites

A. I. Eatah, A. A. Ghani, and E. El Faramawy
Effect of Sintering Temperature on the Electrical Conductivity and Thermoelectric Power of $CuFe_2O_4$
(72.20; S11.2)
(a) Vol. **105**, No. 1, p. 231—233 (1988).

F. Leccabue, O. Arés Muzio, and J. L. Sánchez Llamazares
Shape Anisotropy Effect in Barium Ferrite Interacting Submicronic Particles
(75.30; S11.2)
(a) Vol. **105**, No. 1, p. 285—290 (1988).

A. C. Bódi and I. Ileo
The Effect of Neutron Irradiation on the Cation Distribution in Iron-Deficient Ni–Zn Ferrites
(75.50; S11.2)
(a) Vol. **105**, No. 2, p. 573—577 (1988).

Y. Srinivasava Rao and C. S. Sunandana
Characterization of Diamagnetically Substituted (Na^+, Te^{4+}, Cl^-) Lithium Ferrite
(75.50; 61.60; S11.2)
(a) Vol. **106**, No. 1, p. 217—225 (1988).

R. A. Kosiński
On the Stabilization of a Vertical Bloch Line Pair in a Domain Wall in an Ion-Implanted Garnet Film
(75.60; S11.2)
(a) Vol. **106**, No. 1, p. 227—231 (1988).

E. Z. Katsnelson and A. G. Karoza
On the Irreversibility of the Processes during Photomagnetization of Ferrites
(78.20; 72.40; S11.2)
(a) Vol. **106**, No. 1, p. 277—282 (1988).

A. H. Wafik and S. A. Mazen
Effect of Mg–Fe Replacement in the System $Mg_xZn_{0.3}Fe_{2.7-x}O_{4\pm\delta}$ on Barkhausen Jumps
(75.60; S11.2)
(a) Vol. **106**, No. 1, p. K65—K69 (1988).

S. Ligenza, B. Paluchowska, and M. Konwicki
Neutron Diffraction Studies of 0.2 Titanium Substituted Lithium Ferrite
(75.50; S11.2)
(a) Vol. **106**, No. 1, p. K71—K74 (1988).

Z. Kleszczewski and J. Bodzenta
Phonon–Phonon Interaction in Gadolinium–Gallium Garnet Crystals
(63.20; S11.2)
(b) Vol. **146**, No. 2, p. 467—474 (1988).

F. G. Bass, N. N. Nasonov, and O. V. Naumenko
Solitary Electromagnetic Spin Waves in a Ferrite Waveguide
(75.30; S11.2)
(b) Vol. **146**, No. 2, p. K145—K151 (1988).

M. Guyot, T. Merceron, V. Cagan, and A. Messekher
Mobility and/or Damping of the Domain Wall. Comparison between Experimental and Theoretical Values in Ferrimagnets
(75.60; S11.2)
(a) Vol. **106**, No. 2, p. 595—612 (1988).

V. M. Talanov
Specific Points on Spinel Phase Diagrams
(64.70; S11.2)
(a) Vol. **106**, No. 2, p. K129—K133 (1988).

S. Ram
Magnetic and Magnetic Resonance Studies on Oxygen-Deficient $SrFeO_{3-y}$ Perovskites
(61.40; 75.60; 76.30; S11.2)
(a) Vol. **107**, No. 1, p. 55—62 (1988).

S. E. Sarkisov and A. A. Kaminskii
Optical Phonon Spectroscopy of Heterovalent Disordered $Ca_3(Nb, Ga)_2Ga_3O_{12}$ Crystals with Garnet Structure
(78.30; S11.2)
(a) Vol. **107**, No. 1, p. 365—371 (1988).

A. D. Stalios, P. Delavignette, and R. de Batist
Transmission Electron Microscopy Observation of Crystal Defects in a Natural Diopside Mineral
(61.16 and 61.70; S11.2)
(a) Vol. **107**, No. 2, p. 759—767 (1988).

K. Melzer and A. Martin
A Mössbauer Study of the Reaction between Barium Carbonate and Iron(III) Oxide
(76.80; S11.2)
(a) Vol. **107**, No. 2, p. K163—K168 (1988).

P. V. Reddy
High-Temperature Elastic Behaviour of Mn–Mg Mixed Ferrites
(62.20; 75.50; S11.2)
(a) Vol. **108**, No. 2, p. 607—611 (1988).

C. de Francisco, J. M. Muñoz, J. Ayala, and J. I. Iñiguez
Magnetic Disaccommodation in Ferrous Zinc Ferrites
(75.50; S11.2)
(a) Vol. **108**, No. 2, p. 721—731 (1988).

S. Ligenza and M. Konwicki
Cluster Structure of the Li–Ti–Zn Ferrite System Studied by Neutron Scattering
(75.50; 61.12; S11.2)
(a) Vol. **109**, No. 1, p. 319—327 (1988).

R. Höhne
A Study of Induced Magnetic Anisotropy in Gallium–Iron Ferrites above Room Temperature
(75.30 and 75.50; S11.2)
(a) Vol. **109**, No. 2, p. 575—587 (1988).

H. R. Asatryan
ESR Study of Mo^{3+} Containing Mixed Yttrium–Scandium–Aluminium Garnet Single Crystals
(76.30; S11.2)
(b) Vol. **150**, No. 1, p. 253—259 (1988).

E. Z. Katsnelson, A. G. Karoza, and L. A. Meleshchenko
Photoinduced Magnetization Effect on IR Spectra of Ferrites
(75.50 and 78.30; S11.2)
(a) Vol. **110**, No. 1, p. K51—K56 (1988).

S12. Organic compounds

E. HEINER, S. L. DRECHSLER, and J. MALEK
On the Charge Structure of Polarons in Trans-Polyacetylene
(71.38 and 71.50; S12)
(b) Vol. **145**, No. 1, p. 233—240 (1988).

M. UMAR, M. I. HAQUE, and R. J. SINGH
EPR Study of Mn^{2+} Doped Single Crystal of α-$Ca(HCOO)_2$
(76.30; S12)
(b) Vol. **145**, No. 1, p. K51—K54 (1988).

YUAN-CHAO, T. TSANG, E. RAHIMZADEH, and L. YIN
Solid Phase Structures of Cyano-Biphenyl Liquid Crystals
(61.30; S12)
(a) Vol. **105**, No. 1, p. K1—K6 (1988).

G. BRAUER, W. HERTLE and A. G. BALOGH
γ-Irradiation of Polyethylene Studied by Positron Annihilation
(61.40 and 78.70; 61.80; S12)
(a) Vol. **105**, No. 1, p. K7—K12 (1988).

A. KUCZKOWSKI
The Electrical Properties of Polyester Polymer-Quinoline Salt of Tetracyanoquinodimethane Composites
(72.80; S12)
(a) Vol. **105**, No. 1, p. K61—K65 (1988).

J. WĄSICKI, A. V. BELUSHKIN, V. G. KHOMENKO, I. NATKANIEC, YU. V. TELEZHENKO, and T. WASIUTYŃSKI
Low-Temperature Structure and Lattice Dynamics of 4,4'-Difluorobiphenyl
(63.20; 61.65; S12)
(b) Vol. **145**, No. 2, p. 445—453 (1988).

M. BURKERT, R. LINK, N. PUHLMANN, and W. VOLLMANN
Electron Energy Loss Measurements on Tetracyanoquinodimethane (TCNQ)
(71.25; 71.45; S12)
(b) Vol. **145**, No. 2, p. K121—K123 (1988).

V. M. YARTSEV and C. S. JACOBSEN
Infrared Properties of the Organic Semiconductor $MEM(TCNQ)_2$ for Polarization Perpendicular to the TCNQ Stacks
(78.30; S12)
(b) Vol. **145**, No. 2, p. K149—K153 (1988).

H. VON BERLEPSCH and M. PINNOW
Thermally Stimulated Surface Potential Decay in HDPE and the Cross-Over Phenomenon
(73.40; 73.20; S12)
(a) Vol. **105**, No. 2, p. 485—492 (1988).

G. F. Leal Ferreira and O. N. Oliveira, Jr.
Hole Transport in Corona-Charged 12 μm Sheldahl FEP Samples Interpreted through a Field-Dependent Schubweg Model
(72.20; S12)
(a) Vol. **105**, No. 2, p. 531—539 (1988).

B. Pura, J. Przedmojski, K. Wentowska, R. Dąbrowski, and K. Czupryński
X-Ray Critical Scattering Study of Nematic-Smectic A Phase Transition in 8DBT/120CB Mixture
(64.70; S12)
(a) Vol. **105**, No. 2, p. K107—K110 (1988).

T. Chot, N. N. Xuan, and L. Q. Minh
Barrier Heights and Interface States of Metal/2,4,7-Trinitro-9-Fluorenone Thin Film Contacts
(73.30; S12)
(a) Vol. **105**, No. 2, p. K121—K124 (1988).

S. Dhanuskodi and N. Hariharan
Radiation Damage of Bisglycine Cadmium Chloride Single Crystals. An EPR Study
(61.80; 76.30; S12)
(a) Vol. **106**, No. 1, p. 67—72 (1988).

D. Neogy, A. K. Mukherjee, and T. Purohit
Calculation of Electrostatic Crystal Field Parameters of Rare Earth Hydrated Single Crystals
(71.70; S12)
(a) Vol. **106**, No. 1, p. 173—180 (1988).

G.-H. Wang, M.-K. Teng, D.-X. Shen, C.-Y. Yi, Y.-Y. Zhou, Y.-Y. Lu, H.-W. Wang, Y.-Z. Zhu, and L. Dou
Positron Annihilation Lifetime and Doppler Broadening Studies of Electron-Irradiated Polypropylene
(61.70 and 78.70; S12)
(a) Vol. **106**, No. 1, p. K1—K6 (1988).

H. Kresse and W. Weissflog
Strong Antiparallel Correlation in Liquid Crystalline Esters with Lateral Groups
(77.20; 77.40; S12)
(a) Vol. **106**, No. 1, p. K89—K91 (1988).

W. Possart and I. Müller
The Electric Double Layer at the Polymer–Metal Contact. A Phenomenological Model
(73.40; 73.30; S12)
(a) Vol. **106**, No. 2, p. 525—534 (1988).

A. Graja, G. Sekretarczyk, W. Pukacki, S. Nešpůrek, and O. Ryba
Conduction-Electron Spin Resonance and Thermoelectric Power of the Mixed-Valence Tetrathiafulvalenium Chloride
(72.20 and 76.30; S12)
(b) Vol. **147**, No. 1, p. 401—408 (1988).

I. D. Mikhailov and A. P. Loghinov
Density of Vibrational States of Deuterized Polyethylene with Markovian Ordering
(63.50; S12)
(b) Vol. **147**, No. 1, p. K15—K19 (1988).

A. A. Romanyukha, Yu. N. Shvachko, A. V. Skripov, and A. P. Stepanov
Conduction Electron Spin Resonance in Organic Conductors α- and β-$(ET)_2IBr_2$
(76.30; S12)
(b) Vol. **147**, No. 2, p. K185—K188 (1988).

H. Krikor, P. Nagels, J. Vansummeren, and A. van den Bosch
Static Magnetic Susceptibilities of Poly(phenylethynylcopper) Doped with Iodine
(75.50; S12)
(a) Vol. **107**, No. 2, p. 639—646 (1988).

R. Piasecki, M. Ząbkowska-Wacławek, and W. Wacławek
An Explanation of Electrical Conductivity Data for Copper Phthalocyanine–Acetylene Carbon Black Mixtures
(72.80; S12)
(a) Vol. **108**, No. 1, p. 331—335 (1988).

C. Prieto, R. Ramirez, J. A. Gonzalo, and W. Windsch
Brillouin Scattering Determination of the Elastic Constants of Tris-Sarcosine Calcium Chloride at Room Temperature
(62.20; S12)
(a) Vol. **108**, No. 1, p. K9—K13 (1988).

A. Subbarayan, C. Balusubramanian, and Sa. K. Narayandass
Laser-Induced Damage Study of Vacuum Deposited Perylene Thin Films
(61.80 and 79.20; S12)
(a) Vol. **108**, No. 1, p. K31—K34 (1988).

V. K. Jain and V. S. Yadav
Estimation of T_1 for Co(II) from the EPR Linewidths of Mn(II) in Dibarium Cobalt Formate Tetrahydrate Single Crystals
(76.30; S12)
(b) Vol. **148**, No. 2, p. K157—K160 (1988).

F. X. Cabañas and C. F. Schwerdtfeger
Temperature Dependence of the *g*-Tensor of N-Methyl-N-Ethyl Morpholinium Tetracyanoquinodimethane
(76.30; S12)
(a) Vol. **108**, No. 2, p. 733—737 (1988).

J. A. de Saja and K. Tanaka
Photoelectrochemical Cells with p-Type Poly(3-Methylthiophene)
(73.40; 72.40; 78.40; S12)
(a) Vol. **108**, No. 2, p. K109—K114 (1988).

B. KUMAR and B. AGARWAL
Energy Band Structure of Phenazine N Oxide and Phenazine N Dioxide. Candidate Materials for Organic Photovoltaic Devices
(71.25; 72.20; S12)
(b) Vol. **149**, No. 1, p. 221—229 (1988).

Z. DREGER, J. KALINOWSKI, I. DAVOLI, S. STIZZA, and S. FELIZIANI
Thermoluminescence Study of Charge Traps in Polycrystalline Layers of Phenanthrene
(71.55 and 78.60; S12)
(b) Vol. **149**, No. 1, p. 363—370 (1988).

U. DAHMS, M. BURKERT, M. STARKE, H. J. HÖHNE, R. LINK, and C. HAMANN
Electron Energy Loss Measurements on Metal-Free Phthalocyanine
(78.20; S12)
(b) Vol. **149**, No. 1, p. K85—K88 (1988).

M. V. PÉREZ DIAZ, J. CH. ABBÉ, and G. DUPLÂTRE
Positron Annihilation Studies in Non-Metallic Solids: d-Camphor and Silver Iodide
(78.70; S9; S12)
(a) Vol. **109**, No. 1, p. 337—343 (1988).

K. BIEDRZYCKI
Thermally Stimulated Electron Emission from Triglycine Sulphate Crystals Excited by AC Electric Field
(77.80 and 79.40; S12)
(a) Vol. **109**, No. 1, p. K79—K83 (1988).

P. C. HÄGELE, S. WOLF, and C. SCHMID
Mass Defects in Polyethylene and the Role of Mass Weighting in Phonon Green's Function Theory
(61.70 and 63.20; S12)
(b) Vol. **149**, No. 2, p. 521—532 (1988).

G. M. HEGDE
Effect of the Electrode on the Electrical Conduction Process in Organic Semiconductors
(72.80 and 73.40; S12)
(a) Vol. **109**, No. 2, p. K125—K127 (1988).

J. WĄSICKI, A. M. GUREVICH, V. N. EROPKIN, J. MAYER, I. NATKANIEC, YU. V. TELEZHENKO, and V. G. KHOMENKO
Low-Temperature Heat Capacity and Phonon Spectra of 4,4'-Difluorobiphenyl
(63.20 and 65.40; S12)
(b) Vol. **150**, No. 1, p. 39—45 (1988).

A. A. MAKSIMOV, I. I. TARTAKOVSKII, and T. A. KRIVENKO
Determination of Polariton–Phonon Interaction Parameters in Anthracene Crystals
(71.36; 78.30; 78.55; S12)
(b) Vol. **150**, No. 1, p. 125—132 (1988).

H. PUFF and H. W. STREITWOLF
Optical Absorption and Stability of the Bipolaron Lattice
(71.38; 78.20; S12)
(b) Vol. **150**, No. 1, p. 147—164 (1988).

R. C. ZAMAR and A. H. BRUNETTI
Uniaxial-Stress Dependence of the NQR Frequency of ^{35}Cl in Paradichlorobenzene
(76.60; S12)
(b) Vol. **150**, No. 1, p. 245—252 (1988).

W. POSSART and I. MÜLLER
The Estimation of the Contact Potential Difference from Contact Charging between Polymer and Metal. New Aspects
(73.40; S12)
(a) Vol. **110**, No. 1, p. 205—211 (1988).

S. SCHMITT-RINK
Coherent Nonlinear Optical Processes in Semiconductors. Physics and Applications
(71.35 and 78.20; S7.12; S7.15; S12)
(b) Vol. **150**, No. 2, p. 349—356 (1988).

M. YUDASAKA, M. TANAKA, Y. KUWAE, K. NAKANISHI, and S. KURITA
Uniaxial Stress Effect on Visible and Near-Infrared Absorption Spectra of Si(phthalocyaninato) $(OH)_2$ Epitaxial Films
(78.65; S12)
(a) Vol. **110**, No. 2, p. 645—649 (1988).

M. G. KAPLUNOV, N. D. KUSHCH, and E. B. YAGUBSKII
Optical Properties of the New Organic Superconductor (BEDT-TTF)$_2$Cu(SCN)$_2$
(78.30; 71.38; S12)
(a) Vol. **110**, No. 2, p. K111—K116 (1988).

S12.1. Quasi-one dimensional systems

M. E. Gouvea and A. S. T. Pires
Spin Dynamics in the Classical Easy-Plane Antiferromagnet: Proton Spin–Lattice Relaxation in TMMC
(76.60; S12.1)
(b) Vol. **146**, No. 1, p. 263—268 (1988).

L. N. Kristoforov and A. V. Zolotaryuk
Dynamics of Ionic and Bonding Defects in Quasi-One-Dimensional Hydrogen-Bonded Chains
(61.70 and 63.90; 66.90; 71.45; S12.1)
(b) Vol. **146**, No. 2, p. 487—501 (1988).

L. S. Brizhik, Yu. B. Gaididei, A. A. Vakhnenko, and V. A. Vakhnenko
Soliton Generation in Semi-Infinite Molecular Chains
(71.50; 71.45; S12.1)
(b) Vol. **146**, No. 2, p. 605—612 (1988).

A. S. Davydov
Bisoliton Mechanism of High-Temperature Superconductivity
(71.45 and 74.20; S12.1)
(b) Vol. **146**, No. 2, p. 619—641 (1988).

V. K. Fedyanin and V. A. Osipov
On the Theory of Linear Diatomic Polymers
(71.45; 71.38; S12.1)
(b) Vol. **147**, No. 1, p. 199—209 (1988).

J. Malek, S. L. Drechsler, E. Heiner, and R. Kahnt
Solitons and Polarons in Polyyne. Discrete Lattice and Broken Electron–Hole Symmetry Effects
(71.50; 71.38; 71.45; S12.1)
(b) Vol. **147**, No. 1, p. 281—295 (1988).

Yu. B. Gaididei and V. M. Loktev
On a Theory of Electronic Spectrum and Magnetic Properties of High-T_c Superconductors
(74.20; 71.45; S12.1)
(b) Vol. **147**, No. 1, p. 307—319 (1988).

V. M. Yartsev
Optical Properties of Quasi-1D *n*-Merized Molecular Semiconductors
(72.20 and 78.20; S12.1)
(b) Vol. **149**, No. 1, p. 157—167 (1988).

Ch. Seidel
On the Pinning of Spin-Density Waves by Impurities
(72.15 and 75.30; S12.1)
(b) Vol. **149**, No. 1, p. 327—335 (1988).

R. Świetlik, G. Mihàly, and P. Delhaès
Microwave Conductivity and Thermoelectric Power of Some Ternary Salts of TCNQ and Iodine
(72.20; S12.1)
(a) Vol. **110**, No. 2, p. 593—599 (1988).

Author Index

Attention

To find a paper of your interest please choose first the special entry of the subject index and then use the volume number and page.

A. K. ABASS	(a) **106**, 613 (1988)	78.20; S10
	(a) **108**, K151 (1988)	78.65; S10.1
J. CH. ABBÉ	(b) **146**, 91 (1988)	61.80 and 78.70; S11
	(a) **109**, 337 (1988)	78.70; S9; S12
R. J. ABBUNDI	(a) **108**, 753 (1988)	78.60; S9.12
A. ABDELLAOUI	(a) **109**, 455 (1988)	68.55 and 78.65; S10.1
S. M. ABDEL-MAKSOUD	(b) **147**, 629 (1988)	72.40; S8
	(b) **148**, 629 (1988)	71.55; 72.40; S8
S. S. ABDINBEKOV	(b) **148**, K177 (1988)	78.30; 78.50; S8; S8.15
I. M. ABDUL AZIZ	(a) **109**, K101 (1988)	68.55 and 78.65; S10.1
W. A. ABDUL-GHAFOR	(a) **106**, 479 (1988)	78.65; 71.25; S10.1
A. M. ABDULLAEV	(a) **106**, 315 (1988)	65.40; S8
F. KH. ABDULLAEV	(b) **145**, K131 (1988)	74.50
	(b) **146**, 457 (1988)	63.10; 71.45
A. A. ABDUMALIKOV (b)	(b) **145**, K131 (1988)	74.50
	(b) **146**, 457 (1988)	63.10; 71.45
H. ABE	(a) **107**, 315 (1988)	72.20; S8
A. O. ABOYAN	(a) **106**, 41 (1988)	61.10
M. ABRAHAM	(b) **150**, 867 (1988)	78.20; S7.13
T. ABRITTA	(a) **108**, 795 (1988)	78.55; 61.70; S10
A. J. ABU EL-HAISA	see under letter "E"	
B. N. NARAHARI ACHAR	(b) **147**, K137 (1988)	72.20 and 73.60; S7.12; S7.15
B. S. ACHARYA	(a) **105**, 455 (1988)	68.55; S8
H. N. ACHARYA	(a) **107**, K1 (1988)	61.60 and 65.90; S10
P. K. ACHARYA	(a) **105**, 171 (1988)	68.55 and 78.65; S10.1
B. S. V. S. R. ACHARYULU	(a) **105**, 303 (1988)	78.50 and 78.60; S11.1
A. ADÁN	(a) **107**, K119 (1988)	68.55; 73.60; S9.12
B. L. AGAPOV	(a) **109**, 463 (1988)	73.40; S7.12; S8
B. AGARWAL	(b) **149**, 221 (1988)	71.25; 72.20; S12
A. A. AGASIEV	(b) **149**, K191 (1988)	78.40; S10.1
P. AGHAMKAR	(b) **145**, 343 (1988)	78.20; S7.13
BAL K. AGRAWAL	(b) **147**, 97 (1988)	63.50; S6
R. AGRAWAL	(b) **146**, 97 (1988)	65.50; 62.20; 65.70; S11
	(a) **108**, 803 (1988)	73.40 and 85
N. V. AGRINSKAYA	(b) **146**, K161 (1988)	63.20 and 78.30; S8.13
K. H. AHARONIAN	(b) **150**, 133 (1988)	71.45; 73.60; S8
Z. A. AHMED	(a) **108**, K151 (1988)	78.65; S10.1
NGUYEN AI VIET	see under letter 'N"	
AI-ZHEN LI	see under letter "L"	
T. AKACHI	(a) **105**, 411 (1988)	65.40; S1.2
I. AKAI	(b) **150**, 635 (1988)	71.35 and 78.20; 71.70; 73.40; S9
A. A. AKHMEDOV	(a) **108**, 619 (1988)	72.40 and 73.40; S8
N. D. AKHMEDZADE	(b) **148**, K93 (1988)	63.20 and 78.30; S8.16

S. M. Javed Aghtar	(a) **109**, 255 (1988)	73.60; 79.20; S4
A. I. Akimov	(a) **107**, K139 (1988)	74.70; S10.15
I. Ch. Akopyan	(b) **148**, 185 (1988)	71.35; 78.20; S9
A. Sh. Akramov	(a) **108**, 189 (1988)	64.60; 77.40; S11.1
I. A. Aksenov	(a) **105**, K97 (1988)	64.70; S8.16
	(a) **108**, K63 (1988)	72.80; S8.16
V. L. Aksenov	(b) **146**, K1 (1988)	63.20 and 63.75; S10
	(b) **149**, K7 (1988)	62.80 and 63.75; S10
J. M. Alameda	(a) **107**, 329 (1988)	75.70; S1.2
M. Albani	(a) **105**, 661 (1988)	73.40 and 85; 78.60; S7.15
J. P. Albert	(b) **149**, 641 (1988)	71.35 and 71.55; S8.11; S8.15
J. G. Albornóz	(a) **110**, 575 (1988)	72.20 and 78.20; 71.55; S8.16
E. L. Albuquerque	(b) **145**, 597 (1988)	71.20; 71.55; S7.12; S7.15
	(b) **145**, K5 (1988)	63.20; 76.80
	(b) **146**, 449 (1988)	63.20
	(b) **148**, 543 (1988)	71.20 and 73.40; S7.12; S7.15
Z. T. Al-Dhhan	(b) **145**, 145 (1988)	78.65; 68.55; S10.15
	(b) **149**, K81 (1988)	78.65; S10.1
	(b) **149**, K187 (1988)	78.65; S10.1
	(a) **110**, K39 (1988)	73.60; S10.1
H. M. Al-Doori	(a) **109**, K101 (1988)	68.55 and 78.65; S10.1
K. S. Aleksandrov	(a) **105**, 441 (1988)	64.70; 61.60; 65.40; 77.20; 78.20; S9.16
L. N. Aleksandrov	(a) **106**, K135 (1988)	61.40 and 65.70; S5.11
	(a) **107**, 213 (1988)	68.55; 64.75; 71.55; S5.11
	(a) **109**, K27 (1988)	64.70; S5.11; S10.1
	(a) **110**, 133 (1988)	64.60; 61.50; S5.11
O. V. Aleksandrov	(a) **110**, K61 (1988)	61.70 and 66.30; S5.11
I. P. Aleksandrova	(a) **105**, 145 (1988)	64.70; 76.60; S11.1
M. M. Aleksandrova	(a) **105**, K29 (1988)	61.50 and 62.20; S7.13
O. B. Alekseevskii	(b) **146**, K109 (1988)	61.80 and 72.20
Yu. A. Aleshchenko	(b) **145**, 117 (1988)	63.20; 78.30; S8.16
L. C. R. Alfred	(b) **145**, 203 (1988)	71.10; 71.25; S2
A. R. Ali	(a) **108**, 381 (1988)	75.60; S1.2
	(a) **109**, 313 (1988)	75.60; S1.2
F. A. Ali	(a) **108**, K151 (1988)	78.65; S10.1
A. E. Aliev	(a) **108**, 189 (1988)	64.60; 77.40; S11.1
R. A. Aliev	(a) **109**, 345 (1988)	73.40 and 85; 73.30; S11
F. H. Al-Khateeb	(a) **106**, K143 (1988)	68.55 and 78.65; S10.1
G. Allais	(a) **109**, 111 (1988)	61.60; S7.16
K. R. Allakhverdiev	(b) **147**, K99 (1988)	71.25 and 78.40; S8
	(b) **148**, K93 (1988)	63.20 and 78.30; S8.16
G. R. Allan	(b) **149**, 383 (1988)	78.20; 71.55; S7.13
	(b) **150**, 719 (1988)	78.20; S7.13
J. P. Allemand	(a) **105**, 427 (1988)	62.20; 61.40; 75.80; S1.1
N. A. Al-Liabi	(a) **106**, 613 (1988)	78.20; S10
R. Allub	(b) **147**, K5 (1988)	64.70; S2
N. Suarez Almodovar	see under letter "S"	
R. F. Almukhametov	(a) **108**, 135 (1988)	64.70 and 66.30; S8
K. A. Al-Saleh	(a) **105**, 177 (1988)	61.80 and 64.75; 73.60; S1.3; S5
	(a) **107**, 253 (1988)	61.80 and 73.60; S1.2; S5.11
S. M. Al-Shamari	(a) **106**, K143 (1988)	68.55 and 78.65; S10.1
A. A. Alybakov	(b) **146**, K135 (1988)	76.30 and 78.50; S11.1
T. Amand	(b) **147**, 727 (1988)	78.55; 71.35; S8.11; S8.12
G. Amato	(a) **108**, K25 (1988)	73.25; S5.11
S. Amelinckx	(a) **109**, 67 (1988)	61.14; S11.1
A. Ameziane	(a) **110**, 521 (1988)	73.60; S5; S8

Ch. Ammer	(a) **106**, 493 (1988)	66.30 and 68.48; 68.60; S1.3
	(a) **109**, 525 (1988)	66.30 and 68.48
Nguyen Ba An	see under letter "N"	
An Hu	see under letter "H"	
E. Anastassakis	(b) **149**, K1 (1988)	61.50 and 62.20
W. T. Anderson	(a) **110**, 531 (1988)	61.80 and 73.40; S1.61; S4; S7.12
W. Andrä	(a) **105**, 589 (1988)	75.70; 75.60
	(a) **106**, 589 (1988)	75.70; 61.80; S1.2; S1.61; S1.62
	(a) **109**, K149 (1988)	75.60; S1.2
	(a) **110**, 181 (1988)	68.55 and 78.65; S5.11
C. Andraud	(b) **149**, 759 (1988)	78.60; S9.16
A. V. Andreev	(a) **109**, K67 (1988)	75.50; S1.5
A. V. Andreeva	(a) **106**, 351 (1988)	61.55; 61.70; 72.20; S5.11
A. de Andrés	(b) **150**, 31 (1988)	63.20; 61.70; 78.30; S9.11
A. Anedda	(b) **146**, 757 (1988)	78.30; 63.20; S9.15
M. A. Angadi	(a) **106**, 485 (1988)	73.60; 75.70; S7
R. Angadi	(a) **106**, 485 (1988)	73.60; 75.70; S7
V. Angelov	(b) **146**, 269 (1988)	76.80; S1.1
B. M. Anghelov	(b) **148**, K83 (1988)	71.70 and 78.50; S10
V. N. Anisimova	(a) **105**, 139 (1988)	64.70; 77.20; S11.1
S. Annamalai	(a) **105**, 171 (1988)	68.55 and 78.65; S10.1
Yu. M. Annenkov	(b) **147**, K109 (1988)	61.80; S9.11
A. Antonetti	(b) **150**, 357 (1988)	71.35 and 78.20; S8.11
M. Antonini	(a) **109**, K23 (1988)	64.75; S10.1
G. Antonioli	(b) **145**, 401 (1988)	61.60 and 78.70; S8.16
V. N. Antonov	(b) **147**, 575 (1988)	71.10; 71.25; 71.38; S1.3
Vl. N. Antonov	(b) **147**, 575 (1988)	71.10; 71.25; 71.38; S1.3
B. P. Antonyuk	(b) **145**, 759 (1988)	71.35 and 78.55
M. Anwar	(a) **109**, 469 (1988)	78.65; 68.55; S10.1
S. P. Apanaševich	(b) **150**, 507 (1988)	78.20
P. Apell	(b) **145**, 483 (1988)	73.40
S. S. Aplesnin	(b) **149**, 267 (1988)	75.30
M. Apostol	(b) **146**, K125 (1988)	74.20
A. Apostolov	(a) **105**, K145 (1988)	75.30; S1.4
W. Appel	(a) **105**, 589 (1988)	75.70; 75.60
	(a) **105**, K67 (1988)	75.60 and 75.70; S1.2
S. Arajs	(a) **108**, 343 (1988)	72.15; S1.2
	(a) **108**, 369 (1988)	72.15; 75.80; S1.1
E. T. Arakawa	(a) **105**, 617 (1988)	78.20; 78.65; S1.3
E. D. Arama	(a) **106**, K181 (1988)	78.40; S8.16
	(a) **109**, K55 (1988)	72.40 and 78.55; S8.16
Yu. G. Arapov	(b) **148**, 197 (1988)	71.30; 72.20; S8.15
T. I. Arbuzova	(b) **148**, 361 (1988)	72.20 and 75.30; 76.30; S10.1
O. Arés Muzio	(a) **105**, 243 (1988)	75.50; 76.80; S1.1
	(a) **105**, 285 (1988)	75.30; S11.2
Ariel Fernandez	see under letter "F"	
V. V. Aristov	(a) **108**, 651 (1988)	61.10 and 68.20; S5.11
A. Armigliato	(a) **109**, 53 (1988)	61.16; S5.11
D. A. Aronov	(a) **109**, 211 (1988)	73.40; 73.60; S5.11
I. S. Arora	(b) **150**, K7 (1988)	61.50; S9.11
E. Z. Arshakyan	(a) **106**, 41 (1988)	61.10
G. Aruldhas	(a) **109**, 635 (1988)	78.30; S11
V. V. Arutunyan	(b) **149**, K77 (1988)	78.40; S10.1
S. K. Asadov	(a) **109**, 307 (1988)	75.30; S11
H. R. Asatryan	(b) **150**, 253 (1988)	76.30; S11.2
C. Ascheron	(a) **106**, 73 (1988)	61.80; S7.11
	(a) **106**, 81 (1988)	61.80; 78.70; S5.11; S7.11

B. M. Ashkinadze	(b) **150**, 533 (1988)	78.20; 71.35; S5.11
I. M. Askerov	(a) **105**, K151 (1988)	76.30 and 76.80; S8
H. A. Aslanian	(a) **108**, K85 (1988)	61.10 and 62.30
G. K. Aslanov	(a) **105**, K151 (1988)	76.30 and 76.80; S8
V. G. Aslanyan	(a) **110**, 359 (1988)	61.10
R. Assenov	(a) **108**, 699 (1988)	72.20; S8
V. N. Astratov	(b) **150**, 611 (1988)	78.20; 72.40; S10
P. A. Atanasov	(b) **149**, 739 (1988)	78.20; S8.15
	(b) **150**, 729 (1988)	78.20; S8.15
N. Atarashi	(b) **149**, 477 (1988)	61.70; S5.11; S5.12
A. Audzijonis	(b) **148**, K97 (1988)	78.40; S8.16
	(b) **148**, K169 (1988)	78.20; S8.16
	(b) **150**, K31 (1988)	71.25; S9
M. I. Auslender	(b) **147**, 613 (1988)	72.20; 72.60; S8.16
	(b) **148**, 289 (1988)	72.20; 72.60; S8.1; S10
G. T. Avanesyan	(b) **148**, K29 (1988)	71.36 and 71.38
M. Averous	(b) **147**, 629 (1988)	72.40; S8
	(b) **148**, 629 (1988)	71.55; 72.40; S8
J. E. Avon	(a) **108**, 599 (1988)	64.75; 75.30; S8.16
J. Ayala	(a) **108**, 721 (1988)	75.50; S11.2
X. Aymerich	(a) **109**, 479 (1988)	68.48 and 73.60; S5.11; S10.1
I. M. Abdul Azíz	see under letters "Ab"	
S. A. Azou	(b) **150**, 237 (1988)	72.20; S8
R. Azoulay	(b) **150**, 465 (1988)	68.55 and 78.20; S7.12; S7.15
Nguyen Ba An	see under letter "N"	
V. M. Babich	(a) **109**, 537 (1988)	71.55; 72.80; 76.30; S5.11
I. A. Bachilo	(a) **109**, 231 (1988)	61.80 and 68.55; S1.61; S5.11
A. G. Badalyan	(b) **149**, K31 (1988)	76.30; 78.50; S9.11
A. Badescu	(a) **109**, 61 (1988)	61.55; S1.1; S1.2
S. Bagchi	(b) **147**, K51 (1988)	72.15; S3
E. G. Bagduev	(b) **146**, 75 (1988)	61.70; 78.60; S9.11
V. E. Bagiev	(b) **149**, K191 (1988)	78.40; S10.1
	(a) **109**, 345 (1988)	73.40 and 85; 73.30; S11
N. T. Bagraev	(b) **150**, 19 (1988)	61.70 and 71.55; 72.40; 76.30; 76.60; S5.11
R. Bahadur	(b) **147**, 583 (1988)	67.40; S5.2
A. Bahnev	(a) **108**, 631 (1988)	68.55 and 78.55; 76.30; S7.15
V. N. Baier	(b) **149**, 403 (1988)	61.80 and 78.70
E. Ya. Baiguzin	(a) **108**, 205 (1988)	62.20
B. H. Bairamov	(b) **146**, K161 (1988)	63.20 and 78.30; S8.13
Ya. A. Bairamov	(a) **108**, 619 (1988)	72.40 and 73.40; S8
Baixin Liu	see under letter "L"	
A. S. Bakai	(a) **108**, K101 (1988)	66.30; 62.30; S5.11
R. Z. Bakhtizin	(a) **108**, 251 (1988)	79.70; 68.20; S1; S4
	(b) **150**, 103 (1988)	68.20 and 79.70; S1
N. A. Bakhyshov	(b) **147**, K99 (1988)	71.25 and 78.40; S8
H. Bakker	(b) **145**, 409 (1988)	66.30
	(b) **147**, K105 (1988)	66.30; S1
	(b) **149**, 483 (1988)	66.30; S1
J. Bąk-Misiuk	(a) **106**, 17 (1988)	61.55; 61.10; S1.1; S1.61
	(a) **106**, 451 (1988)	65.40; 61.70; S7.11; S7.12
V. Yu. Balandin	(a) **109**, K27 (1988)	64.70; S5.11; S10.1
	(a) **110**, 133 (1988)	64.60; 61.50; S5.11
C. Balasubramanian	(a) **106**, K41 (1988)	73.60; S10.1
	(a) **108**, K31 (1988)	61.80 and 79.20; S12
T. Balcerzak	(b) **145**, 299 (1988)	75.10 and 75.70

S. Balle	(b) **149**, 585 (1988)	71.20 and 71.25; S1.5; S1.63
A. G. Balogh	(a) **105**, K7 (1988)	61.40 and 78.70; 61.80; S12
I. Balslev	(b) **150**, 413 (1988)	71.35 and 78.20; S7.12; S7.15
R. Baltramiejūnas	(b) **149**, 337 (1988)	63.20 and 78.55; 71.45; S8.12
	(b) **150**, 139 (1988)	71.35; 71.38; 78.55; S8.13
	(b) **150**, 699 (1988)	71.35 and 78.55; S7.12; S7.15
D. Baltrūnas	(a) **106**, K173 (1988)	76.80; S8.15
	(b) **148**, K161 (1988)	76.80; S8
	(a) **108**, 197 (1988)	64.70; 76.80; S8.15
G. V. Bandurkina	(a) **110**, 555 (1988)	72.20; 71.55; S5.12
T. K. Bandyopadhyay	(a) **108**, K119 (1988)	73.60; 72.40; S8.12
J. P. Banerjee	(a) **109**, 359 (1988)	73.40 and 85; S7.11
S. Banerjee	(a) **110**, 331 (1988)	61.55; S1.2
N. S. Banger	(b) **148**, K5 (1988)	63.20; S2
	(b) **149**, 63 (1988)	62.20; S2
R. Banisch	(a) **107**, 281 (1988)	68.55; S5.11
J. Bánkuti	(a) **110**, 549 (1988)	73.60
L. Banyai	(b) **150**, 357 (1988)	71.35 and 78.20; S8.11
	(b) **150**, 393 (1988)	71.35 and 78.20
Tran Thoai Duy Bao	see under letters "Tr"	
Xi-Mao Bao	(a) **108**, 81 (1988)	61.50; 66.30; 72.80; S5.11
H.-P. Bär	(a) **108**, K131 (1988)	61.70 and 72.20; S10.1
	(a) **108**, K135 (1988)	61.70 and 72.20; S10.1
	(a) **110**, 537 (1988)	73.40; 73.60; S10.1
A. F. Barabanov	(b) **147**, K25 (1988)	71.20
M. Baran	(a) **107**, K145 (1988)	74.70 and 75.60; S10.15
N. P. Baran	(a) **109**, 537 (1988)	71.55; 72.80; 76.30; S5.11
A. I. Baranov	(a) **110**, 429 (1988)	64.70; 62.65; 77.20; S11.1
P. G. Baranov	(b) **147**, K171 (1988)	71.55 and 76.30; S9.11
A. Baranowski	(a) **110**, K101 (1988)	78.70; S1.3; S9
U. Barkow	(a) **105**, 627 (1988)	78.20; 61.70; 66.30; 78.50; S5.11
J. Barnaś	(b) **145**, K55 (1988)	75.70; 76.50
K. Bärner	(b) **147**, K189 (1988)	75.30; S1.62
	(a) **110**, 141 (1988)	64.70; 75.40; S7.15
R. G. Barnes	(b) **147**, K175 (1988)	76.60
N. Barniol	(a) **109**, 479 (1988)	68.48 and 73.60; S5.11; S10.1
V. Barone	(b) **145**, K125 (1988)	71.55; S5.11
J. Barrau	(b) **148**, 723 (1988)	78.55; 71.55; S5.11
J. Barthe	(a) **108**, 783 (1988)	78.60; 61.80; S10
C. Barthou	(a) **105**, 637 (1988)	78.60 and 85; 78.65; S8.12
J. Bartoš	(a) **106**, 659 (1988)	73.40 and 85; S7.12
U. Bartuch	(b) **150**, 873 (1988)	78.20
M. Z. Bashirov	(b) **145**, K71 (1988)	78.65; S7
L. A. Bashkirov	(a) **109**, K59 (1988)	76.60; S10.15
F. G. Bass	(b) **146**, K61 (1988)	75.30
	(b) **146**, K145 (1988)	75.30; S11.2
P. Bastie	(a) **107**, 559 (1988)	64.60; 78.20; S10.1
T. J. Bastow	(b) **145**, 719 (1988)	76.60; S8.1
A. N. Basu	(b) **146**, 117 (1988)	62.20 and 77.20; S9.11
P. K. Basu	(b) **147**, 135 (1988)	73.40 and 76.40; S7.12
J. Baszyński	(a) **105**, 501 (1988)	73.60; 75.70; S1.61
	(a) **110**, 227 (1988)	75.70; S1.2
J. Batarūnas	(b) **150**, K31 (1988)	71.25; S9
A. Bath	(a) **108**, K41 (1988)	73.40; S7.11
R. de Batist	(a) **107**, 759 (1988)	61.16 and 61.70; S11.2
G. Battistig	(a) **108**, K35 (1988)	61.70 and 68.48; 78,65; S5.11
C. L. Bauer	(a) **109**, 403 (1988)	61.70; 61.10; S4

G. T. BAUMBACH	(a) **109**, K7 (1988)	61.10 and 68.55; S7.12; S7.15
T. BAUMBACH	(a) **105**, 197 (1988)	61.10 and 68.55; S7.12; S7.15
	(a) **107**, K19 (1988)	68.48; 62.20; S7.12; S7.15
A. V. BAZHENOV	(a) **106**, 351 (1988)	61.55; 61.70; 72.20; S5.11
C. C. BECERRA	(a) **110**, K97 (1988)	75.50; S1.2; S11
F. BECHSTEDT	(b) **148**, 173 (1988)	63.20 and 68.30; 73.40
F. BEERWERTH	(b) **145**, 195 (1988)	71.35; 78.20; S9.11
F. BEHILIL	(b) **148**, 233 (1988)	71.50
F. BEHROOZI	(b) **149**, 495 (1988)	61.50 and 65.40; S5.2
G. BEISTER	(a) **105**, K45 (1988)	73.25; 78.45; S7.15
E. BELAS	(b) **147**, 621 (1988)	72.20; S8.15
V. V. BELKOV	(b) **150**, 533 (1988)	78.20; 71.35; S5.11
V. R. BELOSLUDOV	(b) **149**, 133 (1988)	63.20; S10
A. V. BELUSHKIN	(b) **145**, 445 (1988)	63.20; 61.65; S12
P. BENALLOUL	(a) **105**, 637 (1988)	78.60 and 85; 78.65; S8.12
M. BENDAOUD	(b) **148**, 233 (1988)	71.50
H. BENDER	(a) **107**, 455 (1988)	61.70; S5.11
L. BENEŠ	(a) **109**, 93 (1988)	61.60; 61.70; S8
J. BENOIT	(a) **105**, 637 (1988)	78.60 and 85; 78.65; S8.12
W. BENOIT	(a) **108**, 613 (1988)	62.40 and 64.70; S1.2
M. BEN SALEM	see under letter "S"	
K. BENTE	(a) **106**, K57 (1988)	74.70 and 76.30; S10.15
V. A. BEREZOVETS	(a) **108**, 303 (1988)	73.40; 72.40; 73.25; S7.13
H. BERGNER	(b) **150**, 885 (1988)	72.20; 72.40; S7.12
S. I. BERIL	(b) **145**, 535 (1988)	71.35 and 78.65; S9
	(b) **147**, 163 (1988)	71.38 and 73.40; 71.45
H. VON BERLEPSCH	(a) **105**, 485 (1988)	73.40; 73.20; S12
J. C. BERNÈDE	(a) **110**, 521 (1988)	73.60; S5; S8
G. I. BERSUKER	(b) **148**, 243 (1988)	63.20 and 71.55; S11
H. BERTORELLO	(a) **108**, 577 (1988)	62.40; 61.70; S4
F. BERZ	(b) **145**, K27 (1988)	72.10
H. BETHGE	(a) **106**, 133 (1988)	68.40 and 68.55; 64.75; 79.20; S1.3
	(a) **106**, 493 (1988)	66.30 and 68.48; 68.60; S1.3
	(a) **110**, 189 (1988)	61.70 and 68.55; S1.3
A. VAN DEN BEUKEL	(a) **105**, 235 (1988)	72.15; 61.40; S1.1
	(a) **108**, 97 (1988)	61.50 and 72.15; S4
W. BEYER	(a) **106**, 57 (1988)	61.80 and 68.55; 66.30; S8
J. BEYNON	(a) **106**, 515 (1988)	73.60; 72.20; S10.1
M. BEYSS	(a) **110**, 77 (1988)	61.50; S1.2
N. P. BEZHAN	(a) **109**, K153 (1988)	78.45 and 78.65; S8.15
A. P. BEZIRGANYAN	(a) **105**, 345 (1988)	61.10 and 78.70; 68.20
	(a) **109**, 101 (1988)	61.10 and 78.70; 68.20
P. A. BEZIRGANYAN	(a) **105**, 345 (1988)	61.10 and 78.70; 68.20
	(a) **106**, 41 (1988)	61.10
P. H. BEZIRGANYAN	(a) **110**, 359 (1988)	61.10
S. E. BEZIRGANYAN	(a) **110**, 359 (1988)	61.10
N. N. BEZRYADIN	(a) **109**, 463 (1988)	73.40; S7.12; S8
K. S. BHADORIYA	(b) **147**, 553 (1988)	71.60 and 73.20; S1.2; S1.3
G. C. BHAR	(a) **106**, K93 (1988)	78.20; S7.16; S8.12; S8.16
B. D. BHASIN	(a) **106**, 271 (1988)	78.60; S9.11
V. BHASKARA SASTRY	see under letter "S"	
T. BHATTACHARYA	(b) **145**, 181 (1988)	71.10; 61.55; S1
K. BIAŁAS-BORGIEŁ	(a) **105**, 567 (1988)	75.30 and 76.30; S9
V. T. BICH	(a) **106**, K147 (1988)	73.60; 78.65; S10.1
	(a) **108**, K157 (1988)	78.65; S10.1
K. BIEDRZYCKI	(a) **109**, K79 (1988)	77.80 and 79.40; S12
K. BIELECKI	(b) **148**, K53 (1988)	76.30; 62.50; S11.1

J.-Y. Bigot	(b) **150**, 477 (1988)	78.20; S8.11
T. Yu. Bilgildeeva	(b) **149**, 649 (1988)	72.20; 71.30; S7.12
R. Bindemann	(a) **108**, 285 (1988)	68.55 and 73.60; 76.30; 78.65; S5.11
R. Binder	(b) **150**, 407 (1988)	78.20; 71.35
R. Bindi	(b) **149**, 203 (1988)	79.75; S10.1
Le Thanh Binh	see under letter "L"	
D. J. S. Birch	(b) **145**, 741 (1988)	78.55; 71.35; S9
L. P. Biró	(a) **108**, 233 (1988)	68.55 and 73.60; S8
	(a) **108**, 637 (1988)	73.60 and 78.65; S8
S. K. Biswas	(a) **105**, 467 (1988)	78.65; 68.55; S8
	(a) **109**, 511 (1988)	68.55; S8
V. D. Blank	(a) **105**, K29 (1988)	61.50 and 62.20; S7.13
B. Blanzat	(b) **149**, 759 (1988)	78.60; S9.16
G. Blasek	(a) **110**, K77 (1988)	68.48; S1.2; S4
C. de Blasi	(a) **105**, 101 (1988)	61.70; S8
G. Blasse	(b) **145**, 333 (1988)	78.55; 78.50; S10.1
	(b) **148**, 413 (1988)	78.55; S11.1
	(b) **148**, K77 (1988)	78.55; S9.12
	(b) **149**, 391 (1988)	71.70 and 78.55; S11.1
M. Błaszak	(b) **148**, K147 (1988)	75.25; 75.30
G. L. Bleris	(a) **110**, 383 (1988)	61.16 and 61.70
K. Blum	(b) **146**, K87 (1988)	79.20; S4
H. Blumtritt	(a) **109**, K3 (1988)	61.14; 61.70; S8.13
A. Bobák	(b) **149**, K167 (1988)	75.10
M. Bobeth	(b) **145**, 123 (1988)	62.30
A. I. Bobrysheva	(b) **146**, 329 (1988)	78.20 and 78.30; 71.35; S7.15
	(b) **147**, 711 (1988)	71.35
	(b) **147**, 717 (1988)	78.55; S8
H. Bock	(b) **145**, K79 (1988)	63.20 and 78.30; 71.45; S7.15
S. V. Bodade	(a) **106**, K107 (1988)	78.60; 78.40; S11
A. C. Bódi	(a) **105**, 573 (1988)	75.50; S11.2
I. V. Bodnar	(b) **145**, 117 (1988)	63.20; 78.30; S8.16
J. Bodzenta	(b) **146**, 467 (1988)	63.20; S11.2
E. Boesman	(b) **146**, K149 (1988)	76.30
	(b) **148**, 683 (1988)	71.55 and 76.30; S9.11
	(b) **149**, K25 (1988)	71.70 and 76.30; S9.11
S. V. Bogdanov	(b) **150**, 593 (1988)	78.20; 71.35; S8.11
T. F. Boggess	(b) **150**, 895 (1988)	78.20 and 78.45; 78.55; S7.15
	(b) **150**, 913 (1988)	78.70; S5.11; S7.12
M. C. Böhm	(b) **149**, K135 (1988)	71.10 and 78.70
B. B. Boiko	(a) **107**, K139 (1988)	74.70; S10.15
	(b) **150**, 527 (1988)	78.65
S. A. Boiko	(b) **150**, 557 (1988)	78.20; 78.50; S9.11
	(b) **150**, 907 (1988)	78.20; 78.50; S9.11
B. Bojanowski	(a) **109**, K139 (1988)	76.30; 75.50; S11
G. G. Bokuchava	(b) **146**, 217 (1988)	75.10; 75.50
F. van Bolhuis	(a) **107**, 817 (1988)	61.60; S8
	(a) **107**, 873 (1988)	61.60; S8
Yu. B. Bolkhovityanov	(a) **110**, 489 (1988)	68.45; 68.55; S7.12
R. I. Bolkkhovityanova	(a) **110**, 489 (1988)	68.45; 68.55; S7.12
D. Bonchev	(b) **148**, K101 (1988)	61.55; S5.2
Ts. Bonchev	(b) **146**, 269 (1988)	76.80; S1.1
X. Bonet	(b) **145**, 261 (1988)	72.20; S8
G. Bongiovanni	(b) **146**, 757 (1988)	78.30; 63.20; S9.15
	(b) **150**, 685 (1988)	71.35 and 78.65; 71.45; S7.12; S7.15
A. Bonnet	(b) **150**, 225 (1988)	72.20; 71.25; S10
G. A. Bordovskii	(a) **109**, K51 (1988)	72.80 and 76.80; S8

V. E. Borisenko	(a) **107**, K33 (1988)	61.70 and 68.55; S1.61
	(a) **109**, 395 (1988)	61.70 and 66.30; S5.11
T. Bornath	(b) **147**, 747 (1988)	71.35; S5.12
H. Börner	(b) **147**, K43 (1988)	74.30; 74.70; S10.15
K. Bornkessel	(a) **110**, K77 (1988)	68.48; S1.2; S4
G. Borodi	(a) **108**, 233 (1988)	68.55 and 73.60; S8
G. Boroś	(a) **106**, K197 (1988)	73.30 and 85; S8.16
A. A. Borshch	(b) **150**, 471 (1988)	78.20; S8.11
B. Borzecka-Prokop	(a) **107**, 347 (1988)	76.60; S11
A. van den Bosch	(a) **107**, 639 (1988)	75.50; S12
D. Bose	(a) **106**, 569 (1988)	72.20; S7.12; S7.15
M. K. Bose	(b) **146**, 525 (1988)	71.25 and 73.60; S7.12
	(b) **148**, 165 (1988)	71.25 and 73.60; S7.13
L. Bosi	(b) **149**, K89 (1988)	61.70; 78.50; S9.11
A. Bosio	(a) **109**, K105 (1988)	68.55 and 73.60; S8.15
K. Z. Botros	(a) **106**, 49 (1988)	61.14 and 61.70; S1.2
H. Böttger	(b) **150**, 73 (1988)	63.10 and 63.50
C. Boulesteix	(a) **107**, 469 (1988)	68.48; 61.16; S10
	(a) **109**, 445 (1988)	61.70 and 65.70; 62.20; S8
A. J. Bourdillon	(b) **147**, K153 (1988)	74.70; S10.15
A. Bourret	(a) **107**, 481 (1988)	61.55 and 61.60; 61.16; S5.11; S5.12; S7.1; S8.1
	(a) **109**, 53 (1988)	61.16; S5.11
A. Bouwen	(b) **145**, K135 (1988)	61.70 and 76.30; S9.11
R. Boyn	(b) **148**, 11 (1988)	78.55 and 78.60; 71.55; S8.1
	(b) **148**, 263 (1988)	71.55; 68.60; 78.55; S8.11
	(b) **150**, 191 (1988)	61.70 and 71.55; 71.70; 78.55; S8.11
Bo-Zang Li	see under letter "L"	
A. Brablec	(a) **106**, 167 (1988)	71.55; 72.20
J. Brádler	(a) **106**, 17 (1988)	61.55; 61.10; S1.1; S1.61
T. Brand	(b) **150**, 705 (1988)	78.20; 85; S5.11
J. Brandon	(b) **150**, 465 (1988)	68.55 and 78.20; S7.12; S7.15
N. B. Brandt	(b) **148**, 197 (1988)	71.30; 72.20; S8.15
	(b) **150**, 237 (1988)	72.20; S8
A. Bräuer	(b) **150**, 873 (1988)	78.20
G. Brauer	(a) **105**, K7 (1988)	61.40 and 78.70; 61.80; S12
W. Brauer	(b) **148**, 213 (1988)	79.20; S4
W. Braune	(a) **108**, 303 (1988)	73.40; 72.40; 73.25; S7.13
P. J. Bray	(b) **147**, K175 (1988)	76.60
M. Bredol	(a) **107**, K7 (1988)	61.60; S8.16
C. Breitlow	(b) **147**, K159 (1988)	74.70; S10.15
F. J. Bremer	(a) **110**, 77 (1988)	61.50; S1.2
T. Bremer	(a) **105**, K17 (1988)	66.30 and 79.20; 78.20; S11.1
W. Brenig	(b) **147**, K67 (1988)	72.20
A. Brezini	(b) **148**, 233 (1988)	71.50
J. W. Brigthwell	(a) **110**, 269 (1988)	71.55 and 78.55; 76.30; 78.60; S8
D. T. Britton	(b) **147**, 89 (1988)	61.70 and 78.70; S1.2; S5
L. S. Brizhik	(b) **146**, 605 (1988)	71.50; 71.45; S12.1
D. Broddin	(a) **107**, 597 (1988)	61.55; 61.16; S1.2; S4
A. M. Brodin	(b) **150**, 557 (1988)	78.20; 78.50; S9.11
	(b) **150**, 907 (1988)	78.20; 78.50; S9.11
H. H. Brongersma	(a) **105**, 493 (1988)	73.60; S4
W. Bronowska	(b) **146**, K51 (1988)	75.50; S9
F. Brosens	(b) **145**, 517 (1988)	71.38
	(b) **147**, 173 (1988)	71.10 and 71.45
M. Brousseau	(b) **147**, 727 (1988)	78.55; 71.35; S8.11; S8.12
	(b) **148**, 723 (1988)	78.55; 71.55; S5.11
M. D. Brown	(a) **108**, 753 (1988)	78.60; S9.12

H. Bruchlos	(a) **109**, 503 (1988)	68.55; 74.70; S10.15
V. Brückner	(b) **150**, 885 (1988)	72.20; 72.40; S7.12
V. N. Brudnyi	(a) **105**, K57 (1988)	61.80 and 71.55; S7.12
	(a) **105**, K141 (1988)	61.80 and 72.80; S7.13
H.-G. Brühl	(a) **105**, 197 (1988)	61.10 and 68.55; S7.12; S7.15
	(a) **106**, 451 (1988)	65.40; 61.70; S7.11; S7.12
O. Brümmer	(b) **146**, 287 (1988)	78.70; 71.50; S1.1; S10.1
	(a) **106**, 419 (1988)	61.70; 72.80; 78.55; 78.70; S7.12
	(a) **107**, 111 (1988)	61.80; 78.50; 78.55; 78.70; S7.12
S. Brunet	(b) **147**, 253 (1988)	71.25; S7.12
A. H. Brunetti	(b) **150**, 245 (1988)	76.60; S12
I. Brylowska	(a) **109**, 597 (1988)	78.55 and 78.60; 61.70; S7.12
R. Bubáková	(a) **110**, 35 (1988)	61.10; S5.12
A. V. Budanov	(a) **109**, 463 (1988)	73.40; S7.12; S8
B. A. Budkevich	(a) **110**, 155 (1988)	61.80 and 73.60; 68.60; S10.1
M. Budzyński	(b) **147**, 685 (1988)	76.80; 71.70; S1.1
A. A. Bugaev	(b) **150**, 891 (1988)	72.30 and 78.20; 71.38; S5.11; S7.12
L. A. Bugaev	(b) **150**, 307 (1988)	78.70; 61.10; 71.10; S9.11; S10.1; S11.1
A. A. Bugai	(a) **109**, 537 (1988)	71.55; 72.80; 76.30; S5.11
W. Bührer	(b) **148**, 457 (1988)	61.55 and 64.80; S1.2
L. L. Buishvili	(b) **146**, 217 (1988)	75.10; 75.50
R. Bukowski	(b) **147**, 409 (1988)	78.20
Z. Bukowski	(b) **146**, K103 (1988)	74.70; S10.15
	(b) **146**, K131 (1988)	72.15 and 74.70; S10.15
N. A. Bulenkov	(a) **106**, 351 (1988)	61.55; 61.70; 72.20; S5.11
E. N. Bulgakov	(b) **148**, K139 (1988)	75.30
S. V. Bulyarskii	(b) **146**, 439 (1988)	61.70
D. Bundfuss	(b) **148**, 421 (1988)	73.40; S5.11; S7.1
Yu. A. Burenkov	(a) **105**, K103 (1988)	62.20; S8
P. Burgaud	(a) **110**, 521 (1988)	73.60; S5; S8
E. Burkel	(a) **108**, 129 (1988)	61.10; 68.20; S5.11
M. Burkert	(b) **145**, K121 (1988)	71.25; 71.45; S12
	(b) **149**, K85 (1988)	78.20; S12
M. F. Burkhard	(b) **147**, 589 (1988)	71.45; 79.20; S1.2; S4; S5
L. A. Bursill	(a) **107**, 503 (1988)	68.20; 61.16; 74.70; S10
I. I. Burshta	(b) **146**, 517 (1988)	71.36 and 73.20; S6
V. A. Bushuev	(b) **150**, 9 (1988)	61.10 and 78.70; S5.11
P. Bussemer	(b) **145**, 759 (1988)	71.35 and 78.55
E. R. Butkevich	(b) **146**, 683 (1988)	76.20
M. Z. Butt	(a) **107**, 187 (1988)	62.20; S1.2
D. M. Butusov	(b) **150**, 599 (1988)	78.20; 85; S7.12
E. A. Bychkov	(b) **149**, 427 (1988)	61.60 and 71.55; 76.30; S8.16
F. X. Cabañas	(a) **108**, 733 (1988)	76.30; S12
B. Cabon	(a) **107**, 393 (1988)	73.40 and 85
V. Cagan	(a) **106**, 595 (1988)	75.60; S11.2
P. X. Cai	(b) **148**, 715 (1988)	61.70 and 78.30; 78.50; S10
Shou-Fu Cai	(b) **150**, 797 (1988)	78.20 and 78.65; 71.45; S8
M. Calamiotou	(b) **149**, K157 (1988)	74.70 and 76.30; S10.15
J. Calas	(b) **147**, 629 (1988)	72.40; S8
	(a) **107**, 273 (1988)	66.30 and 78.65; 73.60; S8
	(b) **148**, 629 (1988)	71.55; 72.40; S8
F. Callens	(b) **146**, K149 (1988)	76.30
	(b) **148**, 683 (1988)	71.55 and 76.30; S9.11
	(b) **149**, K25 (1988)	71.70 and 76.30; S9.11
R. Callens	(b) **148**, 683 (1988)	71.55 and 76.30; S9.11

P. E. van Camp	(b) **146**, 573 (1988)	71.25; S6; S7.1
R. M. Cândea	(a) **108**, 233 (1988)	68.55 and 73.60; S8
	(a) **108**, 637 (1988)	73.60 and 78.65; S8
V. Canevari	(a) **109**, K105 (1988)	68.55 and 73.60; S8.15
W. Cao	(a) **107**, 511 (1988)	61.55; 61.14; S1.61
V. Čápek	(b) **147**, K57 (1988)	72.20
	(b) **148**, 319 (1988)	72.20; S8.15
R. Capelletti	(a) **109**, 603 (1988)	78.20; S9.11
P. Čapková	(a) **106**, 107 (1988)	63.20 and 65.70; S1.61
C. Carabatos-Nedelec	(a) **108**, K41 (1988)	73.40; S7.11
F. Carmona	(a) **107**, 329 (1988)	75.70; S1.2
M. C. Carotta	(a) **106**, 583 (1988)	72.20; S5.11
V. M. Castaño	(a) **107**, 845 (1988)	61.14
S. Diaz Castanon	see under letter "D"	
N. J. Castellani	(b) **148**, 121 (1988)	68.40; 79.60; S4
M. Jimenez de Castro	see under letter "J"	
M. Catalano	(a) **105**, 101 (1988)	61.70; S8
P. Cavaliere	(a) **105**, 101 (1988)	61.70; S8
V. Čech	(a) **106**, 167 (1988)	71.55; 72.20
Z. Celiński	(a) **105**, 567 (1988)	75.30 and 76.30; S9
B. Čelustka	(a) **109**, 329 (1988)	78.20; S8.16
G. Chabrier	(a) **105**, 617 (1988)	78.20; 78.65; S1.3
S. B. Chakraborty	(a) **108**, K119 (1988)	73.60; 72.40; S8.12
A. N. Chakravarti	(b) **146**, 525 (1988)	71.25 and 73.60; S7.12
	(b) **147**, 355 (1988)	72.20; S7.13
	(b) **148**, 165 (1988)	71.25 and 73.60; S7.13
	(b) **148**, 559 (1988)	79.60; 73.60; S7.12
	(b) **149**, 565 (1988)	73.60 and 79.60; S7.12
L. Chakravartti	(a) **109**, K89 (1988)	61.10
S. Chandra	(b) **145**, K1 (1988)	61.50; S3
U. Chandra	(b) **147**, 343 (1988)	74.10; 78.70; S11
Changyong Sun	see under letter "S"	
N. V. Chanh	(a) **106**, K147 (1988)	73.60; 78.65; S10.1
Chaorui Li	see under letter "L"	
S. L. Chaplot	(b) **145**, K89 (1988)	61.50; S11.1
St. Charalambous	(a) **105**, K175 (1988)	78.60; S9.11; S9.12
S. Charar	(b) **147**, 253 (1988)	71.25; S7.12
I. Chatterjee	(b) **146**, 717 (1988)	75.30 and 75.40
S. Chatterjee	(b) **148**, 579 (1988)	71.25; S1.4
U. Chatterjee	(a) **107**, K79 (1988)	78.20; S7.15; S8.15
D. Chattopadhyay	(b) **148**, K17 (1988)	72.20 and 73.40; S7.12
S. D. Chaturvedi	(b) **146**, 125 (1988)	64.30; 63.90; S9.11
	(b) **147**, 115 (1988)	61.50 and 63.90; 62.20; 65.70; S9
S. Chaudhuri	(a) **105**, 467 (1988)	78.65; 68.55; S8
	(a) **109**, 511 (1988)	68.55; S8
S. R. Chaudhuri	(a) **108**, K119 (1988)	73.60; 72.40; S8.12
R. N. P. Chaudhury	(a) **108**, K97 (1988)	61.70 and 78.50; S9.12
H. S. Chauhan	(b) **147**, 343 (1988)	74.10; 78.70; S11
S. K. S. Chauhan	(b) **146**, 97 (1988)	65.50; 62.20; 65.70; S11
M. Chávez	(a) **105**, 411 (1988)	65.40; S1.2
G. C. Che	(a) **110**, 55 (1988)	61.60 and 74.70; S11
N. M. Chebotaev	(b) **148**, 361 (1988)	72.20 and 75.30; 76.30; S10.1
V. Z. Chebotaru	(a) **110**, 277 (1988)	72.40 and 78.40; 78.60; S8.16
D.-X. Chen	(a) **107**, K145 (1988)	74.70 and 75.60; S10.15
Guanghua Chen	(a) **105**, K41 (1988)	73.40 and 76.30; S5.11
	(a) **108**, K139 (1988)	76.30 and 78.30; S5.11
	(a) **109**, K39 (1988)	68.55; 76.30; 78.65; S5

Guowei Chen	(a) **109**,	301 (1988)	74.70; S10.15
J. Chen	(a) **107**,	63 (1988)	61.60 and 74.70; S10.15
Jian-min Chen	(a) **110**,	515 (1988)	73.40; 78.30; S5.11; S7.12
Jun Chen	(a) **109**,	171 (1988)	64.70; 77.80; S10
	(a) **109**,	427 (1988)	64.70; 77.80; S11.1
	(a) **110**,	K25 (1988)	68.48; S6; S10.1
Liang-Yao Chen	(b) **148**,	387 (1988)	78.20; S1.3; S4
Li-Rong Chen	(a) **108**,	K15 (1988)	64.10
	(a) **110**,	K19 (1988)	64.60
R. Chen	(b) **149**,	45 (1988)	61.70; 72.20; 78.60; S9.16
T. G. Chen	(a) **110**,	415 (1988)	66.30; S10
B. L. Cheng	(a) **107**,	177 (1988)	62.40; S4
	(a) **108**,	569 (1988)	62.40; S4
Guoan Cheng	(a) **107**,	K29 (1988)	61.80 and 68.48; S1.1; S1.2; S10.1
X. R. Cheng	(a) **110**,	55 (1988)	61.60 and 74.70; S11
P. A. Cheremnykh	(b) **150**,	237 (1988)	72.20; S8
N. Cherkaoui Eddeqaqi	(b) **150**,	477 (1988)	78.20; S8.11
G. B. Cherlov	(b) **148**,	K33 (1988)	71.55; S10.1
M. V. Chernitskii	(b) **145**,	425 (1988)	64.60 and 76.30; S9.16
V. A. Chesnakov	(a) **109**,	373 (1988)	66.90 and 77.20; 64.70; S10
A. Chevy	(b) **145**,	261 (1988)	72.20; S8
A. Chiari	(b) **147**,	421 (1988)	78.55; 73.40; S7.15
R. Chicón	(b) **147**,	335 (1988)	72.10
G. S. Chilana	(a) **110**,	671 (1988)	73.40 and 85
C. Chlupsa	(a) **110**,	197 (1988)	73.60 and 78.60; S7.15
J. Chmist	(b) **145**,	687 (1988)	75.50; S1.2; S1.4
Tong-San Cho	(b) **148**,	K173 (1988)	78.30; S8.16
J. Chojcan	(b) **147**,	K201 (1988)	76.80 and 78.70; S1.1
T. Chot	(a) **105**,	K121 (1988)	73.30; S12
A. Choudhury	(a) **105**,	467 (1988)	78.65; 68.55; S8
	(a) **109**,	511 (1988)	68.55; S8
R. N. P. Chowdhury	(a) **107**,	K1 (1988)	61.60 and 65.90; S10
C. Christodoulides	(a) **109**,	611 (1988)	77.30
V. Christoph	(a) **106**,	K79 (1988)	75.50; S10
	(a) **109**,	K145 (1988)	75.60; S1.1
A. Christou	(a) **110**,	531 (1988)	61.80 and 73.40; S1.61; S4; S7.12
K. K. Christova	(a) **107**,	379 (1988)	77.30 and 78.70; 72.40; S10
Yuan-Chao Chu	(a) **105**,	K1 (1988)	61.30; S12
P. L. Chua	(b) **150**,	719 (1988)	78.20; S7.13
S. M. Chudinov	(b) **148**,	333 (1988)	71.30 and 72.20; S7.13
A. F. Chuiko	(b) **148**,	107 (1988)	71.10 and 73.40; 71.45
V. A. Chumak	(a) **109**,	K153 (1988)	78.45 and 78.65; S8.15
O. V. Chumanova	(b) **145**,	103 (1988)	61.80; S4; S5.11
C. H. Chung	(b) **148**,	K173 (1988)	78.30; S8.16
	(a) **108**,	K77 (1988)	78.20; S8.15
T. A. Chuong	(a) **108**,	643 (1988)	73.60; 64.70; S8.16
Chu Thien Truong	(a) **108**,	103 (1988)	61.55; 64.70; S4
M. Chybicki	(a) **108**,	K105 (1988)	73.25; S10.15
E. Čijauskas	(b) **148**,	K97 (1988)	78.40; S8.16
M. Cinal	(b) **148**,	699 (1988)	75.80; 62.20; S1; S1.1; S1.2
G. Ciobanu	(b) **145**,	283 (1988)	75.10 and 75.30
	(b) **149**,	249 (1988)	74.20
A. Čížek	(a) **105**,	357 (1988)	61.70; 78.70; S1.1; S1.2
L. M. Clarebrough	(a) **105**,	131 (1988)	61.70; S1.1
	(a) **105**,	365 (1988)	61.70; S1.1
N. Clark	(a) **105**,	K125 (1988)	73.60; S4
M. Clausnitzer	(a) **105**,	53 (1988)	61.10; S5.12

Ll. Closas	(a) **105**,	541 (1988)	72.20 and 73.40
W. Coene	(a) **107**,	521 (1988)	61.16; 61.70
	(a) **109**,	67 (1988)	61.14; S11.1
H. Cofta	(b) **146**,	241 (1988)	75.30
D. Colaitis	(a) **107**,	531 (1988)	61.14
J. Collet	(b) **147**,	727 (1988)	78.55; 71.35; S8.11; S8.12
J. H. Collet	(b) **146**,	393 (1988)	78.20; 71.35; S8.12
R. A. Collins	(a) **108**,	537 (1988)	61.70; 73.40; S1.2
M. Colocci	(b) **147**,	421 (1988)	78.55; 73.40; S7.15
F. Comas	(b) **148**,	533 (1988)	71.25 and 78.30; S7.12; S7.15
	(b) **149**,	533 (1988)	71.38 and 73.40; S7.12; S7.15
A. Conan	(b) **150**,	223 (1988)	72.20; 71.25; S10
M. Condat	(a) **109**,	K15 (1988)	61.70; S1.2
H. van Cong	(b) **147**,	253 (1988)	71.25; S7.12
M. C. Contreras	(a) **107**,	329 (1988)	75.70; S1.2
I. Corcotoi	(b) **145**,	283 (1988)	75.10 and 75.30
	(b) **149**,	249 (1988)	74.20
A. Cornet	(b) **147**,	727 (1988)	78.55; 71.35; S8.11; S8.12
B. V. Costa	(b) **146**,	229 (1988)	75.10; 75.40
	(b) **147**,	K79 (1988)	75.10
W. B. Costa	(b) **145**,	597 (1988)	71.20; 71.55; S7.12; S7.15
	(b) **148**,	543 (1988)	71.20 and 73.40; S7.12; S7.15
J. Costa-Quintana	(b) **149**,	585 (1988)	71.20 and 71.25; S1.5; S1.63
G. Couloumiers	(b) **148**,	723 (1988)	78.55; 71.55; S5.11
B. I. Craig	(b) **146**,	149 (1988)	71.10; S5
	(b) **147**,	K31 (1988)	71.10; 62.20; S5.11
J. M. Cowley	(a) **107**,	719 (1988)	68.20; 68.48; S1; S10
A. G. Cui	(b) **150**,	913 (1988)	78.70; S5.11; S7.12
H. L. Cui	(b) **146**,	189 (1988)	72.20 and 72.70
A. E. Curzon	(a) **105**,	335 (1988)	61.60; 61.80; S8.15
	(a) **108**,	529 (1988)	66.30; 61.16; S1.63
R. Cywiński	(a) **109**,	603 (1988)	78.20; S9.11
N. Czajkowski	(a) **110**,	437 (1988)	62.30 and 63.20; S11.1
Z. Czapla	(a) **105**,	K33 (1988)	64.70; 65.70; S11
P. Czarnecki	(b) **147**,	K47 (1988)	72.20; S1.2
J. Czerbniak	(b) **147**,	123 (1988)	61.80 and 68.30; S5.11
A. Czopnik	(a) **107**,	K151 (1988)	75.40; S1.4
	(a) **110**,	601 (1988)	75.40; S1.4
K. Czupryński	(a) **105**,	K107 (1988)	64.70; S12
P. Czurratis	(a) **106**,	407 (1988)	61.70; 61.55; S1.2; S4
	(a) **108**,	171 (1988)	61.70; 61.55; 62.20; S4
	(a) **110**,	123 (1988)	64.70; 61.10; 61.55; S1.2; S4
H. J. van Daal	(a) **105**,	493 (1988)	73.60; S4
R. Dąbrowski	(a) **105**,	K107 (1988)	64.70; S12
W. Dąbrowski	(a) **105**,	511 (1988)	71.55; 73.40; S5.11
D. Dadârlat	(a) **108**,	637 (1988)	73.60 and 78.65; S8
N. Dădârlat	(a) **108**,	233 (1988)	68.55 and 73.60; S8
R. Dafinova	(a) **106**,	K97 (1988)	78.60; S8.11
O. P. Daga	(a) **106**,	K53 (1988)	61.50 and 74.70; S10.15
U. Dahms	(b) **149**,	K85 (1988)	78.20; S12
G. Dalmai	(a) **107**,	731 (1988)	63.20 and 68.30; S10.1
A. A. El-Daly	see under letter "E"		
Yu. K. Danileiko	(a) **106**,	363 (1988)	61.70; 62.20; S5.11
Yu. V. Dankovskii	(a) **106**,	363 (1988)	61.70; 62.20; S5.11
P. Dannberg	(b) **150**,	873 (1988)	78.20

AL. DARABONT	(a) **108**, 233 (1988)	68.55 and 73.60; S8
S. DARDJONOV	(a) **108**, 631 (1988)	68.55 and 78.55; 76.30; S7.15
A. DARGYS	(b) **148**, K125 (1988)	72.10; 71.25; S5.12
G. DAS	(b) **149**, 355 (1988)	78.70; S2; S4
RATNADAS	(a) **110**, 347 (1988)	61.25; S9.11
N. DASS	(b) **146**, 105 (1988)	64.70; 62.20; S2; S5.2; S9
A. DAUNOIS	(b) **150**, 477 (1988)	78.20; S8.11
A. DAVIDSON	(b) **149**, K81 (1988)	78.65; S10.1
I. DAVOLI	(b) **149**, 363 (1988)	71.55 and 78.60; S12
A. S. DAVYDOV	(b) **146**, 619 (1988)	71.45 and 74.20; S12.1
	(b) **148**, 305 (1988)	74.20; 71.45
DAYU YANG	see under letter "Y"	
A. DE	(a) **105**, 297 (1988)	78.50; 78.60; S9.12
	(a) **106**, K101 (1988)	77.20; 77.40; S10.1
	(a) **108**, 457 (1988)	77.20 and 77.40; S9.12
E. DĘBOWSKA	(b) **148**, 227 (1988)	71.25; 71.60; S1.2
M. DĘBOWSKA	(b) **146**, 91 (1988)	61.80 and 78.70; S11
B. DECAMPS	(a) **109**, K15 (1988)	61.70; S1.2
DE-CHENG TIAN	see under letter "T"	
DE-GANG ZHANG	see under letter "Z"	
I. YA. DEKHTYAR	(a) **105**, 357 (1988)	61.70; 78.70; S1.1; S1.2
L. DELAEY	(b) **146**, 427 (1988)	61.50; 62.20
	(a) **107**, 597 (1988)	61.55; 61.16; S1.2; S4
P. DELAVIGNETTE	(a) **107**, 551 (1988)	61.70; 61.16
	(a) **107**, 745 (1988)	61.55 and 61.70; 61.16; S1.2
	(a) **107**, 759 (1988)	61.16 and 61.70; S11.2
P. DELHAÈS	(a) **110**, 593 (1988)	72.20; S12.1
G. B. DEMISHEV	(a) **108**, 89 (1988)	61.10; 64.70; S8
S. E. DEMYANOV	(a) **107**, K139 (1988)	74.70; S10.15
ZICHENG DENG	(b) **147**, 51 (1988)	61.70
DESHENG JIANG	see under letter "J"	
B. T. DESHMUKH	(a) **106**, K107 (1988)	78.60; 78.40; S11
A. P. DESHPANDE	(b) **145**, 77 (1988)	61.10; S1.2; S8.16
D. DESNICA	(a) **109**, 329 (1988)	78.20; S8.16
P. DEUS	(a) **108**, 225 (1988)	65.70; S7.16
DEVENDRA KUMAR	see under letter "K"	
DEVENDRA PRAKASH	see under letter "P"	
J. T. DEVREESE	(b) **145**, 517 (1988)	71.38
	(b) **146**, 573 (1988)	71.25; S6; S7.1
	(b) **147**, 173 (1988)	71.10 and 71.45
I. DÉZSI	(a) **105**, 219 (1988)	71.55; 76.80; S11.1
A. H. G. EL-DHAHER	see under letter "E"	
S. DHANUSKODI	(a) **106**, 67 (1988)	61.80; 76.30; S12
Z. T. AL-DHHAN	see under letter "A"	
E. M. DIANOV	(b) **147**, 261 (1988)	71.55; S10.1
	(a) **110**, K107 (1988)	78.50; 76.30; S10.15
M. V. PÉREZ DIAZ	see under letter "P"	
S. DÍAZ CASTAÑÓN	(a) **105**. 243 (1988)	75.50; 76.80; S1.1
	(a) **109**, 569 (1988)	75.30; 61.60; S1.1
A. DIEDERING	(a) **107**, 873 (1988)	61.60; S8
N. DE DIEGO	(a) **110**, 409 (1988)	61.70; 78.70; S1.2
E. DIEGUEZ	(a) **108**, K145 (1988)	78.55 and 78.60; S10
G. DIENER	(b) **145**, 123 (1988)	62.30
M. P. DIKSHIT	(b) **149**, 69 (1988)	63.20; S1.4
V. I. DIMITROV	(b) **145**, 65 (1988)	61.50; 64.75
DI-HUA DING	(b) **147**, 51 (1988)	61.70
S. Y. DING	(a) **107**, 63 (1988)	61.60 and 74.70; S10.15

Xingzhao Ding	(a) **110**, K67 (1988)	64.70 and 65.50; S1.3
N. N. Dinh	(a) **106**, K147 (1988)	73.60; 78.65; S10.1
	(a) **108**, K157 (1988)	78.65; S10.1
J. M. B. Fernandes Diniz	see under letter "F"	
M. de Dios Leyva	(b) **145**, K155 (1988)	71.25; S7.13; S7.15
	(b) **148**, K113 (1988)	71.25 and 73.40
D. J. Diskett	(a) **110**, 531 (1988)	61.80 and 73.40; S1.61; S4; S7.12
R. C. Dixit	(b) **147**, 115 (1988)	61.50 and 63.90; 62.20; 65.70; S9
R. P. Djalić-Jovanović	(b) **148**, 81 (1988)	64.60; 77.80; S11.1
A. Dlubek	(a) **106**, 419 (1988)	61.70; 72.80; 78.55; 78.70; S7.12
	(a) **107**, 111 (1988)	61.80; 78.50; 78.55; 78.70; S7.12
G. Dlubek	(a) **106**, 81 (1988)	61.80; 78.70; S5.11; S7.11
	(a) **106**, 419 (1988)	61.70; 72.80; 78.55; 78.70; S7.12
	(a) **107**, 111 (1988)	61.80; 78.50; 78.55; 78.70; S7.12
Yu. Dmitriev	(a) **106**, K7 (1988)	61.80; 72.80; S8
V. S. Dneprovskii	(b) **146**, 341 (1988)	71.35 and 78.20; S8; S8.11
	(b) **150**, 839 (1988)	78.20; S8.11
L. Dobiášová	(a) **106**, 107 (1988)	63.20 and 65.70; S1.61
D. D. Dobrev	(a) **109**, K85 (1988)	61.55; S5
I. I. Dobynde	(b) **147**, 717 (1988)	78.55; S8
A. E. Dolbak	(a) **110**, 489 (1988)	68.45; 68.55; S7.12
K. Dolecki	(a) **110**, 231 (1988)	64.70 and 75.60; S1.2
G. Dolino	(a) **107**, 559 (1988)	64.60; 78.20; S10.1
D. S. Domanevskii	(a) **106**, 249 (1988)	78.55; 71.55; S7.12
	(a) **110**, 221 (1988)	71.55; 78.55; S7.12
B. Domengès	(a) **107**, 73 (1988)	61.60; 64.80; S10.15
J. Dominec	(a) **107**, K135 (1988)	74.70; S10.15
E. Domínguez	(a) **106**, 11 (1988)	61.40 and 68.55; 78.30; S5.11
F. Domínguez-Adame	(a) **108**, K81 (1988)	61.70 and 78.60; S1.61
I. A. Domoryad	(a) **106**, 333 (1988)	61.40; 62.20; S8
Yuanda Dong	(a) **110**, 317 (1988)	61.40; 64.70; S1.2
Do Ngoc Uan	(a) **110**, K29 (1988)	68.55; S1.3
E. G. Doni	(a) **110**, 383 (1988)	61.16 and 61.70
H. Donker	(b) **145**, 333 (1988)	78.55; 78.50; S10.1
	(b) **148**, 413 (1988)	78.55; S11.1
A. Donnadieu	(a) **109**, 455 (1988)	68.55 and 78.65; S10.1
H. M. Al-Doori	see under letter "A"	
R. Dorbez	(a) **110**, 43 (1988)	61.14; S10.1
V. E. van Doren	(b) **146**, 573 (1988)	71.25; S6; S7.1
C. Dörnfeld	(b) **150**, 387 (1988)	71.35 and 78.20; S8.12; S8.15
L. Dou	(a) **106**, K1 (1988)	61.70 and 78.70; S12
S. X. Dou	(b) **147**, K153 (1988)	74.70; S10.15
L. Dozsa	(a) **105**, 521 (1988)	71.55; 72.40; 78.50; S7.12
G. Dräger	(b) **146**, 287 (1988)	78.70; 71.50; S1.1; S10.1
V. Drchal	(b) **146**, 597 (1988)	71.55 and 79.20; S1
	(b) **147**, 215 (1988)	71.10 and 79.20; S1.2
	(b) **148**, K23 (1988)	71.20; 71.55; S1.3
S. L. Drechsler	(b) **145**, 233 (1988)	71.38 and 71.50; S12
	(b) **147**, 281 (1988)	71.50; 71.38; 71.45; S12.1
Z. Dreger	(b) **149**, 363 (1988)	71.55 and 78.60; S12
M. Drozdowski	(a) **110**, 437 (1988)	62.30 and 63.20; S11.1
T. Drüsedau	(a) **108**, 285 (1988)	68.55 and 73.60; 76.30; 78.65; S5.11
Duan Feng	see under letter "F"	
J. Dubard	(b) **150**, 913 (1988)	78.70; S5.11; S7.12
A. Duda	(b) **150**, 273 (1988)	75.40 and 75.70
P. Duhaj	(a) **105**, 319 (1988)	61.40; 61.50
	(b) **146**, K17 (1988)	71.25 and 79.60; S1.2

F. Dujardin	(b) **150**, 201 (1988)	71.35 and 71.55
W. W. Duley	(b) **148**, 729 (1988)	78.55; 71.35; 71.55; S10.1
S. M. Dunaevskii	(b) **146**, 511 (1988)	73.40; 71.25; S7.15
T. T. Dung	(a) **106**, 201 (1988)	75.30; S1.2; S1.4
K. Dunphy	(b) **148**, 729 (1988)	78.55; 71.35; 71.55; S10.1
G. Duplâtre	(b) **146**, 91 (1988)	61.80 and 78.70; S11
	(a) **109**, 337 (1988)	78.70; S9, S12
V. M. Durasov	(b) **150**, 901 (1988)	66.70 and 78.35; S6
K. Durczewski	(b) **147**, 321 (1988)	72.15; 71.70; S1.4
Ch. Durga Prasad	see under letter "P"	
K.-D. Durst	(a) **108**, 403 (1988)	75.30; 75.50; 75.60; S1.2
	(a) **108**, 705 (1988)	75.30; 75.50; 75.60; S1.1; S1.2
Tran Thoai Duy Bao	see under letters "Tr"	
V. Dvořák	(b) **146**, K167 (1988)	63.20 and 64.70; S10
	(b) **149**, K61 (1988)	77.80
A. V. Dvurechenskii	(a) **109**, K21 (1988)	64.70; S5.11; S10.1
	(a) **110**, 133 (1988)	64.60; 61.50; S5.11
H. Dwelk	(b) **146**, K25 (1988)	74.70; S10.15
	(b) **147**, K159 (1988)	74.70; S10.15
R. Dwivedi	(a) **108**, 803 (1988)	73.40 and 85
S. V. Dyachenko	(a) **105**, 87 (1988)	61.80; 78.70; S9.11
	(b) **146**, 295 (1988)	78.55; 61.70; S9.11
V. V. Dyakin	(b) **148**, 501 (1988)	71.25 and 73.40
V. V. Dyakonov	(b) **147**, K171 (1988)	71.55 and 76.30; S9.11
D. van Dyck	(a) **107**, 521 (1988)	61.16; 61.70
V. P. Dymont	(a) **107**, K89 (1988)	61.60; S7

K. E. Easterling	(b) **147**, K153 (1988)	74.70; S10.15
A. I. Eatah	(a) **105**, 231 (1988)	72.20; S11.2
L. Eaves	(b) **150**, 177 (1988)	71.55; 72.20; 78.50; 78.55; S7.11
M. Eckler	(a) **108**, 285 (1988)	68.55 and 73.60; 76.30; 78.65; S5.11
N. Cherkaoui Eddeqaqi	see under letter "C"	
Sh. M. Efendiev	(a) **109**, 345 (1988)	73.40 and 85; 73.30; S11
I. K. Efendieva	(b) **147**, K99 (1988)	71.25 and 78.40; S8
A. L. Efros	(b) **150**, 627 (1988)	71.35 and 78.20; 78.60
K. Eftaxias	(a) **105**, K87 (1988)	66.30; S5.11
	(b) **147**, 83 (1988)	61.70; S9.11
	(a) **107**, K109 (1988)	66.30; S1.1
V. D. Egorov	(b) **146**, 351 (1988)	71.35 and 78.20; S8.11; S8.15
	(b) **148**, 407 (1988)	78.20; S8.11
	(b) **150**, 519 (1988)	78.20; S8.11
V. M. Egorov	(a) **105**, K93 (1988)	62.20 and 64.70; S8
A. V. Egranov	(b) **147**, 267 (1988)	61.70 and 71.35; 78.60; S9.11
K. H. G. Egziabher	(a) **106**, 399 (1988)	66.30; 61.70; 64.75; S1.1; S1.2
J. Ehlert	(a) **106**, 239 (1988)	75.60
P. Ehrhart	(b) **145**, 351 (1988)	61.80 and 63.20; S4
O. Eibl	(a) **108**, 495 (1988)	61.70; S11.1
F. Eichhorn	(a) **107**, 387 (1988)	78.70; S5.12
H. J. Eichler	(b) **150**, 705 (1988)	78.20; 85; S5.11
T. Eick	(a) **109**, 503 (1988)	68.55; 74.70; S10.15
G. Eidžiūnas	(a) **107**, K75 (1988)	78.30; S8.15
J. Eigemann	(a) **110**, 141 (1988)	64.70; 75.40; S7.15
C. W. E. van Eijk	(b) **148**, K77 (1988)	78.55; S9.12
H. M. Eissa	(a) **106**, 387 (1988)	61.70
E. A. Eivazov	(a) **110**, K45 (1988)	65.70; 75.50; S8
A. I. Ekimov	(b) **150**, 627 (1988)	71.35 and 78.20; 78.60

S. Ekmekçi	(b) **148**, 377 (1988)	75.30 and 75.40
A. A. El-Daly	(a) **110**, K13 (1988)	62.20; S1.2
A. H. G. El-Dhaher	(a) **106**, 49 (1988)	61.14 and 61.70; S1.2
D. Elefant	(b) **146**, 207 (1988)	74.10; S1
	(b) **146**, K29 (1988)	74.70; S10.15
E. El Faramawy	(a) **105**, 231 (1988)	72.20; S11.2
M. El-Fazani	(b) **147**, 601 (1988)	72.15; 75.30; S1.2; S1.3
	(b) **150**, K45 (1988)	75.30
A. J. Abu El-Haija	(a) **107**, 253 (1988)	61.80 and 73.60; S1.2; S5.11
M. El-Hiti	(a) **108**, 241 (1988)	73.60; 72.15; S1.3
	(a) **109**, 517 (1988)	72.10 and 73.60; S1.3
A. P. Eliseev	(a) **105**, K169 (1988)	78.55; 71.70; S10
A. I. Elizarov	(b) **148**, 197 (1988)	71.30; 72.20; S8.15
K. Elk	(b) **150**, 289 (1988)	75.60
A. M. Elkorashy	(b) **146**, 279 (1988)	78.20; S8
	(b) **149**, 747 (1988)	78.20; S8
C. Ell	(b) **150**, 357 (1988)	71.35 and 78.20; S8.11
	(b) **150**, 393 (1988)	71.35 and 78.20
K. El Sayed	(b) **150**, 393 (1988)	71.35 and 78.20
A. S. Em	(a) **107**, 261 (1988)	61.80 and 68.55; 64.70; 78.20; S5.11; S5.12
O. Emile	(b) **150**, 465 (1988)	68.55 and 78.20; S7.12; S7.15
R. Enderlein	(b) **145**, 167 (1988)	78.20; 71.25; 71.35; S7.12; S7.15
	(b) **148**, 173 (1988)	63.20 and 68.30; 73.40
	(a) **109**, 11 (1988)	61.50 and 68.48; 68.20; S5.11; S7.12; S8.13
	(b) **150**, 85 (1988)	63.20 and 68.30
P. Enders	(b) **149**, 447 (1988)	66.30; S7.1
S. Endo	(b) **148**, 673 (1988)	72.20; 71.25; S8.16
H. Engelmann	(a) **105**, 219 (1988)	71.55; 76.80; S11.1
I. R. Entin	(a) **106**, 25 (1988)	61.10
	(a) **106**, 339 (1988)	61.10; 62.60; S5.11
	(b) **147**, 459 (1988)	61.10
M. I. Eremets	(b) **147**, 297 (1988)	71.55 and 78.50; 72.40; 76.40; S5
H. Erhard	(a) **106**, 81 (1988)	61.80; 78.70; S5.11; S7.11
G. N. Eritsyan	(b) **149**, K77 (1988)	78.40; S10.1
H. L. Erknapetian	(b) **150**, 133 (1988)	71.45; 73.60; S8
S. Erkoc	(b) **149**, 81 (1988)	63.10; 61.50; 68.40; S5.11
V. N. Ermakov	(b) **145**, 545 (1988)	79.80
	(b) **148**, 305 (1988)	74.20; 71.45
V. N. Eropkin	(b) **150**, 39 (1988)	63.20 and 65.40; S12
W. Ervens	(a) **108**, 403 (1988)	75.30; 75.50; 75.60; S1.2
	(a) **108**, 705 (1988)	75.30; 75.50; 75.60; S1.1; S1.2
Er-wang Mao	see under letter "M"	
H. J. Erzgräber	(a) **106**, K215 (1988)	73.40 and 85; 72.40; S5.11
H. Eschrig	(b) **146**, K29 (1988)	74.70; S10.15
R. Escudero	(a) **105**, 411 (1988)	65.40; S1.2
S. A. Eshraghi	(a) **105**, 563 (1988)	72.80 and 78.20; S8.16
S. E. Esipov	(b) **150**, 501 (1988)	78.20; S8.12; S8.15
B. Esser	(b) **148**, 619 (1988)	71.10 and 71.35
	(b) **150**, K35 (1988)	71.10 and 71.35
E. I. Estrin	(a) **105**, K29 (1988)	61.50 and 62.20; S7.13
B. Etlinger	(a) **109**, 329 (1988)	78.20; S8.16
	(a) **109**, K55 (1988)	72.40 and 78.55; S8.16
P. C. Euthymiou	(a) **108**, K41 (1988)	73.40; S7.11
B. L. Evans	(a) **110**, 165 (1988)	73.60; S5
R. A. Evarestov	(b) **145**, 391 (1988)	61.50; 71.50
B. S. Evseev	(a) **109**, K31 (1988)	64.70; 66.70; 68.45
R. Ewertowski	(a) **110**, K101 (1988)	78.70; S1.3; S9

A. EXTREMERA	(a) **105**, 281 (1988)	75.30; 61.55; S1.3
L. EYRING	(a) **107**, 469 (1988)	68.48; 61.16; S10
R. K. EZOYAN	(b) **149**, K77 (1988)	78.40; S10.1
M. A. FAHIM	(a) **108**, 381 (1988)	75.60; S1.2
	(a) **109**, 313 (1988)	75.60; S1.2
G. FALCI	(b) **146**, K125 (1988)	74.20
HANJIE FAN	(a) **107**, 45 (1988)	61.55; S1.1; S1.4
XUDONG FAN	(a) **107**, 503 (1988)	68.20; 61.16; 74.70; S10
YUDIAN FAN	(a) **107**, K123 (1988)	68.55 and 78.65; S8
C. S. FANG	(b) **147**, 791 (1988)	78.30; 63.20; S11
GUO-PING FANG	(a) **110**, 515 (1988)	73.40; 78.30; S5.11; S7.12
FANGHUA LI	see under letter "L"	
FANGQING ZHANG	see under letter "Z"	
D. EL-FARAMAWY	see under letter "E"	
S. C. FARMER	(a) **107**, 571 (1988)	68.20; 64.60; S10
E. FARRÉS	(a) **109**, 479 (1988)	68.48 and 73.60; S5.11; S10.1
G. H. FATAHALAH	(a) **109**, K101 (1988)	68.55 and 78.65; S10.1
FATANG WANG	see under letter "W"	
P. M. FAUCHET	(b) **148**, K71 (1988)	78.20
A. FAWZY	(a) **107**, 171 (1988)	62.40; S4
	(a) **110**, 481 (1988)	62.40; S4
A. B. FAZAKAS	(b) **145**, 283 (1988)	75.10 and 75.30
	(b) **149**, 249 (1988)	74.20
M. EL-FAZANI	see under letter "E"	
R. FAZIO	(b) **146**, K125 (1988)	74.20
B. FECHNER	(b) **148**, K147 (1988)	75.25; 75.30
A. A. FEDOROV	(a) **107**, K19 (1988)	68.48; 62.20; S7.12; S7.15
A. V. FEDOROV	(b) **150**, 545 (1988)	78.20
V. K. FEDYANIN	(b) **147**, 199 (1988)	71.45; 71.38; S12.1
A. FEHER	(b) **150**, K63 (1988)	75.30 and 75.40; S11
FEI HAO SHENG	see under letter "S"	
I. V. FEKESHGAZI	(b) **146**, 723 (1988)	78.40; S11
M. S. FELDMAN	(b) **145**, K71 (1988)	78.65; S7
I. FELDVARI	(b) **150**, 907 (1988)	78.20; 78.50; S9.11
S. FELIZIANI	(b) **149**, 363 (1988)	71.55 and 78.60; S12
M. FELLER-KNIEPMEIER	(a) **110**, 107 (1988)	66.30; 75.50; S1.1
B. FELLMUTH	(b) **146**, 207 (1988)	74.10; S1
A. FELTZ	(a) **108**, 149 (1988)	66.30; 72.20; S8
D. FENG	(a) **107**, 63 (1988)	61.60 and 74.70; S10.15
DUANG FENG	(a) **106**, K117 (1988)	61.50; S11.1
	(a) **107**, 153 (1988)	63.20 and 68.30; 61.55; S1.2
	(a) **109**, 171 (1988)	64.70; 77.80; S10
	(a) **109**, 427 (1988)	64.70; 77.80; S11.1
	(a) **109**, 435 (1988)	64.70; 61.16; S11
	(a) **110**, K25 (1988)	68.48; S6; S10.1
JIAYOU FENG	(a) **107**, K123 (1988)	68.55 and 78.65; S8
ZHAO FENG	(b) **150**, 663 (1988)	78.20; S8.15
S. B. FEODOSEV	(b) **150**, K19 (1988)	65.70
N. A. FEOKTISTOV	(a) **108**, 323 (1988)	72.20 and 73.40; S5.11
K. FERENCZ	(b) **150**, 811 (1988)	78.20; S8.15
J. R. FERGUSON	(b) **147**, K137 (1988)	72.20 and 73.60; S7.12; S7.15
F. FERMI	(b) **147**, 421 (1988)	78.55; 73.40; S7.15
J. M. B. FERNANDES DINIZ	(a) **105**, 123 (1988)	66.30 and 72.20; 61.70; S10
	(a) **108**, 155 (1988)	66.30 and 72.10; S10
ARIEL FERNÁNDEZ	(b) **149**, 127 (1988)	64.10

L. Fernández	(a) **108**, K163 (1988)	78.60; S10.1
P. Fernández	(a) **107**, 197 (1988)	62.20; 78.60; S10.1
J. Fernández Rubio	(a) **105**, 541 (1988)	72.20 and 73.40
G. F. Leal Ferreira	(a) **105**, 531 (1988)	72.20; S12
J. R. Faleiro Ferreira	(b) **148**, 709 (1988)	75.10
	(b) **150**, 281 (1988)	75.10
	(b) **150**, K51 (1988)	75.10
L. N. Fershtat	(a) **108**, 189 (1988)	64.60; 77.40; S11.1
L. Fiermans	(a) **107**, 625 (1988)	68.20; 79.60; S1.2; S10
H. Figiel	(b) **145**, 687 (1988)	75.50; S1.2; S1.4
A. P. Filippov	(a) **110**, 83 (1988)	61.70
Yu. A. Firsov	(b) **145**, 625 (1988)	72.10; 73.40; S5.12
A. A. Firsova	(a) **106**, 351 (1988)	61.55; 61.70; 72.20; S5.11
H. J. Fischbeck	(b) **146**, 555 (1988)	71.20; 64.80; 71.38; 71.45
F. Fischer	(b) **146**, K71 (1988)	77.30; 61.70; S9.11
	(b) **147**, 811 (1988)	78.55; 63.20; S9.11; S9.12
	(b) **149**, 371 (1988)	78.30; S9.11
P. Fitori	(a) **108**, 233 (1988)	68.55 and 73.60; S8
I. P. Fitting	(a) **108**, 265 (1988)	73.60 and 79.20; S10.1
I. P. Fittipaldi	(b) **150**, 261 (1988)	75.10 and 75.40
F. Fizzotti	(a) **108**, K25 (1988)	73.25; S5.11
H. Fjellvag	(a) **110**, 141 (1988)	64.70; 75.40; S7.15
S. Flach	(b) **146**, K1 (1988)	63.20 and 63.75; S10
	(b) **149**, K7 (1988)	62.80 and 63.75; S10
T. Flade	(a) **108**, 689 (1988)	71.55 and 72.20; S7.12
R. Flagmeyer	(a) **107**, K19 (1988)	68.48; 62.20; S7.12; S7.15
	(a) **108**, 123 (1988)	61.60; S7.16
	(a) **108**, 257 (1988)	73.40; 68.55; 78.65; S7.1; S9.15
J. Flahaut	(a) **107**, 775 (1988)	61.60; 61.10; 61.16; S8
J. G. Fleming	(a) **105**, 77 (1988)	66.30; 68.55; S8.15
I. N. Flerov	(a) **105**, 441 (1988)	64.70; 61.60; 65.40; 77.20; 78.20; S9.16
N. K. Flevaris	(a) **107**, 579 (1988)	68.55; S1.2
I. Flint	(b) **150**, 623 (1988)	78.20
M. Fliyou	(b) **149**, 615 (1988)	71.35; S5.11; S5.12; S7.1; S8.1
P. Flögel	(b) **146**, 351 (1988)	71.35 and 78.20; S8.11; S8.15
V. Florescu	(a) **109**, 61 (1988)	61.55; S1.1; S1.2
B. D. Fluegel	(b) **150**, 357 (1988)	71.35 and 78.20; S8.11
H.-J. Föckel	(a) **105**, K111 (1988)	62.20
M. V. Fok	(a) **108**, 657 (1988)	73.40 and 78.60; S7.1
I. Földvári	(a) **107**, 161 (1988)	62.20 and 62.65; S11.1
O. Foltin	(b) **149**, K73 (1988)	78.20; 71.20
V. M. Fomin	(b) **145**, 535 (1988)	71.35 and 78.65; S9
	(b) **147**, 163 (1988)	71.38 and 73.40; 71.45
I. Fonseca Morena	see under letter "M"	
M. D. Fontana	(b) **147**, 441 (1988)	77.80 and 78.30; 63.20; S11.1
A. Forchel	(b) **150**, 679 (1988)	71.35 and 78.20; S7.12; S7.15
A. Forkl	(a) **105**, 597 (1988)	75.80; 78.20; S1.1
R. Fornari	(a) **105**, 521 (1988)	71.55; 72.40; 78.50; S7.12
E. Förster	(a) **107**, 85 (1988)	61.10; 65.70; 68.20; S5.11
E. Fortin	(b) **146**, 757 (1988)	78.30; 63.20; S9.15
C. T. Forwood	(a) **105**, 131 (1988)	61.70; S1.1
	(a) **105**, 365 (1988)	61.70; S1.1
F. Fouquet	(a) **105**, 427 (1988)	62.20; 61.40; 75.80; S1.1
S. Frabboni	(a) **109**, 53 (1988)	61.16; S5.11
R. Frahm	(b) **146**, 287 (1988)	78.70; 71.50; S1.1; S10.1
Z. Frait	(b) **146**, K119 (1988)	74.30 and 76.30; S10.15
D. Fraitová	(b) **146**, K119 (1988)	74.30 and 76.30; S10.15

J. FRANC	(b) **147**, 621 (1988)	72.20; S8.15
C. DE FRANCISCO	(a) **108**, 721 (1988)	75.50; S11.2
P. FRANCOIS	(a) **109**, 187 (1988)	62.20; S7.12
N. FRANGIS	(a) **107**, 589 (1988)	61.70; S8.16
T. S. FRANGULIAN	(b) **147**, K109 (1988)	61.80; S9.11
F. C. FRANK	(a) **105**, K21 (1988)	61.70
S. P. FREIDMAN	(b) **148**, K33 (1988)	71.55; S10.1
W. FRENTRUP	(a) **106**, 73 (1988)	61.80; S7.11
H. FREY	(a) **108**, 123 (1988)	61.60; S7.16
M. FRIED	(a) **107**, K15 (1988)	61.80 and 68.48; S1.3; S7.12
	(a) **108**, K35 (1988)	61.70 and 68.48; 78.65; S5.11
K. FRIEDLAND	(a) **106**, 419 (1988)	61.70; 72.80; 78.55; 78.70; S7.12
	(b) **150**, 177 (1988)	71.55; 72.20; 78.50; 78.55; S7.11
K. J. FRIEDLAND	(b) **146**, 199 (1988)	72.20; 72.60; S7.12
S. FRIES	(a) **108**, 449 (1988)	78.65; S11.1
M. FRINDI	(b) **149**, 775 (1988)	78.20; S9
B. FRIT	(a) **108**, 111 (1988)	61.14 and 64.80; S10.15
H.-G. FRITSCHE	(b) **148**, 155 (1988)	68.40; 64.10; S5.2
	(b) **148**, K101 (1988)	61.55; S5.2
V. FRITZSCHE	(b) **147**, 485 (1988)	79.60; 61.10; 68.20
	(b) **148**, 395 (1988)	78.20; S1.2
D. FRÖHLICH	(b) **145**, 195 (1988)	71.35; 78.20; S9.11
V. V. FROLOV	(b) **148**, K105 (1988)	61.70; 78.30; S5.11
Y. FU	(b) **147**, 767 (1988)	78.30; 63.20; S8.15
YAO-XIAN FU	(a) **109**, 135 (1988)	61.70; 74.70; 78.70; S10.15
	(a) **109**, K133 (1988)	74.70 and 78.70; S10.15
J. F. FUERTES	(a) **107**, 339 (1988)	75.60; S1.1
H. FUESS	(a) **109**, 589 (1988)	75.25; S11.1
I. YA. FUGOL	(b) **147**, 757 (1988)	78.30; 63.20; S5.2
W. FUHS	(a) **109**, 245 (1988)	72.40 and 73.40; 61.40; 71.55; S5.11
H. FUJII	(b) **145**, 649 (1988)	72.60 and 74.10; S10.15
S. FUJITA	(b) **147**, 649 (1988)	72.10
	(a) **109**, 383 (1988)	61.70 and 62.20; S7.12
H. FUJIWARA	(b) **147**, 559 (1988)	61.60 and 72.20; S10
Y. FUJIWARA	(b) **145**, K141 (1988)	75.10
P. FULCO	(b) **146**, 449 (1988)	63.20
M. FUNKE	(b) **148**, K151 (1988)	75.40; 64.60
FUQUAN JI	see under letter "J"	
A. S. FURMAN	(b) **150**, 611 (1988)	78.20; 72.40; S10
A. I. FURTICHEV	(b) **146**, 341 (1988)	71.35 and 78.20; S8; S8.11
	(b) **150**, 839 (1988)	78.20; S8.11
W. FÜSSEL	(a) **106**, K215 (1988)	73.40 and 85; 72.40; S5.11

R. J. GABORIAUD	(a) **107**, 867 (1988)	68.55; 61.16; S1.2
R. G. GABRIELYAN	(a) **108**, K85 (1988)	61.10 and 62.30
YU. B. GAIDIDEI	(b) **146**, 605 (1988)	71.50; 71.45; S12.1
	(b) **146**, 643 (1988)	72.10; 73.40
	(b) **147**, 307 (1988)	74.20; 71.45; S12.1
P. I. GAIDUK	(a) **107**, K105 (1988)	61.80; S5.11
A. V. GAINER	(b) **150**, 539 (1988)	78.20
	(b) **150**, 605 (1988)	78.20; 78.45
A. GALDIKAS	(a) **107**, K53 (1988)	72.20; S8.16
A. GALECKAS	(b) **150**, 743 (1988)	78.20; 61.80; S5.11
G. GALIBERT	(a) **107**, 273 (1988)	66.30 and 78.65; 73.60; S8
J. M. GALLIGAN	(a) **109**, K19 (1988)	62.20; S8.11
	(a) **110**, 565 (1988)	72.20; 61.70; S8.15

N. Ganesan	(a) **105**, 461 (1988)	73.60; 68.55; S8
N. Ganev	(a) **106**, 467 (1988)	61.10 and 62.20; 68.25
M. Ganne	(b) **150**, 225 (1988)	72.20; 71.25; S10
V. A. Ganshin	(a) **110**, 397 (1988)	61.70 and 78.20; 61.10; S11.1
J. Ganster	(b) **147**, 191 (1988)	71.10
S. V. Gaponenko	(b) **148**, K65 (1988)	71.35 and 78.20; S8.12
	(b) **150**, 653 (1988)	78.65; S8.15
	(b) **150**, 761 (1988)	71.35 and 78.65; S8.12
V. I. Garasim	(b) **148**, K161 (1988)	76.80; S8
S. García García	(a) **105**, 243 (1988)	75.50; 76.80; S1.1
	(a) **109**, 569 (1988)	75.30; 61.60; S1.1
J. V. García-Ramos	(a) **106**, 11 (1988)	61.40 and 68.55; 78.30; S5.11
A. K. Garg	(a) **110**, 467 (1988)	64.70; 62.50; 78.30; S10
K. B. Garg	(b) **145**, 269 (1988)	72.20; S5.11
	(b) **147**, 343 (1988)	74.10; 78.70; S11
R. K. Gartia	(a) **106**, 291 (1988)	78.60
	(a) **106**, 635 (1988)	78.60
G. I. Gashimov	(a) **107**, K129 (1988)	72.80; S8
J. Gasiot	(a) **109**, K109 (1988)	68.45 and 78.60; S11.1
J. Gastaldi	(a) **109**, 403 (1988)	61.70; 61.10; S4
I. K. Gasymov	(a) **107**, K49 (1988)	72.40; S8.16
V. I. Gatalskaya	(a) **107**, K139 (1988)	74.70; S10.15
G. V. Gatalskii	(a) **107**, K139 (1988)	74.70; S10.15
W. Gatzweiler	(a) **105**, 219 (1988)	71.55; 76.80; S11.1
E. Gaulke	(b) **147**, K143 (1988)	71.35 and 72.40; 78.20; S5.12
D. K. Gautam	(b) **145**, 269 (1988)	72.20; S5.11
N. P. Gavaleshko	(b) **146**, 671 (1988)	72.10; 72.20; S5
J. R. Gavarri	(a) **109**, 445 (1988)	61.70 and 65.70; 62.20; S8
V. I. Gavrilenko	(b) **145**, 209 (1988)	71.25; 78.20; S5
T. A. Gavrilova	(a) **110**, 489 (1988)	68.45; 68.55; S7.12
W. Gawalek	(a) **109**, 503 (1988)	68.55; 74.70; S10.15
Tingsui Ge	(a) **105**, 447 (1988)	62.40; 64.70; S1.2
E. D. Gehain	(b) **149**, 465 (1988)	61.70; 79.20; S5.11
W. Gehlhoff	(b) **145**, 609 (1988)	71.55 and 76.30; 61.70; S5.11
	(b) **150**, K57 (1988)	76.30; S7.11
R. Geick	(b) **148**, 445 (1988)	61.12 and 61.60; 62.50; 65.90; S8
J. Geiger	(a) **108**, 275 (1988)	68.55; 78.30; S5.11
D. Geiss	(b) **147**, 191 (1988)	71.10
A. L. Gentile	(a) **105**, 563 (1988)	72.80 and 78.20; S8.16
D. Genzow	(a) **106**, 509 (1988)	68.55 and 78.55; S8
A. Geoffroy	(a) **105**, 637 (1988)	78.60 and 85; 78.65; S8.12
J. George	(a) **106**, 123 (1988)	68.55 and 78.65; S8
G. Gerecke	(b) **148**, 173 (1988)	63.20 and 68.30; 73.40
V. Gerold	(a) **106**, 641 (1988)	61.55 and 78.20; S1.2
I. A. Ges	(a) **110**, 155 (1988)	61.80 and 73.60; 68.60; S10.1
V. A. Gevorkyan	(b) **149**, K77 (1988)	78.40; S10.1
A. A. Ghani	(a) **105**, 231 (1988)	72.20; S11.2
K. P. Ghatak	(b) **146**, K97 (1988)	72.30; S4
	(b) **147**, K179 (1988)	75.20
	(b) **148**, 645 (1988)	72.20; S8.15
	(b) **149**, K117 (1988)	72.20 and 73.40; S7.12; S7.15
	(b) **150**, K67 (1988)	79.60; S4
I. M. Ghauri	(a) **107**, 187 (1988)	62.20; S1.2
G. Ghibaudo	(a) **107**, 393 (1988)	73.40 and 85
A. A. Ghilarducci de Salva	(a) **108**, 577 (1988)	62.40; 61.70; S4
A. Ghosal	(b) **150**, K67 (1988)	79.60; S4

A. Ghosh	(a) **110**, 651 (1988)	78.20; S10.15
A. K. Ghosh	(b) **145**, 181 (1988)	71.10; 61.55; S1
B. K. Ghosh	(b) **147**, 97 (1988)	63.50; S6
D. K. Ghosh	(a) **107**, K79 (1988)	78.20; S7.15; S8.15
N. P. Ghosh	(a) **106**, K93 (1988)	78.20; S7.16; S8.12; S8.16
S. Ghosh	(b) **147**, 355 (1988)	72.20; S7.13
G. Giaquinta	(b) **146**, K125 (1988)	74.20
Z. Gierak	(a) **107**, 299 (1988)	72.15; S1.1; S1.2
R. Gilde	(a) **105**, 493 (1988)	73.60; S4
C. Trallero Giner	(b) **148**, 533 (1988)	71.25 and 78.30; S7.12; S7.15
J. Giordano	(b) **147**, 537 (1988)	63.20 and 78.30; S9
W. Giriat	(b) **147**, 767 (1988)	78.30; 63.20; S8.15
B. A. Gizhevskii	(b) **148**, 361 (1988)	72.20 and 75.30; 76.30; S10.1
P. S. Gladkov	(a) **108**, K125 (1988)	71.55 and 72.40; 61.70; S7.12
S. J. Gładysz	(b) **147**, K87 (1988)	75.10; 75.50
H. Glaeske	(b) **146**, 385 (1988)	72.10; 78.20
P. Glas	(a) **107**, 85 (1988)	61.10; 65.70; 68.20; S5.11
V. A. Gley	(a) **107**, K11 (1988)	62.20; S1
M. Gliński	(b) **148**, 567 (1988)	73.40; S7.13
M. Glogarova	(b) **149**, K61 (1988)	77.80
M. Glotz	(b) **150**, 705 (1988)	78.20; 85; S5.11
E. Gmelin	(a) **110**, 141 (1988)	64.70; 75.40; S7.15
E. O. Göbel	(b) **150**, 749 (1988)	78.20; 78.55; S8.15
G. Gobsch	(b) **147**, K125 (1988)	71.25 and 73.40; S7.13
	(b) **148**, 349 (1988)	73.25 and 73.40; S7.13
	(b) **148**, 611 (1988)	71.25 and 73.40; S8.15
S. Godoy	(b) **147**, 649 (1988)	72.10
K. Godwod	(a) **106**, 17 (1988)	61.55; 61.10; S1.1; S1.61
O. Goede	(b) **146**, 11 (1988)	78.20; 71.35; 78.55; S8.1
	(b) **146**, K65 (1988)	78.40 and 78.55; 71.35; S8.11
	(b) **146**, K157 (1988)	63.20 and 78.30; S8.11
	(a) **108**, 443 (1988)	78.20; S8.13
K. Goetz	(a) **107**, 85 (1988)	61.10; 65.70; 68.20; S5.11
W. M. Goho	(a) **106**, 641 (1988)	61.55 and 78.20; S1.2
	(b) **147**, 417 (1988)	71.50 and 78.20; S1.2
I. A. Goichuk	(b) **149**, 55 (1988)	63.20; 75.10
Th. Goldberg	(a) **105**, K111 (1988)	62.20
O. A. Golikova	(a) **108**, 323 (1988)	72.20 and 73.40; S5.11
A. E. Golobokov	(a) **105**, K29 (1988)	61.50 and 62.20; S7.13
E. M. Gololobov	(b) **149**, K145 (1988)	61.70 and 74.70; S10.15
V. P. Goltsev	(a) **108**, K1 (1988)	61.70 and 62.20; S4
A. V. Golubkov	(a) **105**, K93 (1988)	62.20 and 64.70; S8
	(a) **105**, K103 (1988)	62.20; S8
M. Golz	(b) **150**, 867 (1988)	78.20; S7.13
A. Gómez	(a) **107**, 845 (1988)	61.14
F. Gömöry	(a) **109**, 205 (1988)	73.60 and 75.30; 74.70; S10.15
Gong Yu	see under letter "Y"	
U. Gonser	(a) **105**, 219 (1988)	71.55; 76.80; S11.1
J. A. Gonzalo	(a) **108**, K9 (1988)	62.20; S12
R. Gopal	(a) **108**, 803 (1988)	73.40 and 85
R. V. Gopala Rao	see under letter "R"	
V. V. Gorbach	(b) **149**, K49 (1988)	75.30; 76.50
V. V. Gorbachev	(b) **150**, 901 (1988)	66.70 and 78.35; S6
D. A. Gorbachevskii	(a) **106**, K23 (1988)	61.80 and 68.48; 66.30; S1.3; S4
I. N. Gorbatyuk	(b) **148**, 197 (1988)	71.30; 72.20; S8.15
E. A. Goremychkin	(b) **145**, K15 (1988)	71.70; S1.4
P. N. Gorley	(b) **146**, 671 (1988)	72.10; 72.20; S5

S. Ya. Goroshchenko	(b) **145**, 187 (1988)	71.25; 72.10; 73.40
	(b) **145**, K107 (1988)	73.40
V. G. Gorshkov	(a) **106**, 363 (1988)	61.70; 62.20; S5.11
G. Gosmanová	(a) **109**, 199 (1988)	62.20; 68.20
I. A. Gospodarev	(b) **150**, K19 (1988)	65.70
G. G. Gotsadze	(b) **150**, 599 (1988)	78.20; 85; S7.12
V. Gottschalch	(a) **108**, 123 (1988)	61.60; S7.16
G. Götz	(a) **106**, 589 (1988)	75.70; 61.80; S1.2; S1.61; S1.62
	(a) **110**, 181 (1988)	68.55 and 78.65; S5.11
J. P. Goudonnet	(a) **105**, 617 (1988)	78.20; 78.65; S1.3
M. E. Gouvea	(b) **146**, 263 (1988)	76.60; S12.1
R. P. Goyal	(b) **146**, 97 (1988)	65.50; 62.20; 65.70; S11
	(b) **146**, 475 (1988)	64.30; 62.20; 65.50; S9.12; S11
S. C. Goyal	(b) **145**, 435 (1988)	62.20; S9.12
V. K. Goyal	(a) **107**, K83 (1988)	73.40 and 85; 72.40; S5.11
B. C. Grabmeier	(a) **106**, 89 (1988)	61.70; 72.40; 78.20; S11.1
M. de Graef	(b) **146**, 427 (1988)	61.50; 62.20
	(a) **107**, 597 (1988)	61.55; 61.16; S1.2; S4
J. Graefenstein	(b) **147**, 575 (1988)	71.10; 71.25; 71.38; S1.3
G. Graiss	(a) **107**, 171 (1988)	62.40; S4
	(a) **110**, 481 (1988)	62.40; S4
A. Graja	(b) **147**, 401 (1988)	72.20 and 76.30; S12
J. Grammatikakis	(a) **109**, 421 (1988)	61.80; 71.55; S5.11
G. Grange	(a) **109**, 403 (1988)	61.70; 61.10; S4
D. Gratias	(a) **107**, 739 (1988)	61.16
S. Grebinskii	(a) **107**, K53 (1988)	72.20; S8.16
R. M. Grechishki	(a) **110**, 259 (1988)	78.20; 75.60
A. G. J. Green	(a) **110**, 269 (1988)	71.55 and 78.55; 76.30; 78.60; S8
G. S. Green	(a) **110**, 323 (1988)	61.10; 61.70; S5
B. Greń	(b) **146**, K103 (1988)	74.70; S10.15
M. Grendel	(a) **106**, 659 (1988)	73.40 and 85; S7.12
V. P. Gribkovskii	(b) **148**, K65 (1988)	71.35 and 78.20; S8.12
	(b) **150**, 647 (1988)	78.20; S8.15
	(b) **150**, 761 (1988)	71.35 and 78.65; S8.12
V. I. Gridnev	(b) **149**, K129 (1988)	71.25; S5.11
M. Griepentrog	(a) **106**, 73 (1988)	61.80; S7.11
J. Griesche	(a) **105**, 189 (1988)	68.55; S8.13
	(a) **106**, 509 (1988)	68.55 and 78.55; S8
	(a) **109**, 11 (1988)	61.50 and 68.48; 68.20; S5.11; S7.12; S8.13
J. Grigas	(a) **110**, 429 (1988)	64.70; 62.65; 77.20; S11.1
V. G. Grigoryan	(b) **149**, K123 (1988)	71.45 and 73.40
R. Grill	(b) **145**, 637 (1988)	72.20; S8.15
	(b) **146**, K35 (1988)	72.10; 72.20
	(b) **147**, 621 (1988)	72.20; S8.15
	(b) **148**, 319 (1988)	72.20; S8.15
P. Grima	(a) **107**, 165 (1988)	61.55 and 64.70; S8.15
	(a) **107**, 205 (1988)	64.70; S8.15
	(a) **108**, 599 (1988)	64.75; 75.30; S8.16
M. Grinberg	(b) **145**, 619 (1988)	71.55; S8.12
K.-O. E. Groeneveld	(b) **147**, 589 (1988)	71.45; 79.20; S1.2; S4; S5
D. N. Gromov	(b) **148**, 185 (1988)	71.35; 78.20; S9
R. Gronsky	(a) **105**, 207 (1988)	68.20; 61.16; S10
R. Grötzschel	(a) **106**, 57 (1988)	61.80 and 68.55; 66.30; S8
	(a) **109**, 493 (1988)	68.55; 65.70; S9.12
K. I. Grozdev	(b) **148**, K143 (1988)	75.10
G. Grummt	(a) **107**, 405 (1988)	73.40; 71.55; 72.40; 78.60; S7.11; S7.15
	(a) **107**, 419 (1988)	73.40 and 78.60; 85; S7.12

	(a) **108**, 813 (1988)	73.40
	(a) **110**, 687 (1988)	73.40 and 85
J. B. GRUN	(b) **150**, 825 (1988)	78.20; S9
H. F. GU	(b) **148**, 715 (1988)	61.70 and 78.30; 78.50; S10
QUING GU	(a) **108**, 81 (1988)	61.50; 66.30; 72.80; S5.11
Z. G. GU	(b) **145**, 151 (1988)	71.36 and 73.20; 78.30; S9.12
GUANGHUA CHEN	see under letter "C"	
GUANGTIAN ZOU	see under letter "Z"	
V. A. GUBANOV	(b) **148**, K33 (1988)	71.55; S10.1
	(b) **149**, 175 (1988)	71.25; S1.61
V. A. GUBANOVA	(b) **146**, K135 (1988)	76.30 and 78.50; S11.1
B. DE GUILLEBON	(a) **105**, 427 (1988)	62.20; 61.40; 75.80; S1.1
M. GUITTARD	(a) **107**, 775 (1988)	61.60; 61.10; 61.16; S8
H. GÜNTHER	(b) **149**, 101 (1988)	62.20
GUOAN CHENG	see under letter "C"	
GUOAN WU	see under letter "W"	
L. GUOBA	(a) **109**, 239 (1988)	68.48; 79.20; S7.11; S7.12
GUO-PING FANG	see under letter "F"	
GUOQUING PAN	see under letter "P"	
GUOSHENG SUN	see under letter "S"	
GUOWEI CHEN	see under letter "C"	
GUO-XIN XIE	see under letter "X"	
B. R. K. GUPTA	(b) **146**, 97 (1988)	65.50; 62.20; 65.70; S11
	(b) **146**, 475 (1988)	64.30; 62.20; 65.50; S9.12; S11
	(b) **150**, 59 (1988)	62.20; S8
D. C. GUPTA	(b) **149**, 121 (1988)	62.20; S8
H. C. GUPTA	(b) **146**, 481 (1988)	63.20; S1.2
O. P. GUPTA	(b) **146**, 475 (1988)	64.30; 62.20; 65.50; S9.12; S11
R. S. GUPTA	(a) **110**, 671 (1988)	73.40 and 85
R. SEN GUPTA	(b) **148**, 579 (1988)	71.25; S1.4
S. N. GUPTA	(a) **105**, 275 (1988)	76.80; S1; S1.1
V. B. GUPTA	(b) **146**, 481 (1988)	63.20; S1.2
V. K. GUPTA	(b) **147**, 553 (1988)	71.60 and 73.20; S1.2; S1.3
A. M. GUREVICH	(b) **150**, 39 (1988)	63.20 and 65.40; S12
A. S. GUREVICH	(b) **149**, K145 (1988)	61.70 and 74.70; S10.15
A. V. GUREVICH	(b) **145**, K47 (1988)	74.60
A. N. GURYANOV	(a) **110**, K107 (1988)	78.50; 76.30; S10.15
G. M. GUSAKOV	(a) **107**, 261 (1988)	61.80 and 68.55; 64.70; 78.20; S5.11; S5.12
A. N. GUSATINSKII	(b) **146**, K21 (1988)	71.25 and 78.70; S9
A. G. GUSEINOV	(a) **106**, 315 (1988)	65.40; S8
S. S. GUSEINOV	(b) **147**, K99 (1988)	71.25 and 78.40; S8
YA. YU. GUSEINOV	(b) **149**, K191 (1988)	78.40; S10.1
A. I. GUSEV	(a) **106**, 459 (1988)	64.60; 75.30; S1.61
N. A. GUSHCHINA	(b) **146**, 511 (1988)	73.40; 71.25; S7.15
N. GUSKOS	(b) **147**, 349 (1988)	72.20; S9.12
	(b) **149**, K157 (1988)	74.70 and 76.30; S10.15
J. GUTOWSKI	(b) **146**, 589 (1988)	71.35 and 71.55; 78.45; S8.11
	(b) **150**, 833 (1988)	78.20; S8.11
I. GUTZOW	(a) **107**, K99 (1988)	61.80 and 78.60; S9.12
M. GUYMONT	(a) **107**, 739 (1988)	61.16
	(a) **107**, 775 (1988)	61.60; 61.10; 61.16; S8
M. GUYOT	(a) **106**, 595 (1988)	75.60; S11.2
J. GWIAZDA	(a) **108**, K67 (1988)	76.80; 75.50; S1.1
J. GYULAI	(a) **107**, K15 (1988)	61.80 and 68.48; S1.3; S7.12
	(a) **108**, K35 (1988)	61.70 and 68.48; 78.65; S5.11
I. GYÚRÓ	(a) **105**, K129 (1988)	72.80; S7.13
K. GZARA	(b) **148**, K71 (1988)	78.20

R. J. Haange	(a) **107**, 817 (1988)	61.60; S8
	(a) **107**, 873 (1988)	61.60; S8
M. Haas	(b) **149**, 283 (1988)	76.80
P. Haasen	(a) **108**, 553 (1988)	61.70 and 62.40; S5.12
	(a) **109**, K19 (1988)	62.20; S8.11
A. Hachimi	(b) **149**, 711 (1988)	78.20 and 78.50; 61.70; 78.55; S9.12
M. Hada	(a) **108**, 295 (1988)	68.55 and 75.70; 76.80; S1.1
V. Hadjicontis	(a) **105**, K87 (1988)	66.30; S5.11
	(b) **147**, 83 (1988)	61.70; S9.11
	(a) **107**, K109 (1988)	66.30; S1.1
A. A. A. Hady	(a) **106**, K191 (1988)	78.40; S8.15
	(a) **107**, K25 (1988)	73.60; 68.55; S8.15
H. Haefner	(a) **107**, 405 (1988)	73.40; 71.55; 72.40; 78.60; S7.11; S7.15
	(a) **107**, 419 (1988)	73.40 and 78.60; 85; S7.12
P. C. Hägele	(b) **149**, 521 (1988)	61.70 and 63.20; S12
N. T. Q. Hai	(a) **109**, K115 (1988)	71.55; 73.40; 85; S7.11
S. Hajaig	(a) **106**, 49 (1988)	61.14 and 61.70; S1.2
S. A. Hakeem	(a) **106**, 635 (1988)	78.60
T. Halicioglu	(b) **149**, 81 (1988)	63.10; 61.50; 68.40; S5.11
N. A. Halim	(a) **107**, 253 (1988)	61.80 and 73.60; S1.2; S5.11
G. C. Hallam	(a) **107**, K157 (1988)	75.60; S1.1
D. P. Halliday	(b) **150**, 177 (1988)	71.55; 72.20; 78.50; 78.55; S7.11
C. Hamaguchi	(a) **106**, 283 (1988)	78.55; S8.12
A. Hamaïdia	(b) **149**, 711 (1988)	78.20 and 78.50; 61.70; 78.55; S9.12
	(b) **150**, 323 (1988)	78.20 and 78.50; 61.70; 76.30; S9.12
C. Hamann	(b) **149**, K85 (1988)	78.20; S12
	(a) **110**, K29 (1988)	68.55; S1.3
M. Hamerský	(a) **107**, K11 (1988)	62.20; S1
S. Handschack	(b) **145**, 157 (1988)	73.40; S7.13
	(b) **147**, 823 (1988)	73.40; S7.1
	(b) **147**, K125 (1988)	71.25 and 73.40; S7.13
Hanjie Fan	see under letter "F"	
Han Li	see under letter "L"	
Hanmin Jin	see under letter "J"	
D. Hans	(b) **147**, 529 (1988)	62.20; 63.90; 77.20; S9.11
	(b) **150**, 47 (1988)	62.20; S9.12; S10.1
M. I. Haque	(b) **145**, K51 (1988)	76.30; S12
G. Harburn	(a) **108**, 111 (1988)	61.14 and 64.80; S10.15
A. Harendt	(b) **148**, 407 (1988)	78.20; S8.11
T. M. Haridasan	(b) **145**, 139 (1988)	63.20; 61.70; S8.13
	(b) **147**, 495 (1988)	66.30; S9
N. Hariharan	(a) **106**, 67 (1988)	61.80; 76.30; S12
H. J. Haritonidis	(b) **148**, 653 (1988)	72.20
M. Hartmann	(b) **146**, 357 (1988)	71.25 and 72.20; 72.10
	(b) **150**, 365 (1988)	71.35 and 78.20
U. Hartmann	(a) **108**, 387 (1988)	75.60; S1.1
	(a) **110**, 247 (1988)	75.60; S1.1
H. W. den Hartog	(b) **148**, K77 (1988)	78.55; S9.12
J. Härtwig	(b) **145**, 363 (1988)	61.10 and 61.50; 61.70; S5.11
	(a) **105**, 61 (1988)	61.10 and 61.70; S10.1
	(a) **109**, 79 (1988)	61.10; 61.70; 62.20; S10.1
S. Häseler	(b) **148**, K93 (1988)	63.20 and 78.30; S8.16
H. Hashimoto	(a) **107**, 611 (1988)	61.14
I. Hashimoto	(a) **107**, 123 (1988)	61.70; 62.20; 72.15; S4
M. Hashimoto	(a) **108**, 163 (1988)	61.70; 64.75; S1.2; S1.3
A. R. Hassan	(b) **147**, 211 (1988)	71.25 and 71.35; S7.11; S9

M. A. Hassan	(a) **105**, 609 (1988)	72.20 and 78.20; 78.30; S10.15
	(a) **106**, 191 (1988)	72.80; 76.30; 78.30; S10.15
T. Hattori	(b) **148**, K47 (1988)	75.30; S9
H. Haug	(b) **150**, 357 (1988)	71.35 and 78.20; S8.11
	(b) **150**, 393 (1988)	71.35 and 78.20
R. Haupt	(b) **149**, K123 (1988)	71.45 and 73.40
K. Havancsak	(a) **106**, 371 (1988)	61.80; 62.20; S1.2
Ha Vinh Tan	(b) **147**, K63 (1988)	74.20
Y. Hayashi	(a) **107**, 619 (1988)	61.55 and 66.30; S1.2
H. He	(b) **147**, 185 (1988)	61.40 and 71.20
Jiaqing He	(a) **110**, 317 (1988)	61.40; 64.70; S1.2
Yizhen He	(a) **105**, 325 (1988)	61.40; 65.40; S1.1; S1.2
	(a) **109**, 39 (1988)	61.40 and 64.70; S1.1; S1.2
	(a) **109**, K1 (1988)	61.40; S1.61
	(a) **110**, 317 (1988)	61.40; 64.70; S1.2
	(a) **110**, K67 (1988)	64.70 and 65.50; S1.3
Zhigang He	(a) **107**, K123 (1988)	68.55 and 78.65; S8
D. Hecht	(a) **108**, 265 (1988)	73.60 and 79.20; S10.1
G. M. Hedge	(a) **109**, K125 (1988)	72.80 and 73.40; S12
L. Hedman	(a) **106**, 233 (1988)	75.50; S1.2; S1.4
E. Hegenbarth	(b) **147**, 103 (1988)	66.70; S5.11
K. Hehl	(b) **149**, 111 (1988)	63.20; 68.30; 71.35
M. Heider	(a) **105**, 661 (1988)	73.40 and 85; 78.60; S7.15
W. Heiland	(a) **105**, K17 (1988)	66.30 and 79.20; 78.20; S11.1
A. Heilmann	(a) **110**, K29 (1988)	68.55; S1.3
R. Heilmann	(a) **106**, K113 (1988)	78.45 and 85; 73.40; S7.11; S7.16
	(a) **107**, 405 (1988)	73.40; 71.55; 72.40; 78.60; S7.11; S7.15
W. Heimbrodt	(b) **146**, 11 (1988)	78.20; 71.35; 78.55; S8.1
	(b) **146**, K65 (1988)	78.40 and 78.55; 71.35; S8.11
	(a) **108**, 443 (1988)	78.20; S8.13
E. Heiner	(b) **145**, 233 (1988)	71.38 and 71.50; S12
	(b) **146**, 655 (1988)	72.10; 72.20
	(b) **147**, 281 (1988)	71.50; 71.38; 71.45; S12.1
	(b) **148**, 599 (1988)	71.10; 65.40; 71.38; S1.3
	(b) **149**, K153 (1988)	72.10; 72.20
H. Hejase	(b) **149**, 685 (1988)	75.70; S1.2
B. Hejda	(b) **148**, K129 (1988)	72.10; S7.12
K. Helmling	(b) **150**, K1 (1988)	61.10; 61.12; 61.14
H. Hempel	(a) **110**, 459 (1988)	64.70 and 77.80; S11
Hengde Li	see under letter "L"	
Z. Henkie	(b) **146**, K131 (1988)	72.15 and 74.70; S10.15
F. Henneberger	(b) **145**, K83 (1988)	78.20; S8.12
	(b) **150**, 419 (1988)	71.35 and 78.20; S8.11; S8.12; S8.15
	(b) **150**, 495 (1988)	78.20; S8.12
	(b) **150**, 641 (1988)	78.20; 71.35; S8.1; S9.11
	(b) **150**, 735 (1988)	78.20; 78.65; S7.15
	(b) **150**, 769 (1988)	71.35 and 78.20; S8.11
K. Henneberger	(b) **147**, 733 (1988)	71.35 and 78.20; 73.40; S7.15
	(b) **150**, 425 (1988)	71.35 and 78.20; S8.11
	(b) **150**, 691 (1988)	71.35 and 78.20; S7.12; S7.15
K. Henniger	(a) **109**, K93 (1988)	62.20
U. Henniger	(a) **108**, 469 (1988)	73.40 and 85
A. Henry	(a) **107**, 101 (1988)	61.70; 72.80; 78.30; S5.11
W. Hergert	(a) **109**, 261 (1988)	73.40; 78.60; S7.12
	(a) **109**, 625 (1988)	78.55 and 78.60; 72.10
	(a) **110**, 283 (1988)	78.60; S7.12; S7.15
R. Hergt	(a) **109**, 503 (1988)	68.55; 74.70; S10.15

M. Herms	(b) **145**, K79 (1988)	63.20 and 78.30; 71.45; S7.15
K. Herrman	(a) **106**, 509 (1988)	68.55 and 78.55; S8
R. Herrmann	(b) **145**, 157 (1988)	73.40; S7.13
	(b) **146**, K25 (1988)	74.70; S10.15
	(b) **147**, 297 (1988)	71.55 and 78.50; 72.40; 76.40; S5
	(b) **147**, K159 (1988)	74.70; S10.15
	(b) **148**, 143 (1988)	76.40 and 78.20; 73.40; S7.13
	(b) **148**, 333 (1988)	71.30 and 72.20; S7.13
	(b) **148**, 349 (1988)	73.25 and 73.40; S7.13
	(b) **148**, 567 (1988)	73.40; S7.13
	(b) **150**, 119 (1988)	73.40 and 76.40; 72.10; S7.13
P. Hertel	(a) **108**, 449 (1988)	78.65; S11.1
W. Hertle	(a) **105**, K7 (1988)	61.40 and 78.70; 61.80; S12
M. Hervieu	(a) **107**, 73 (1988)	61.60; 64.80; S10.15
G. Hess	(b) **145**, 729 (1988)	78.20; 64.60; S11.1
D. Hesse	(a) **109**, 217 (1988)	68.48 and 68.55; S1.61
A. H. Heuer	(a) **107**, 571 (1988)	68.20; 64.60; S10
J. Heydenreich	(a) **107**, 807 (1988)	61.60; 61.16; S11
	(a) **109**, K3 (1988)	61.14; 61.70; S8.13
M. Heyting	(a) **108**, 97 (1988)	61.50 and 72.15; S4
M. Hidaka	(b) **145**, 649 (1988)	72.60 and 74.10; S10.15
T. D. Hien	(a) **106**, 201 (1988)	75.30; S1.2; S1.4
Nguyen Van Hieu	see under letter "N"	
T. Higashimura	(b) **150**, 431 (1988)	78.20; 71.35
S. Hildebrandt	(a) **109**, 625 (1988)	78.55 and 78.60; 72.10
	(a) **110**, 283 (1988)	78.60; S7.12; S7.15
K. Hiratsuka	(a) **108**, 509 (1988)	61.70; 61.16; S1.2
	(a) **109**, 119 (1988)	61.14
C. Hirlimann	(b) **147**, 779 (1988)	78.30 and 78.55; S7.11
A. A. Hirsch	(a) **105**, 249 (1988)	75.30; 76.80; S1.1; S1.2
M. El-Hiti	see under letter "E"	
V. Hizhnyakov	(b) **149**, 283 (1988)	76.80
T. X. Hoai	(a) **106**, K147 (1988)	73.60; 78.65; S10.1
	(b) **147**, K165 (1988)	72.40 and 78.55; S8.12
N. H. Hoang	(a) **106**, K147 (1988)	73.60; 78.65; S10.1
	(a) **108**, K157 (1988)	78.65; S10.1
Hoang Xuan Nguyen	see under letter "N"	
H. Hobert	(a) **110**, 181 (1988)	68.55 and 78.65; S5.11
H.-R. Höche	(a) **105**, 53 (1988)	61.10; S5.12
	(a) **108**, 489 (1988)	61.10
S. A. Hodorowicz	(a) **107**, 347 (1988)	76.60; S11
C. A. Hogarth	(b) **145**, 145 (1988)	78.65; 68.55; S10.15
	(a) **105**, 609 (1988)	72.20 and 78.20; 78.30; S10.15
	(a) **106**, 191 (1988)	72.80; 76.30; 78.30; S10.15
	(b) **149**, K81 (1988)	78.65; S10.1
	(b) **149**, K187 (1988)	78.65; S10.1
	(a) **109**, 469 (1988)	78.65; 68.55; S10.1
	(a) **110**, K39 (1988)	73.60; S10.1
H. J. Höhne	(b) **149**, K85 (1988)	78.20; S12
R. Höhne	(a) **109**, 575 (1988)	75.30 and 75.50; S11.2
J. M. Holender	(a) **110**, 123 (1988)	64.70; 61.10; 61.55; S1.2; S4
L. Holtmann	(a) **106**, 89 (1988)	61.70; 72.40; 78.20; S11.1
V. Holý	(b) **145**, 363 (1988)	61.10 and 61.50; 61.70; S5.11
	(a) **105**, 61 (1988)	61.10 and 61.70; S10.1
R. Holzberger	(b) **147**, K159 (1988)	74.70; S10.15
W. Holzhäuser	(b) **146**, K29 (1988)	74.70; S10.15

B. HÖNERLAGE	(b) **149**, 775 (1988)	78.20; S9
	(b) **150**, 825 (1988)	78.20; S9
N. M. HONG	(a) **106**, 201 (1988)	75.30; S1.2; S1.4
NGUYEN VAN HONG	see under letter "N"	
HONGJIAN LIU	see under letter "L"	
HONG-RUI ZHANG	see under letter "Z"	
HONG WANG	see under letter "W"	
HONG ZHANG	see under letter "Z"	
T. HÖNIG	(b) **146**, 589 (1988)	71.35 and 71.55; 78.45; S8.11
	(b) **150**, 833 (1988)	78.20; S8.11
R. HOOGEWIJS	(a) **107**, 625 (1988)	68.20; 79.60; S1.2; S10
J. HORÁK	(b) **147**, 501 (1988)	61.50 and 61.70; 72.80; S8; S8.15
	(a) **109**, 93 (1988)	61.60; 61.70; S8
	(b) **150**, 237 (1988)	72.20; S8
Y. HORIE	(b) **145**, 649 (1988)	72.60 and 74.10; S10.15
N. J. M. HORING	(b) **146**, 189 (1988)	72.20 and 72.70
S. HORISAKA	(b) **148**, 659 (1988)	72.20; S10
G. HORVÁTH	(a) **110**, 549 (1988)	73.60
R. HORÝN	(b) **146**, K103 (1988)	74.70; S10.15
	(b) **146**, K131 (1988)	72.15 and 74.70; S10.15
P. HÖSCHL	(b) **145**, 637 (1988)	72.20; S8.15
	(b) **147**, 621 (1988)	72.20; S8.15
	(b) **148**, 319 (1988)	72.20; S8.15
T. J. C. HOSEA	(b) **146**, K153 (1988)	78.50; S10.1
J. HREBÍČEK	(a) **106**, 167 (1988)	71.55; 72.20
M. HROBÁR	(a) **107**, 235 (1988)	73.40; 77.55; S10.15
YUANFU HSIA	(a) **110**, K89 (1988)	64.70 and 74.70; S10.15
AN HU	(a) **107**, 153 (1988)	63.20 and 68.30; 61.55; S1.2
MEI-SHEN HU	(a) **109**, 435 (1988)	64.70; 61.16; S11
T. B. HU	(b) **147**, 471 (1988)	61.14; 61.90
J. HUANG	(a) **110**, 415 (1988)	66.30; S10
J. Q. HUANG	(a) **110**, 55 (1988)	61.60 and 74.70; S11
L. J. HUANG	(a) **110**, 443 (1988)	64.75; 71.25; S1.3
HUANG MING	see under letter "M"	
HUAQUIN WANG	see under letter "W"	
F. K. HÜBNER	(a) **106**, 239 (1988)	75.60
HUILIN JIANG	see under letter "J"	
HUIMIN SHAO	see under letter "H"	
HUI SHEN	see under letter "S"	
E. HUIZER	(a) **105**, 235 (1988)	72.15; 61.40; S1.1
D. HULIN	(b) **150**, 357 (1988)	71.35 and 78.20; S8.11
R. E. HUMMEL	(a) **106**, 641 (1988)	61.55 and 78.20; S1.2
	(b) **147**, 417 (1988)	71.50 and 78.20; S1.2
	(a) **107**, K175 (1988)	66.30 and 85; S4
N. D. HUNG	(b) **147**, K165 (1988)	72.40 and 78.55; S8.12
VU VAN HUNG	see under letters "VU"	
J. J. HUNTER	(b) **149**, 383 (1988)	78.20; 71.55; S7.13
	(b) **150**, 719 (1988)	78.20; S7.13
H. D. HUSSEINOV	(b) **148**, K177 (1988)	78.30; 78.50; S8; S8.15
S. H. HUSSEINOV	(b) **148**, K177 (1988)	78.30; 78.50; S8; S8.15
J. M. HVAM	(b) **150**, 387 (1988)	71.35 and 78.20; S8.12; S8.15
S. HYODO	(a) **109**, 383 (1988)	61.70 and 62.20; S7.12

P. IACCONI	(a) **108**, 783 (1988)	78.60; 61.80; S10
A. IBARRA	(a) **105**, 601 (1988)	77.30 and 77.40

T. D. Ibragimov	(b) **145**, K103 (1988)	63.20 and 64.70; 78.30; S8.16
	(b) **148**, K89 (1988)	63.20 and 78.30; S8.16
T. Idogaki	(b) **148**, K47 (1988)	75.30; S9
B. Idzikowski	(b) **147**, K47 (1988)	72.20; S1.2
	(a) **108**, 375 (1988)	75.60; 75.80; S1.1; S1.4
	(a) **108**, 747 (1988)	75.50; 61.80; S1.1; S1.2
M. Igalson	(a) **108**, K169 (1988)	73.40 and 85; 72.40; S8.11
J. R. Iglesias	(b) **145**, 525 (1988)	71.20
T. Iida	(b) **150**, 431 (1988)	78.20; 71.35
T. Ikari	(a) **108**, K53 (1988)	71.55 and 72.80; S8
I. Ileo	(a) **105**, 573 (1988)	75.50; S11.2
A. V. Ilinskii	(b) **150**, 611 (1988)	78.20; 72.40; S10
V. Ilkovič	(b) **150**, K55 (1988)	75.10
K. Imai	(b) **149**, K183 (1988)	63.20 and 78.30; S8
T. Imura	(a) **108**, 163 (1988)	61.70; 64.75; S1.2; S1.3
K. Inabe	(b) **149**, K67 (1988)	61.70 and 78.50; S9.15
	(b) **149**, K101 (1988)	61.70; 78.50; S9.15
T. Inagaki	(a) **105**, 617 (1988)	78.20; 78.65; S1.3
J. I. Iñiguez	(a) **108**, 721 (1988)	75.50; S11.2
M. Inoue	(b) **147**, 559 (1988)	61.60 and 72.20; S10
	(b) **148**, 659 (1988)	72.20; S10
A. N. Ionov	(b) **146**, 199 (1988)	72.20; 72.60; S7.12
T. Irie	(b) **148**, 673 (1988)	72.20; 71.25; S8.16
G. Irmer	(b) **145**, K79 (1988)	63.20 and 78.30; 71.45; S7.15
	(a) **105**, 549 (1988)	61.70 and 72.20; 78.30; S7.11
	(b) **146**, K161 (1988)	63.20 and 78.30; S8.13
S. Iselborn	(a) **108**, 275 (1988)	68.55; 78.30; S5.11
H. Ishikawa	(b) **146**, 661 (1988)	72.10 and 72.80; 75.30; S1.63
	(b) **149**, 257 (1988)	72.20 and 75.30; S1.63; S8.16
S. Ishio	(a) **107**, 355 (1988)	75.50; 75.60; S1.1
A. O. Ismailov	(a) **110**, K45 (1988)	65.70; 75.50; S8
M. Isomae	(b) **148**, 673 (1988)	72.20; 71.25; S8.16
T. Itoh	(b) **145**, 567 (1988)	61.10 and 71.35; 78.20; S9
	(b) **146**, 531 (1988)	71.35; 78.55; S9
E. V. Ivakin	(b) **150**, 901 (1988)	66.70 and 78.35; S6
A. L. Ivanov	(b) **150**, 443 (1988)	78.20; 71.35; 71.36; S8.11
V. A. Ivanshin	(b) **148**, K61 (1988)	76.30 and 77.80; S11.1
E. L. Ivchenko	(b) **150**, 673 (1988)	71.35 and 78.20; 71.25; S7.12; S7.15
G. D. Ivlev	(a) **106**, K123 (1988)	64.70; S5.11
	(a) **110**, 155 (1988)	61.80 and 73.60; 68.60; S10.1
Y. Iwabuchi	(b) **145**, 567 (1988)	61.10 and 71.35; 78.20; S9
	(b) **146**, 531 (1988)	71.35; 78.55; S9
M. Iwamatsu	(b) **150**, K39 (1988)	71.25; S8

A. Jabłoński	(a) **110**, 495 (1988)	68.20; 79.20; S1.1; S1.2
J. Jabłoński	(a) **105**, 113 (1988)	61.70; S5.11
I. J. Jabr	(a) **105**, 177 (1988)	61.80 and 64.75; 73.60; S1.3; S5
A. Jacobi	(b) **147**, K159 (1988)	74.70; S10.15
G. Jacobi	(a) **108**, 171 (1988)	61.70; 61.55; 62.20; S4
B. Jacobs	(a) **105**, 661 (1988)	73.40 and 85; 78.60; S7.15
	(a) **107**, 419 (1988)	73.40 and 78.60; 85; S7.12
C. S. Jacobsen	(b) **145**, K149 (1988)	78.30; S12
C. Jagadish	(a) **110**, 671 (1988)	73.40 and 85
E. Jäger	(b) **149**, 111 (1988)	63.20; 68.30; 71.35
H. U. Jäger	(a) **105**, 387 (1988)	61.80; S1.62; S5.11

L. Jahn	(a) **106**, K79 (1988)	75.50; S10
	(b) **150**, 289 (1988)	75.60
E. Jahne	(b) **146**, K157 (1988)	63.20 and 78.30; S8.11
F. Jahnke	(b) **150**, 425 (1988)	71.35 and 78.20; S8.11
D. C. Jain	(b) **147**, 343 (1988)	74.10; 78.70; S11
J. D. Jain	(a) **106**, K53 (1988)	61.50 and 74.70; S10.15
V. K. Jain	(b) **148**, K157 (1988)	76.30; S12
P. C. Jana	(a) **110**, K93 (1988)	74.70; S10.15
I. Jánossy	(b) **150**, 783 (1988)	78.20; 78.65; S8
L. Jansa	(b) **146**, K75 (1988)	78.30; S8.16
J. Janszky	(a) **107**, 161 (1988)	62.20 and 62.65; S11.1
F. Januschewski	(a) **106**, K215 (1988)	73.40 and 85; 72.40; S5.11
K. Jarašiūnas	(b) **150**, 879 (1988)	78.20; S5.11; S7.12; S7.13; S8.12
F. Jarka	(b) **150**, 357 (1988)	71.35 and 78.20; S8.11
E. Jároli	(a) **107**, K15 (1988)	61.80 and 68.48; S1.3; S7.12
	(a) **108**, K35 (1988)	61.70 and 68.48; 78.65; S5.11
B. Jassemnejad	(a) **108**, 753 (1988)	78.60; S9.12
S. M. Javed Akhtar	see under letter "A"	
V. S. Jayakumar	(a) **109**, 635 (1988)	78.30; S11
G. Jeandel	(a) **108**, 183 (1988)	66.30 and 77.50; S9.12
R. J. Jelitto	(b) **147**, 391 (1988)	75.10
S. U. Jen	(a) **106**, 99 (1988)	65.70; 75.80; S1.61
U. Jendrich	(a) **108**, 553 (1988)	61.70 and 62.40; S5.12
T. A. Jenkins	(a) **108**, 343 (1988)	72.15; S1.2
H. W. Jeon	(a) **108**, K77 (1988)	78.20; S8.15
K. S. Jerath	(b) **147**, 343 (1988)	74.10; 78.70; S11
D. N. Jha	(b) **145**, 277 (1988)	72.10; 72.15; S1.1; S1.2; S1.3; S1.4
Fuquan Ji	(a) **110**, K25 (1988)	68.48; S6; S10.1
Desheng Jiang	(b) **145**, 167 (1988)	78.20; 71.25; 71.35; S7.12; S7.15
Huilin Jiang	(a) **109**, K129 (1988)	74.70 and 78.70; S10.15
P. H. Jiang	(b) **145**, K111 (1988)	73.40; S7.12; S7.15
S. S. Jiang	(a) **110**, 323 (1988)	61.10; 61.70; S5
Jian-Min Chen	see under letter "C"	
Jianshi Zhou	see under letter "Z"	
A. Jianu	(a) **109**, 61 (1988)	61.55; S1.1; S1.2
Jiaquing He	see under letter "H"	
Jiayou Feng	see under letter "F"	
Lu Jie	(b) **150**, 749 (1988)	78.20; 78.55; S8.15
D. C. Jiles	(a) **108**, 417 (1988)	75.60; 62.20; S1.1
I. Jiménez	(a) **106**, 11 (1988)	61.40 and 68.55; 78.30; S5.11
M. Jiménez de Castro	(a) **105**, 601 (1988)	77.30 and 77.40
Hanmin Jin	(b) **145**, 679 (1988)	75.30; 71.70; S1.1; S1.4
Xin Jin	(a) **110**, K89 (1988)	64.70 and 74.70; S10.15
Zhun-Zhi Jin	(b) **145**, 73 (1988)	61.25; S1; S1.2
	(b) **147**, 45 (1988)	61.25; S1; S1.2
X. Jing	(a) **107**, 63 (1988)	61.60 and 74.70; S10.15
Jinghua Tian	see under letter "T"	
Jingkul Liang	see under letter "L"	
Jin-Song Zhu	see under letter "Z"	
S. Jitsukawa	(a) **108**, 163 (1988)	61.70; 64.75; S1.2; S1.3
M. Joffre	(b) **150**, 357 (1988)	71.35 and 78.20; S8.11
J. Jõgi	(b) **149**, 283 (1988)	76.80
St. Joksch	(a) **107**, 85 (1988)	61.10; 65.70; 68.20; S5.11
A. F. de Jong	(a) **107**, 521 (1988)	61.16; 61.70
K. S. Joseph	(a) **106**, 123 (1988)	68.55 and 78.65; S8
J. P. Jouart	(b) **149**, 633 (1988)	71.55; 78.50; 78.55; S9.12
C. Jourdan	(a) **109**, 403 (1988)	61.70; 61.10; S4

JUN CHEN	see under letter "C"	
G. JUNGK	(b) **146**, 335 (1988)	78.20
	(b) **150**, 483 (1988)	78.20
P. JUNGWIRTH	(b) **147**, K149 (1988)	71.55
S. JURŠENAS	(b) **150**, 699 (1988)	71.35 and 78.55; S7.12; S7.15
N. M. KABACHNIK	(b) **145**, 103 (1988)	61.80; S4; S5.11
S. S. KABALKINA	(a) **108**, 89 (1988)	61.10; 64.70; S8
R. KABULOV	(a) **109**, 211 (1988)	73.40; 73.60; S5.11
C. M. KACHHAVA	(b) **147**, 583 (1988)	67.40; S5.2
G. A. KACHURIN	(a) **109**, 141 (1988)	61.80 and 66.30; S5.11
V. N. KADANTSEV	(b) **147**, 155 (1988)	71.45
S. KADEČKOVÁ	(a) **105**, K37 (1988)	62.40; 61.70; S1.1
	(a) **106**, 17 (1988)	61.55; 61.10; S1.1; S1.61
H.-MATSUO KAGAYA	(b) **147**, 109 (1988)	64.75; 62.50; S6
	(b) **149**, 477 (1988)	61.70; S5.11; S5.12
H. G. KAHLE	(b) **145**, 729 (1988)	78.20; 64.60; S11.1
	(b) **147**, 383 (1988)	75.50; 75.25; 75.30; S11.1
H. KAHNT	(a) **108**, 149 (1988)	66.30; 72.20; S8
R. KAHNT	(b) **147**, 281 (1988)	71.50; 71.38; 71.45; S12.1
Y. KAIFU	(b) **150**, 635 (1988)	71.35 and 78.20; 71.70; 73.40; S9
Z. KĄKOL	(b) **145**, 687 (1988)	75.50; S1.2; S1.4
J. KALINOWSKI	(b) **149**, 363 (1988)	71.55 and 78.60; S12
J. A. KALOMIROS	(a) **107**, 633 (1988)	78.20; S8.16
H. KALT	(b) **150**, 895 (1988)	78.20 and 78.45; 78.55; S7.15
K. KAMADA	(b) **147**, 109 (1988)	64.75; 62.50; S6
S. KAMBA	(b) **146**, 743 (1988)	63.20 and 78.30; 74.70; S10.15
Y. KAMEDA	(b) **149**, 257 (1988)	72.20 and 75.30; S1.63; S8.16
I. V. KAMENSKAYA	(a) **105**, K141 (1988)	61.80 and 72.80; S7.13
A. A. KAMINSKII	(a) **105**, K155 (1988)	78.45; 78.55; S10
	(a) **107**, 365 (1988)	78.30; S11.2
H. KAMLEH	(a) **105**, 661 (1988)	73.40 and 85; 78.60; S7.15
T. KANADANI	(a) **110**, K9 (1988)	64.75; S4
T. KANEYOSHI	(b) **146**, 253 (1988)	75.30; 75.50
	(b) **150**, 261 (1988)	75.10 and 75.40
	(b) **150**, 297 (1988)	75.50
Z. KANG	(a) **107**, 469 (1988)	68.48; 61.16; S10
YU. O. KANTER	(a) **107**, K19 (1988)	68.48; 62.20; S7.12; S7.15
	(a) **110**, 61 (1988)	61.14 and 68.55; S1.2
M. G. KAPLUNOV	(a) **110**, K111 (1988)	78.30; 71.38; S12
N. KAPOOR	(a) **105**, 403 (1988)	66.30 and 72.15; S1.2
J. KARABA	(b) **149**, K167 (1988)	75.10
P. M. KARAGEORGY-ALKALAEV	(a) **108**, 323 (1988)	72.20 and 73.40; S5.11
TH. KARAKOSTAS	(a) **107**, 551 (1988)	61.70; 61.16
	(a) **107**, 559 (1988)	68.55; S1.2
T. KARASAWA	(b) **150**, 635 (1988)	71.35 and 78.20; 71.70; 73.40; S
A. M. KARGUPPIKAR	(a) **109**, 549 (1988)	72.80 and 78.50; S1.63
A. G. KAROZA	(a) **106**, 277 (1988)	78.20; 72.40; S11.2
	(a) **110**, K51 (1988)	75.50 and 78.30; S11.2
F. V. KARPUSHKO	(b) **150**, 791 (1988)	73.40 and 78.20; S7.12; S8.12
V. V. KARYAGIN	(b) **148**, 501 (1988)	71.25 and 73.40
M. KASCHKE	(b) **146**, 351 (1988)	71.35 and 78.20; S8.11; S8.15
R. KASCHNER	(b) **145**, K9 (1988)	68.25 and 68.40; 71.10
D. KASHCHIEV	(a) **110**, 61 (1988)	61.14 and 68.55; S1.2

M. KASNER	(b) **148**, 273 (1988)	71.50
	(b) **148**, 635 (1988)	71.50
A. KASTEN	(b) **145**, 729 (1988)	78.20; 64.60; S11.1
G. KÄSTNER	(b) **145**, 373 (1988)	61.14
	(b) **146**, 63 (1988)	61.14
T. K. KASUMOV	(a) **107**, K49 (1988)	72.40; S8.16
M. KATAOKA	(b) **145**, 567 (1988)	61.10 and 71.35; 78.20; S9
V. N. KATERINCHUK	(a) **108**, K115 (1988)	73.40; 85; S8; S10.1
S. P. KATHURIA	(a) **106**, 271 (1988)	78.60; S9.11
V. M. KATKOV	(b) **149**, 403 (1988)	61.80 and 78.70
E. Z. KATSNELSON	(a) **106**, 277 (1988)	78.20; 72.40; S11.2
	(a) **110**, K51 (1988)	75.50 and 78.30; S11.2
J. KAVALIAUSKAS	(a) **107**, K75 (1988)	78.30; S8.15
K. KAWABATA	(a) **106**, 181 (1988)	71.55 and 79.75; 78.60; S10.1
R. KAWASHIMA	(a) **108**, K73 (1988)	77.80; S11.1
A. E. KAZACHENKO	(b) **150**, 647 (1988)	78.20; S8.15
L. A. KAZAKEVICH	(a) **105**, 97 (1988)	61.80; S5.11
A. A. KAZAKOV	(b) **145**, 291 (1988)	75.30
	(b) **145**, 659 (1988)	75.10; 75.30; 75.40; S1.4
M. M. KAZANIN	(a) **108**, 323 (1988)	72.20 and 73.40; S5.11
A. YU. KAZIMIROV	(b) **150**, 9 (1988)	61.10 and 78.70; S5.11
T. S. KÊ	(a) **105**, 447 (1988)	62.40; 64.70; S1.2
	(a) **107**, 177 (1988)	62.40; S4
	(a) **108**, 569 (1988)	62.40; S4
S. W. S. MCKEEVER	see under letter "M"	
R. KEIPER	(b) **148**, K41 (1988)	72.20
N. P. KEKELIDSE	(a) **108**, K93 (1988)	61.80; 78.50; S7.11; S7.12
KELIN WANG	see under letter "W"	
P. KELLER	(a) **108**, 783 (1988)	78.60; 61.80; S10
M. A. KENAWY	(a) **107**, 171 (1988)	62.40; S4
H. A. MACKENZIE	see under letter "M"	
KEQIN XIAO	see under letter "X"	
F. R. KESSLER	(a) **105**, 627 (1988)	78.20; 61.70; 66.30; 78.50; S5.11
	(b) **147**, K143 (1988)	71.35 and 72.40; 78.20; S5.12
U. R. KEVORKYAN	(a) **106**, 379 (1988)	61.70; S1.1
P. K. KHABIBULLAEV	(a) **108**, 189 (1988)	64.60; 77.40; S11.1
P. I. KHADZHI	(b) **147**, 721 (1988)	71.35 and 78.20
	(b) **147**, 741 (1988)	78.20; 71.35; 73.40
	(b) **150**, 513 (1988)	78.20
J. M. KHALIFEH	(a) **107**, 253 (1988)	61.80 and 73.60; S1.2; S5.11
S. KH. KHALILOV	(a) **108**, 619 (1988)	72.40 and 73.40; S8
M. A. KHAN	(b) **149**, 615 (1988)	71.35; S5.11; S5.12; S7.1; S8.1
M. MEHAR KHAN	see under letter "M"	
M. N. KHAN	(a) **105**, 609 (1988)	72.20 and 78.20; 78.30; S10.15
	(a) **106**, 191 (1988)	72.80; 76.30; 78.30; S10.15
D. D. KHANG	(a) **109**, K115 (1988)	71.55; 73.40; 85; S7.11
N. Q. KHANH	(a) **108**, K35 (1988)	61.70 and 68.48; 78.65; S5.11
Z. KHÁS	(b) **148**, K129 (1988)	72.10; S7.12
A. G. KHASABOV	(b) **146**, K21 (1988)	71.25 and 78.70; S9
L. V. KHASHCHINA	(b) **147**, 757 (1988)	78.30; 63.20; S5.2
F. H. AL-KHATEEB	see under letter "A"	
F. KHATER	(a) **108**, 241 (1988)	73.60; 72.15; S1.3
	(a) **109**, 517 (1988)	72.20 and 73.60; S1.3
V. G. KHLABUTIN	(b) **149**, K129 (1988)	71.25; S5.11
W. I. KHLEIF	(b) **149**, K187 (1988)	78.65; S10.1
G. V. KHODOVA	(b) **150**, 545 (1988)	78.20

W. S. Khokle	(b) **145**,	269 (1988)	72.20; S5.11
	(a) **106**,	K53 (1988)	61.50 and 74.70; S10.15
V. G. Khomenko	(b) **145**,	445 (1988)	63.20; 61.65; S12
	(b) **150**,	39 (1988)	63.20 and 65.40; S12
V. F. Khopin	(a) **110**,	K107 (1988)	78.50; 76.30; S10.15
V. A. Khramtsov	(a) **109**,	127 (1988)	61.80; 76.30; S5.11
A. G. Khrapak	(b) **145**,	455 (1988)	68.10; 73.30; S2
B. I. Khudik	(b) **148**,	297 (1988)	74.20
V. P. Khvostikov	(b) **150**,	863 (1988)	71.35 and 78.20; S7.12; S7.15
S. Kianian	(a) **105**,	563 (1988)	72.80 and 78.20; S8.16
M. Kiknadze	(b) **146**,	K141 (1988)	76.70; S9.12
I. Kikuchi	(a) **108**,	K89 (1988)	61.40; S8
Y. Kikuchi	(a) **108**,	509 (1988)	61.70; 61.16; S1.2
	(a) **109**,	119 (1988)	61.14
Chang-Dae Kim	(b) **148**,	K173 (1988)	78.30; S8.16
H. K. Kim	(b) **148**,	K173 (1988)	78.30; S8.16
	(a) **108**,	K77 (1988)	78.20; S8.15
H. S. Kim	(a) **105**,	45 (1988)	61.14
K. M. Kim	(a) **108**,	K77 (1988)	78.20; S8.15
V. M. Kim	(a) **110**,	K107 (1988)	78.50; 76.30; S10.15
Wha-Tek Kim	(b) **148**,	K173 (1988)	78.30; S8.16
D. Kindl	(a) **106**,	297 (1988)	73.30 and 85; 73.60; S8.13
G. Kiriakidis	(b) **147**,	349 (1988)	72.20; S9.12
T. Kirihara	(b) **146**,	531 (1988)	71.35; 78.55; S9
K. Kirouni	(b) **148**,	723 (1988)	78.55; 71.55; S5.11
F.-G. Kirscht	(a) **109**,	279 (1988)	71.55; S5.11
Y. Kirsh	(b) **149**,	45 (1988)	61.70; 72.20; 78.60; S9.16
I. V. Kiryushin	(a) **109**,	161 (1988)	62.20; S5.11
E. S. Kiseleva	(b) **147**,	741 (1988)	78.20; 71.35; 73.40
V. S. Kishan Kumar	see under letters "Ku"		
S. Kishida	(a) **105**,	K165 (1988)	78.55; S8.12
	(a) **106**,	283 (1988)	78.55; S8.12
	(a) **109**,	617 (1988)	78.55; 71.55; S8.12
V. G. Kistol	(b) **148**,	K37 (1988)	72.15; S4
M. Kitamura	(b) **149**,	791 (1988)	78.70; S9
G. Kitis	(a) **105**,	K175 (1988)	78.60; S9.11; S9.12
R. Kittner	(a) **105**,	61 (1988)	61.10 and 61.70; S10.1
M. Kiwaki	(b) **147**,	K1 (1988)	61.70; 76.30; S10.1
A. Kjekshus	(a) **110**,	141 (1988)	64.70; 75.40; S7.15
R. Klabes	(a) **106**,	57 (1988)	61.80 and 68.55; 66.30; S8
	(a) **110**,	K81 (1988)	68.60; S8
J. Klamut	(b) **146**,	K103 (1988)	74.70; S10.15
	(b) **146**,	K131 (1988)	72.15 and 74.70; S10.15
M. Klaua	(a) **106**,	133 (1988)	68.40 and 68.55; 64.75; 79.20; S1.3
	(a) **106**,	493 (1988)	66.30 and 68.48; 68.60; S1.3
	(a) **109**,	525 (1988)	66.30 and 68.48
	(a) **109**,	K97 (1988)	66.30 and 68.48; 68.60; S1.3
	(a) **110**,	189 (1988)	61.70 and 68.55; S1.3
P. Kleinert	(b) **147**,	237 (1988)	71.20; 71.50
	(b) **148**,	325 (1988)	74.10; 74.20
	(b) **149**,	K111 (1988)	63.10
A. Klekamp	(a) **105**,	K17 (1988)	66.30 and 79.20; 78.20; S11.1
S. M. Kleshchev	(b) **145**,	291 (1988)	75.30
	(b) **145**,	659 (1988)	75.10; 75.30; 75.40; S1.4
Z. Kleszczewski	(b) **146**,	467 (1988)	63.20; S11.2
	(b) **147**,	409 (1988)	78.20
Z. Kletowski	(a) **108**,	363 (1988)	72.10; 72.15; S1.4

J. Klikorka	(b) **146**,	K75 (1988)	78.30; S8.16
	(b) **147**,	501 (1988)	61.50 and 61.70; 72.80; S8; S8.15
D. Klimm	(a) **106**,	81 (1988)	61.80; 78.70; S5.11; S7.11
V. I. Klimov	(b) **146**,	341 (1988)	71.35 and 78.20; S8; S8.11
	(b) **150**,	839 (1988)	78.20; S8.11
C. Klingshirn	(b) **150**,	477 (1988)	78.20; S8.11
	(b) **150**,	749 (1988)	78.20; 78.55; S8.15
A. A. Klochikhin	(b) **147**,	727 (1988)	78.55; 71.35; S8.11; S8.12
	(b) **148**,	185 (1988)	71.35; 78.20; S9
	(b) **150**,	777 (1988)	71.35 and 78.55; S8.11
H. A. Klose	(a) **106**,	73 (1988)	61.80; S7.11
M. Kłosek	(a) **106**,	K13 (1988)	62.40; S1.1; S1.2
G. Kluge	(a) **106**,	57 (1988)	61.80 and 68.55; 66.30; S8
	(a) **106**,	K47 (1988)	72.40; S8.16
	(a) **110**,	K81 (1988)	68.60; S8
N. I. Klyui	(b) **145**,	209 (1988)	71.25; 78.20; S5
A. Knauer	(a) **107**,	321 (1988)	72.20; 78.50; S7.11
S. Y. Knyazev	(a) **107**,	213 (1988)	68.55; 64.75; 71.55; S5.11
M. Kobayashi	(b) **147**,	141 (1988)	71.36 and 71.45; 68.30; 73.40
T. Kobayashi	(a) **150**,	667 (1988)	73.40 and 78.55; S7.12; S7.15
	(a) **110**,	375 (1988)	61.60; 61.16; S1.63; S9
M. Kobiyama	(a) **108**,	219 (1988)	62.40; S4
F. Koch	(a) **109**,	261 (1988)	73.40; 78.60; S7.12
	(a) **110**,	197 (1988)	73.60 and 78.60; S7.15
S. W. Koch	(b) **150**,	357 (1988)	71.35 and 78.20; S8.11
	(b) **150**,	379 (1988)	71.35 and 78.20; S8.11
V. V. Kochelab	(a) **108**,	67 (1988)	61.10; 61.70; S5.12
V. A. Kochelap	(b) **146**,	311 (1988)	78.20
	(b) **146**,	319 (1988)	78.20; S8.12
	(b) **150**,	489 (1988)	78.20; S8.11
A. S. Kochemirovskii	(a) **107**,	11 (1988)	73.25; 68.45; 72.40; 73.20; S8
V. P. Kochereshko	(b) **150**,	673 (1988)	71.35 and 78.20; 71.25; S7.12; S7.15
D. D. Koditsa	(a) **107**,	K169 (1988)	78.30; S8.16
Y. Koga	(a) **108**,	K53 (1988)	71.55 and 72.80; S8
E. M. Kogan	(b) **147**,	613 (1988)	72.20; 72.60; S8.16
	(b) **148**,	289 (1988)	72.20; 72.60; S8.1; S10
D. Köhler	(a) **105**,	377 (1988)	66.30; 61.70; 78.60; S10.1
H.-J. Köhler	(a) **106**,	K165 (1988)	74.50; S4
J. Köhler	(b) **147**,	797 (1988)	78.55; 71.35; 71.55; S11
V. G. Kohn	(a) **106**,	31 (1988)	61.10
S. Kokai	(a) **108**,	K89 (1988)	61.40; S8
E. Kokmeijer	(a) **105**,	235 (1988)	72.15; 61.40; S1.1
I. T. Kokov	(a) **105**,	441 (1988)	64.70; 61.60; 65.40; 77.20; 78.20; S9.16
J. Kolenda	(b) **150**,	735 (1988)	78.20; 78.65; S7.15
V. M. Koleshko	(a) **109**,	161 (1988)	62.20; S5.11
E. Kolley	(b) **145**,	585 (1988)	71.10; 71.45
W. Kolley	(b) **145**,	585 (1988)	71.10; 71.45
T. N. Kolobyanina	(a) **108**,	89 (1988)	61.10; 64.70; S8
A. Kołodziejczyk	(b) **145**,	687 (1988)	75.50; S1.2; S1.4
A. B. Kolpachev	(b) **148**,	205 (1988)	71.20; 74.70; S1.61; S1.62
A. A. Komarnitskii	(a) **107**,	261 (1988)	61.80 and 68.55; 64.70; 78.20; S5.11; S5.12
F. F. Komarov	(a) **107**,	K105 (1988)	61.80; S5.11
	(a) **109**,	231 (1988)	61.80 and 68.55; S1.61; S5.11
T. Komatsu	(b) **150**,	431 (1988)	78.20; 71.35
	(b) **150**,	635 (1988)	71.35 and 78.20; 71.70; 73.40; S9
K. Konan	(a) **107**,	273 (1988)	66.30 and 78.65; 73.60; S8
A. A. Konchits	(a) **109**,	537 (1988)	71.55; 72.80; 76.30; S5.11

V. N. Kondratev	(b) **145**, 103 (1988)	61.80; S4; S5.11
V. K. Kononenko	(b) **150**, 695 (1988)	78.20; S7.12
P. Konsin	(b) **149**, 11 (1988)	63.75 and 77.80; 64.70; S8; S11
M. Konwicki	(a) **106**, K71 (1988)	75.50; S11.2
	(a) **109**, 319 (1988)	75.50; 61.12; S11.2
Yu. S. Konyaev	(a) **105**, K29 (1988)	61.50 and 62.20; S7.13
T. K. Kopeć	(b) **147**, K37 (1988)	74.20; S10.15
P. S. Kopev	(b) **150**, 673 (1988)	71.35 and 78.20; 71.25; S7.12; S7.15
Th. Köpp	(a) **108**, 443 (1988)	78.20; S8.13
D. V. Korbutyak	(b) **145**, 535 (1988)	71.35 and 78.65; S9
Yu. N. Korkishko	(a) **110**, 397 (1988)	61.70 and 78.20; 61.10; S11.1
W. Korneta	(b) **146**, 697 (1988)	75.10 and 75.70
E. Korngiebel	(a) **108**, 171 (1988)	61.70; 61.55; 62.20; S4
A. A. Kornilova	(b) **150**, 25 (1988)	63.20 and 76.80; 62.50; S4
V. M. Korol	(a) **110**, 9 (1988)	61.80; 66.30; 72.80; 73.40; S5.11
V. S. Korolkov	(b) **149**, 231 (1988)	71.70; 78.50; S9.12
S. Yu. Korostelev	(b) **148**, 483 (1988)	61.50 and 62.50; S1.1
E. V. Korovkin	(b) **146**, 75 (1988)	61.70; 78.60; S9.11
F. P. Korshunov	(a) **107**, K139 (1988)	74.70; S10.15
V. S. Kortov	(b) **148**, K33 (1988)	71.55; S10.1
I. Kosacki	(a) **108**, 765 (1988)	77.20 and 77.40; S9
V. V. Kosarev	(a) **107**, K43 (1988)	71.55; S7.13
G. Koschek	(a) **105**, 377 (1988)	66.30; 61.70; 78.60; S10.1
	(a) **106**, 651 (1988)	78.60; 71.55; S7.12
	(a) **108**, 683 (1988)	71.55; 78.60; S7.12
B. Kościelniak-Mucha	(a) **105**, K135 (1988)	72.40 and 73.30; S8.16
	(a) **106**, K197 (1988)	73.30 and 85; S8.16
V. Koshkin	(a) **106**, K7 (1988)	61.80; 72.80; S8
R. A. Kosiński	(a) **106**, 227 (1988)	75.60; S11.2
L. N. Kostadinov	(a) **109**, K85 (1988)	61.55; S5
G. Kostorz	(b) **148**, 457 (1988)	61.55 and 64.80; S1.2
N. S. Kostyukov	(b) **146**, K109 (1988)	61.80 and 72.20
H. S. Kothari	(a) **106**, K53 (1988)	61.50 and 74.70; S10.15
S. S. Kotsur	(b) **149**, 347 (1988)	78.20
J. Kováč	(a) **107**, 405 (1988)	73.40; 71.55; 72.40; 78.60; S7.11; S7.15
V. T. Kovachev	(b) **145**, K47 (1988)	74.60
M. V. Kovalchuk	(b) **150**, 9 (1988)	61.10 and 78.70; S5.11
V. B. Kovalchuk	(a) **109**, 537 (1988)	71.55; 72.80; 76.30; S5.11
Z. D. Kovalyuk	(a) **108**, K115 (1988)	73.40; 85; S8; S10.1
A. P. Kovchavtsev	(a) **106**, 669 (1988)	73.40 and 85; S5.11
A. Kowalczyk	(a) **107**, K61 (1988)	75.50; S1.1
	(a) **110**, 241 (1988)	75.30; 61.55; S1.1; S1.2
M. Koyano	(b) **147**, 559 (1988)	61.60 and 72.20; S10
	(b) **148**, 659 (1988)	72.20; S10
M. Kozielski	(a) **110**, 437 (1988)	62.30 and 63.20; S11.1
P. Kozioł	(a) **108**, 739 (1988)	76.60; 62.50; S11
V. V. Kozlovskii	(a) **110**, K61 (1988)	61.70 and 66.30; S5.11
U. F. Kozmidis-Luburi	(b) **148**, 81 (1988)	64.60; 77.80; S11.1
W. Kraak	(b) **145**, 157 (1988)	73.40; S7.13
	(b) **148**, 333 (1988)	71.30 and 72.20; S7.13
	(b) **148**, 349 (1988)	73.25 and 73.40; S7.13
	(b) **148**, 567 (1988)	73.40; S7.13
G. Krabbes	(b) **146**, K29 (1988)	74.70; S10.15
T. Krajewski	(a) **109**, 193 (1988)	64.70; 76.30; S11.1
K. Král	(b) **148**, K129 (1988)	72.10; S7.12
V. M. Kramar	(b) **146**, 613 (1988)	71.70 and 78.50; S8.1
M. Kranjčec	(a) **109**, 329 (1988)	78.20; S8.16

A. KRAPF	(b) **146**,	K25 (1988)	74.70; S10.15
E. KRÄTZIG	(a) **105**,	K17 (1988)	66.30 and 79.20; 78.20; S11.1
	(a) **106**,	89 (1988)	61.70; 72.40; 78.20; S11.1
I. KRAUS	(a) **106**,	467 (1988)	61.10 and 62.20; 68.25
	(a) **109**,	199 (1988)	62.20; 68.20
L. KRAUS	(a) **109**,	K71 (1988)	75.80
E. KRAUSE	(a) **110**,	K33 (1988)	71.55 and 78.60; S8.12
H. KRAUSE	(a) **108**,	K131 (1988)	61.70 and 72.20; S10.1
	(a) **108**,	K135 (1988)	61.70 and 72.20; S10.1
	(a) **110**,	537 (1988)	73.40; 73.60; S10.1
R. KRAUSE	(a) **106**,	81 (1988)	61.80; 78.70; S5.11; S7.11
	(a) **106**,	419 (1988)	61.70; 72.80; 78.55; 78.70; S7.12
	(a) **107**,	111 (1988)	61.80; 78.50; 78.55; 78.70; S7.12
H. KRAUTZ	(a) **110**,	K77 (1988)	68.48; S1.2; S4
I. N. KRAVTSOVA	(b) **147**,	K91 (1988)	76.30
O. V. KRECHIVSKII	(b) **148**,	K29 (1988)	71.36 and 71.38
	(b) **148**,	K165 (1988)	71.35 and 78.20
	(b) **150**,	561 (1988)	71.35 and 78.20
U. KREISSIG	(a) **105**,	387 (1988)	61.80; S1.62; S5.11
J. KREISSL	(b) **145**,	609 (1988)	71.55 and 76.30; 61.70; S5.11
	(b) **150**,	K57 (1988)	76.30; S7.11
D. KREMP	(b) **147**,	747 (1988)	71.35; S5.12
H. KRESSE	(a) **106**,	K89 (1988)	77.20; 77.40; S12
M. KRETZSCHMAR	(b) **150**,	769 (1988)	71.35 and 78.20; S8.11
H. KRIKOR	(a) **107**,	639 (1988)	75.50; S12
R. M. KRISHNA	(b) **148**,	K119 (1988)	71.70; 76.30; 78.50; S9.12
P. KRISPIN	(a) **109**,	K115 (1988)	71.55; 73.40; 85; S7.11
N. KRISTIANPOLLER	(b) **149**,	45 (1988)	61.70; 72.20; 78.60; S9.16
N. KRISTOFFEL	(b) **149**,	11 (1988)	63.75 and 77.80; 64.70; S8; S11
	(b) **150**,	617 (1988)	72.40 and 78.20
L. N. KRISTOFOROV	(b) **146**,	487 (1988)	61.70 and 63.90; 66.90; 71.45; S12.1
G. KRIVAITĖ	(a) **107**,	K75 (1988)	78.30; S8.15
T. A. KRIVENKO	(b) **150**,	125 (1988)	71.36; 78.30; 78.55; S12
R. KROGGEL	(a) **106**,	407 (1988)	61.70; 61.55; S1.2; S4
H. KRONMÜLLER	(a) **105**,	597 (1988)	75.80; 78.20; S1.1
	(b) **147**,	361 (1988)	74.60
	(a) **108**,	403 (1988)	75.30; 75.50; 75.60; S1.2
	(a) **108**,	705 (1988)	75.30; 75.50; 75.60; S1.1; S1.2
A. I. KROTKII	(b) **145**,	425 (1988)	64.60 and 76.30; S9.16
E. KRÜGER	(b) **149**,	669 (1988)	74.20
H. KRÜGER	(b) **147**,	K159 (1988)	74.70; S10.15
	(b) **148**,	349 (1988)	73.25 and 73.40; S7.13
A. I. KRUGLIK	(a) **105**,	441 (1988)	64.70; 61.60; 65.40; 77.20; 78.20; S9.16
E. S. KRYACHKO	(b) **147**,	69 (1988)	61.50 and 61.70
J. KRYNICKI	(a) **109**,	531 (1988)	61.70 and 71.55; S5.11
C. KRYSCHI	(b) **147**,	797 (1988)	78.55; 71.35; 71.55; S11
T. KRZYSZTON	(b) **145**,	K41 (1988)	74.60
E. KUBALEK	(a) **105**,	377 (1988)	66.30; 61.70; 78.60; S10.1
	(a) **106**,	651 (1988)	78.60; 71.55; S7.12
	(a) **108**,	683 (1988)	71.55; 78.60; S7.12
J. KUBĚNA	(a) **105**,	61 (1988)	61.10 and 61.70; S10.1
N. KUBICKI	(b) **146**,	K25 (1988)	74.70; S10.15
	(a) **108**,	303 (1988)	73.40; 72.40; 73.25; S7.13
K. KUCHARSKI	(a) **105**,	113 (1988)	61.70; S5.11
A. KUCZKOWSKI	(a) **105**,	K61 (1988)	72.80; S12
K. KUDABAEV	(b) **146**,	K135 (1988)	76.30 and 78.50; S11.1
J. KUDRNOVSKÝ	(b) **146**,	597 (1988)	71.55 and 79.20; S1

	(b) **147**,	215 (1988)	71.10 and 79.20; S1.2
	(b) **148**,	K23 (1988)	71.20; 71.55; S1.3
G. E. KUGEL	(b) **147**,	441 (1988)	77.80 and 78.30; 63.20; S11.1
H.-A. KUHN	(a) **110**,	449 (1988)	62.20; S1.2
G. KÜHNEL	(a) **108**,	689 (1988)	71.55 and 72.20; S7.12
R. KÜHNERT	(a) **109**,	K93 (1988)	62.20
W. F. KUHS	(b) **148**,	445 (1988)	61.12 and 61.60; 62.50; 65.90; S8
V. KUKERMANN	(b) **147**,	661 (1988)	72.15; S1.3
V. V. KUKHTIN	(b) **149**,	55 (1988)	63.20; 75.10
T. N. KULAGINA	(b) **147**,	149 (1988)	71.38; 72.10; S4
K. KUŁAKOWSKI	(a) **106**,	K169 (1988)	75.80; S1.1; S1.2
V. A. KULBACHINSKII	(b) **148**,	197 (1988)	71.30; 72.20; S8.15
	(b) **148**,	333 (1988)	71.30 and 72.20; S7.13
	(b) **150**,	237 (1988)	72.20; S8
	(b) **150**,	K25 (1988)	73.40; S7.15
O. V. KULIKOVA	(a) **107**,	373 (1988)	78.30 and 78.55; S8.16
	(a) **107**,	K57 (1988)	71.55; 72.40; S7.11
	(a) **109**,	K153 (1988)	78.45 and 78.65; S8.15
N. R. KULISH	(b) **146**,	319 (1988)	78.20; S8.12
U. D. KULKARNI	(a) **110**,	331 (1988)	61.55; S1.2
W. KULLMANN	(b) **148**,	445 (1988)	61.12 and 61.60; 62.50; 65.90; S8
O. A. KULYASOVA	(a) **109**,	K27 (1988)	64.70; S5.11; S10.1
L. L. KULYUK	(a) **107**,	373 (1988)	78.30 and 78.55; S8.16
	(a) **107**,	K57 (1988)	71.55; 72.40; S7.11
	(a) **109**,	K153 (1988)	78.45 and 78.65; S8.15
A. KUMAO	(a) **107**,	611 (1988)	61.14
A. KUMAR	(b) **148**,	549 (1988)	71.25 and 73.40; S5.12; S7.12; S8.12
B. KUMAR	(b) **149**,	221 (1988)	71.25; 72.20; S12
DEVENDRA KUMAR	(a) **106**,	627 (1988)	77.20 and 77.40; 77.80; S10.15
M. KUMAR	(b) **146**,	125 (1988)	64.30; 63.90; S9.11
	(b) **147**,	529 (1988)	62.20; 63.90; 77.20; S9.11
	(b) **150**,	K7 (1988)	61.50; S9.11
V. KUMAR	(a) **109**,	273 (1988)	71.55; S5.11
V. S. KISHAN KUMAR	(a) **105**,	303 (1988)	78.50 and 78.60; S11.1
M. KUMARI	(b) **146**,	105 (1988)	64.70; 62.20; S2; S5.2; S9
K. KUMAZAKI	(b) **149**,	K183 (1988)	63.20 and 78.30; S8
M. KUMEDA	(b) **147**,	K1 (1988)	61.70; 76.30; S10.1
S. KUNDU	(b) **147**,	135 (1988)	73.40 and 76.40; S7.12
K. KUNZE	(b) **150**,	K1 (1988)	61.10; 61.12; 61.14
K. H. KUO	(a) **107**,	511 (1988)	61.55; 61.14; S1.61
M. H. KUOK	(b) **147**,	K195 (1988)	78.30; 63.20; S9.16
K. KURBANOV	(a) **105**,	K155 (1988)	78.45; 78.55; S10
J. KURIATA	(a) **109**,	K139 (1988)	76.30; 75.50; S11
G. L. KURISHEV	(a) **106**,	669 (1988)	73.40 and 85; S5.11
S. KURITA	(a) **110**,	645 (1988)	78.65; S12
I. N. KURKIN	(b) **148**,	K61 (1988)	76.30 and 77.80; S11.1
M. KURZAWA	(a) **109**,	K139 (1988)	76.30; 75.50; S11
N. D. KUSHCH	(a) **110**,	K111 (1988)	78.30; 71.38; S12
D. B. KUSHEV	(b) **150**,	729 (1988)	78.20; S8.15
S. S. KUSHWAHA	(b) **149**,	69 (1988)	63.20; S1.4
D. KUSNICK	(b) **145**,	157 (1988)	73.40; S7.13
R. KUSZELEWICZ	(b) **150**,	465 (1988)	68.55 and 78.20; S7.12; S7.15
Y. KUWAE	(a) **110**,	645 (1988)	78.65; S12
R. O. KUZIAN	(b) **147**,	K25 (1988)	71.20
O. S. KUZMIN	(a) **109**,	K11 (1988)	61.70 and 62.20; S1.1
A. V. KUZNETSOV	(b) **150**,	489 (1988)	78.20; S8.11

P. I. Kuznetsov	(b) **148**, K65 (1988)	71.35 and 78.20; S8.12
	(b) **150**, 761 (1988)	71.35 and 78.65; S8.12
V. I. Kuznetsov	(a) **105**, 97 (1988)	61.80; S5.11
	(a) **108**, 143 (1988)	61.80; 71.55; S5.11
B. Kużnicka	(a) **108**, K5 (1988)	64.75 and 66.30; S1.2
P. Kwapuliński	(a) **107**, 299 (1988)	72.15; S1.1; S1.2
A. G. Kyazymzade	(a) **108**, 619 (1988)	72.40 and 73.40; S8
D. S. Kyriakos	(b) **148**, 653 (1988)	72.20
D. Lab	(b) **150**, 419 (1988)	71.35 and 78.20; S8.11; S8.12; S8.15
E. Lach	(b) **150**, 679 (1988)	71.35 and 78.20; S7.12; S7.15
H. K. Lachowicz	(a) **105**, 597 (1988)	75.80; 78.20; S1.1
Z. Laczik	(a) **108**, K35 (1988)	61.70 and 68.48; 78.65; S5.11
M. Lago-Izquierdo	(b) **145**, 501 (1988)	71.10 and 73.40
Z. Y. Lai	(b) **145**, 151 (1988)	71.36 and 73.20; 78.30; S9.12
M. Laitinen	(a) **106**, 107 (1988)	63.20 and 65.70; S1.61
H. Lakner	(a) **106**, 651 (1988)	78.60; 71.55; S7.12
	(a) **108**, 683 (1988)	71.55; 78.60; S7.12
S. V. J. Lakshman	(b) **148**, K119 (1988)	71.70; 76.30; 78.50; S9.12
S. Lakshmi Raghavan	see under letter "R"	
M. Lamla	(b) **146**, K65 (1988)	78.40 and 78.55; 71.35; S8.11
G. Lamarche	(a) **108**, 599 (1988)	64.75; 75.30; S8.16
J. van Landuyt	(a) **107**, 521 (1988)	61.16; 61.70
	(a) **109**, 67 (1988)	61.14; S11.1
	(a) **110**, K1 (1988)	61.16; S1
U. Langbein	(b) **150**, 805 (1988)	78.65
Lanping Yue	see under letter "Y"	
H. Lanteri	(b) **149**, 203 (1988)	79.75; S10.1
Wanming Lao	(a) **109**, 301 (1988)	74.70; S10.15
D. Lapraz	(a) **108**, 783 (1988)	78.60; 61.80; S10
V. R. Larionov	(b) **150**, 599 (1988)	78.20; 85; S7.12
E. G. Lashkevich	(b) **145**, 535 (1988)	71.35 and 78.65; S9
V. Latinis	(b) **150**, 699 (1988)	71.35 and 78.55; S7.12; S7.15
H. Lauter	(b) **145**, 351 (1988)	61.80 and 63.20; S4
M. Yu. Lavrentiev	(b) **149**, 133 (1988)	63.20; S10
N. P. Lazarev	(a) **108**, K101 (1988)	66.30; 62.30; S5.11
M. Lazaridou	(a) **105**, K13 (1988)	61.70; S9.12
V. N. Lazukin	(b) **149**, 427 (1988)	61.60 and 71.55; 76.30; S8.16
G. F. Leal Ferreira	see under letter "F"	
V. Yu. Lebed	(b) **150**, 653 (1988)	78.65; S8.15
A. B. Lebedev	(a) **105**, K103 (1988)	62.20; S8
Ya. D. Lebedev	(a) **108**, 657 (1988)	73.40 and 78.60; S7.1
F. Leccabue	(a) **105**, 285 (1988)	75.30; S11.2
F. Lederer	(b) **150**, 805 (1988)	78.65
J. Ledzion	(a) **107**, K115 (1988)	73.60; S1.3
A. Lefebvre	(a) **107**, 647 (1988)	61.70 and 62.20; S7.12
	(a) **109**, 187 (1988)	62.20; S7.12
A. Lehmann	(b) **148**, 401 (1988)	78.30 and 78.65
G. Lehmann	(b) **147**, 691 (1988)	61.70 and 76.30; S11
X. L. Lei	(b) **146**, 189 (1988)	72.20 and 72.70
A. Yu. Leiderman	(a) **108**, 323 (1988)	72.20 and 73.40; S5.11
L. Leine	(b) **145**, 759 (1988)	71.35 and 78.55
V. Leinweber	(b) **145**, 195 (1988)	71.35; 78.20; S9.11
R. B. McLellan	see under letter "M"	

K. LENKEIT	(a) **107**, K19 (1988)	68.48; 62.20; S7.12; S7.15
M. LENZNER	(b) **150**, 885 (1988)	72.20; 72.40; S7.12
K. LEO	(b) **145**, K31 (1988)	72.30; 71.38
	(b) **149**, 215 (1988)	71.38; S7.12
H. LEON	(b) **149**, 533 (1988)	71.38 and 73.40; S7.12; S7.15
B. LEPLEY	(a) **108**, K41 (1988)	73.40; S7.11
V. LERCHE	(a) **105**, 61 (1988)	61.10 and 61.70; S10.1
	(a) **109**, 79 (1988)	61.10; 61.70; 62.20; S10.1
D. B. LEROY	(b) **148**, 121 (1988)	68.40; 79.60; S4
LE THANH BINH	(a) **105**, K51 (1988)	71.55; S7.15
V. LEUTE	(a) **107**, K7 (1988)	61.60; S8.16
M. LEVALOIS	(a) **109**, 111 (1988)	61.60; S7.16
TH. LEVENTOURI	(b) **149**, K157 (1988)	74.70 and 76.30; S10.15
I. G. LEVIN	(b) **150**, 307 (1988)	78.70; 61.10; 71.10; S9.11; S10.1; S11.1
R. LÉVY	(b) **150**, 825 (1988)	78.20; S9
M. DE DIOS LEYVA	see under letter "D"	
AI-ZHEN LI	(a) **110**, 515 (1988)	73.40; 78.30; S5.11; S7.12
BO-ZANG LI	(b) **149**, K173 (1988)	75.60
CHAORUI LI	(a) **109**, 301 (1988)	74.70; S10.15
FANGHUA LI	(a) **107**, 45 (1988)	61.55; S1.1; S1.4
HAN LI	(b) **150**, 663 (1988)	78.20; S8.15
HENGDE LI	(a) **107**, K29 (1988)	61.80 and 68.48; S1.1; S1.2; S10.1
	(a) **107**, K123 (1988)	68.55 and 78.65; S8
SHEN LI	(b) **150**, 839 (1988)	78.20; S8.11
SHUHONG LI	(b) **147**, 51 (1988)	61.70
WU LI	(a) **109**, 301 (1988)	74.70; S10.15
YAN-MIN LI	(b) **150**, 211 (1988)	74.20 and 75.30; 71.38
Y. F. LI	(a) **106**, 233 (1988)	75.50; S1.2; S1.4
YUZHANG LI	(b) **147**, 421 (1988)	78.55; 73.40; S7.15
ZHEN-YA LI	(b) **148**, 369 (1988)	75.10; 75.30
ZHONGRONG LI	(a) **109**, 301 (1988)	74.70; S10.15
ZONGQUAN LI	(a) **109**, K1 (1988)	61.40; S1.61
Z. Y. LI	(b) **146**, 223 (1988)	75.10; 75.50
	(b) **148**, K133 (1988)	75.10 and 75.50
N. A. AL-LIABI	see under letter "A"	
JINGKUI LIANG	(a) **107**, 45 (1988)	61.55; S1.1; S1.4
J. K. LIANG	(a) **110**, 55 (1988)	61.60 and 74.70; S11
Y. H. LIANG	(a) **107**, 63 (1988)	61.60 and 74.70; S10.15
LIANGPING WANG	see under letter "W"	
LIANG-YAO CHEN	see under letter "C"	
E. LIAROKAPIS	(b) **149**, K1 (1988)	61.50 and 62.20
A. E. LIGACHEV	(a) **109**, K11 (1988)	61.70 and 62.20; S1.1
S. LIGENZA	(a) **106**, K71 (1988)	75.50; S11.2
	(a) **109**, 319 (1988)	75.50; 61.12; S11.2
W. LIH	(b) **147**, 791 (1988)	78.30; 63.20; S11
XIANDUAN LIN	(a) **109**, 301 (1988)	74.70; S10.15
ZIJING LIN	(b) **148**, 525 (1988)	63.20 and 68.30; S5
	(b) **149**, 143 (1988)	63.20 and 68.30; S5.11
M. LINDBERG	(b) **150**, 357 (1988)	71.35 and 78.20; S8.11
	(b) **150**, 379 (1988)	71.35 and 78.20; S8.11
S. LINDEROTH	(b) **149**, 785 (1988)	78.70; S9.12
R. LINK	(b) **145**, K121 (1988)	71.25; 71.45; S12
	(b) **149**, K85 (1988)	78.20; S12
V. N. LIPATNIKOV	(a) **106**, 459 (1988)	64.60; 75.30; S1.61
J. LIPAVIČIUS	(b) **148**, K97 (1988)	78.40; S8.16
	(b) **148**, K169 (1988)	78.20; S8.16

B. Lippold	(b) **147**, K43 (1988)	74.30; 74.70; S10.15
Li-Rong Chen	see under letter "C"	
M. P. Lisitsa	(b) **146**, 319 (1988)	78.20; S8.12
	(b) **146**, 723 (1988)	78.40; S11
	(b) **150**, 907 (1988)	78.20; 78.50; S9.11
Litian Wang	see under letter "W"	
V. G. Litovshenko	(b) **145**, 209 (1988)	71.25; 78.20; S5
	(b) **145**, 535 (1988)	71.35 and 78.65; S9
V. V. Litvinov	(a) **108**, 311 (1988)	71.55; 61.70; S5.12
Baixin Liu	(a) **107**, K29 (1988)	61.80 and 68.48; S1.1; S1.2; S10.1
B. X. Liu	(a) **110**, 443 (1988)	64.75; 71.25; S1.3
H. K. Liu	(b) **147**, K153 (1988)	74.70; S10.15
Hongjian Liu	(a) **108**, 337 (1988)	74.70; 75.20; S10.15
Rongchuan Liu	(a) **110**, K89 (1988)	64.70 and 74.70; S10.15
J. Liwo	(a) **108**, K105 (1988)	73.25; S10.15
Li-Yuan Zhang	see under letter "Z"	
Lizhong Wang	see under letter "W"	
J. L. Sanchez Llamazares	see under letter "S"	
J. Llopis	(a) **107**, 197 (1988)	62.20; 78.60; S10.1
	(a) **108**, K163 (1988)	78.60; S10.1
K. Löbe	(b) **148**, 263 (1988)	71.55; 68.60; 78.55; S8.11
	(b) **150**, 191 (1988)	61.70 and 71.55; 71.70; 78.55; S8.11
P. Lobotka	(a) **109**, 205 (1988)	73.60 and 75.30; 74.70; S10.15
P. I. Loeff	(b) **147**, K105 (1988)	66.30; S1
H. Löffler	(a) **106**, 407 (1988)	61.70; 61.55; S1.2; S4
	(a) **108**, 103 (1988)	61.55; 64.70; S4
	(a) **108**, 171 (1988)	61.70; 61.55; 62.20; S4
A. P. Loghinov	(b) **147**, K15 (1988)	63.50; S12
T. Lohner	(a) **107**, K15 (1988)	61.80 and 68.48; S1.3; S7.12
	(a) **108**, K35 (1988)	61.70 and 68.48; 78.65; S5.11
A. Loiseau	(a) **107**, 655 (1988)	64.70; 61.16; S1.2; S4
V. M. Loktev	(b) **147**, 307 (1988)	74.20; 71.45; S12.1
V. N. Lomasov	(a) **109**, 127 (1988)	61.80; 76.30; S5.11
C. A. Londos	(a) **105**, K87 (1988)	66.30; S5.11
	(a) **109**, 421 (1988)	61.80; 71.55; S5.11
Qiwei Long	(a) **109**, K129 (1988)	74.70 and 78.70; S10.15
Longshu Wang	see under letter "W"	
F. López	(a) **107**, 329 (1988)	75.70; S1.2
F. López-Aguilar	(a) **149**, 585 (1988)	71.20 and 71.25; S1.5; S1.63
J. López-Gondar	(b) **145**, 501 (1988)	71.10 and 73.40
	(b) **145**, K155 (1988)	71.25; S7.13; S7.15
	(b) **148**, K113 (1988)	71.25 and 73.40
M. H. Loretto	(a) **107**, 673 (1988)	61.16
P. Lošťák	(b) **146**, K75 (1988)	78.30; S8.16
	(a) **106**, 619 (1988)	78.20; 78.30; S8
	(b) **150**, 237 (1988)	72.20; S8
W. Lottermoser	(a) **109**, 589 (1988)	75.25; S11.1
P. P. Lottici	(b) **145**, 401 (1988)	61.60 and 78.70; S8.16
	(b) **146**, 503 (1988)	63.20; 78.30; S8.16
	(b) **146**, K81 (1988)	63.50; S5
S. V. Lozovskii	(a) **107**, 213 (1988)	68.55; 64.75; 71.55; S5.11
V. Z. Lozovsky	(b) **148**, 297 (1988)	74.20
Ping Lu	(a) **107**, 681 (1988)	68.20; 61.16; S8.13
W. Lu	(b) **147**, 767 (1988)	78.30; 63.20; S8.15
Y.-Y. Lu	(a) **106**, K1 (1988)	61.70 and 78.70; S12
Zhong-nan Lu	(a) **107**, 153 (1988)	63.20 and 68.30; 61.55; S1.2
T. Luciński	(a) **105**, 501 (1988)	73.60; 75.70; S1.61

J. Łuczka	(b) **149**, K35 (1988)	76.20; 77.40
F. Ludwig	(b) **147**, 297 (1988)	71.55 and 78.50; 72.40; 76.40; S5
	(b) **148**, 143 (1988)	76.40 and 78.20; 73.40; S7.13
	(b) **150**, 119 (1988)	73.40 and 76.40; 72.10; S7.13
P. F. Lugakov	(a) **105**, 97 (1988)	61.80; S5.11
	(a) **108**, 143 (1988)	61.80; 71.55; S5.11
	(a) **110**, 403 (1988)	61.80; 72.80; S5.11
Lu Jie	see under letter "J"	
P. Lukáč	(a) **107**, K11 (1988)	62.20; S1
T. A. Lukashevich	(a) **110**, 403 (1988)	61.80; 72.80; S5.11
F. Lukeš	(a) **106**, K27 (1988)	68.55; 78.65; S8
	(a) **107**, 239 (1988)	78.65; S7.12
V. V. Lukyanitsa	(a) **108**, 143 (1988)	61.80; 71.55; S5.11
G. Lulli	(a) **108**, 503 (1988)	61.70; 72.20; S5.11
Lun-Biao Xu	see under letter "X"	
Ningsheng Luo	(b) **147**, 511 (1988)	63.20; 68.30; S1.2
L. N. Lupichev	(b) **147**, 155 (1988)	71.45
V. G. Luppov	(b) **149**, 697 (1988)	76.60; 67.90
F. Lüty	(b) **146**, 303 (1988)	78.30; 78.50; S9.11
A. V. Lyakhnovich	(b) **150**, 507 (1988)	78.20
I. I. Lyapilin	(b) **148**, 501 (1988)	71.25 and 73.40
S. G. Lyapin	(b) **147**, 297 (1988)	71.55 and 78.50; 72.40; 76.40; S5
D. W. Lynch	(b) **148**, 387 (1988)	78.20; S1.3; S4
V. G. Lyssenko	(b) **150**, 593 (1988)	78.20; 71.35; S8.11
E. Ma	(a) **110**, 443 (1988)	64.75; 71.25; S1.3
	(a) **110**, 509 (1988)	68.48 and 68.55; S4
H. Maack	(a) **110**, 459 (1988)	64.70 and 77.80; S11
R. Mach	(a) **110**, 657 (1988)	73.40 and 78.60; 85; S8.11
H. A. MacKenzie	(b) **149**, 383 (1988)	78.20; 71.55; S7.13
	(b) **150**, 719 (1988)	78.20; S7.13
J. Maćkowiak	(b) **145**, 419 (1988)	64.60; 61.20
M. Maćkowiak	(a) **108**, 739 (1988)	76.60; 62.50; S11
M. P. Madan	(b) **149**, K55 (1988)	77.30; S9.11
E. G. Madatova	(a) **105**, 357 (1988)	61.70; 78.70; S1.1; S1.2
M. Mäder	(a) **107**, K119 (1988)	68.55; 73.60; S9.12
	(a) **109**, 493 (1988)	68.55; 65.70; S9.12
H. Mädge	(a) **107**, K151 (1988)	75.40; S1.4
	(a) **110**, 601 (1988)	75.40; S1.4
Y. Madhavan	(b) **145**, 139 (1988)	63.20; 61.70; S8.13
K. N. Madhusoodanan	(a) **108**, 775 (1988)	78.20; S5
K. Maeda	(a) **109**, 383 (1988)	61.70 and 62.20; S7.12
J. Maege	(a) **105**, K45 (1988)	73.25; 78.45; S7.15
E. L. Mager	(b) **149**, K145 (1988)	61.70 and 74.70; S10.15
N. Magnea	(a) **107**, 101 (1988)	61.70; 72.80; 78.30; S5.11
D. P. Mahapatra	(b) **149**, 355 (1988)	78.70; S2; S4
J. Mahy	(a) **107**, 873 (1988)	61.60; S8
A. Maignan	(a) **107**, 73 (1988)	61.60; 64.80; S10.15
A. B. Maity	(b) **146**, 525 (1988)	71.25 and 73.60; S7.12
	(b) **148**, 165 (1988)	71.25 and 73.60; S7.13
	(b) **148**, 559 (1988)	79.60; 73.60; S7.12
	(b) **149**, 565 (1988)	73.60 and 79.60; S7.12
J. Majhi	(a) **106**, K209 (1988)	73.30 and 85; S5.11
E. Majková	(b) **146**, K17 (1988)	71.25 and 79.60; S1.2

C. Majumdar	(b) **146**,	525 (1988)	71.25 and 73.60; S7.12
	(b) **148**,	165 (1988)	71.25 and 73.60; S7.13
	(b) **148**,	559 (1988)	79.60; 73.60; S7.12
	(b) **149**,	565 (1988)	73.60 and 79.60; S7.12
F. A. Majumder	(b) **150**,	749 (1988)	78.20; 78.55; S8.15
L. F. Makarenko	(b) **145**,	241 (1988)	71.55; 72.40; S5.11
	(a) **106**,	K153 (1988)	71.55; 72.40; S5.11
A. G. Makhanek	(b) **149**,	231 (1988)	71.70; 78.50; S9.12
A. V. Makhankov	(b) **145**,	669 (1988)	75.10
V. G. Makhankov	(b) **145**,	669 (1988)	75.10
L. A. Makovetskaya	(a) **105**,	K97 (1988)	64.70; S8.16
	(a) **108**,	K63 (1988)	72.80; S8.16
G. I. Makovetskii	(a) **107**,	K89 (1988)	61.60; S7
V. M. Maksimenko	(a) **109**,	537 (1988)	71.55; 72.80; 76.30; S5.11
P. A. Maksimiyuk	(a) **107**,	K11 (1988)	62.20; S1
A. A. Maksimov	(b) **150**,	125 (1988)	71.36; 78.30; 78.55; S12
T. I. Maksimova	(b) **149**,	K31 (1988)	76.30; 78.50; S9.11
J. Malek	(b) **145**,	233 (1988)	71.38 and 71.50; S12
	(b) **147**,	281 (1988)	71.50; 71.38; 71.45; S12.1
V. L. Malevich	(a) **106**,	K123 (1988)	64.70; S5.11
G. L. Malhotra	(a) **109**,	K45 (1988)	68.55; 78.65; S7.16
I. E. Malinovskii	(b) **148**,	K65 (1988)	71.35 and 78.20; S8.12
	(b) **150**,	761 (1988)	71.35 and 78.65; S8.12
W. Malinowski	(a) **106**,	371 (1988)	61.80; 62.20; S1.2
N. I. Malysh	(b) **146**,	319 (1988)	78.20; S8.12
V. S. Malyshevskii	(b) **147**,	11 (1988)	61.80; 61.70; 63.20; 78.70
A. M. Mamedov	(b) 149,	K191 (1988)	78.40; S10.1
F. I. Mamedov	(a) **107**,	K49 (1988)	72.40; S8.16
K. K. Mamedov	(a) **106**,	315 (1988)	65.40; S8
N. T. Mamedov	(b) **145**,	K103 (1988)	63.20 and 64.70; 78.30; S8.16
	(b) **148**,	K89 (1988)	63.20 and 78.30; S8.16
	(a) **110**,	K5 (1988)	61.60; S8.16
T. G. Mamedov	(b) **147**,	K99 (1988)	71.25 and 78.40; S8
T. N. Mamontova	(a) **107**,	11 (1988)	73.25; 68.45; 72.40; 73.20; S8
Yun Dun Man	see under letter "Y"		
R. Manaila	(a) **109**,	61 (1988)	61.55; S1.1; S1.2
V. A. Manasson	(a) **108**,	K115 (1988)	73.40; 85; S8; S10.1
C. Mande	(b) **145**,	77 (1988)	61.10; S1.2; S8.16
I. Mandjukov	(b) **146**,	269 (1988)	76.80; S1.1
M. Manfredi	(a) **109**,	603 (1988)	78.20; S9.11
C. Manfredotti	(a) **108**,	K25 (1988)	73.25; S5.11
Yu. I. Mankov	(b) **149**,	241 (1988)	72.15; 75.60; S1.1
A. Manna	(b) **145**,	181 (1988)	71.10; 61.55; S1
I. Manna	(a) **108**,	K19 (1988)	64.75 and 68.48; S4
C. Manolikas	(a) **107**,	589 (1988)	61.70; S8.16
N. B. Manson	(b) **146**,	K153 (1988)	78.50; S10.1
G. Manzke	(b) **147**,	733 (1988)	71.35 and 78.20; 73.40; S7.15
	(b) **150**,	691 (1988)	71.35 and 78.20; S7.12; S7.15
Er-wang Mao	(a) **110**,	515 (1988)	73.40; 78.30; S5.11; S7.12
Ming Mao	(a) **105**,	325 (1988)	61.40; 65.40; S1.1; S1.2
	(a) **109**,	39 (1988)	61.40 and 64.70; S1.1; S1.2
I. V. Marchishin	(a) **106**,	153 (1988)	71.55; 73.40
	(a) **106**,	557 (1988)	71.55 and 73.40; 61.80
M. J. Marcinkowski	(a) **105**,	K25 (1988)	61.70
I. F. Margolych	(b) **149**,	301 (1988)	75.30 and 75.50
E. Mariańska	(a) **108**,	K67 (1988)	76.80; 75.50; S1.1

P. J. Markowski	(b) **146**, K131 (1988)	72.15 and 74.70; S10.15
	(b) **147**, 321 (1988)	72.15; 71.70; S1.4
F. Marlow	(b) **148**, 619 (1988)	71.10 and 71.35
	(b) **150**, K35 (1988)	71.10 and 71.35
S. A. Marshall	(b) **149**, 691 (1988)	76.30; S10.1
A. Martin	(a) **107**, K163 (1988)	76.80; S11.2
D. Martin	(b) **150**, 685 (1988)	71.35 and 78.65; 71.45; S7.12; S7.15
L. Martin	(a) **109**, 455 (1988)	68.55 and 78.65; S10.1
E. K. Martynchuk	(b) **145**, 309 (1988)	78.20; S8; S8.15
S. N. Martynov	(b) **146**, 691 (1988)	76.30
	(b) **149**, K41 (1988)	71.55 and 76.30; S9
M. Marvan	(b) **149**, K61 (1988)	77.80
G. Mary	(b) **149**, 683 (1988)	71.55; 78.50; 78.55; S9.12
S. Mase	(b) **145**, 649 (1988)	72.60 and 74.10; S10.15
V. M. Mashinskii	(a) **110**, K107 (1988)	78.50; 76.30; S10.15
Lj. D. Mašković	(b) **148**, 81 (1988)	64.60; 77.80; S11.1
T. Massalha	(a) **105**, 249 (1988)	75.30; 76.80; S1.1; S1.2
V. F. Masterov	(a) **107**, 291 (1988)	71.55; 76.80; S8
B. Maťáko	(a) **106**, 659 (1988)	73.40 and 85; S7.12
G. Materlik	(a) **105**, 53 (1988)	61.10; S5.12
	(b) **146**, 287 (1988)	78.70; 71.50; S1.1; S10.1
T. Matsuda	(a) **108**, K73 (1988)	77.80; S11.1
A. Matsuoka	(a) **105**, K165 (1988)	78.55; S8.12
K. Matsuura	(a) **105**, K165 (1988)	78.55; S8.12
	(a) **106**, 283 (1988)	78.55; S8.12
	(a) **109**, 617 (1988)	78.55; 71.55; S8.12
R. Mattheis	(a) **109**, 217 (1988)	68.48 and 68.55; S1.61
S. Matthies	(b) **150**, K1 (1988)	61.10; 61.12; 61.14
P. Matthys	(b) **146**, K149 (1988)	76.30
	(b) **148**, 683 (1988)	71.55 and 76.30; S9.11
	(b) **149**, K25 (1988)	71.70 and 76.30; S9.11
P. W. Mauriz	(b) **145**, K5 (1988)	63.20; 76.80
J. Mayer	(b) **150**, 39 (1988)	63.20 and 65.40; S12
A. Mayr	(b) **147**, 383 (1988)	75.70; 75.25; 75.30; S11.1
S. A. Mazen	(a) **106**, K65 (1988)	75.60; S11.2
P. Mazurek	(a) **109**, 597 (1988)	78.55 and 78.60; 61.70; S7.12
P. S. Mazumdar	(a) **106**, 291 (1988)	78.60
S. W. S. McKeever	(a) **108**, 545 (1988)	61.70; 78.50; 78.60; S9.11
	(a) **108**, 753 (1988)	78.60; S9.12
R. B. McLellan	(a) **110**, 421 (1988)	61.70; S1.1
B. E. Mdivanyan	(a) **107**, 131 (1988)	62.20; 61.80; S7.12
N. I. Medvedeva	(b) **149**, 175 (1988)	71.25; S1.61
G. A. Medvedkin	(a) **110**, 631 (1988)	78.20; 72.40; S8.16
M. Mehar Khan	(b) **147**, K51 (1988)	72.15; S3
K. Meinel	(a) **106**, 133 (1988)	68.40 and 68.55; 64.75; 79.20; S1.3
	(a) **106**, 493 (1988)	66.30 and 68.48; 68.60; S1.3
	(a) **109**, 525 (1988)	66.30 and 68.48
	(a) **109**, K97 (1988)	66.30 and 68.48; 68.60; S1.3
	(a) **110**, 189 (1988)	61.70 and 68.55; S1.3
A. Meisel	(b) **145**, 603 (1988)	71.50; 79.20; S5.11; S6
Mei-shen Hu	see under letter "H"	
O. Mekenyan	(b) **148**, K101 (1988)	61.55; S5.2
A. O. Mekhrabov	(a) **105**, K151 (1988)	76.30 and 76.80; S8
L. A. Meleshchenko	(a) **110**, K51 (1988)	75.50 and 78.30; S11.2
A. I. Melker	(a) **108**, 205 (1988)	62.20
S. V. Melnichuk	(b) **146**, 613 (1988)	71.70 and 78.50; S8.1
A. A. Melnikov	(a) **108**, 431 (1988)	78.60; S7.1

S. V. Melnikova	(a) **105**, 441 (1988)	64.70; 61.60; 65.40; 77.20; 78.20; S9.16
K. Melzer	(a) **107**, K161 (1988)	76.80; S11.2
B. Mendez	(a) **108**, K81 (1988)	61.70 and 78.60; S1.61
J. Menon	(a) **107**, 693 (1988)	61.55; S1.2; S4
M. Menyhárd	(a) **110**, 495 (1988)	68.20; 79.20; S1.1; S1.2
H. P. Menzler	(a) **108**, 449 (1988)	78.65; S11.1
T. Merceron	(a) **106**, 595 (1988)	75.60; S11.2
D. Mercurio	(a) **108**, 111 (1988)	61.14 and 64.80; S10.15
M. Merisalo	(a) **106**, 107 (1988)	63.20 and 65.70; S1.61
M. Merli	(a) **106**. 583 (1988)	72.20; S5.11
	(a) **108**, 503 (1988)	61.70; 72.20; S5.11
I. Mertig	(b) **147**, 575 (1988)	71.10; 71.25; 71.38; S1.3
M. Mertig	(b) **149**, 697 (1988)	76.60; 67.90
A. Messekher	(a) **106**, 595 (1988)	75.60; S11.2
H. Metzger	(a) **108**, 129 (1988)	61.10; 68.20; S5.11
J. D. Meyer	(a) **105**, 177 (1988)	61.80 and 64.75; 73.60; S1.3; S5
M. M. Mezdrogina	(a) **108**, 323 (1988)	72.20 and 73.40; S5.11
Z. V. Mezhevich	(b) **149**, K145 (1988)	61.70 and 74.70; S10.15
N. I. Michalev	(a) **106**, 333 (1988)	61.40; 62.20; S8
W. Michalke	(a) **109**, 503 (1988)	68.55; 74.70; S10.15
B. Michel	(a) **109**, K93 (1988)	62.20
C. Michel	(a) **107**, 73 (1988)	61.60; 64.80; S10.15
J. C. Michel	(b) **150**, 465 (1988)	68.55 and 78.20; S7.12; S7.15
R. Mickevičius	(b) **150**, 437 (1988)	78.20; S7.12
S. Mickevičius	(a) **107**, K53 (1988)	72.20; S8.16
T. R. Middya	(b) **146**, 117 (1988)	62.20 and 77.20; S9.11
J. Mielnicki	(b) **145**, 299 (1988)	75.10 and 75.70
A. Migus	(b) **150**, 357 (1988)	71.35 and 78.20; S8.11
G. Mihály	(a) **110**, 593 (1988)	72.20; S12.1
K. Mika	(b) **149**, 457 (1988)	61.70 and 65.50
A. I. Mikhailin	(a) **108**, 205 (1988)	62.20
G. V. Mikhailov	(b) **150**, 777 (1988)	71.35 and 78.55; S8.11
I. D. Mikhailov	(b) **147**, K15 (1988)	63.50; S12
E. V. Mikhailovskaya	(b) **145**, 535 (1988)	71.35 and 78.65; S9
M. Mikhov	(a) **105**, K145 (1988)	75.30; S1.4
J. Miletic	(b) **149**, 775 (1988)	78.20; S9
J. Mimkes	(b) **148**, 91 (1988)	65.70; 65.50; S1.2; S1.3; S2
B. F. Minaev	(b) **148**, 689 (1988)	71.70 and 76.30; S5.11
Huang Ming	(b) **145**, K21 (1988)	71.70; S9.16
J. Minge	(a) **109**, 193 (1988)	64.70; 76.30; S11.1
Ming Mao	see under letters "Ma"	
L. Q. Minh	(a) **105**, K121 (1988)	73.30; S12
	(a) **108**, K157 (1988)	78.65; S10.1
C. Minichino	(b) **145**, K125 (1988)	71.55; S5.11
H. Minowa	(b) **145**, 89 (1988)	61.80
R. Mirianashvili	(b) **146**, K141 (1988)	76.70; S9.12
D. Lj. Mirjanic	(b) **148**, 593 (1988)	71.45; 77.20
R. H. Misho	(a) **106**, K143 (1988)	68.55 and 78.65; S10.1
	(a) **109**, K101 (1988)	68.55 and 78.65; S10.1
L. K. Mishra	(b) **148**, 585 (1988)	71.25; S2; S4
S. K. Mishra	(b) **149**, 503 (1988)	63.20; S1; S1.1
E. Misiak	(a) **110**, 231 (1988)	64.70 and 75.60; S1.2
J. Misiewicz	(a) **106**, 551 (1988)	72.40 and 73.30; S7
	(a) **107**, K65 (1988)	71.55 and 78.55; S7
S. K. Misra	(b) **147**, 677 (1988)	76.30; S11
V. Misyuchenko	(a) **110**, 181 (1988)	68.55 and 78.65; S5.11
R. Mitdank	(a) **110**, 197 (1988)	73.60 and 78.60; S7.15

B. MITRA	(b) **149**, K117 (1988)	72.20 and 73.40; S7.12; S7.15
	(b) **150**, K67 (1988)	79.60; S4
G. B. MITRA	(a) **109**, K83 (1988)	61.10
T. MIYAZAKI	(a) **105**, 263 (1988)	75.50; 75.80; S1.1
	(a) **107**, 355 (1988)	75.50; 75.60; S1.1
R. MIZERIS	(a) **110**, 429 (1988)	64.70; 62.65; 77.20; S11.1
Y. MIZUGUCHI	(a) **109**, 617 (1988)	78.55; 71.55; S8.12
I. M. MLADENOV	(b) **145**, K115 (1988)	71.10 and 71.25
S. MOCHIZUKI	(b) **145**, K75 (1988)	78.30; S10.1
H. MOELLER	(a) **108**, 469 (1988)	73.40 and 85
F. MOGEON	(a) **107**, 559 (1988)	64.60; 78.20; S10.1
A. Z. MOHAMED	(a) **110**, K13 (1988)	62.20; S1.2
M. T. MOHAMMAD	(a) **106**, 479 (1988)	78.65; 71.25; S10.1
S. MOHAN	(a) **105**, K181 (1988)	62.20 and 73.60; S1.2
S. V. MOHARIL	(a) **106**, 271 (1988)	78.60; S9.11
	(a) **106**, K107 (1988)	78.60; 78.40; S11
P. MOHAZZABI	(b) **149**, 495 (1988)	61.50 and 65.40; S5.2
I. MOJZES	(a) **107**, K15 (1988)	61.80 and 68.48; S1.3; S7.12
R. MOKRANI	(a) **109**, 445 (1988)	61.70 and 65.70; 62.20; S8
M. MOLDOVANOVA	(a) **108**, 699 (1988)	72.20; S8
N. A. MOLDOVYAN	(a) **106**, K181 (1988)	78.40; S8.16
	(a) **110**, 277 (1988)	72.40 and 78.40; 78.60; S8.16
K.-P. MÖLLMANN	(b) **149**, 663 (1988)	72.40; S8.16
	(a) **109**, K121 (1988)	71.55; 72.80; S8
V. B. MOLODKIN	(a) **108**, 67 (1988)	61.10; 61.70; S5.12
M. MONDAL	(b) **146**, K97 (1988)	72.30; S4
	(b) **147**, K179 (1988)	75.20
	(b) **148**, 645 (1988)	72.20; S8.15
J. MONECKE	(b) **145**, K79 (1988)	63.20 and 78.30; 71.45; S7.15
	(b) **146**, K161 (1988)	63.20 and 78.30; S8.13
M. MONETA	(b) **147**, 123 (1988)	61.80 and 68.30; S5.11
C. MONTEVERDI	(a) **108**, 503 (1988)	61.70; 72.20; S5.11
F. MONZER	(a) **105**, 661 (1988)	73.40 and 85; 78.60; S7.15
P. MORAVEC	(b) **145**, 637 (1988)	72.20; S8.15
	(b) **147**, 621 (1988)	72.20; S8.15
	(b) **148**, 319 (1988)	72.20; S8.15
I. FONSECA MORENO	(b) **145**, 25 (1988)	71.20
D. V. MORGAN	(a) **110**, 531 (1988)	61.80 and 73.40; S1.61; S4; S7.12
H. MORI	(a) **106**, 283 (1988)	78.55; S8.12
	(a) **109**, 617 (1988)	78.55; 71.55; S8.12
G. MORLOT	(a) **108**, 183 (1988)	66.30 and 77.50; S9.12
J. W. MOROŃ	(a) **106**, K13 (1988)	62.40; S1.1; S1.2
T. V. MOROZOVA	(a) **110**, 397 (1988)	61.70 and 78.20; 61.10; S11.1
M. MORSLI	(b) **150**, 225 (1988)	72.20; 71.25; S10
M. MORVIC	(a) **106**, 659 (1988)	73.40 and 85; S7.12
N. MOSER	(a) **105**, 597 (1988)	75.80; 78.20; S1.1
E. MOSINIEWICZ-SZABLEWSKA	(a) **108**, 747 (1988)	75.50; 61.80; S1.1; S1.2
S. A. MOSKALENKO	(b) **147**, 711 (1988)	71.35
	(b) **149**, 187 (1988)	71.36
	(b) **150**, 401 (1988)	71.35 and 78.20
YU. N. MOSKVICH	(a) **105**, 145 (1988)	64.70; 76.60; S11.1
M. MOSSA	(b) **149**, 111 (1988)	63.20; 68.30; 71.35
M. M. MOSTAFA	(a) **110**, K13 (1988)	62.20; S1.2
S. MOTIEJŪNAS	(a) **106**, K173 (1988)	76.80; S8.15
	(b) **148**, K161 (1988)	76.80; S8
	(a) **108**, 197 (1988)	64.70; 76.80; S8.15

E. MROSAN	(b) **147**, 575 (1988)	71.10; 71.25; 71.38; S1.3
J. MRÓZ	(a) **105**, K33 (1988)	64.70; 65.70; S11
P. MROZEK	(a) **110**, 495 (1988)	68.20; 79.20; S1.1; S1.2
E. MÜHLE	(b) **145**, K15 (1988)	71.70; S1.4
S. M. MUJIBUR RAHMAN	see under letter "R"	
B. N. MUKASHEV	(b) **148**, K105 (1988)	61.70; 78.30; S5.11
N. N. MUKHAMADEEVA	(a) **108**, 135 (1988)	64.70 and 66.30; S8
A. A. MUKHAMEDOV	(b) **145**, 759 (1988)	71.35 and 78.55
A. K. MUKHERJEE	(b) **146**, K45 (1988)	71.70 and 75.50; S11
	(a) **106**, 173 (1988)	71.70; S12
K. K. MUKHERJEE	(b) **147**, K51 (1988)	72.15; S3
S. MUKHOPADHYAY	(b) **146**, 717 (1988)	75.30 and 75.40
E. MÜLLER	(b) **148**, 155 (1988)	68.40; 64.10; S5.2
F. MÜLLER	(a) **106**, K165 (1988)	74.50; S4
G. O. MÜLLER	(a) **110**, 657 (1988)	73.40 and 78.60; 85; S8.11
H. MÜLLER	(b) **148**, K11 (1988)	73.30; 73.40
H.-U. MÜLLER	(b) **146**, K25 (1988)	74.70; S10.15
	(b) **147**, 297 (1988)	71.55 and 78.50; 72.40; 76.40; S5
	(b) **148**, 143 (1988)	76.40 and 78.20; 73.40; S7.13
	(b) **150**, 119 (1988)	73.40 and 76.40; 72.10; S7.13
I. MÜLLER	(a) **106**, 525 (1988)	73.40; 73.30; S12
	(a) **110**, 205 (1988)	73.40; S12
J. F. MÜLLER	(b) **150**, 393 (1988)	71.35 and 78.20
M. MÜLLER	(b) **150**, 641 (1988)	78.20; 71.35; S8.1; S9.11
R. MÜLLER	(b) **150**, 587 (1988)	78.20
J. M. MUÑOZ	(a) **108**, 721 (1988)	75.50; S11.2
G. MUNSCHY	(b) **150**, 201 (1988)	71.35 and 71.55
F. M. MUNTYANU	(b) **148**, K37 (1988)	72.15; S4
W. A. MURAD	(a) **106**, K143 (1988)	68.55 and 78.65; S10.1
	(a) **109**, K101 (1988)	68.55 and 78.65; S10.1
S. MURAMATSU	(b) **149**, 791 (1988)	78.70; S9
	(b) **149**, K19 (1988)	71.55 and 78.50; 61.70; S9.11; S9.12
K. R. MURALI	(a) **105**, 477 (1988)	68.55; 73.60; 78.65; S7
S. MURALIDHAR	(a) **110**, 331 (1988)	61.55; S1.2
G. MURALIDHARAN	(b) **146**, 727 (1988)	78.60; 61.70; S9.11
	(b) **150**, 315 (1988)	78.60; 61.70; S9.11
L. I. MURIN	(b) **145**, 241 (1988)	71.55; 72.40; S5.11
K. SATYANARAYANA MURTHY	(a) **110**, K5 (1988)	61.60; S8.16
T. S. N. MURTHY	(a) **106**, K185 (1988)	78.30; S11.1
V. R. K. MURTHY	(a) **105**, K71 (1988)	77.20 and 77.40; S9
	(a) **109**, K75 (1988)	77.30 and 77.40
Y. V. G. S. MURTI	(a) **105**, 397 (1988)	61.70; 77.30; 77.40; 78.50; S9.11
A. A. MUSTAFAEV	(a) **107**, K129 (1988)	72.80; S8
E. R. MUSTAFAEV	(a) **109**, 345 (1988)	73.40 and 85; 73.30; S11
O. ARÉS MUZIO	see under letter "A"	
N. MYKOLAITIENĖ	(b) **150**, K31 (1988)	71.25; S9
V. MYSLÍK	(a) **106**, K35 (1988)	68.45 and 78.70; S7.11; S7.12
G. NACHTWEI	(b) **147**, K159 (1988)	74.70; S10.15
	(b) **148**, 349 (1988)	73.25 and 73.40; S7.13
	(b) **148**, 567 (1988)	73.40; S7.13
V. NÁDAŽDY	(a) **106**, 659 (1988)	73.40 and 85; S7.12
B. R. NAG	(a) **106**, 569 (1988)	72.20; S7.12; S7.15
E. L. NAGAEV	(b) **145**, 11 (1988)	75.30 and 78.20; 71.25; 75.60
N. NAGASAWA	(b) **150**, 817 (1988)	71.35 and 78.20; S9

P. NAGELS	(a) **107**, 639 (1988)	75.50; S12
	(a) **107**, 769 (1988)	61.40; 64.60; S8
L. E. NAGLI	(a) **105**, 87 (1988)	61.80; 78.70; S9.11
	(a) **106**, 295 (1988)	78.55; 61.70; S9.11
S. NAITOH	(a) **108**, 509 (1988)	61.70; 61.16; S1.2
M. NAKAGAWA	(a) **107**, 709 (1988)	61.55 and 64.70; S1.3
Y. NAKAGAWA	(a) **108**, 463 (1988)	78.60; S9.11
S. NAKAMURA	(a) **108**, 463 (1988)	78.60; S9.11
	(b) **149**, K67 (1988)	61.70 and 78.50; S9.15
	(b) **149**, K101 (1988)	61.70; 78.50; S9.15
Y. NAKAMURA	(a) **105**, 291 (1988)	75.70; 61.10; 76.80; S1.1
	(a) **107**, 315 (1988)	72.20; S8
	(a) **108**, 295 (1988)	68.55 and 75.70; 76.80; S1.1
H. NAKANISHI	(b) **148**, 673 (1988)	72.20; 71.25; S8.16
K. NAKANISHI	(a) **110**, 645 (1988)	78.65; S12
K. NAKATA	(a) **108**, 219 (1988)	62.40; S4
E. P. NAKHMEDOV	(b) **145**, 625 (1988)	72.10; 73.40; S5.12
S. M. NAKHMETOV	(a) **105**, K151 (1988)	76.30 and 76.80; S8
A. D. NAMAZOV	(a) **107**, K129 (1988)	72.80; S8
J. NANDI	(b) **149**, 275 (1988)	71.70 and 75.30; 75.40; S11
S. NARA	(b) **149**, 555 (1988)	68.55; 61.70; 73.20; S1.3
B. N. NARAHARI ACHAR	see under letter "A"	
B. NARASHIMHA	(a) **108**, K97 (1988)	61.70 and 78.50; S9.12
SA. K. NARAYANDASS	(a) **106**, K41 (1988)	73.60; S10.1
	(a) **108**, K31 (1988)	61.80 and 79.20; S12
N. M. NARTYA	(a) **107**, K57 (1988)	71.55; 72.40; S7.11
N. N. NASONOV	(b) **146**, K61 (1988)	75.30
	(b) **146**, K145 (1988)	75.30; S11.2
F. S. NASREDINOV	(a) **107**, 291 (1988)	71.55; 76.80; S8
G. Y. NASSER	(a) **110**, 165 (1988)	73.60; S5
I. NATKANIEC	(b) **145**, 445 (1988)	63.20; 61.65; S12
	(b) **145**, K15 (1988)	71.40; S1.4
	(b) **150**, 39 (1988)	63.20 and 65.40; S12
V. D. NATSIK	(a) **105**, K37 (1988)	62.40; 61.70; S1.1
O. V. NAUMENKO	(b) **146**, K61 (1988)	75.30
	(b) **146**, K145 (1988)	75.30; S11.2
J. NAVASQUILLO	(a) **110**, 585 (1988)	72.20
J. NAVRÁTIL	(a) **106**, 619 (1988)	78.20; 78.30; S8
B. B. NAYAK	(a) **105**, 455 (1988)	68.55; S8
A. B. NAZARENKO	(b) **148**, 297 (1988)	74.20
H. N. NAZARENO	(b) **145**, 525 (1988)	71.20
	(b) **146**, 341 (1988)	71.35 and 78.20; S8; S8.11
E. V. NAZVANOVA	(b) **148**, 407 (1988)	78.20; S8.11
	(b) **150**, 839 (1988)	78.20; S8.11
E. NEBAUER	(a) **109**, K35 (1988)	73.40; S7.12
YU. S. NECHAEV	(a) **106**, 399 (1988)	66.30; 61.70; 64.75; S1.1; S1.2
H. NEGISHI	(b) **147**, 559 (1988)	61.60 and 72.20; S10
	(b) **148**, 659 (1988)	72.20; S10
D. K. NELSON	(b) **147**, 727 (1988)	78.55; 71.35; S8.11; S8.12
	(b) **150**, 777 (1988)	71.35 and 78.55; S8.11
K. NEMOTO	(a) **107**, 429 (1988)	61.80 and 73.40; 85; S5.11
	(a) **110**, 301 (1988)	61.80 and 73.40; 85; S5.11
	(a) **110**, 677 (1988)	61.70 and 73.40; 85; S5.11
D. NEOGY	(b) **146**, 181 (1988)	71.70; 78.50; S9
	(b) **146**, K45 (1988)	71.70 and 75.50; S11
	(b) **149**, 275 (1988)	71.70 and 75.30; 75.40; S11
	(a) **106**, 173 (1988)	71.70; S12

A. I. Nepomnyachikh	(b) **147**, 267 (1988)	61.70 and 71.35; 78.60; S9.11
S. Nešpůrek	(b) **147**, 401 (1988)	72.20 and 76.30; S12
V. Netiksis	(b) **150**, 743 (1988)	78.20; 61.80; S5.11
J. M. Neto	(a) **108**, 795 (1988)	78.55; 61.70; S10
J. Neugebauer	(b) **145**, 579 (1988)	71.35; 78.20; S9
H. Neuhäuser	(a) **109**, 149 (1988)	61.70; 62.20; S1.2
G. Neumann	(b) **149**, 483 (1988)	66.30; S1
H. Neumann	(a) **108**, 225 (1988)	65.70; S7.16
V. B. Neustruev	(a) **110**, K107 (1988)	78.50; 76.30; S10.15
N. T. Nghi	(a) **109**, K115 (1988)	71.55; 73.40; 85; S7.11
Hoang Xuan Nguyen	(b) **146**, 351 (1988)	71.35 and 78.20; S8.11; S8.15
H. X. Nguyen	(b) **148**, 407 (1988)	78.20; S8.11
	(b) **150**, 519 (1988)	78.20; S8.11
Nguyen Ai Viet	(b) **147**, K63 (1988)	74.20
Nguyen Ba An	(b) **150**, 845 (1988)	71.35; 78.20
Nguyen Tang	(b) **149**, 511 (1988)	65.50
Nguyen Toan Thang	(b) **147**, K63 (1988)	74.20
Nguyen Van Hieu	(b) **147**, K63 (1988)	74.20
Nguyen Van Hung	(b) **148**, 49 (1988)	61.10 and 78.70; 71.10
Nguyen Van Trong	(b) **147**, 567 (1988)	71.45
T. H. Nhung	(b) **147**, K165 (1988)	72.40 and 78.55; S8.12
P. V. Nickles	(a) **107**, 85 (1988)	61.10; 65.70; 68.20; S5.11
M-A. Nicolet	(a) **110**, 509 (1988)	68.48 and 68.55; S4
J. Nieber	(a) **105**, 53 (1988)	61.10; S5.12
	(a) **108**, 489 (1988)	61.10
J. Nieder	(b) **147**, K143 (1988)	71.35 and 72.40; 78.20; S5.12
R. Nies	(a) **105**, 627 (1988)	78.20; 61.70; 66.30; 78.50; S5.11
	(b) **147**, K143 (1988)	71.35 and 72.40; 78.20; S5.12
H. Niezgoda	(b) **147**, 685 (1988)	76.80; 71.70; S1.1
A. N. Nigam	(a) **105**, 403 (1988)	66.30 and 72.15; S1.2
G. Nihoul	(a) **110**, 43 (1988)	61.14; S10.1
S. P. Nikanorov	(a) **105**, K103 (1988)	62.20; S8
A. E. Nikiforov	(b) **145**, 425 (1988)	64.60 and 76.30; S9.16
I. Ya. Nikiforov	(b) **148**, 205 (1988)	71.20; 74.70; S1.61; S1.62
L. A. Nikiforova	(b) **147**, K25 (1988)	71.20
M. Nikl	(b) **145**, 741 (1988)	78.55; 71.35; S9
A. Yu. Nikulin	(a) **108**, 651 (1988)	61.10 and 68.20; S5.11
V. K. Nikulin	(b) **146**, 511 (1988)	73.40; 71.25; S7.15
M. Nimis	(b) **149**, K89 (1988)	61.70; 78.50; S9.11
Ningsheng Luo	see under letter "L"	
A. K. Nishad	(a) **106**, K177 (1988)	77.30; S9.11
L. Nistor	(a) **110**, K1 (1988)	61.16; S1
S. V. Nistor	(b) **145**, K135 (1988)	61.70 and 76.30; S9.11
B. M. Nitsovich	(b) **148**, K29 (1988)	71.36 and 71.38
	(b) **148**, K165 (1988)	71.35 and 78.20
	(b) **150**, 561 (1988)	71.35 and 78.20
H. Nitta	(b) **149**, 41 (1988)	61.80
W. Nitzsche	(a) **107**, 851 (1988)	61.70; S5.11
M. A. Nizametdinova	(b) **147**, K99 (1988)	71.25 and 78.40; S8
G. Noll	(b) **150**, 749 (1988)	78.20; 78.55; S8.15
R. Nolte	(b) **148**, K41 (1988)	72.20
V. Nolting	(b) **149**, 313 (1988)	71.20 and 75.30; 75.10
W. Nolting	(b) **149**, 313 (1988)	71.20 and 75.30; 75.10
M. Nomura	(b) **147**, 559 (1988)	61.60 and 72.20; S10
A. P. Novikov	(a) **109**, 231 (1988)	61.80 and 68.55; S1.61; S5.11
B. V. Novikov	(b) **148**, 185 (1988)	71.35; 78.20; S9

D. L. Novikov	(b) **149**, 175 (1988)	71.25; S1.61
R. Novotný	(b) **146**, K75 (1988)	78.30; S8.16
	(a) **106**, 619 (1988)	78.20; 78.30; S8
G. Nouet	(a) **107**, 551 (1988)	61.70; 61.16
W. A. Oates	(b) **149**, 457 (1988)	61.70 and 65.50
J. Oberle	(b) **150**, 477 (1988)	78.20; S8.11
Z. Obuszko	(a) **106**, K169 (1988)	75.80; S1.1; S1.2
G. Oelgart	(a) **106**, K113 (1988)	78.45 and 85; 73.40; S7.11; S7.16
	(a) **107**, 419 (1988)	73.40 and 78.60; 85; S7.12
	(a) **109**, 261 (1988)	73.40; 78.60; S7.12
	(a) **110**, 197 (1988)	73.60 and 78.60; S7.15
S. G. Ogloblin	(b) **148**, 185 (1988)	71.35; 78.20; S9
S. Ōhara	(b) **147**, 559 (1988)	61.60 and 72.20; S10
	(b) **148**, 659 (1988)	72.20; S10
T. Ohgaku	(a) **105**, 153 (1988)	62.20; 62.40; S9.11
	(a) **106**, K19 (1988)	62.20; S9.11
S. Ohtaki	(a) **107**, 123 (1988)	61.70; 62.20; 72.15; S4
Y. H. Ohtsuki	(b) **149**, 41 (1988)	61.80
H. Ohyama	(a) **107**, 429 (1988)	61.80 and 73.40; 85; S5.11
	(a) **110**, 301 (1988)	61.80 and 73.40; 85; S5.11
	(a) **110**, 677 (1988)	61.70 and 73.40; 85; S5.11
S. Ohzawa	(b) **149**, 41 (1988)	61.80
D. K. Okorokov	(b) **146**, 341 (1988)	71.35 and 78.20; S8; S8.11
	(b) **150**, 839 (1988)	78.20; S8.11
S. Okuda	(a) **106**, 181 (1988)	71.55 and 79.75; 78.60; S10.1
V. P. Oleinikov	(b) **146**, 439 (1988)	61.70
A. I. Olekhnovich	(a) **108**, 431 (1988)	78.60; S7.1
J. Olfe	(a) **109**, 149 (1988)	61.70; 62.20; S1.2
S. I. Olikhovskii	(a) **108**, 67 (1988)	61.10; 61.70; S5.12
L. E. Oliveira	(b) **147**, 223 (1988)	71.25; 71.45; S8
O. N. Oliveira, Jr.	(a) **105**, 531 (1988)	72.20; S12
M. Olko	(a) **109**, K63 (1988)	75.70
S. Olszewski	(b) **148**, 699 (1988)	75.80; 62.20; S1; S1.1; S1.2
Om Parkash	see under letter "P"	
B. N. Onwuagba	(b) **146**, 131 (1988)	63.20; S1.4
Z. Onyszkiewicz	(b) **150**, 109 (1988)	68.40
A. A. Opalenko	(b) **150**, 25 (1988)	63.20 and 76.80; 62.50; S4
A. Opanowicz	(a) **105**, K135 (1988)	72.40 and 73.30; S8.16
	(a) **106**, K197 (1988)	73.30 and 85; S8.16
	(a) **108**, K47 (1988)	71.55; 78.60; S8.12
Ch. Opitz	(b) **148**, K11 (1988)	73.30; 73.40
H. Opyrchał	(a) **109**, 603 (1988)	78.20; S9.11
A. Orendáčová	(b) **150**, K63 (1988)	75.30 and 75.40; S11
T. S. Orlova	(a) **105**, K93 (1988)	62.20 and 64.70; S8
E. Orozco	(b) **147**, 431 (1988)	78.30; S9.11
C. Ortiz	(a) **109**, 413 (1988)	61.70 and 78.60; S9.11
V. V. Osiko	(a) **106**, 363 (1988)	61.70; 62.20; S5.11
M. E. Osinovskii	(a) **108**, 67 (1988)	61.10; 61.70; S5.12
V. A. Osipov	(b) **147**, 199 (1988)	71.45; 71.38; S12.1
P. M. Ossi	(a) **108**, 587 (1988)	62.30 and 64.70; S1.2; S1.3
B. Ostrom	(a) **105**, 563 (1988)	72.80 and 78.20; S8.16
J. Otto	(b) **148**, 489 (1988)	63.20; 78.30; S9.11
H.-J. Ou	(a) **107**, 719 (1988)	68.20; 68.48; S1; S10
J. L. Oudar	(b) **150**, 465 (1988)	68.55 and 78.20; S7.12; S7.15

K. L. OVANESYAN	(a) **105**, K155 (1988)	78.45; 78.55; S10
A. OVERBERG	(b) **147**, 811 (1988)	78.55; 63.20; S9.11; S9.12
V. N. OVSYUK	(a) **106**, 153 (1988)	71.55; 73.40
	(a) **106**, 557 (1988)	71.55 and 73.40; 61.80
K. B. OZANYAN	(a) **108**, K125 (1988)	71.55 and 72.40; 61.70; S7.12
G. PAASCH	(b) **147**, 823 (1988)	73.40; S7.1
	(b) **147**, K125 (1988)	71.25 and 73.40; S7.13
	(b) **148**, 349 (1988)	73.25 and 73.40; S7.13
	(b) **148**, 421 (1988)	73.40; S5.11; S7.1
	(b) **148**, 611 (1988)	71.25 and 73.40; S8.15
S. K. PABI	(a) **108**, K19 (1988)	64.75 and 68.48; S4
S. PAČESOVÁ	(b) **146**, 743 (1988)	63.20 and 78.30; 74.70; S10.15
A. K. PACHAURI	(b) **146**, 125 (1988)	64.30; 63.90; S9.11
O. PACHEROVÁ	(a) **110**, 35 (1988)	61.10; S5.12
G. K. PADAM	(a) **109**, K45 (1988)	68.55; 78.65; S7.16
H. C. PADHI	(b) **149**, 355 (1988)	78.70; S2; S4
A. PADMAJA	(b) **148**, K1 (1988)	62.20; S2
H. PAGNIA	(a) **108**, 11 (1988)	73.40 and 85; S10
K. F. PAI	(b) **147**, 791 (1988)	78.30; 63.20; S11
P. PAJÓN-SUAREZ	(b) **147**, 127 (1988)	71.25 and 73.40; S7.12; S7.15
	(b) **147**, 547 (1988)	71.25 and 73.40; S7.12
S. PAKEVA	(a) **106**, K97 (1988)	78.60; S8.11
R. PAL	(a) **107**, K83 (1988)	73.40 and 85; 72.40; S5.11
G. V. PALCHIK	(a) **108**, 311 (1988)	71.55; 61.70; S5.12
T. I. PALSON	(a) **106**, 123 (1988)	68.55 and 78.65; S8
B. PALUCHOWSKA	(a) **106**, K71 (1988)	75.50; S11.2
P. P. PAL-VAL	(a) **105**, K37 (1988)	62.40; 61.70; S1.1
H. O. PAMUK	(b) **149**, 81 (1988)	63.10; 61.50; 68.40; S5.11
GUOQING PAN	(a) **109**, K129 (1988)	74.70 and 78.70; S10.15
H. B. PAN	(b) **148**, 129 (1988)	68.30 and 77.50; 71.38
J. S. PAN	(b) **148**, 129 (1988)	68.30 and 77.50; 71.38
X. Q. PAN	(a) **107**, 63 (1988)	61.60 and 74.70; S10.15
XIAOQING PAN	(a) **106**, K117 (1988)	61.50; S11.1
B. K. PANDA	(b) **149**, 355 (1988)	78.70; S2; S4
V. PANDIAN	(a) **109**, 273 (1988)	71.55; S5.11
A. G. PANFILOV	(b) **150**, 777 (1988)	71.35 and 78.55; S8.11
G. N. PANIN	(a) **109**, K3 (1988)	61.14; 61.70; S8.13
V. E. PANIN	(b) **148**, 483 (1988)	61.50 and 62.50; S1.1
D. K. PAPADEMITRIOU	(b) **147**, K95 (1988)	71.45 and 78.30; S5
G. J. PAPAIOANNOU	(a) **108**, K41 (1988)	73.40; S7.11
M. P. PARDO	(a) **107**, 775 (1988)	61.60; 61.10; 61.16; S8
R. V. PARFENEV	(a) **105**, K129 (1988)	72.80; S7.13
	(b) **148**, K173 (1988)	78.30; S8.16
A. PARISINI	(a) **109**, 53 (1988)	61.16; S5.11
V. PARIZEK	(a) **108**, 643 (1988)	73.60; 64.70; S8.16
H. L. PARK	(a) **108**, K77 (1988)	78.20; S8.15
OM PARKASH	(a) **106**, 627 (1988)	77.20 and 77.40; 77.80; S10.15
J. L. PARRA-SANTIAGO	(b) **147**, 127 (1988)	71.25 and 73.40; S7.12; S7.15
B. H. PARRY	(a) **108**, 111 (1988)	61.14 and 64.80; S10.15
L. PARTHIER	(a) **108**, 699 (1988)	72.20; S8
YU. A. PASECHNIK	(b) **146**, 517 (1988)	71.36 and 73.20; S6
P. P. PASKOV	(b) **149**, 739 (1988)	78.20; S8.15
	(b) **150**, 729 (1988)	78.20; S8.15
L. PASSARI	(a) **106**, 583 (1988)	72.20; S5.11
G. PASTOR	(a) **106**, 11 (1988)	61.40 and 68.55; 78.30; S5.11

W. Paszkowicz	(a) **106**, 451 (1988)	65.40; 61.70; S7.11; S7.12
F. Pászti	(a) **107**, K15 (1988)	61.80 and 68.48; S1.3; S7.12
P. C. Pathak	(b) **146**, 97 (1988)	65.50; 62.20; 65.70; S11
V. M. Pathak	(a) **110**, 293 (1988)	72.40 and 73.40; S8
S. P. Pati	(a) **109**, 359 (1988)	73.40 and 85; S7.11
M. Pattabi	(a) **106**, 145 (1988)	73.60; S1.3; S10.1
S. Patu	(a) **105**, 419 (1988)	62.20; 61.70; S1.2
W. Paul	(b) **147**, 383 (1988)	75.50; 75.25; 75.30; S11.1
J. L. Pautrat	(a) **107**, 101 (1988)	61.70; 72.80; 78.30; S5.11
L. I. Pavlov	(b) **149**, 739 (1988)	78.20; S8.15
	(b) **150**, 729 (1988)	78.20; S8.15
S. T. Pavlov	(b) **148**, 533 (1988)	71.25 and 78.30; S7.12; S7.15
M. Pawłowska	(b) **146**, K51 (1988)	75.50; S9
Z. Pawlowska	(b) **145**, 555 (1988)	71.10; 71.25; S5.11
B. Pawłowski	(b) **147**, 123 (1988)	61.80 and 68.30; S5.11
C. A. Payling	(b) **150**, 177 (1988)	71.55; 72.20; 78.50; 78.55; S7.11
B. Pécz	(a) **107**, K15 (1988)	61.80 and 68.48; S1.3; S7.12
J. Peisl	(a) **108**, 129 (1988)	61.10; 68.20; S5.11
F. Pelle	(b) **149**, 759 (1988)	78.60; S9.16
J. Pellegrino	(a) **110**, 565 (1988)	72.20; 61.70; S8.15
N. I. Pepik	(a) **107**, K43 (1988)	71.55; S7.13
G. I. Pereseda	(b) **146**, 75 (1988)	61.70; 78.60; S9.11
S. E. Urreta de Pereyra	see under letter "U"	
G. S. Pérez	(a) **107**, 165 (1988)	61.55 and 64.70; S8.15
	(a) **107**, 205 (1988)	64.70; S8.15
J. Perez	(a) **105**, 427 (1988)	62.20; 61.40; 75.80; S1.1
R. Pérez-Alvarez	(b) **145**, 493 (1988)	71.10 and 73.40
	(b) **145**, 501 (1988)	71.10 and 73.40
	(b) **147**, 127 (1988)	71.25 and 73.40; S7.12; S7.15
	(b) **147**, 547 (1988)	71.25 and 73.40; S7.12
M. V. Pérez Diaz	(a) **109**, 337 (1988)	78.70; S9; S12
Th. Peschel	(b) **150**, 657 (1988)	78.20; 71.35; S8.1
V. V. Peschev	(a) **105**, K57 (1988)	61.80 and 71.55; S7.12
P. D. Pester	(a) **105**, 649 (1988)	73.30 and 85; 72.40
	(a) **106**, 577 (1988)	72.40 and 73.25
G. M. Pestryakov	(b) **148**, K29 (1988)	71.36 and 71.38
	(b) **148**, K165 (1988)	71.35 and 78.20
	(b) **150**, 561 (1988)	71.35 and 78.20
Á. Péter	(a) **107**, 161 (1988)	62.20 and 62.65; S11.1
G. Petermann	(a) **106**, 535 (1988)	71.20 and 73.40; S5.11
M. Petrauskas	(b) **150**, 735 (1988)	78.20; 78.65; S7.15
	(b) **150**, 743 (1988)	78.20; 61.80; S5.11
V. F. Petrenko	(a) **109**, 373 (1988)	66.90 and 77.20; 64.70; S10
A. G. Petrosyan	(a) **105**, K155 (1988)	78.45; 78.55; S10
E. G. Petrov	(b) **149**, 55 (1988)	63.20; 75.10
N. S. Petrov	(b) **150**, 527 (1988)	78.65
S. A. Petrov	(a) **109**, 231 (1988)	61.80 and 68.55; S1.61; S5.11
S. B. Petrov	(b) **149**, K41 (1988)	71.55 and 76.30; S9
V. I. Petrov	(a) **110**, 283 (1988)	78.60; S7.12; S7.15
B. Petrović	(a) **109**, K55 (1988)	72.40 and 78.55; S8.16
S. Petrowa	(b) **145**, 603 (1988)	71.50; 79.20; S5.11; S6
J. Petzelt	(b) **146**, 743 (1988)	63.20 and 78.30; 74.70; S10.15
N. Peyghambarian	(b) **150**, 357 (1988)	71.35 and 78.20; S8.11
S. Pfeffer	(a) **105**, K115 (1988)	68.55; S8.15
H. Pfeiffer	(a) **106**, K161 (1988)	64.70 and 74.70; S10.15
J. Philip	(a) **108**, 775 (1988)	78.20; S5
R. Piasecki	(a) **108**, 331 (1988)	72.80; S12

J. Pichet	(a) **109**, K139 (1988)	76.30; 75.50; S11
R. Pickenhain	(a) **107**, 419 (1988)	73.40 and 78.60; 85; S7.12
C. R. Pidgeon	(b) **150**, 719 (1988)	78.20; S7.13
U. Pietsch	(a) **105**, 197 (1988)	61.10 and 68.55; S7.12; S7.15
	(a) **106**, 451 (1988)	65.40; 61.70; S7.11; S7.12
	(b) **147**, 517 (1988)	61.10 and 62.20; S5.12; S7.12
	(a) **107**, K93 (1988)	68.55; 61.10; S7.15
	(a) **109**, K7 (1988)	61.10 and 68.55; S7.12; S7.15
B. Pilawa	(b) **145**, 729 (1988)	78.20; 64.60; S11.1
Yu. K. Pilipenko	(b) **149**, 697 (1988)	76.60; 67.90
V. A. Pilipovich	(a) **110**, 155 (1988)	61.80 and 73.60; 68.60; S10.1
Ya. Ya. Pilkevich	(a) **109**, 127 (1988)	61.80; 76.30; S5.11
O. Pilla	(b) **149**, 759 (1988)	78.60; S9.16
E. Pinčík	(a) **106**, 659 (1988)	73.40 and 85; S7.12
Ping Lu	see under letter "L"	
M. Pinnow	(a) **105**, 485 (1988)	73.40; 73.20; S12
E. Piotrowski	(b) **149**, K35 (1988)	76.20; 77.40
J. Piqueras	(a) **107**, 197 (1988)	62.20; 78.60; S10.1
	(a) **108**, K81 (1988)	61.70 and 78.60; S1.61
A. S. T. Pires	(b) **145**, 711 (1988)	75.10
	(b) **146**, 229 (1988)	75.10; 75.40
	(b) **146**, 263 (1988)	76.60; S12.1
F. V. Pirogov	(a) **105**, 87 (1988)	61.80; 78.70; S9.11
Yu. V. Pisarevskii	(a) **107**, 161 (1988)	62.20 and 62.65; S11.1
M. H. Pischedda	(a) **110**, 43 (1988)	61.14; S10.1
A. Pishchev	(b) **150**, 617 (1988)	72.40 and 78.20
L. V. Pivovarova	(a) **107**, 11 (1988)	73.25; 68.45; 72.40; 73.20; S8
M. Piwowarski	(b) **147**, 123 (1988)	61.80 and 68.30; S5.11
I. Placencia	(a) **109**, 479 (1988)	68.48 and 73.60; S5.11; S10.1
N. M. Plakida	(b) **146**, K1 (1988)	63.20 and 63.75; S10
	(b) **148**, 473 (1988)	64.70; 61.50; S11
	(b) **149**, K7 (1988)	62.80 and 63.75; S10
K. Plankiewicz	(a) **108**, K169 (1988)	73.40 and 85; 72.40; S8.11
B. Pödör	(a) **105**, K129 (1988)	72.80; S7.13
	(b) **149**, K149 (1988)	72.20
K. M. Podurets	(a) **107**, 387 (1988)	78.70; S5.12
H. Poelman	(a) **107**, 731 (1988)	63.20 and 68.30; S10.1
A. Poggi	(a) **108**, 503 (1988)	61.70; 72.20; S5.11
V. V. Pogosov	(b) **145**, 455 (1988)	68.10; 73.30; S2
E. P. Pokatilov	(b) **145**, 535 (1988)	71.35 and 78.65; S9
	(b) **147**, 163 (1988)	71.38 and 73.40; 71.45
S. I. Pokutnyi	(b) **147**, K21 (1988)	73.60 and 78.55
K. Polák	(b) **145**, 741 (1988)	78.55; 71.35; S9
K. Polański	(a) **110**, 231 (1988)	64.70 and 75.60; S1.2
H. Polat	(b) **148**, 377 (1988)	75.30 and 75.40
M. Polcarová	(a) **106**, 17 (1988)	61.55; 61.10; S1.1; S1.61
I. V. Polikarpov	(b) **149**, 435 (1988)	61.10
E. Pollert	(b) **146**, 743 (1988)	63.20 and 78.30; 74.70; S10.15
	(b) **146**, K119 (1988)	74.30 and 76.30; S10.15
I. S. Polovtsev	(b) **150**, 19 (1988)	61.70 and 71.55; 72.40; 76.30; 76.60; S5.11
T. A. Polyanskaya	(b) **149**, 649 (1988)	72.20; 71.30; S7.12
F. Pomer	(b) **145**, 261 (1988)	72.20; S8
	(a) **110**, 585 (1988)	72.20
G. Pompe	(b) **147**, 103 (1988)	66.70; S5.11
E. A. Ponezha	(b) **145**, 545 (1988)	79.80
P. Pongratz	(a) **108**, 495 (1988)	61.70; S11.1

K. Ponnuraju	(a) **105**, 161 (1988)	68.55 and 73.60; S8.15
E. G. Ponyatovskii	(b) **146**, 75 (1988)	61.70; 78.60; S9.11
G. P. Popelnyuk	(a) **105**, K97 (1988)	64.70; S8.16
A. Popov	(a) **108**, 631 (1988)	68.55 and 78.55; 76.30; S7.15
A. I. Popov	(a) **106**, 333 (1988)	61.40; 62.20; S8
I. A. Popov	(b) **148**, K37 (1988)	72.15; S4
V. V. Popov	(a) **105**, K129 (1988)	72.80; S7.13
	(a) **107**, K43 (1988)	71.55; S7.13
	(a) **110**, K61 (1988)	61.70 and 66.30; S5.11
N. S. Popovich	(a) **107**, K169 (1988)	78.30; S8.16
A. Pöppl	(a) **109**, 295 (1988)	71.55; 76.30; S10.1
R. Poprawski	(a) **110**, K73 (1988)	64.70 and 77.80; S11.1
G. Portal	(a) **108**, 783 (1988)	78.60; 61.80; S10
R. Portier	(a) **107**, 739 (1988)	61.16
W. Possart	(a) **106**, 525 (1988)	73.40; 73.30; S12
	(a) **110**, 205 (1988)	73.40; S12
T. Possner	(b) **150**, 873 (1988)	78.20
K. O. Postnikov	(a) **106**, 669 (1988)	73.40 and 85; S5.11
V. S. Postnikov	(a) **109**, 463 (1988)	73.40; S7.12; S8
G. V. Potemkin	(a) **109**, K11 (1988)	61.70 and 62.20; S1.1
R. Pott	(a) **110**, 601 (1988)	75.40; S1.4
J. K. Poželа	(a) **110**, 555 (1988)	72.20; 71.55; S5.12
N. V. K. Prabhakar	(b) **146**, 111 (1988)	62.20 and 64.30; S8.15
Prabhavathy	(a) **110**, K5 (1988)	61.60; S8.16
R. B. Prabhu	(b) **149**, 623 (1988)	71.70
B. Pradeep	(a) **106**, 123 (1988)	68.55 and 78.65; S8
A. K. Pradhan	(a) **110**, K93 (1988)	74.70; S10.15
Devendra Prakash	(b) **148**, K109 (1988)	63.20; S1.2
J. Prakash	(b) **146**, K71 (1988)	77.30; 61.70; S9.11
	(a) **106**, K177 (1988)	77.30; S9.11
L. Pranevičius	(a) **109**, 239 (1988)	68.48; 79.20; S7.11; S7.12
Ch. Durga Prasad	(a) **106**, 627 (1988)	77.20 and 77.40; 77.80; S10.15
A. Pratap	(b) **149**, 93 (1988)	63.50; S2
Chr. Pratsch	(b) **149**, 547 (1988)	73.40; S7.12; S7.15
B. Prevot	(b) **150**, 65 (1988)	63.20 and 78.30; S7.12
C. Prieto	(a) **108**, K9 (1988)	62.20; S12
V. N. Prigodin	(b) **145**, 625 (1988)	72.10; 73.40; S5.12
E. Prise	(b) **150**, 447 (1988)	78.20
I. M. Pritula	(b) **147**, 757 (1988)	78.30; 63.20; S5.2
N. L. Prokhorenko	(a) **107**, K105 (1988)	61.80; S5.11
V. K. Prokopenko	(a) **109**, K59 (1988)	76.60; S10.15
M. V. Prokopenya	(a) **106**, 249 (1988)	78.55; 71.55; S7.12
	(a) **110**, 221 (1988)	71.55; 78.55; S7.12
V. Prosser	(b) **145**, 637 (1988)	72.20; S8.15
M. Prudenziati	(a) **109**, K23 (1988)	64.75; S10.1
N. Pruss	(b) **146**, K25 (1988)	74.70; S10.15
N. A. Pryatkova	(b) **149**, K145 (1988)	61.70 and 74.70; S10.15
W. Prystasz	(b) **146**, 707 (1988)	75.30
J. Przedmojski	(a) **105**, K107 (1988)	64.70; S12
P. Przysłupski	(a) **107**, K145 (1988)	74.70 and 75.60; S10.15
S. G. Psakhe	(b) **148**, 483 (1988)	61.50 and 62.50; S1.1
H. Puff	(b) **150**, 147 (1988)	71.38; 78.20; S12
M. Pugnet	(b) **146**, 393 (1988)	78.20; 71.35; S8.12
	(b) **147**, 727 (1988)	78.55; 71.35; S8.11; S8.12
N. Puhlmann	(b) **145**, K121 (1988)	71.25; 71.45; S12
	(a) **109**, 261 (1988)	73.40; 78.60; S7.12
W. Pukacki	(b) **147**, 401 (1988)	72.20 and 76.30; S12

J. Puls	(b) **150**, 419 (1988)	71.35 and 78.20; S8.11; S8.12; S8.15
	(b) **150**, 769 (1988)	71.35 and 78.20; S8.11
B. Pura	(a) **105**, K107 (1988)	64.70; S12
T. Purohit	(b) **146**, 181 (1988)	71.70; 78.50; S9
	(a) **106**, 173 (1988)	71.70; S12
T. Purtak	(b) **148**, K57 (1988)	75.10
L. Půst	(b) **146**, K119 (1988)	74.30 and 76.30; S10.15
M. Pützler	(b) **149**, 371 (1988)	78.30; S9.11
R. Puźniak	(a) **107**, K145 (1988)	74.70 and 75.60; S10.15
H. Pykacz	(a) **105**, K33 (1988)	64.70; 65.70; S11
N. B. Pyshnaya	(a) **108**, K59 (1988)	71.55 and 72.80; S7.11
Z. Pytel	(b) **146**, 697 (1988)	75.10 and 75.70
Qirui Zhang	see under letter "Z"	
Y. Qiu	(a) **110**, 323 (1988)	61.10; 61.70; S5
Qiwei Long	see under letter "L"	
P. Quenon	(b) **149**, 465 (1988)	61.70; 79.20; S5.11
R. Querzoli	(b) **147**, 421 (1988)	78.55; 73.40; S7.15
Quing Gu	see under letter "G"	
Quing Shen	see under letter "S"	
M. Quintero	(a) **107**, 165 (1988)	61.55 and 64.70; S8.15
	(a) **107**, 205 (1988)	64.70; S8.15
	(a) **108**, 599 (1988)	64.75; 75.30; S8.16
V. N. Quynh	(a) **106**, K147 (1988)	73.60; 78.65; S10.1
S. I. Radautsan	(a) **106**, K181 (1988)	78.40; S8.16
	(a) **107**, 373 (1988)	78.30 and 78.55; S8.16
	(a) **107**, K57 (1988)	71.55; 72.40; S7.11
	(a) **108**, K59 (1988)	71.55 and 72.80; S7.11
	(a) **109**, K55 (1988)	72.40 and 78.55; S8.16
	(a) **109**, K153 (1988)	78.45 and 78.65; S8.16
	(a) **110**, 277 (1988)	72.40 and 78.40; 78.60; S8.16
E. Radzhabov	(b) **149**, 441 (1988)	61.70 and 71.55; 78.50; S9.11
S. M. Rafique	(b) **147**, K51 (1988)	72.15; S3
S. Lakshmi Raghavan	(a) **108**, 771 (1988)	78.20; S9.11
E. Rahimzadeh	(a) **105**, K1 (1988)	61.30; S12
S. M. Mujibur Rahman	(b) **147**, K5 (1988)	64.70; S2
P. Rajagopal	(a) **109**, 635 (1988)	78.30; S11
K. Rajanna	(a) **105**, K181 (1988)	62.20 and 73.60; S1.2
O. S. Rajora	(a) **105**, 335 (1988)	61.60; 61.80; S8.15
	(a) **108**, 529 (1988)	66.30; 61.16; S1.63
A. Rajput	(b) **147**, K11 (1988)	63.20; S1.2; S1.3
A. E. Rakhshani	(a) **105**, 183 (1988)	68.55; S10.1
P. C. Rakshit	(b) **148**, K17 (1988)	72.20 and 73.40; S7.12
S. Ram	(a) **107**, 55 (1988)	61.40; 75.60; 76.30; S11.2
K. Ramachandran	(b) **145**, 139 (1988)	63.20; 61.70; S8.13
R. Raman	(a) **109**, K75 (1988)	77.30 and 77.40
N. Ramesh	(a) **108**, 771 (1988)	78.20; S9.11
R. Ramírez	(a) **108**, K9 (1988)	62.20; S12
	(b) **149**, K135 (1988)	71.10 and 78.70
M. Rani	(b) **149**, 93 (1988)	63.50; S2
G. H. Rao	(a) **110**, 55 (1988)	61.60 and 74.70; S11
G. S. T. Rao	(a) **106**, K53 (1988)	61.50 and 74.70; S10.15

J. L. Rao	(b) **145**, 255 (1988)	71.70; 76.30; S9.16
	(b) **148**, K119 (1988)	71.70; 76.30; 78.50; S9.12
K. V. Rao	(a) **105**, 297 (1988)	78.50; 78.60; S9.12
	(a) **106**, K101 (1988)	77.20; 77.40; S10.1
	(a) **107**, K145 (1988)	74.70 and 75.60; S10.15
	(a) **108**, 457 (1988)	77.20 and 77.40; S9.12
	(a) **108**, K97 (1988)	61.70 and 78.50; S9.12
K. V. R. Rao	(b) **147**, 343 (1988)	74.10; 78.70; S11
R. Ramji Rao	(b) **147**, K11 (1988)	63.20; S1.2; S1.3
	(b) **148**, K1 (1988)	62.20; S2
R. V. Gopala Rao	(b) **150**, 53 (1988)	65.50; S9.11
	(a) **110**, 347 (1988)	61.25; S9.11
S. U. M. Rao	(a) **109**, K45 (1988)	68.55; 78.65; S7.16
V. V. Rao	(a) **110**, 293 (1988)	72.40 and 73.40; S8
Y. Srinivasa Rao	(a) **106**, 217 (1988)	75.50; 61.60; S11.2
A. Raoufi	(a) **108**, 369 (1988)	72.15; 75.80; S1.1
Ö. Rapp	(a) **106**, 233 (1988)	75.50; S1.2; S1.4
J. Rasek	(a) **107**, 299 (1988)	72.15; S1.1; S1.2
V. M. Ratag	(b) **147**, 243 (1988)	71.25; S5.12
P. J. Ratcliffe	(a) **108**, 537 (1988)	61.70; 73.40; S1.2
Ratna Das	see under letter "D"	
A. M. Ratner	(b) **147**, 757 (1988)	78.30; 63.20; S5.2
S. A. Ratseev	(a) **107**, 373 (1988)	78.30 and 78.55; S8.16
H. Rauh	(b) **148**, 445 (1988)	61.12 and 61.60; 62.50; 65.90; S8
B. Raveau	(a) **107**, 73 (1988)	61.60; 64.80; S10.15
S. Ravelet	(a) **108**, K41 (1988)	73.40; S7.11
B. Ray	(a) **110**, 269 (1988)	71.55 and 78.55; 76.30; 78.60; S8
B. S. Razbirin	(b) **147**, 727 (1988)	78.55; 71.35; S8.11; S8.12
	(b) **150**, 777 (1988)	71.35 and 78.55; S8.11
I. M. Razdobreev	(b) **147**, 717 (1988)	78.55; S8
C. Razzetti	(b) **145**, 401 (1988)	61.60 and 78.70; S8.16
E. Realo	(b) **149**, 283 (1988)	76.80
I. K. Rebane	(b) **145**, 749 (1988)	78.50
E. Recknagel	(b) **145**, 151 (1988)	71.36 and 73.20; 78.30; S9.12
P. V. Reddy	(a) **108**, 607 (1988)	62.20; 75.50; S11.2
S. V. Redkin	(a) **108**, 651 (1988)	61.10 and 68.20; S5.11
C. Regnaut	(b) **146**, 413 (1988)	61.20; 61.25
W. Reich	(b) **149**, K163 (1988)	75.40
M. Reiche	(a) **107**, 851 (1988)	61.70; S5.11
J. Reichel	(a) **107**, 851 (1988)	61.70; S5.11
L. Reinhard	(b) **148**, 457 (1988)	61.55 and 64.80; S1.2
J. Reinhold	(b) **145**, 603 (1988)	71.50; 79.20; S5.11; S6
A. Reklaitis	(b) **150**, 437 (1988)	78.20; S7.12
D. S. Remenko	(a) **106**, K181 (1988)	78.40; S8.16
A. A. Rempel	(a) **106**, 459 (1988)	64.60; 75.30; S1.61
H. D. Rempel	(b) **148**, 689 (1988)	71.70 and 76.30; S5.11
R. Renner	(b) **150**, 749 (1988)	78.20; 78.55; S8.15
P. Rennert	(b) **148**, 49 (1988)	61.10 and 78.70; 71.10
R. Rentzsch	(b) **146**, 199 (1988)	72.20; 72.60; S7.12
	(a) **106**, 419 (1988)	61.70; 72.80; 78.55; 78.70; S7.12
I. I. Reshina	(a) **106**, 261 (1988)	78.30; 63.50; S7.11; S7.12
H. Rhan	(a) **107**, K93 (1988)	68.55; 61.10; S7.15
	(a) **109**, K7 (1988)	61.10 and 68.55; S7.12; S7.15
B. Rheinländer	(a) **106**, K113 (1988)	78.45 and 85; 73.40; S7.11; S7.16
	(a) **107**, 405 (1988)	73.40; 71.55; 72.40; 78.60; S7.11; S7.15
R. M. Riberto-Teixeira	(b) **145**, 525 (1988)	71.20
P. C. Rice-Evans	(b) **147**, 89 (1988)	61.70 and 78.70; S1.2; S5

G. Richter	(a) **105**, K45 (1988)	73.25; 78.45; S7.15
H. H. Richter	(a) **107**, 281 (1988)	68.55; S5.11
A. Ridah	(b) **147**, 441 (1988)	77.80 and 78.30; 63.20; S11.1
N. Riddleston	(b) **145**, 145 (1988)	78.65; 68.55; S10.15
V. Riede	(b) **150**, 165 (1988)	71.25; 72.20; 72.30; 78.20; S1.63
R. Riera	(b) **148**, 533 (1988)	71.25 and 78.30; S7.12; S7.15
J. Ringeissen	(b) **149**, 615 (1988)	71.35; S5.11; S5.12; S7.1; S8.1
D. Ríos-Jara	(a) **105**, 411 (1988)	65.40; S1.2
D. Ristau	(a) **109**, 255 (1988)	73.60; 79.20; S4
M. Ritschel	(b) **146**, K29 (1988)	74.70; S10.15
J. P. Rivière	(a) **107**, 867 (1988)	68.55; 61.16; S1.2
A. Rizzo	(a) **105**, 101 (1988)	61.70; S8
D. Yu. Rodichev	(b) **150**, K25 (1988)	73.40; S7.15
L. Rodríguez	(a) **105**, 411 (1988)	65.40; S1.2
V. D. Rodriguez	(a) **109**, 413 (1988)	61.70 and 78.60; S9.11
H. Rodriguez-Coppola	(b) **145**, 493 (1988)	71.10 and 73.40
	(b) **145**, 501 (1988)	71.10 and 73.40
K. Rogacki	(a) **146**, K103 (1988)	74.70; S10.15
N. G. Romanov	(b) **147**, K171 (1988)	71.55 and 76.30; S9.11
S. Romanowski	(b) **145**, 467 (1988)	73.30; 71.10; S1.2
	(b) **148**, 515 (1988)	78.65; S1.2
	(b) **148**, K11 (1988)	73.30; 73.40
A. A. Romanyukha	(b) **147**, K185 (1988)	76.30; S12
N. Romeo	(a) **109**, K105 (1988)	68.55 and 73.60; S8.15
Rongchuan Liu	see under letter "L"	
J. Rosa	(b) **149**, K31 (1988)	76.30; 78.50; S9.11
N. N. Rosanov	(b) **150**, 545 (1988)	78.20
A. Röseler	(a) **107**, K69 (1988)	78.30
J. Röseler	(b) **145**, 579 (1988)	71.35; 78.20; S9
S. B. Rosenkov	(b) **149**, 427 (1988)	61.60 and 71.55; 76.30; S8.16
M. Rösler	(b) **148**, 213 (1988)	79.20; S4
H. Rossmann	(b) **145**, K83 (1988)	78.20; S8.12
	(b) **150**, 495 (1988)	78.20; S8.12
P. Rostaing	(b) **149**, 203 (1988)	79.25; S10.1
A. H. Rotaru	(b) **149**, 187 (1988)	71.36
	(b) **150**, 401 (1988)	71.35 and 78.20
A. Kh. Rotaru	(b) **147**, 721 (1988)	71.35 and 78.20
	(b) **150**, 513 (1988)	78.20
S. Roth	(a) **106**, K61 (1988)	75.50; 75.60; S1.1; S1.2
H. Rothard	(b) **147**, 589 (1988)	71.45; 79.20; S1.2; S4; S5
L. Rothkirch	(b) **146**, K25 (1988)	74.70; S10.15
B. K. Roul	(a) **110**, K93 (1988)	74.70; S10.15
R. Rousina	(a) **110**, K57 (1988)	78.65; 61.40; S8
J. L. Rouvière	(a) **107**, 481 (1988)	61.55 and 61.60; 61.16; S5.11; S5.12; S7.1; S8.1
D. Roy	(b) **146**, 117 (1988)	62.20 and 77.20; S9.11
M. Roy	(a) **107**, K1 (1988)	61.60 and 65.90; S10
S. K. Roy	(a) **109**, 359 (1988)	73.40 and 85; S7.11
P. Royer	(a) **105**, 617 (1988)	78.20; 78.65; S1.3
O. V. Rozanov	(a) **105**, 145 (1988)	64.70; 76.60; S11.1
E. I. Rozum	(b) **149**, K129 (1988)	71.25; S5.11
A. S. Rubanov	(b) **150**, 901 (1988)	66.70 and 78.35; S6
N. M. Rubinina	(a) **108**, 437 (1988)	78.50; S11.1
R. Ruckh	(b) **147**, 229 (1988)	71.38
I. Rückmann	(b) **150**, 735 (1988)	78.20; 78.65; S7.15
Yu. V. Rud	(a) **110**, 631 (1988)	78.20; 72.40; S8.16
N. Rudolph	(b) **150**, 419 (1988)	71.35 and 78.20; S8.11; S8.12; S8.15
C. Rudowicz	(b) **147**, 677 (1988)	76.30; S11

J. Rudzińska-Girulska	(b) **148**, 227 (1988)	71.25; 71.60; S1.2
E. Ruedl	(a) **107**, 745 (1988)	61.55 and 61.70; 61.16; S1.2
W. W. Rühle	(b) **149**, 215 (1988)	71.38; S7.12
C. Ruiz-Mejía	(b) **147**, 593 (1988)	71.55; 61.70; S9.11
V. Rusanov	(b) **146**, 269 (1988)	76.80; S1.1
N. Russo	(b) **145**, K125 (1988)	71.55; S5.11
E. V. Russu	(a) **107**, K57 (1988)	71.55; 72.40; S7.11
S. S. Russu	(b) **146**, 329 (1988)	78.20 and 78.30; 71.35; S7.15
A. G. Rustamov	(a) **107**, K129 (1988)	72.80; S8
O. Ryba	(b) **147**, 401 (1988)	72.20 and 76.30; S12
V. Rybka	(a) **106**, K35 (1988)	68.45 and 78.70; S7.11; S7.12
B. S. Ryvkin	(b) **150**, 567 (1988)	78.20; 72.40; S7.12
	(b) **150**, 599 (1988)	78.20; 85; S7.12
V. M. Ryzhkovskii	(a) **107**, K89 (1988)	61.60; S7
H. Rzewuski	(a) **109**, 531 (1988)	61.70 and 71.55; S5.11
G. Saad	(a) **107**, 171 (1988)	62.40; S4
	(a) **110**, 481 (1988)	62.40; S4
J. Sabin del Valle	(b) **145**, K155 (1988)	71.25; S7.13; S7.15
	(b) **148**, K113 (1988)	71.25 and 73.40
R. Kh. Sabirov	(b) **146**, 683 (1988)	76.20
M. Sacchi	(a) **109**, K23 (1988)	64.75; S10.1
A. Sachs	(b) **148**, 101 (1988)	62.20; S4
F. Sachslehner	(a) **107**, 305 (1988)	72.15; S1.2
L. Sadłowski	(a) **109**, K139 (1988)	76.30; 75.50; S11
A. F. Safarov	(a) **110**, K45 (1988)	65.70; 75.50; S8
G. Safoula	(a) **110**, 521 (1988)	73.60; S5; S8
S. F. Sagnowski	(a) **107**, 347 (1988)	76.60; S11
Ch. S. Saidov	(a) **107**, 291 (1988)	71.55; 76.80; S8
A. G. Saif	(b) **149**, 657 (1988)	74.60
J. A. de Saja	(a) **108**, K109 (1988)	73.40; 72.40; 78.40; S12
V. D. Sajfert	(b) **148**, 81 (1988)	64.60; 77.80; S11.1
H. Saka	(a) **108**, 163 (1988)	61.70; 64.75; S1.2; S1.3
A. Sakakibara	(a) **110**, K9 (1988)	64.75; S4
M. S. Sakr	(a) **110**, K13 (1988)	62.20; S1.2
M. Salagram	(a) **105**, K161 (1988)	78.30; S11
	(a) **106**, K185 (1988)	78.30; S11.1
K. A. Al-Saleh	see under letter "A"	
S. M. Saleh	(a) **106**, 49 (1988)	61.14 and 61.70; S1.2
N. S. Saleh	(a) **105**, 177 (1988)	61.80 and 64.75; 73.60; S1.3; S5
	(a) **107**, 253 (1988)	61.80 and 73.60; S1.2; S5.11
W. Salejda	(b) **148**, 473 (1988)	64.70; 61.50; S11
M. Ben Salem	(a) **107**, 469 (1988)	68.48; 61.16; S10
	(a) **110**, 43 (1988)	61.14; S10.1
E. G. Salman	(a) **108**, 625 (1988)	73.40 and 77.30; 77.50; S10.1
A. A. Ghilarducci de Salva	see under letter "G"	
L. Šamaj	(b) **149**, 291 (1988)	75.10
	(b) **149**, 675 (1988)	75.10 and 75.40
L. K. Samanta	(a) **107**, K79 (1988)	78.20; S7.15; S8.15
E. A. Samedov	(b) **146**, K161 (1988)	63.20 and 78.30; S8.13
K. Saminadayar	(a) **107**, 101 (1988)	61.70; 72.80; 78.30; S5.11
E. G. Samoilova	(a) **105**, K169 (1988)	78.55; 71.70; S10
A. A. Samokhvalov	(b) **148**, 361 (1988)	72.20 and 75.30; 76.30; S10.1
B. E. Samorukov	(a) **110**, K61 (1988)	61.70 and 66.30; S5.11
V. A. Samuilov	(a) **110**, 181 (1988)	68.55 and 78.65; S5.11

V. SAMULIONIS	(a) **110**, 429 (1988)	64.70; 62.65; 77.20; S11.1
T. SANADZE	(b) **146**, K141 (1988)	76.70; S9.12
J. L. SÀNCHEZ LLAMAZARES	(a) **105**, 243 (1988)	75.50; 76.80; S1.1
	(a) **105**, 285 (1988)	75.30; S11.2
	(a) **109**, 569 (1988)	75.30; 61.60; S1.1
B. SANDOW	(a) **109**, K35 (1988)	73.40; S7.12
A. A. S. SANGACHIN	(b) **147**, 115 (1988)	61.50 and 63.90; 62.20; 65.70; S9
	(b) **147**, 529 (1988)	62.20; 63.90; 77.20; S9.11
M. P. DOS SANTOS	(b) **147**, 779 (1988)	78.30 and 78.55; S7.11
S. P. SANYAL	(b) **149**, 121 (1988)	62.20; S8
J. A. SANZ-GARCÍA	(a) **108**, K145 (1988)	78.55 and 78.60; S10
V. B. SAPRE	(b) **145**, 77 (1988)	61.10; S1.2; S8.16
V. I. SAPTSOV	(b) **150**, 605 (1988)	78.20; 78.45
V. SARGŪNAS	(a) **109**, 239 (1988)	68.48; 79.20; S7.11; S7.12
C. K. SARKAR	(b) **147**, 135 (1988)	73.40 and 76.40; S7.12
	(b) **148**, K17 (1988)	72.20 and 73.40; S7.12
S. E. SARKISOV	(a) **107**, 365 (1988)	78.30; S11.2
E. F. SARMENTO	(b) **150**, 261 (1988)	75.10 and 75.40
C. SARRAZIN	(a) **107**, 867 (1988)	68.55; 61.16; S1.2
M. SASAKI	(b) **147**, 559 (1988)	61.60 and 72.20; S10
N. SASI	(a) **106**, K41 (1988)	73.60; S10.1
M. S. M. SASTRY	(a) **106**, 145 (1988)	73.60; S1.3; S10.1
	(a) **107**, 225 (1988)	73.60; 68.55; S4
S. B. S. SASTRY	(b) **146**, 727 (1988)	78.60; 61.70; S9.11
	(b) **150**, 315 (1988)	78.60; 61.70; S9.11
S. S. SASTRY	(a) **105**, K71 (1988)	77.20 and 77.40; S9
V. BHASKARA SASTRY	(a) **110**, K5 (1988)	61.60; S8.16
F. SATO	(a) **105**, 263 (1988)	75.50; 75.80; S1.1
	(a) **107**, 355 (1988)	75.50; 75.60; S1.1
T. SATOW	(b) **146**, 661 (1988)	72.10 and 72.80; 75.30; S1.63
	(a) **108**, K89 (1988)	61.40; S8
	(b) **149**, 257 (1988)	72.20 and 75.30; S1.63; S8.16
A. SATPATHY	(b) **150**, 53 (1988)	65.50; S9.11
S. SATPATHY	(b) **145**, 555 (1988)	71.10; 71.25; S5.11
G. SATYANADAM	(a) **105**, K71 (1988)	77.20 and 77.40; S9
K. SATYANARAYANA MURTHY	see under letter "M"	
V. A. SAUTENKOV	(b) **150**, 605 (1988)	78.20; 78.45;
J. L. SAUVAJOL	(b) **147**, 537 (1988)	63.20 and 78.30; S9
V. A. SAVCHUK	(a) **108**, K63 (1988)	72.80; S8.16
A. V. SAVIN	(b) **147**, 155 (1988)	71.45
N. A. SAVINOVA	(a) **109**, K51 (1988)	72.80 and 76.80; S8
N. SAVVIDES	(b) **147**, K153 (1988)	74.70; S10.15
Y. SAYADI	(a) **108**, 783 (1988)	78.60; 61.80; S10
K. EL SAYED	see under letter "E"	
N. S. SAXENA	(b) **149**, 93 (1988)	63.50; S2
U. SCHADE	(b) **149**, 447 (1988)	66.30; S7.1
H. SCHAEBEN	(b) **148**, 63 (1988)	61.50; 61.10; 61.12
W. SCHÄFER	(b) **147**, 699 (1988)	71.35 and 78.20
	(b) **150**, 407 (1988)	78.20; 71.35
D. SCHALCH	(a) **105**, K81 (1988)	61.80 and 78.65; S6
R. SCHALLER	(a) **108**, 613 (1988)	62.40 and 64.70; S1.2
F. W. SCHAPINK	(a) **106**, 433 (1988)	61.70
	(a) **107**, 551 (1988)	61.70; 61.16
A. SCHARMANN	(a) **105**, K81 (1988)	61.80 and 78.65; S6
M. SCHELL	(b) **150**, 575 (1988)	78.45
M. SCHELUDKO	(a) **105**, K145 (1988)	75.30; S1.4

D. Schikora	(a) **105**, 189 (1988)	68.55; S8.13
	(a) **106**, 509 (1988)	68.55 and 78.55; S8
	(a) **109**, 11 (1988)	61.50 and 68.48; 68.20; S5.11; S7.12; S8.13
V. Schiller	(a) **108**, K35 (1988)	61.70 and 68.48; 78.65; S5.11
W. Schiller	(b) **145**, K35 (1988)	72.15
W. Schilling	(b) **145**, 351 (1988)	61.80 and 63.20; S4
L. Schimmele	(b) **147**, 361 (1988)	74.60
F. Schirrmeister	(a) **108**, 149 (1988)	66.30; 72.20; S8
M. Schlanges	(b) **147**, 747 (1988)	71.35; S5.12
H. Schlegel	(b) **149**, K111 (1988)	63.10
J. Schmal	(a) **106**, K47 (1988)	72.40; S8.16
K. Schmalz	(a) **109**, 279 (1988)	71.55; S5.11
H. Schmelz	(a) **108**, 495 (1988)	61.70; S11.1
C. Schmid	(b) **149**, 521 (1988)	61.70 and 63.20; S12
D. Schmid	(b) **147**, 797 (1988)	78.55; 71.35; 71.55; S11
B. Schmidt	(a) **108**, 469 (1988)	73.40 and 85
G. Schmidt	(a) **109**, 503 (1988)	68.55; 74.70; S10.15
Th. Schmidt	(b) **147**, 797 (1988)	78.55; 71.35; 71.55; S11
H. Schmiedel	(b) **147**, K43 (1988)	74.30; 74.70; S10.15
S. Schmitt-Rink	(b) **150**, 349 (1988)	71.35 and 78.20; S7.12; S7.15; S12
B. Schnackenburg	(b) **145**, 157 (1988)	73.40; S7.13
H. A. Schneider	(a) **108**, 689 (1988)	71.55 and 72.20; S7.12
Ch. Schnittler	(a) **108**, 519 (1988)	61.70; 64.75; S5.11
M. Schnürer	(a) **107**, 85 (1988)	61.10; 65.70; 68.20; S5.11
H. R. Schober	(b) **145**, 351 (1988)	61.80 and 63.20; S4
D. Schoemaker	(b) **145**, K135 (1988)	61.70 and 76.30; S9.11
E. Schöll	(b) **150**, 575 (1988)	78.45
B. Schönfeld	(b) **148**, 457 (1988)	61.55 and 64.80; S1.2
P. Schotanus	(b) **148**, K77 (1988)	78.55; S9.12
J. Schreiber	(b) **145**, 695 (1988)	75.40
	(a) **110**, 97 (1988)	61.70 and 71.55; 72.40; 78.50; 78.55; S8.1
	(a) **110**, 283 (1988)	78.60; S7.12; S7.15
A. Schröder	(a) **110**, 107 (1988)	66.30; 75.50; S1.1
B. Schröder	(a) **108**, 275 (1988)	68.55; 78.30; S5.11
K. Schröder	(b) **149**, 685 (1988)	75.70; S1.2
B. Schubert	(a) **105**, K115 (1988)	68.55; S8.15
M. Schubert	(b) **146**, 385 (1988)	72.10; 78.20
K.-H. Schuldt	(b) **147**, 699 (1988)	71.35 and 78.20
	(b) **150**, 407 (1988)	78.20; 71.35
D. Schulz	(b) **145**, 695 (1988)	75.40
G. Schulz	(a) **110**, 657 (1988)	73.40 and 78.60; 85; S8.11
D. Schulze	(b) **148**, 349 (1988)	73.25 and 73.40; S7.13
H. Schulze	(a) **108**, K131 (1988)	61.70 and 72.20; S10.1
	(a) **108**, K135 (1988)	61.70 and 72.20; S10.1
A. Schülzgen	(b) **145**, K83 (1988)	78.20; S8.12
	(b) **150**, 495 (1988)	78.20; S8.12
B. Schumann	(a) **108**, 257 (1988)	73.40; 68.55; 78.65; S7.1; S9.15
J.-W. Schünemann	(a) **110**, 141 (1988)	64.70; 75.40; S7.15
Th. Schurig	(b) **145**, 157 (1988)	73.40; S7.13
C. Schwab	(b) **150**, 65 (1988)	63.20 and 78.30; S7.12
H. Schwab	(b) **150**, 387 (1988)	71.30 and 78.20; S8.12; S8.15
C. F. Schwerdtfeger	(a) **108**, 733 (1988)	76.30; S12
D. Schwitalla	(b) **145**, K35 (1988)	72.15
H. Scibior	(a) **109**, 597 (1988)	78.55 and 78.60; 61.70; S7.12
M. Sebbani	(b) **148**, 233 (1988)	71.50
K. Seetharam	(a) **106**, K185 (1988)	78.30; S11.1
M. Segnini	(a) **105**, K125 (1988)	73.60; S4

A. SEGURA	(b) **145**,	261 (1988)	72.20; S8
CH. SEIDEL	(b) **149**,	327 (1988)	72.15 and 75.30; S12.1
R. P. SEISYAN	(b) **150**,	863 (1988)	71.35 and 78.20; S7.12; S7.15
T. SEIYAMA	(a) **106**,	181 (1988)	71.55 and 79.75; 78.60; S10.1
G. SEKAR	(a) **109**,	635 (1988)	78.30; S11
P. SEKAR	(b) **147**,	495 (1988)	66.30; S9
G. SEKRETARCZYK	(b) **147**,	401 (1988)	72.20 and 76.30; S12
B. SELLE	(a) **110**,	657 (1988)	73.40 and 78.60; 85; S8.11
D. I. SEMENTSOV	(a) **110**,	259 (1988)	78.20; 75.60
K. SEN	(a) **107**,	K83 (1988)	73.40 and 85; 72.40; S5.11
P. SEN	(b) **145**,	343 (1988)	78.20; S7.13
P. K. SEN	(b) **145**,	343 (1988)	78.20; S7.13
S. SEN	(a) **105**,	171 (1988)	68.55 and 78.65; S10.1
M. SENCE	(b) **150**,	749 (1988)	78.20; 78.55; S8.15
R. SEN GUPTA	see under letter "G"		
C. A. C. SEQUEIRA	(a) **105**,	123 (1988)	66.30 and 72.20; 61.70; S10
	(a) **108**,	155 (1988)	66.30 and 72.10; S10
YU. A. SEREBRENNIKOV	(b) **148**,	689 (1988)	71.70 and 76.30; S5.11
P. P. SEREGIN	(a) **107**,	291 (1988)	71.55; 76.80; S8
V. F. SEREGIN	(b) **150**,	605 (1988)	78.20; 78.45
L. N. SEREGINA	(a) **109**,	K51 (1988)	72.80 and 76.80; S8
V. M. SERGEEVA	(a) **105**,	K93 (1988)	62.20 and 64.70; S8
A. SERPI	(a) **108**,	351 (1988)	72.40; 71.55; S8.16
A. SERRA	(a) **110**,	409 (1988)	61.70; 78.70; S1.2
J. P. ŠETRAJČIĆ	(b) **148**,	593 (1988)	71.45; 77.20
S. B. SEVASTIANOV	(a) **106**,	153 (1988)	71.55; 73.40
	(a) **106**,	557 (1988)	71.55 and 73.40; 61.80
A. L. SHABALOV	(b) **145**,	K71 (1988)	78.65; S7
I. N. SHABANOVA	(b) **149**,	K177 (1988)	79.60; S1.4
S. J. SHAFFER	(a) **107**,	799 (1988)	61.70 and 64.80; S10
M. D. SHAGALOV	(a) **108**,	657 (1988)	73.40 and 78.60; S7.1
V. A. SHAKIN	(b) **150**,	527 (1988)	78.65
S. M. AL-SHAMARI	see under letter "A"		
B. D. SHANINA	(a) **109**,	537 (1988)	71.55; 72.80; 76.30; S5.11
J. SHANKER	(b) **150**,	K7 (1988)	61.50; S9.11
HUIMIN SHAO	(a) **110**,	K89 (1988)	64.70 and 74.70; S10.15
A. K. SHARMA	(b) **145**,	435 (1988)	62.20; S9.12
	(b) **146**,	125 (1988)	64.30; 63.90; S9.11
M. N. SHARMA	(b) **149**,	K55 (1988)	77.30; S9.11
K. SHARSHEEV	(b) **146**,	K135 (1988)	76.30 and 78.50; S11.1
V. A. SHCHEGOLEV	(a) **106**,	371 (1988)	61.80; 62.20; S1.2
A. W. SHEIKH	(b) **147**,	601 (1988)	72.15; 75.30; S1.2; S1.3
	(b) **150**,	K95 (1988)	75.30
S. S. SHEININ	(a) **105**,	45 (1988)	61.14
M. K. SHEINKMAN	(a) **110**,	213 (1988)	71.55; 73.40; 78.55; S7.15
R. I. SHEKHMAMETEV	(b) **145**,	579 (1988)	71.35; 78.20; S9
YU. G. SHEKUN	(b) **147**,	717 (1988)	78.55; S8
A. A. SHEMYAKOV	(a) **109**,	K59 (1988)	76.60; S10.15
D.-X. SHEN	(a) **106**,	K1 (1988)	61.70 and 78.70; S12
HUI SHEN	(a) **109**,	K1 (1988)	61.40; S1.61
QING SHEN	(a) **107**,	153 (1988)	63.20 and 68.30; 61.55; S1.2
S. C. SHEN	(b) **147**,	767 (1988)	78.30; 63.20; S8.15
SHUECHU SHEN	(b) **147**,	511 (1988)	63.20; 68.30; S1.2
V. A. SHENDEROVSKII	(b) **146**,	671 (1988)	72.10; 72.20; S5
	(a) **107**,	95 (1988)	61.70; 72.20; 75.20; S8.13
FEI HAO SHENG	(b) **150**,	663 (1988)	78.20; S8.15

Y. Q. Sheng	(b) **145**, 151 (1988)	71.36 and 73.20; 78.30; S9.12
Shen Li	see under letter "L"	
G. A. Shepelskii	(a) **110**, 97 (1988)	61.70 and 71.55; 72.40; 78.50; 78.55; S8.1
A. V. Sherman	(b) **145**, 319 (1988)	71.35 and 78.40
	(b) **149**, 725 (1988)	71.35 and 78.30
T. S. Shi	(b) **148**, 715 (1988)	61.70 and 78.30; 78.50; S10
G. D. Shibarshina	(b) **147**, 721 (1988)	71.35 and 78.20
	(b) **150**, 513 (1988)	78.20
S. Shigetomi	(a) **108**, K53 (1988)	71.55 and 72.80; S8
M. Sh. Shikhsaidov	(b) **146**, 75 (1988)	61.70; 78.60; S9.11
	(a) **107**, 131 (1988)	62.20; 61.80; S7.12
V. B. Shikin	(a) **108**, 669 (1988)	61.70 and 71.55; S5.11; S5.12
N. I. Shikina	(a) **108**, 669 (1988)	61.70 and 71.55; S5.11; S5.12
N. Shiktorov	(a) **108**, 197 (1988)	64.70; 76.80; S8.15
N. A. Shilkova	(b) **146**, 173 (1988)	71.10; 71.25
	(b) **149**, 195 (1988)	71.10
	(b) **149**, 571 (1988)	71.10; 71.25
I. P. Shilovich	(a) **105**, K97 (1988)	64.70; S8.16
S. Sh. Shilstein	(a) **107**, 387 (1988)	78.70; S5.12
N. M. Shimanskaya	(b) **149**, K145 (1988)	61.70 and 74.70; S10.15
S. Shimanuki	(b) **148**, 251 (1988)	71.55 and 78.55; S9.11
T. Shimizu	(b) **147**, K1 (1988)	61.70; 76.30; S10.1
S. Shindo	(b) **145**, 89 (1988)	61.80
T. Shiozawa	(a) **110**, 375 (1988)	61.60; 61.16; S1.63; S9
V. B. Shipilo	(a) **104**, 431 (1988)	78.60; S7.1
M. M. Shirinov	(b) **148**, K93 (1988)	63.20 and 78.30; S8.16
A. M. Shirokov	(b) **147**, 297 (1988)	71.55 and 78.50; 72.40; 76.40; S5
V. P. Shirokovskii	(b) **146**, 173 (1988)	71.10; 71.25
	(b) **149**, 195 (1988)	71.10
	(b) **149**, 571 (1988)	71.10; 71.25
A. V. Shishkin	(b) **149**, 785 (1988)	78.70; S9.12
E. M. Shishonok	(a) **108**, 431 (1988)	78.60; S7.1
G. K. Shivakumar	(a) **110**, K57 (1988)	78.65; 61.40; S8
B. L. Shklyar	(a) **106**, 363 (1988)	61.70; 62.20; S5.11
Yu. K. Shlyk	(a) **109**, 463 (1988)	73.40; S7.12; S8
A. A. Shmatov	(a) **110**, 213 (1988)	71.55; 73.40; 78.55; S7.15
G. M. Shmelev	(b) **147**, K131 (1988)	73.40
Y. Shono	(a) **106**, 181 (1988)	71.55 and 79.75; 78.60; S10.1
A. P. Shotov	(a) **108**, 637 (1988)	73.60 and 78.65; S8
Shou-Fu Cai	see under letter "C"	
A. M. Shpak	(b) **150**, 907 (1988)	78.20; 78.50; S9.11
E. M. Shpilevskii	(a) **106**, K23 (1988)	61.80 and 68.48; 66.30; S1.3; S4
L. G. Shpinkova	(b) **147**, K75 (1988)	75.30 and 76.80; S1.1
S. B. Shrivastava	(b) **147**, 553 (1988)	71.60 and 73.20; S1.2; S1.3
T. V. Shubina	(b) **150**, 863 (1988)	71.35 and 78.20; S7.12; S7.15
Shuhong Li	see under letter "L"	
L. I. Shulga	(a) **109**, K153 (1988)	78.45 and 78.65; S8.15
L. A. Shuvalov	(a) **110**, 429 (1988)	64.70; 62.65; 77.20; S11.1
Shu-yi Zhang	see under letter "Z"	
Yu. N. Shvachko	(b) **147**, K185 (1988)	76.30; S12
Yu. M. Shvera	(b) **147**, 711 (1988)	71.35
	(b) **149**, 187 (1988)	71.36
P. Siddambary	(a) **106**, K185 (1988)	78.30; S11.1
A. V. Sidorin	(a) **106**, 363 (1988)	61.70; 62.20; S5.11
D. Siebert	(a) **106**, K57 (1988)	74.70 and 76.30; S10.15
W. Siegel	(a) **105**, 549 (1988)	61.70 and 72.20; 78.30; S7.11
	(a) **108**, 689 (1988)	71.55 and 72.20; S7.12

A. Siemko	(a) **105**, 597 (1988)	75.80; 78.20; S1.1
W. Sienicki	(a) **106**, K157 (1988)	72.80; 61.70; S1.63
K. Sierański	(a) **106**, 473 (1988)	73.30; 72.40; S7.1
	(a) **106**, 551 (1988)	72.40 and 73.30; S7
	(a) **106**, K31 (1988)	72.40 and 73.30; S7
E. Sigmund	(b) **147**, 229 (1988)	71.38
A. Šileika	(a) **107**, K75 (1988)	78.30; S8.15
V. P. Silin	(b) **147**, 373 (1988)	75.10; 75.50
P. R. Silva	(b) **147**, K79 (1988)	75.10
I. M. Silvestrova	(a) **107**, 161 (1988)	62.20 and 62.65; S11.1
M. Simerska	(a) **106**, 407 (1988)	61.70; 61.55; S1.2; S4
A. V. Siminel	(a) **107**, K57 (1988)	71.55; 72.40; S7.11
	(a) **109**, K153 (1988)	78.45 and 78.65; S8.15
Siming Zheng	see under letter "Z"	
A. Šimůnek	(b) **149**, 765 (1988)	78.70; 71.20; S6
A. Singh	(a) **106**, K139 (1988)	68.55 and 73.40; S10.1
B. R. Singh	(a) **106**, K53 (1988)	61.50 and 74.70; S10.15
I. Singh	(b) **148**, 549 (1988)	71.25 and 73.40; S5.12; S7.12; S8.12
J. Singh	(b) **147**, 343 (1988)	74.10; 78.70; S11
J. K. Singh	(a) **106**, K53 (1988)	61.50 and 74.70; S10.15
N. Singh	(b) **148**, K5 (1988)	63.20; S2
	(b) **149**, 63 (1988)	62.20; S2
R. J. Singh	(b) **145**, K51 (1988)	76.30; S12
R. K. Singh	(b) **146**, 111 (1988)	62.20 and 64.30; S8.15
	(b) **149**, 121 (1988)	62.20; S8
S. J. Singh	(a) **106**, 291 (1988)	78.60
S. K. Singh	(a) **105**, 455 (1988)	68.55; S8
S. P. Singh	(b) **148**, K5 (1988)	63.20; S2
	(b) **149**, 63 (1988)	62.20; S2
T. N. Singh	(b) **149**, 503 (1988)	63.20; S1; S1.1
R. K. Singhal	(b) **147**, 343 (1988)	74.10; 78.70; S11
A. Sinha	(b) **146**, 141 (1988)	68.30 and 71.45; S7.1
G. V. Sinitsyn	(b) **150**, 455 (1988)	78.20; 78.65; S8.1
D. B. Sirdeshmukh	(b) **150**, K11 (1988)	63.70; S9.11
V. Sivaramakrishnan	(a) **105**, 461 (1988)	73.60; 68.55; S8
	(a) **106**, 145 (1988)	73.60; S1.3; S10.1
F. F. Sizov	(b) **145**, 309 (1988)	78.20; S8; S8.15
V. V. Skadorov	(b) **149**, 435 (1988)	61.10
L. Skála	(b) **147**, K149 (1988)	71.55
	(b) **148**, K11 (1988)	73.30; 73.40
P. Skalicky	(a) **108**, 495 (1988)	61.70; S11.1
W. Skorupa	(a) **105**, 387 (1988)	61.80; S1.62; S5.11
I. Škorvánek	(a) **108**, 747 (1988)	75.50; 61.80; S1.1; S1.2
A. V. Skripov	(b) **147**, K185 (1988)	76.30; S12
V. Skritski	(a) **110**, 429 (1988)	64.70; 62.65; 77.20; S11.1
D. Skrzypek	(a) **105**, 567 (1988)	75.30 and 76.30; S9
V. Skumriev	(a) **105**, K145 (1988)	75.30; S1.4
V. A. Skuratov	(a) **106**, 371 (1988)	61.80; 62.20; S1.2
Yu. D. Slavov	(b) **147**, 721 (1988)	71.35 and 78.20
	(b) **150**, 513 (1988)	78.20
B. Smandek	(b) **150**, 705 (1988)	78.20; 85; S5.11
L. Smardz	(a) **110**, 227 (1988)	75.70; S1.2
A. L. Smirl	(b) **150**, 895 (1988)	78.20 and 78.45; 78.55; S7.15
	(b) **150**, 913 (1988)	78.70; S5.11; S7.12
A. O. Smirnov	(a) **108**, 303 (1988)	73.40; 72.40; 73.25; S7.13
B. I. Smirnov	(a) **105**, K93 (1988)	62.20 and 64.70; S8
	(a) **110**, 89 (1988)	61.70; 62.20; 72.20; S8

I. A. Smirnov	(a) **105**, K93 (1988)	62.20 and 64.70; S8
	(a) **105**, K103 (1988)	62.20; S8
	(a) **110**, 89 (1988)	61.70; 62.20; 72.20; S8
V. P. Smirnov	(b) **145**, 391 (1988)	61.50; 71.50
G. F. Smirnova	(b) **145**, 177 (1988)	63.20; 78.30; S8.16
I. A. Smirnova	(a) **106**, 339 (1988)	61.10; 62.60; S5.11
	(b) **147**, 459 (1988)	61.10
T. V. Smirnova	(b) **145**, 117 (1988)	63.20; 78.30; S8.16
W. M. A. Smit	(b) **145**, 333 (1988)	78.55; 78.50; S10.1
	(b) **148**, 413 (1988)	78.55; S11.1
David J. Smith	(a) **107**, 681 (1988)	68.20; 61.16; S8.13
P. V. Smith	(b) **146**, 149 (1988)	71.10; S5
	(b) **147**, K31 (1988)	71.10; 62.20; S5.11
L. Smrčka	(a) **107**, K135 (1988)	74.70; S10.15
O. Smrčková	(b) **146**, 743 (1988)	63.20 and 78.30; 74.70; S10.15
	(a) **107**, K135 (1988)	74.70; S10.15
A. A. Snigirev	(a) **108**, 651 (1988)	61.10 and 68.20; S5.11
O. V. Snitko	(b) **146**, 517 (1988)	71.36 and 73.20; S6
M. Šob	(a) **105**, 357 (1988)	61.70; 78.70; S1.1; S1.2
E. Sobeslavsky	(a) **105**, 387 (1988)	61.80; S1.62; S5.11
J. Sobhanadri	(a) **105**, 161 (1988)	68.55 and 73.60; S8.15
P. Sobkowicz	(b) **146**, K7 (1988)	71.25 and 73.40
H. Sobotta	(b) **150**, 165 (1988)	71.25; 72.20; 72.30; 78.20; S1.63
H. G. Sockel	(a) **110**, 449 (1988)	62.20; S1.2
J. P. Sokoloff	(b) **150**, 357 (1988)	71.35 and 78.20; S8.11
R. A. Sokolov	(a) **106**, 669 (1988)	73.40 and 85; S5.11
V. N. Sokolov	(b) **146**, 311 (1988)	78.20
	(b) **146**, 319 (1988)	78.20; S8.12
	(b) **150**, 489 (1988)	78.20; S8.11
V. O. Sokolov	(b) **147**, 261 (1988)	71.55; S10.1
H. Solbrig	(b) **146**, 733 (1988)	78.70
A. V. Soldatov	(b) **146**, K21 (1988)	71.25 and 78.70; S9
V. F. Solodkov	(b) **150**, 605 (1988)	78.20; 78.45
A. Z. Solontson	(b) **147**, 373 (1988)	75.10; 75.50
V. S. Solovev	(a) **107**, K105 (1988)	61.80; S5.11
J. Sołtys	(a) **110**, 123 (1988)	64.70; 61.10; 61.55; S1.2; S4
T. Soma	(b) **147**, 109 (1988)	64.75; 62.50; S6
	(b) **149**, 477 (1988)	61.70; S5.11; S5.12
R. Sommerfeldt	(a) **106**, 89 (1988)	61.70; 72.40; 78.20; S11.1
U. Sondermann	(a) **110**, 141 (1988)	64.70; 75.40; S7.15
L. M. Song	(b) **149**, 595 (1988)	71.45 and 73.20
Wei-ji Song	(b) **148**, 369 (1988)	75.10; 75.30
G. Sorge	(a) **110**, 459 (1988)	64.70 and 77.80; S11
A. A. Sorokin	(b) **147**, K75 (1988)	75.30 and 76.80; S1.1
K. L. Sorokina	(a) **108**, 323 (1988)	72.20 and 73.40; S5.11
C. C. Sorrell	(b) **147**, K153 (1988)	74.70; S10.15
J. Sosnowski	(a) **105**, 555 (1988)	74.30
N. Sotnik	(a) **108**, 11 (1988)	73.40 and 85; S10
J. Spendeler	(a) **107**, 481 (1988)	61.55 and 61.60; 61.16; S5.11; S5.12; S7.1; S8.1
W. Sperber	(a) **106**, 239 (1988)	75.60
Ch. Spiegelberg	(b) **150**, 769 (1988)	71.35 and 78.20; S8.11
F. H. M. Spit	(b) **147**, K105 (1988)	66.30; S1
H. Spustek	(b) **147**, 685 (1988)	76.80; 71.70; S1.1
J. Spyridelis	(a) **107**, 633 (1988)	78.20; S8.16
J. Šrámek	(b) **146**, 743 (1988)	63.20 and 78.30; 74.70; S10.15
J. Šrámková	(a) **106**, 619 (1988)	78.20; 78.30; S8
K. Srinivasan	(a) **108**, 771 (1988)	78.20; S9.11

Y. Srinivasa Rao	see under letter "R"		
A. Srivastava	(a) **106**,	K203 (1988)	73.40 and 85; S5.11
G. P. Srivastava	(a) **110**,	671 (1988)	73.40 and 85
K. S. Srivastava	(b) **146**,	141 (1988)	68.30 and 71.45; S7.1
R. Srivastava	(b) **146**,	141 (1988)	68.30 and 71.45; S7.1
	(a) **110**,	293 (1988)	72.40 and 73.40; S8
S. K. Srivastava	(a) **108**,	803 (1988)	73.40 and 85
	(b) **149**,	69 (1988)	63.20; S1.4
F. Šrobár	(b) **150**,	581 (1988)	78.20 and 78.45
V. A. Stadnik	(b) **150**,	501 (1988)	78.20; S8.12; S8.15
J. L. Staehli	(b) **150**,	685 (1988)	71.35 and 78.65; 71.45; S7.12; S7.15
O. M. Stafsudd	(a) **105**,	563 (1988)	72.80 and 78.20; S8.16
A. Stahl	(b) **150**,	413 (1988)	71.35 and 78.20; S7.12; S7.15
B. Stalinski	(a) **107**,	K151 (1988)	75.40; S1.4
	(a) **110**,	601 (1988)	75.40; S1.4
A. D. Stalios	(a) **107**,	759 (1988)	61.16 and 61.70; S11.2
P. M. Starik	(a) **106**,	K173 (1988)	76.80; S8.15
	(a) **107**,	95 (1988)	61.70; 72.20; 75.20; S8.13
	(b) **148**,	K161 (1988)	76.80; S8
M. Starke	(b) **149**,	K85 (1988)	78.20; S12
Z. Starý	(b) **147**,	501 (1988)	61.50 and 61.70; 72.80; S8; S8.15
	(a) **109**,	93 (1988)	61.60; 61.70; S8
I. V. Stasyuk	(b) **149**,	347 (1988)	78.20
B. Stébé	(b) **150**,	201 (1988)	71.35 and 71.55
C. B. Steele	(a) **106**,	515 (1988)	73.60; 72.20; S10.1
P. Stefański	(a) **107**,	K61 (1988)	75.50; S1.1
P. Stefányi	(b) **150**,	K63 (1988)	75.30 and 75.40; S11
R. Stegmann	(a) **105**,	661 (1988)	73.40 and 85; 78.60; S7.15
U. Steigenberger	(b) **148**,	445 (1988)	61.12 and 61.60; 62.50; 65.90; S8
T. Steinkopff	(b) **150**,	K1 (1988)	61.10; 61.12; 61.14
V. Stelmakh	(a) **110**,	181 (1988)	68.55 and 78.65; S5.11
V. Stepankevičius	(b) **150**,	139 (1988)	71.35; 71.38; 78.55; S8.13
A. P. Stepanov	(b) **147**,	K185 (1988)	76.30; S12
G. Stephani	(a) **106**,	K61 (1988)	75.50; 75.60; S1.1; S1.2
J. D. Stephenson	(a) **106**,	441 (1988)	61.70; 64.60; S10.1
A. Stesmans	(a) **110**,	77 (1988)	76.30; 61.70; S10.1
D. A. Stevenson	(a) **105**,	77 (1988)	66.30; 68.55; S8.15
S. Stizza	(b) **149**,	363 (1988)	71.55 and 78.60; S12
J. Stoemenos	(a) **107**,	579 (1988)	68.55; S1.2
V. F. Stolyarenko	(b) **147**,	K109 (1988)	61.80; S9.11
H. Stolz	(b) **146**,	357 (1988)	71.25 and 72.20; 72.10
M. Stölzer	(b) **150**,	165 (1988)	71.25; 72.20; 72.30; 78.20; S1.63
M. Stordeur	(b) **150**,	165 (1988)	71.25; 72.20; 72.30; 78.20; S1.63
V. M. Strakhovenko	(b) **149**,	403 (1988)	61.80 and 78.70
N. Streibl	(b) **150**,	447 (1988)	78.20
H. W. Streitwolf	(b) **149**,	K13 (1988)	71.38
	(b) **150**,	147 (1988)	71.38; 78.20; S12
Ya. M. Strelniker	(b) **149**,	603 (1988)	71.35; 73.20
V. E. Strelnitskii	(b) **145**,	209 (1988)	71.25; 78.20; S5
P. Streubel	(b) **145**,	603 (1988)	71.50; 79.20; S5.11; S6
E. K. Stribuk	(a) **107**,	K139 (1988)	74.70; S10.15
R. Strobel	(b) **150**,	885 (1988)	72.20; 72.40; S7.12
E. E. Strumban	(a) **107**,	373 (1988)	78.30 and 78.55; S8.16
	(a) **107**,	K57 (1988)	71.55; 72.40; S7.11
	(a) **109**,	K153 (1988)	78.45 and 78.65; S8.15
H. Strusny	(a) **109**,	K35 (1988)	73.40; S7.12
S. N. Stuart	(b) **145**,	719 (1988)	76.60; S8.1

M. Stutzmann	(b) **149**,	K95 (1988)	61.70 and 63.20
Wenhui Su	(a) **108**,	337 (1988)	74.70; 75.20; S10.15
N. Suarez Almodovar	(a) **109**,	569 (1988)	75.30; 61.60; S1.1
A. V. Subashiev	(b) **150**,	533 (1988)	78.20; 71.35; S5.11
A. Subbarayan	(a) **108**,	K31 (1988)	61.80 and 79.20; S12
I. M. Subbotin	(a) **106**,	669 (1988)	73.40 and 85; S5.11
K. G. Subhadra	(b) **150**,	K11 (1988)	63.70; S9.11
M. Subotowicz	(b) **147**,	685 (1988)	76.80; 71.70; S1.1
	(a) **109**,	597 (1988)	78.55 and 78.60; 61.70; S7.12
A. Subrahmanyam	(a) **105**,	K71 (1988)	77.20 and 77.40; S9
C. Sugiura	(b) **149**,	791 (1988)	78.70; S9
M. Suhrke	(b) **149**,	547 (1988)	73.40; S7.12; S7.15
Yu. V. Sukhetskii	(b) **146**,	K21 (1988)	71.25 and 78.70; S9
A. A. Sukhovskii	(a) **105**,	145 (1988)	64.70; 76.60; S11.1
V. B. Sulimov	(b) **147**,	261 (1988)	71.55; S10.1
C. Sułkowski	(b) **146**,	K103 (1988)	74.70; S10.15
K. Sumiyama	(a) **105**,	291 (1988)	75.70; 61.10; 76.80; S1.1
	(a) **108**,	295 (1988)	68.55 and 75.70; 76.80; S1.1
Changyong Sun	(a) **108**,	337 (1988)	74.70; 75.20; S10.15
D. Z. Sun	(b) **145**,	K111 (1988)	73.40; S7.12; S7.15
Guosheng Sun	(a) **105**,	K41 (1988)	73.40 and 76.30; S5.11
	(a) **108**,	K139 (1988)	76.30 and 78.30; S5.11
X. Y. Sun	(b) **147**,	K153 (1988)	74.70; S10.15
C. S. Sunandana	(a) **105**,	11 (1988)	78.20
	(a) **106**,	217 (1988)	75.50; 61.60; S11.2
	(a) **110**,	K85 (1988)	74.70; S10.15
M. Sundaram	(a) **105**,	477 (1988)	68.55; 73.60; 78.65; S7
J. Suñé	(a) **109**,	479 (1988)	68.48 and 73.60; S5.11; S10.1
I. T. Suprun	(a) **107**,	141 (1988)	62.40; 66.30
P. Süptitz	(a) **106**,	57 (1988)	61.80 and 68.55; 66.30; S8
G. I. Surdutovich	(b) **150**,	539 (1988)	78.20
	(b) **150**,	605 (1988)	78.20; 78.45
C. Suryanarayana	(a) **107**,	693 (1988)	61.55; S1.2; S4
E. Susi	(a) **106**,	583 (1988)	72.20; S5.11
	(a) **108**,	503 (1988)	61.70; 72.20; S5.11
K. Suzuki	(a) **108**,	163 (1988)	61.70; 64.75; S1.2; S1.3
P. Švec	(a) **105**,	319 (1988)	61.40; 61.50
S. V. Svirida	(b) **149**,	785 (1988)	78.70; S9.12
P. Svoboda	(a) **106**,	115 (1988)	66.70; 72.15; S1.1
	(b) **147**,	K159 (1988)	74.70; S10.15
V. Švorčík	(a) **106**,	K35 (1988)	68.45 and 78.70; S7.11; S7.12
K. Svozil	(b) **147**,	635 (1988)	74.20
W. Świątkowski	(a) **110**,	K101 (1988)	78.70; S1.3; S9
J. Świerczek	(a) **109**,	559 (1988)	75.50; S1.1; S1.2
R. Świetlik	(a) **110**,	593 (1988)	72.20; S12.1
H.-E. Swoboda	(b) **150**,	749 (1988)	78.20; 78.55; S8.15
H. K. Sy	(b) **149**,	595 (1988)	71.45 and 73.20
D. Sýkorová	(b) **146**,	743 (1988)	63.20 and 78.30; 74.70; S10.15
	(a) **107**,	K135 (1988)	74.70; S10.15
V. Syneček	(a) **106**,	407 (1988)	61.70; 61.55; S1.2; S4
E. S. Syrkin	(b) **150**,	K19 (1988)	65.70
S. A. Syskin	(b) **149**,	133 (1988)	63.20; S10
B. I. Sysoev	(a) **109**,	463 (1988)	73.40; S7.12; S8
J. Szablewski	(a) **108**,	K5 (1988)	64.75 and 66.30; S1.2
J. Szatkowski	(a) **106**,	473 (1988)	73.30; 72.40; S7.1
	(a) **106**,	551 (1988)	72.40 and 73.30; S7
	(a) **106**,	K31 (1988)	72.40 and 73.30; S7

G. SZENES	(a) **106**, 371 (1988)	61.80; 62.20; S1.2
R. SZIPÖCS	(b) **150**, 811 (1988)	78.20; S8.15
V. SZÖCS	(b) **145**, 637 (1988)	72.20; S8.15
A. E. SZUKIEL	(b) **147**, 321 (1988)	72.15; 71.70; S1.4
M. SZUSZKIEWICZ	(b) **147**, K201 (1988)	76.80 and 78.70; S1.1
J. SZYMAŃSKI	(b) **146**, 241 (1988)	75.30
H. SZYMCZAK	(a) **107**, K145 (1988)	74.70 and 75.60; S10.15
S. SZYMURA	(a) **109**, 559 (1988)	75.50; S1.1; S1.2
B. G. TAGIEV	(a) **105**, K151 (1988)	76.30 and 76.80; S8
M. M. TAGIEV	(b) **148**, K93 (1988)	63.20 and 78.30; S8.16
A. TAGLIACOZZO	(b) **145**, 483 (1988)	73.40
W. A. TAHA	(a) **106**, 613 (1988)	78.20; S10
M. A. TAIROV	(a) **110**, 631 (1988)	78.20; 72.40; S8.16
M. TAKAHASHI	(a) **105**, 263 (1988)	75.50; 75.80; S1.1
S. TAKAMURA	(a) **108**, 219 (1988)	62.40; S4
K. TAKARABE	(b) **145**, 219 (1988)	71.25; 78.40; S8
N. TAKEUCHI	(a) **105**, 153 (1988)	62.20; 62.40; S9.11
	(a) **106**, K19 (1988)	62.20; S9.11
	(a) **108**, 463 (1988)	78.60; S9.11
	(b) **149**, K67 (1988)	61.70 and 78.50; S9.15
	(b) **149**, K101 (1988)	61.70; 78.50; S9.15
T. TAKEUCHI	(b) **147**, K9 (1988)	63.90; S9.16
	(b) **149**, 579 (1988)	71.25; S9
V. M. TALANOV	(a) **106**, K129 (1988)	64.70; S11.2
S. L. TALIM	(b) **145**, 711 (1988)	75.10
Z. N. TAMAŠEVIČIENĖ	(a) **110**, 555 (1988)	72.20; 71.55; S5.12
A. V. TAMAŠEVIČIUS	(a) **110**, 555 (1988)	72.20; 71.55; S5.12
G. TAMULAITIS	(b) **150**, 139 (1988)	71.35; 71.38; 78.55; S8.13
HA VINH TAN	see under letter "H"	
K. TANAKA	(a) **108**, K109 (0988)	73.40; 72.40; 78.40; S12
M. TANAKA	(b) **145**, K141 (1988)	75.10
	(b) **149**, K45 (1988)	76.30; S9.15
	(a) **110**, 645 (1988)	78.65; S12
A. TANDON	(b) **146**, 141 (1988)	68.30 and 71.45; S7.1
NGUYEN TANG	see under letter "N"	
S. H. TANG	(b) **147**, K195 (1988)	78.30; 63.20; S9.16
YINSHENG TANG	(b) **145**, 167 (1988)	78.20; 71.25; 71.35; S7.12; S7.15
R. B. TANGSALI	(b) **149**, 623 (1988)	71.70
W. TAŃSKA-KRUPA	(b) **147**, 685 (1988)	76.80; 71.70; S1.1
YE TAO	(a) **109**, 435 (1988)	64.70; 61.16; S11
V. V. TARBA	(a) **108**, K93 (1988)	61.80; 78.50; S7.11; S7.12
N. I. TARBAEV	(a) **110**, 97 (1988)	61.70 and 71.55; 72.40; 78.50; 78.55; S8.1
G. G. TARASOV	(b) **150**, 557 (1988)	78.20; 78.50; S9.11
	(b) **150**, 907 (1988)	78.20; 78.50; S9.11
I. I. TARASYUK	(b) **146**, 671 (1988)	72.10; 72.20; S5
I. I. TARTAKOVSKII	(b) **150**, 125 (1988)	71.36; 78.30; 78.55; S12
J. TATARKIEWICZ	(b) **145**, K95 (1988)	61.70 and 63.20; 78.30; S8.13
	(b) **149**, K95 (1988)	61.70 and 63.20
N. A. TATYANINA	(b) **150**, 901 (1988)	66.70 and 78.35; S6
E. TAUTE	(b) **146**, K87 (1988)	79.20; S4
K. P. TCHAKPELE	(b) **149**, 641 (1988)	71.35 and 71.55; S8.11; S8.15
P. TCHOLAKOV	(a) **105**, K145 (1988)	75.30; S1.4
H. TEICHLER	(b) **147**, 361 (1988)	74.60
G. TEICHMANN	(a) **108**, 519 (1988)	61.70; 64.75; S5.11
M. TEJEDOR	(a) **107**, 339 (1988)	75.60; S1.1

P. TEJEDOR	(a) **106**, 11 (1988)	61.40 and 68.55; 78.30; S5.11
YU. V. TELEZHENKO	(b) **145**, 445 (1988)	63.20; 61.65; S12
	(b) **150**, 39 (1988)	63.20 and 65.40; S12
A. TEMPEL	(a) **109**, 493 (1988)	68.55; 65.70; S9.12
R. TEMPEL	(a) **108**, 469 (1988)	73.40 and 85
G. VAN TENDELOO	(a) **107**, 521 (1988)	61.16; 61.70
	(a) **107**, 785 (1988)	61.70; 62.20; S1.2; S4
	(a) **107**, 799 (1988)	61.70 and 64.80; S10
	(a) **109**, 67 (1988)	61.14; S11.1
M.-K. TENG	(a) **106**, K1 (1988)	61.70 and 78.70; S12
H. TERAUCHI	(a) **105**, 197 (1988)	61.10 and 68.55; S7.12; S7.15
V. E. TEZLEVAN	(a) **107**, 373 (1988)	78.30 and 78.55; S8.16
NGUYEN TOAN THANG	see under letter "N"	
D. X. THANH	(b) **147**, K165 (1988)	72.40 and 78.55; S8.12
M. THEUERKAUF	(b) **150**, 73 (1988)	63.10 and 63.50
W. THIELEMANN	(a) **107**, 419 (1988)	73.40 and 78.60; 85; S7.12
B. J. THIJSSE	(a) **105**, 235 (1988)	72.15; 61.40; S1.1
	(a) **108**, 97 (1988)	61.50 and 72.15; S4
A. THOMAS	(a) **106**, 57 (1988)	61.80 and 68.55; 66.30; S8
	(a) **110**, K81 (1988)	68.60
G. THOMAS	(a) **105**, 207 (1988)	68.20; 61.16; S10
	(a) **107**, 825 (1988)	61.16 and 61.70; 74.70; S10.15
H. THOMAS	(a) **110**, 531 (1988)	61.80 and 73.40; S1.61; S4; S7.12
V. THOMAS	(b) **148**, 723 (1988)	78.55; 71.55; S5.11
P. THOMPSON	(a) **110**, 531 (1988)	61.80 and 73.40; S1.61; S4; S7.12
G. THUMMES	(b) **147**, 661 (1988)	72.15; S1.3
I. THURZO	(a) **106**, 659 (1988)	73.40 and 85; S7.12
N. P. THUY	(a) **106**, 201 (1988)	75.30; S1.2; S1.4
DE-CHENG TIAN	(b) **145**, 73 (1988)	61.25; S1; S1.2
JINGHUA TIAN	(a) **107**, 45 (1988)	61.55; S1.1; S1.4
F. D. TICHELAAR	(a) **106**, 433 (1988)	61.70
L. TICHÝ	(a) **107**, 769 (1988)	61.40; 64.60; S8
TIESONG ZHAO	see under letter "Z"	
I. M. TIGINYANU	(a) **108**, K59 (1988)	71.55 and 72.80; S7.11
	(a) **109**, K55 (1988)	72.40 and 78.55; S8.16
S. A. TIKHOMIROV	(b) **150**, 647 (1988)	78.20; S8.15
V. A. TIKHOMIROV	(a) **110**, K107 (1988)	78.50; 76.30; S10.15
B. TILLACK	(a) **107**, 281 (1988)	68.55; S5.11
D. R. TILLEY	(b) **146**, 449 (1988)	63.20
	(b) **150**, 133 (1988)	71.45; 73.60; S8
R. J. D. TILLEY	(a) **108**, 111 (1988)	61.14 and 64.80; S10.15
A. N. TIMOSHEVSKII	(b) **146**, 161 (1988)	71.20 and 71.25; S11
T. C. T. TING	(b) **146**, 81 (1988)	61.70; 62.20
TINGSUI GE	see under letter "G"	
L. S. TITOV	(b) **149**, 241 (1988)	72.15; 75.60; S1.1
K. TITTELBACH-HELMRICH	(a) **109**, 279 (1988)	71.55; S5.11
M. D. TIWARI	(b) **147**, 661 (1988)	72.15; S1.3
B. M. TODRIS	(a) **109**, 307 (1988)	75.30; S11
V. V. TOKAREV	(a) **107**, K33 (1988)	61.70 and 68.55; S1.61
E. TOKUNAGA	(b) **150**, 667 (1988)	73.40 and 78.55; S7.12; S7.15
A. TOMAS	(a) **107**, 775 (1988)	61.60; 61.10; 61.16; S8
P. M. TOMCHUK	(b) **146**, 671 (1988)	72.10; 72.20; S5
ZH. M. TOMILO	(b) **149**, K145 (1988)	61.70 and 74.70; S10.15
W. J. TOMLINSON	(b) **150**, 851 (1988)	78.20; 78.30; 78.35

J. W. TOMM	(a) **106**, 509 (1988)	68.55 and 78.55; S8
	(a) **106**, K83 (1988)	78.55; S8.15
T. TOMOYOSE	(b) **147**, 141 (1988)	71.36 and 71.45; 68.30; 73.40
TONG-SAN CHO	see under letter "C"	
V. V. TOPOROV	(b) **146**, K161 (1988)	63.20 and 78.30; S8.13
T. V. TORCHINSKAYA	(a) **110**, 213 (1988)	71.55; 73.40; 78.55; S7.15
D. R. TORGESON	(b) **147**, K175 (1988)	76.60
A. A. TOROPOV	(b) **150**, 863 (1988)	71.35 and 78.20; S7.12; S7.15
M. TORRES	(a) **106**, 11 (1988)	61.40 and 68.55; 78.30; S5.11
G. TORRES-VILLASENOR	(a) **107**, 785 (1988)	61.70; 62.20; S1.2; S4
M. TOSCANO	(b) **145**, K125 (1988)	71.55; S5.11
B. S. TOŠIĆ	(b) **148**, 593 (1988)	71.45; 77.20
L. TÓTH	(b) **149**, K167 (1988)	75.10
M. TOURNOUX	(b) **150**, 225 (1988)	72.20; 71.25; S10
J. TOUŠEK	(a) **110**, 687 (1988)	73.40 and 85
J. TOUŠKOVÁ	(a) **106**, 297 (1988)	73.30 and 85; 73.60; S8.13
R. TOVAR	(a) **107**, 165 (1988)	61.55 and 64.70; S8.15
	(a) **107**, 205 (1988)	64.70; S8.15
K. D. TOVSTYUK	(b) **146**, 613 (1988)	71.70 and 78.50; S8.1
C. TRALLERO GINER	see under letter "G"	
G. TRÄNKLE	(b) **150**, 679 (1988)	71.35 and 78.20; S7.12; S7.15
D. B. TRAN THOAI	(b) **146**, 137 (1988)	71.45 and 73.20; S1.3; S4
	(a) **107**, 791 (1988)	61.80 and 73.20; 71.45; S4
	(b) **149**, 169 (1988)	71.10 and 73.25
TRAN THOAI DUY BAO	(b) **145**, 509 (1988)	71.45 and 73.20; 61.80; S4
I. TRAVĚNEC	(a) **107**, 235 (1988)	73.40; 77.55; S10.15
S. V. TRETYAKOV	(b) **148**, 289 (1988)	72.20; 72.60; S8.1; S10
J. TREUSCH	(b) **147**, 699 (1988)	71.35 and 78.20
G. P. TRIBERIS	(b) **147**, 349 (1988)	72.20; S9.12
B. TRIEMAN	(b) **149**, 45 (1988)	61.70; 72.20; 78.60; S9.16
S. A. TRIGGER	(b) **145**, 455 (1988)	68.10; 73.30; S2
CH. TRIKALINOS	(b) **149**, K157 (1988)	74.70 and 76.30; S10.15
S. TRIMPER	(b) **148**, K151 (1988)	75.40; 64.60
D. N. TRIPATHY	(b) **148**, 585 (1988)	71.25; S2; S4
N. N. TROITSKAYA	(a) **107**, 291 (1988)	71.55; 76.80; S8
Z. TROJANOVÁ	(a) **107**, K11 (1988)	62.20; S1
E. TROJNAR	(b) **146**, K103 (1988)	74.70; S10.15
NGUYEN VAN TRONG	see under letter "N"	
U. TROPPENZ	(b) **148**, 333 (1988)	71.30 and 72.20; S7.13
I. O. TROYANCHUK	(a) **109**, K59 (1988)	76.60; S10.15
CHU THIEN TRUONG	see under letter "C"	
U. TRUTSCHEL	(b) **150**, 805 (1988)	78.65
	(b) **150**, 867 (1988)	78.20; S7.13
R. TRYKOZKO	(b) **108**, K169 (1988)	73.40 and 85; 72.40; S8.11
B. TRYZNA	(a) **110**, 687 (1988)	73.40 and 85
K. TRZEBIATOWSKI	(a) **108**, K105 (1988)	73.25; S10.15
T. TSANG	(a) **105**, K1 (1988)	61.30; S12
H. U. TSCHIRNER	(a) **107**, 291 (1988)	71.55; 76.80; S8
W. S. TSE	(b) **147**, 791 (1988)	78.30; 63.20; S11
I. M. TSIDILKOVSKII	(b) **148**, 197 (1988)	71.30; 72.20; S8.15
V. I. TSITSANU	(a) **107**, 373 (1988)	78.30 and 78.55; S8.16
E. N. TSOI	(b) **146**, 457 (1988)	63.10; 71.45
T. TSUBOI	(b) **145**, K145 (1988)	75.25 and 76.50; S9.16
	(b) **146**, K11 (1988)	71.35 and 78.40; S9; S9.12
R. TSUMURA	(a) **105**, 411 (1988)	65.40; S1.2
G. I. TSURKAN	(b) **147**, K131 (1988)	73.40

I. TSURUMI	(a) **105**, K165 (1988)	78.55; S8.12
	(a) **106**, 283 (1988)	78.55; S8.12
	(a) **109**, 617 (1988)	78.55; 71.55; S8.12
D. A. TUAN	(a) **109**, K115 (1988)	71.55; 73.40; 85; S7.11
A. I. TUGUSHI	(b) **146**, 217 (1988)	75.10; 75.50
C. TUIJN	(b) **149**, 483 (1988)	66.30; S1
R. TURCU	(a) **108**, 233 (1988)	68.55 and 73.60; S8
	(a) **108**, 637 (1988)	73.60 and 78.65; S8
A. TURKOVIĆ	(a) **109**, K55 (1988)	72.40 and 78.55; S8.16
D. M. TURTSEVICH	(b) **149**, K145 (1988)	61.70 and 74.70; S10.15
I. D. TURYANITSA	(b) **146**, 723 (1988)	78.40; S11
I. E. TYSHCHENKO	(a) **109**, 141 (1988)	61.80 and 66.30; S5.11
G. TYULIEV	(a) **110**, 495 (1988)	68.20; 79.20; S1.1; S1.2
TZU-TUNG YEN	see under letter "Y"	
DO NGOC UAN	see under letter "D"	
H. ÜBENSEE	(b) **147**, 823 (1988)	73.40; S7.1
	(b) **148**, 421 (1988)	73.40; S5.11; S7.1
	(b) **148**, 611 (1988)	71.25 and 73.40; S8.15
H. UCHIKI	(b) **150**, 667 (1988)	73.40 and 78.55; S7.12; S7.15
O. UEMURA	(b) **146**, 661 (1988)	72.10 and 72.80; 75.30; S1.63
	(a) **108**, K89 (1988)	61.40; S8
	(b) **149**, 257 (1988)	72.20 and 75.30; S1.63; S8.16
I. I. UKRAINSKII	(b) **145**, 187 (1988)	71.25; 72.10; 73.40
	(b) **145**, K107 (1988)	73.40
J. K. ULBIKAS	(a) **110**, 555 (1988)	72.20; 71.55; S5.12
U. ULMANIS	(a) **106**, K7 (1988)	61.80; 72.80; S8
W. ULRICI	(b) **150**, 177 (1988)	71.55; 72.20; 78.50; 78.55; S7.11
	(b) **150**, K57 (1988)	76.30; S7.11
M. UMAR	(b) **145**, K51 (1988)	76.30; S12
K. UNGER	(b) **149**, K141 (1988)	72.20
A. UNKROTH	(a) **106**, 589 (1988)	75.70; 61.80; S1.2; S1.61; S1.62
	(a) **109**, K149 (1988)	75.60; S1.2
N. UNZNER	(a) **105**, 627 (1988)	78.20; 61.70; 66.30; 78.50; S5.11
J. C. UPADHYAYA	(b) **148**, K109 (1988)	63.20; S1.2
I. N. URALTSEV	(b) **150**, 673 (1988)	71.35 and 78.20; 71.25; S7.12; S7.15
R. URBAN	(b) **145**, 351 (1988)	61.80 and 63.20; S4
A. URBANIAK-KUCHARCZYK	(b) **147**, K71 (1988)	75.30 and 78.70
	(b) **147**, K119 (1988)	73.60 and 75.70
	(b) **149**, 543 (1988)	73.60; S1.4
V. I. URENEV	(a) **108**, 311 (1988)	71.55; 61.70; S5.12
N. B. URLI	(a) **109**, K55 (1988)	72.40 and 78.55; S8.16
S. E. URRETA DE PEREYRA	(a) **108**, 577 (1988)	62.40; 61.70; S4
V. URUMOV	(b) **145**, K59 (1988)	75.10
N. URYÛ	(b) **148**, K47 (1988)	75.30; S9
G. VACQUIER	(a) **109**, 445 (1988)	61.70 and 65.70; 62.20; S8
J. VAITKUS	(b) **150**, 743 (1988)	78.20; 61.80; S5.11
	(b) **150**, 749 (1988)	78.20; 78.55; S8.15
	(b) **150**, 879 (1988)	78.20; S5.11; S7.12; S7.13; S8.12
I. A. VAKARCHUK	(b) **149**, 301 (1988)	75.30 and 75.50
A. A. VAKHNENKO	(b) **146**, 605 (1988)	71.50; 71.45; S12.1
V. A. VAKHNENKO	(b) **146**, 605 (1988)	71.50; 71.45; S12.1
M. YA. VALAKH	(b) **150**, 557 (1988)	78.20; 78.50; S9.11
	(b) **150**, 907 (1988)	78.20; 78.50; S9.11

G. Valdrè	(a) **107**, 745 (1988)	61.55 and 61.70; 61.16; S1.2
U. Valdrè	(a) **107**, 745 (1988)	61.55 and 61.70; 61.16; S1.2
V. G. Valeev	(a) **108**, 251 (1988)	79.70; 68.20; S1; S4
M. V. Valeiko	(a) **108**, 637 (1988)	73.60 and 78.65; S8
J. Sabin del Valle	see under letter "S"	
G. C. Valley	(b) **150**, 913 (1988)	78.70; S5.11; S7.12
V. Valvoda	(a) **106**, 107 (1988)	63.20 and 65.70; S1.61
G. Vanderschaeve	(a) **107**, 647 (1988)	61.70 and 62.20; S7.12
	(a) **109**, 187 (1988)	62.20; S7.12
D. M. Vanderwalker	(a) **105**, K77 (1988)	61.50; 61.70; S1.2
U. V. Vandishev	(b) **146**, 341 (1988)	71.35 and 78.20; S8; S8.11
	(b) **150**, 839 (1988)	78.20; S8.11
J. Vanhellemont	(a) **107**, 455 (1988)	61.70; S5.11
C. Vannuffel	(a) **107**, 655 (1988)	64.70; 61.16; S1.2; S4
J. Vansummeren	(a) **107**, 639 (1988)	75.50; S12
J. Varghese	(a) **105**, 183 (1988)	68.55; S10.1
C. Varotsos	(a) **105**, K13 (1988)	61.70; S9.12
	(b) **147**, 83 (1988)	61.70; S9.11
	(a) **107**, K109 (1988)	66.30; S1.1
P. Vašek	(a) **106**, 115 (1988)	66.70; 72.15; S1.1
	(a) **107**, K135 (1988)	74.70; S10.15
A. D. Vasilev	(a) **105**, 441 (1988)	64.70; 61.60; 65.40; 77.20; 78.20; S9.16
A. B. Vassilikou-Dova	(b) **147**, 691 (1988)	61.70 and 76.30; S11
L. Vatskichev	(a) **106**, K75 (1988)	72.15 and 75.70; S1.2; S1.4
M. Vatskicheva	(a) **106**, K75 (1988)	72.15 and 75.70; S1.2; S1.4
A. F. Vavrek	(a) **107**, 379 (1988)	77.30 and 78.70; 72.40; S10
P. R. Vaya	(a) **105**, 161 (1988)	68.55 and 73.60; S8.15
	(a) **105**, 477 (1988)	68.55; 73.60; 78.65; S7
	(a) **106**, K209 (1988)	73.30 and 85; S5.11
A. G. Vedeshwar	(a) **109**, 549 (1988)	72.80 and 78.50; S1.63
R. V. Vedrinskii	(b) **150**, 307 (1988)	78.70; 61.10; 71.10; S9.11; S10.1; S11.1
A. de Veirman	(a) **110**, K1 (1988)	61.16; S1
A. N. Veis	(a) **107**, K169 (1988)	78.30; S8.16
G. Vekeman	(b) **149**, K25 (1988)	71.70 and 76.30; S9.11
V. S. Vekshina	(a) **107**, K43 (1988)	71.55; S7.13
V. L. Velichanskii	(b) **150**, 605 (1988)	78.20; 78.45
B. Vengalis	(a) **108**, 197 (1988)	64.70; 76.80; S8.15
E. F. Venger	(b) **146**, 517 (1988)	71.36 and 73.20; S6
J. Vennik	(a) **107**, 625 (1988)	68.20; 79.60; S1.2; S10
	(a) **107**, 731 (1988)	63.20 and 68.30; S10.1
R. Veresegyházy	(a) **107**, K15 (1988)	61.80 and 68.48; S1.3; S7.12
H. C. Verma	(b) **145**, K1 (1988)	61.50; S3
	(a) **105**, 275 (1988)	76.80; S1; S1.1
V. N. Vertoprakhov	(a) **108**, 625 (1988)	73.40 and 77.30; 77.55; S10.1
M. Verwerft	(a) **109**, 67 (1988)	61.14; S11.1
N. V. Veselovskaya	(a) **106**, 363 (1988)	61.70; 62.20; S5.11
V. A. Vetrov	(b) **147**, K171 (1988)	71.55 and 76.30; S9.11
Nguyen Ai Viet	see under letter "N"	
N. A. Viglin	(b) **148**, 361 (1988)	72.20 and 75.30; 76.30; S10.1
C. Vijayan	(a) **105**, 397 (1988)	61.70; 77.30; 77.40; 78.50; S9.11
R. Vila	(a) **105**, 601 (1988)	77.30 and 77.40
A. Vinattieri	(b) **147**, 421 (1988)	78.55; 73.40; S7.15
I. V. F. Viney	(a) **110**, 269 (1988)	71.55 and 78.55; 76.30; 78.60; S8
H. Vinzelberg	(b) **146**, K29 (1988)	74.70; S10.15
L. S. Vlasenko	(a) **109**, 127 (1988)	61.80; 76.30; S5.11
M. P. Vlasenko	(a) **109**, 127 (1988)	61.80; 76.30; S5.11

J. P. M. VAN VLIET	(b) **149**,	391 (1988)	71.70 and 78.55; S11.1
L. K. VODOPYANOV	(b) **145**,	117 (1988)	63.20; 78.30; S8.16
W. VOGELSBERGER	(b) **148**,	155 (1988)	68.40; 64.10; S5.2
B. VOIGTSBERGER	(a) **109**,	295 (1988)	71.55; 76.30; S10.1
U. VOLAND	(a) **108**,	225 (1988)	65.70; S7.16
P. DE VOLDER	(a) **107**,	625 (1988)	68.20; 79.60; S1.2; S10
T. R. VOLK	(a) **108**,	437 (1988)	78.50; S11.1
G. VÖLKEL	(b) **148**,	K61 (1988)	76.30 and 77.80; S11.1
	(a) **109**,	295 (1988)	71.55; 76.30; S10.1
R. A. VOLKOV	(b) **148**,	107 (1988)	71.10 and 73.40; 71.45
V. I. VOLKOV	(b) **150**,	471 (1988)	78.20; S8.11
W. VOLLMANN	(b) **145**,	K121 (1988)	71.25; 71.45; S12
S. A. VOROBIEV	(b) **146**,	K109 (1988)	61.80 and 72.20
G. DE VOS	(a) **110**,	615 (1988)	76.30; 61.70; S10.1
R. VOSKA	(b) **150**,	907 (1988)	78.20; 78.50; S9.11
R. VOSZKA	(a) **107**,	161 (1988)	62.20 and 62.65; S11.1
A. J. DE VRIES	(b) **149**,	391 (1988)	71.70 and 78.55; S11.1
G. DE VRIES	(b) **149**,	483 (1988)	66.30; S1
VU VAN HUNG	(b) **149**,	511 (1988)	65.50
G. S. VYGOVSKII	(b) **150**,	443 (1988)	78.20; 71.35; 71.36; S8.11
W. WACŁAWEK	(a) **108**,	331 (1988)	72.80; S12
A. H. WAFIK	(a) **106**,	K65 (1988)	75.60; S11.2
	(a) **107**,	K157 (1988)	75.60; S1.1
J. WALCZAK	(a) **109**,	K139 (1988)	76.30; 75.50; S11
A. C. WALKER	(b) **150**,	719 (1988)	78.20; S7.13
G. WALLNER	(a) **108**,	129 (1988)	61.10; 68.20; S5.11
M. WALTHER	(b) **150**,	679 (1988)	71.35 and 78.20; S7.12; S7.15
FATANG WANG	(a) **108**,	337 (1988)	74.70; 75.20; S10.15
G.-H. WANG	(a) **106**,	K1 (1988)	61.70 and 78.70; S12
G. Q. WANG	(a) **107**,	63 (1988)	61.60 and 74.70; S10.15
H.-W. WANG	(a) **106**,	K1 (1988)	61.70 and 78.70; S12
HONG WANG	(a) **109**,	301 (1988)	74.70; S10.15
HUAQIN WANG	(a) **110**,	K89 (1988)	64.70 and 74.70; S10.15
KELIN WANG	(b) **148**,	525 (1988)	63.20 and 68.30; S5
	(b) **149**,	143 (1988)	63.20 and 68.30; S5.11
L. C. WANG	(a) **107**,	63 (1988)	61.60 and 74.70; S10.15
LIANPING WANG	(b) **146**,	K115 (1988)	74.20
LITIAN WANG	(a) **105**,	447 (1988)	62.40; 64.70; S1.2
LIZHONG WANG	(b) **146**,	K115 (1988)	74.20
LONGSHU WANG	(a) **110**,	317 (1988)	61.40; 64.70; S1.2
XIAO-GANG WANG	(a) **109**,	135 (1988)	61.70; 74.70; 78.70; S10.15
	(a) **109**,	K133 (1988)	74.70 and 78.70; S10.15
YUANSHENG WANG	(a) **109**,	39 (1988)	61.40 and 64.70; S1.1; S1.2
Z. WANG	(b) **150**,	K15 (1988)	63.50
Z. G. WANG	(a) **105**,	419 (1988)	62.20; 61.70; S1.2
WAN-LUN YU	see under letter "Y"		
WANMING LAO	see under letter "L"		
A. WARIN	(a) **108**,	183 (1988)	66.30 and 77.50; S9.12
J. WĄSICKI	(b) **145**,	445 (1988)	63.20; 61.65; S12
	(b) **150**,	39 (1988)	63.20 and 65.40; S12
R. WASIEWICZ	(b) **147**,	685 (1988)	76.80; 71.70; S1.1
S. M. WASIM	(a) **110**,	575 (1988)	72.20 and 78.20; 71.55; S8.16
T. WASIUTYŃSKI	(b) **145**,	445 (1988)	63.20; 61.65; S12
M. L. WASZ	(a) **110**,	421 (1988)	61.70; S1.1

K. WATANABE	(a) **108**, 509 (1988)	61.70; 61.16; S1.2
	(a) **109**, 119 (1988)	61.14
M. WAUTELET	(b) **149**, 465 (1988)	61.70; 79.20; S5.11
C. WEBER	(b) **150**, 749 (1988)	78.20; 78.55; S8.15
M. WEGENER	(b) **150** 477 (1988)	78.20; S8.11
A. WĘGRZYN	(a) **106**, K169 (1988)	75.80; S1.1; S1.2
M. WEI	(a) **107**, 63 (1988)	61.60 and 74.70; S10.15
A. WEIDINGER	(b) **145**, 151 (1988)	71.36 and 73.20; 78.30; S9.12
WEIHUA ZHUANG	see under letter "Z"	
WEI-JI SONG	see under letter "S"	
G. WEIMANN	(b) **150**, 679 (1988)	71.35 and 78.20; S7.12; S7.15
H. WEINERT	(a) **108**, 257 (1988)	73.40; 68.55; 78.65; S7.1; S9.15
	(b) **150**, 699 (1988)	71.35 and 78.55; S7.12; S7.15
V. WEINHOLD	(b) **146**, K65 (1988)	78.40 and 78.55; 71.35; S8.11
	(b) **146**, K157 (1988)	63.20 and 78.30; S8.11
CH. WEISE	(a) **109**, 141 (1988)	61.80 and 66.30; S5.11
W. WEISSFLOG	(a) **106**, K89 (1988)	77.20; 77.40; S12
W. WELLER	(b) **145**, 625 (1988)	72.10; 73.40; S5.12
	(b) **148**, 273 (1988)	71.50
	(b) **148**, 635 (1988)	71.50
WEN-CHEN ZHENG	see under letter "Z"	
E. WENDLER	(a) **108**, K93 (1988)	61.80; 78.50; S7.11; S7.12
L. WENDLER	(b) **149**, K123 (1988)	71.45 and 73.40
G. WENDROCK	(a) **108**, 103 (1988)	61.55; 64.70; S4
	(a) **110**, 123 (1988)	64.70; 61.10; 61.55; S1.2; S4
C. J. WENG	(a) **106**, 99 (1988)	65.70; 75.80; S1.61
WENHUI SU	see under letter "S"	
H. R. WENK	(a) **107**, 799 (1988)	61.70 and 64.80; S10
WENLAN XU	see under letter "X"	
WEN-QIN ZHAO	see under letter "Z"	
K. WENTOWSKA	(a) **105**, K107 (1988)	64.70; S12
CH. WENZEL	(a) **110**, K77 (1988)	68.48; S1.2; S4
H. WENZL	(a) **110**, 77 (1988)	61.50; S1.2
L. WERNER	(a) **106**, 509 (1988)	68.55 and 78.55; S8
	(a) **106**, K83 (1988)	78.55; S8.15
P. WERNER	(a) **107**, 807 (1988)	61.60; 61.16; S11
W. WESCH	(a) **108**, K93 (1988)	61.80; 78.50; S7.11; S7.12
J. M. WESSELINOWA	(b) **145**, K65 (1988)	75.10; 63.10; S1.5
	(b) **146**, K55 (1988)	75.10
J. P. A. WESTERVELD	(b) **145**, 409 (1988)	66.30
C. H. WESTPHAL	(a) **110**, K97 (1988)	75.50; S1.2; S11
WHA-TEK KIM	see under letter "K"	
B. S. WHERRETT	(b) **149**, 383 (1988)	78.20; 71.55; S7.13
G. WIATROWSKI	(b) **145**, 299 (1988)	75.10 and 75.70
G. WIECH	(b) **149**, 765 (1988)	78.70; 71.20; S6
C. WIECKO	(b) **148**, 73 (1988)	63.50
H. WIECZOREK	(a) **109**, 245 (1988)	72.40 and 73.40; 61.40; 71.55; S5.11
G. A. WIEGERS	(a) **107**, 817 (1988)	61.60; S8
	(a) **107**, 873 (1988)	61.60; S8
E. WIESER	(a) **109**, 141 (1988)	61.80 and 66.30; S5.11
U. WIESNER	(b) **146**, K28 (1988)	74.70; S10.15
S. R. WILK	(b) **146**, 303 (1988)	78.30; 78.50; S9.11
I. WILL	(a) **107**, 85 (1988)	61.10; 65.70; 68.20; S5.11
R. P. WILLIAMS	(a) **108**, 111 (1988)	61.14 and 64.80; S10.15
T. WILSON	(a) **105**, 649 (1988)	73.30 and 85; 72.40
	(a) **106**, 577 (1988)	72.40 and 73.25
W. WINDSCH	(a) **108**, K9 (1988)	62.20; S12

U. Winter	(a) **108**, 651 (1988)	61.10 and 68.20; S5.11
A. M. Witowski	(b) **145**, K95 (1988)	61.70 and 63.20; 78.30; S8.13
A. Witt	(b) **150**, 477 (1988)	78.20; S8.11
H. Witte	(a) **108**, 683 (1988)	71.55 and 72.20; S7.12
J. J. Wnuk	(b) **146**, K91 (1988)	73.40
U. Woggon	(b) **150**, 641 (1988)	78.20; 71.35; S8.1; S9.11
J. Wojciechowski	(a) **105**, 113 (1988)	61.70; S5.11
W. Wojciechowski	(b) **146**, K51 (1988)	75.50; S9
B. Wolf	(a) **107**, K37 (1988)	68.55
	(b) **149**, 149 (1988)	68.20; 68.30; S7.13
E. Wolf	(b) **146**, K29 (1988)	74.70; S10.15
S. Wolf	(b) **149**, 521 (1988)	61.70 and 63.20; S12
R. Wolfrat	(a) **105**, K81 (1988)	61.80 and 78.65; S6
K. Wollschläger	(a) **105**, 387 (1988)	61.80; S1.62; S5.11
J. C. Woolley	(a) **107**, 165 (1988)	61.55 and 64.70; S8.15
	(a) **107**, 205 (1988)	64.70; S8.15
	(a) **108**, 599 (1988)	64.75; 75.30; S8.16
P. Wróbel	(b) **145**, K41 (1988)	74.60
D. Wruck	(a) **107**, 321 (1988)	72.20; 78.50; S7.11
A. Wrzeciono	(b) **147**, K47 (1988)	72.20; S1.2
	(a) **107**, K61 (1988)	75.50; S1.1
	(a) **108**, 375 (1988)	75.60; 75.80; S1.1; S1.4
	(a) **110**, 241 (1988)	75.30; 61.55; S1.1; S1.2
Guoan Wu	(a) **110**, 317 (1988)	61.40; 64.70; S1.2
Wu Li	see under letter "L"	
K. I. Wysokiński	(b) **147**, K67 (1988)	72.20

Xianduan Lin	see under letter "L"	
Xiang-yang Yuan	see under letter "Y"	
Xian-Wu Zou	see under letter "Z"	
Xianyi Zhou	see under letter "Z"	
Keqin Xiao	(a) **110**, 317 (1988)	61.40; 64.70; S1.2
Xiaobo Zhang	see under letter "Z"	
Xiao-gang Wang	see under letter "W"	
Xiaquing Pan	see under letter "P"	
Guo-Xin Xie	(b) **150**, 797 (1988)	78.20 and 78.65; 71.45; S8
L. M. Xie	(b) **148**, 715 (1988)	61.70 and 78.30; 78.50; S10
S. S. Xie	(a) **110**, 55 (1988)	61.60 and 74.70; S11
X. M. Xie	(a) **110**, 415 (1988)	66.30; S10
Xi-Mao Bao	see under letter "B"	
Xingzhao Ding	see under letter "D"	
Xin Jin	see under letter "J"	
G. Xiong	(b) **150**, 765 (1988)	78.20; S8.11
Lun-Biao Xu	(b) **150**, 797 (1988)	78.20 and 78.65; 71.45; S8
M. H. Xu	(a) **105**, 419 (1988)	62.20; 61.70; S1.2
Wenlan Xu	(b) **147**, 511 (1988)	63.20; 68.30; S1.2
W. L. Xu	(b) **147**, 767 (1988)	78.30; 63.20; S8.15
X. Xu	(b) **150**, 765 (1988)	78.20; S8.11
Yuhuan Xu	(a) **109**, 301 (1988)	74.70; S10.15
N. N. Xuan	(a) **105**, K121 (1988)	73.30; S12
Hoang Xuan Nguyen	see under letter "N"	
XuDong Fan	see under letter "F"	

V. S. Yadav	(b) **148**, K157 (1988)	76.30; S12
Yafei Zhang	see under letter "Z"	

E. B. YAGUBSKII	(a) **110**, K111 (1988)	78.30; 71.38; S12
I. I. YAKIMENKO	(b) **146**, 643 (1988)	72.10; 73.40
E. B. YAKIMOV	(a) **109**, K3 (1988)	61.14; 61.70; S8.13
D. R. YAKOVLEV	(b) **150**, 673 (1988)	71.35 and 78.20; 71.25; S7.12; S7.15
R. A. YAKSHIBAEV	(a) **108**, 135 (1988)	64.70 and 66.30; S8
I. T. YAKUBOV	(b) **145**, 455 (1988)	68.10; 73.30; S2
G. G. YAKUSHCHEVA	(b) **148**, K65 (1988)	71.35 and 78.20; S8.12
	(b) **150**, 761 (1988)	71.35 and 78.65; S8.12
E. D. YAKUSHKIN	(a) **105**, 139 (1988)	64.70; 77.20; S11.1
H. YAMAGUCHI	(a) **108**, 509 (1988)	61.70; 61.16; S1.2
	(a) **109**, 119 (1988)	61.14
H. YAMAMOTO	(b) **147**, K115 (1988)	73.40
M. YAMASHIRO	(b) **147**, 141 (1988)	71.36 and 71.45; 68.30; 73.40
YONG YAN	(a) **110**, K25 (1988)	68.48; S6; S10.1
T. YANAGAWA	(a) **106**, 283 (1988)	78.55; S8.12
C. Z. YANG	(b) **146**, 223 (1988)	75.10; 75.50
	(b) **148**, K133 (1988)	75.10 and 75.50
DAYU YANG	(a) **107**, 45 (1988)	61.55; S1.1; S1.4
F. YANG	(b) **150**, 765 (1988)	78.20; S8.11
A. YU. YANGIROV	(a) **106**, 315 (1988)	65.40; S8
B. YANGUI	(a) **107**, 469 (1988)	68.48; 61.16; S10
	(a) **110**, 43 (1988)	61.14; S10.1
YAN-MIN LI	see under letter "L"	
O. E. YANOVITSKII	(b) **147**, 69 (1988)	61.50 and 61.70
V. P. YANOVSKII	(a) **109**, K11 (1988)	61.70 and 62.20; S1.1
YAN-ZHONG ZHANG	see under letter "Z"	
YONG-QIAO YAO	(b) **147**, 45 (1988)	61.25; S1; S1.2
YAO-XIAN FU	see under letter "F"	
V. M. YARTSEV	(b) **145**, K149 (1988)	78.30; S12
	(b) **149**, 157 (1988)	72.20 and 78.20; S12.1
F. YASSIN	(a) **106**, K79 (1988)	75.50; S10
	(a) **109**, K145 (1988)	75.60; S1.1
K. YASUDA	(a) **107**, 709 (1988)	61.55 and 64.70; S1.3
H. J. YE	(b) **147**, 767 (1988)	78.30; 63.20; S8.15
H. Q. YE	(a) **107**, 511 (1988)	61.55; 61.14; S1.61
YE TAO	see under letter "T"	
TZU-TUNG YEN	(a) **108**, K15 (1988)	64.10
C.-Y. YI	(a) **106**, K1 (1988)	61.70 and 78.70; S12
Y. YIAN	(a) **107**, 63 (1988)	61.60 and 74.70; S10.15
L. YIN	(a) **105**, K1 (1988)	61.30; S12
YINSHENG TANG	see under letter "T"	
YI-YANG ZHOU	see under letter "Z"	
YIZHEN HE	see under letter "H"	
K. YONASHIRO	(b) **147**, 141 (1988)	71.36 and 71.45; 68.30; 73.40
YONG ZHU	see under letter "Z"	
J. Q. YOU	(b) **147**, 471 (1988)	61.14; 61.90
C. YU	(b) **149**, 691 (1988)	76.30; S10.1
GONG YU	(a) **109**, K39 (1988)	68.55; 76.30; 78.65; S5
WAN-LUN YU	(b) **145**, 255 (1988)	71.70; 76.30; S9.16
Z. Y. YU	(b) **147**, 767 (1988)	78.30; 63.20; S8.15
YU. M. YUABOV	(a) **109**, 211 (1988)	73.40; 73.60; S5.11
X. L. YUAN	(a) **108**, 545 (1988)	61.70; 78.50; 78.60; S9.11
XIANG-YANG YUAN	(a) **107**, 153 (1988)	63.20 and 68.30; 61.55; S1.2
YUAN-CHAO CHU	see under letter "C"	
YUANDA DONG	see under letter "D"	
YUANFU HSIA	see under letter "H"	
YUANSHENG WANG	see under letter "W"	

M. YUDASAKA	(a) **110**, 645 (1988)	78.65; S12
S. G. YUDIN	(a) **109**, 395 (1988)	61.70 and 66.30; S5.11
YUDIAN FAN	see under letter "F"	
LANPING YUE	(a) **109**, K1 (1988)	61.40; S1.61
L. A. YUGURYAN	(b) **149**, 231 (1988)	71.70; 78.50; S9.12
YUHUAN XU	see under letter "X"	
YU. M. YUMAGUZIN	(b) **150**, 103 (1988)	68.20 and 79.70; S1
ZHANG YUN	(b) **150**, 663 (1988)	78.20; S8.15
YUN DUN MAN	(a) **106**, 371 (1988)	61.80; 62.20; S1.2
V. A. YUNKIN	(a) **108**, 651 (1988)	61.10 and 68.20; S5.11
YUPU ZHANG	see under letter "Z"	
SH. N. YURAEV	(b) **148**, K177 (1988)	78.30; 78.50; S8; S8.15
A. M. YURKIN	(a) **105**, K169 (1988)	78.55; 71.70; S10
YUZHANG LI	see under letter "L"	
M. ZĄBKOWSKA-WACŁAWEK	(a) **108**, 331 (1988)	72.80; S12
A. ZAGÒRSKI	(b) **148**, K57 (1988)	75.10
	(a) **109**, K63 (1988)	75.70
A. M. ZAITSEV	(a) **108**, 431 (1988)	78.60; S7.1
B. P. ZAKHARCHENYA	(b) **150**, 891 (1988)	72.30 and 78.20; 71.38; S5.11; S7.12
YU. V. ZAKHAROV	(b) **149**, 241 (1988)	72.15; 75.60; S1.1
C. ZALDO	(b) **147**, 431 (1988)	78.30; S9.11
	(a) **108**, K145 (1988)	78.55 and 78.60; S10
A. J. ZALESKI	(b) **146**, K103 (1988)	74.70; S10.15
K. ZALIBOWSKI	(a) **108**, 765 (1988)	77.20 and 77.40; S9
V. A. ZALOI	(b) **146**, 329 (1988)	78.20 and 78.30; 71.35; S7.15
	(b) **149**, 187 (1988)	71.36
	(b) **150**, 401 (1988)	71.35 and 78.20
R. C. ZAMAR	(b) **150**, 245 (1988)	76.60; S12
H. W. ZANDBERGEN	(a) **105**, 207 (1988)	68.20; 61.16; S10
	(a) **107**, 825 (1988)	61.16 and 61.70; 74.70; S10.15
R. ZÂNGARO	(a) **109**, K109 (1988)	68.45 and 78.60; S11.1
A. V. ZARETSKII	(a) **109**, 373 (1988)	66.90 and 77.20; 64.70; S10
D. I. ZAROVSKII	(a) **107**, K33 (1988)	61.70 and 68.55; S1.61
I. ZASADA	(a) **110**, 231 (1988)	64.70 and 75.60; S1.2
I. I. ZASAVITSKII	(a) **108**, 637 (1988)	73.60 and 78.65; S8
A. F. ZATSEPIN	(b) **148**, K33 (1988)	71.55; S10.1
P. ZAUMSEIL	(a) **108**, 651 (1988)	61.10 and 68.20; S5.11
E. A. ZAVADSKII	(a) **109**, 307 (1988)	75.30; S11
YU. D. ZAVOROTNEV	(a) **109**, 307 (1988)	75.30; S11
D. M. ZAYACHUK	(a) **106**, K173 (1988)	76.80; S8.15
	(a) **107**, 95 (1988)	61.70; 72.20; 75.20; S8.13
K. ŽĎÁNSKÝ	(a) **105**, K51 (1988)	71.55; S7.15
A. ZEHE	(a) **107**, K37 (1988)	68.55
	(a) **107**, K119 (1988)	68.55; 73.60; S9.12
	(b) **149**, 149 (1988)	68.20; 68.30; S7.13
	(a) **109**, 493 (1988)	68.55; 65.70; S9.12
E. ZEITLER	(b) **146**, 137 (1988)	71.45 and 73.20; S1.3; S4
	(a) **107**, 791 (1988)	61.80 and 73.20; 71.45; S4
	(b) **149**, 169 (1988)	71.10 and 73.25
S. ZEKENG	(b) **150**, 65 (1988)	63.20 and 78.30; S7.12
M. ZELENÝ	(a) **109**, K67 (1988)	75.50; S1.5
Y. P. ZENG	(b) **145**, K111 (1988)	73.40; S7.12; S7.15
A. ZENTKO	(a) **108**, 747 (1988)	75.50; 61.80; S1.1; S1.2
A. YU. ZERR	(a) **105**, K29 (1988)	61.50 and 62.20; S7.13
R. B. ZEZIN	(b) **150**, 901 (1988)	66.70 and 78.35; S6

De-Gang Zhang	(b) **149**, K173 (1988)	75.60
Fangqing Zhang	(a) **105**, K41 (1988)	73.40 and 76.30; S5.11
	(a) **108**, K139 (1988)	76.30 and 78.30; S5.11
	(b) **109**, K39 (1988)	68.55; 76.30; 78.65; S5
Hong Zhang	(a) **109**, 135 (1988)	61.70; 74.70; 78.70; S10.15
	(a) **109**, K133 (1988)	74.70 and 78.70; S10.15
Hong-rui Zhang	(a) **110**, 515 (1988)	73.40; 78.30; S5.11; S7.12
L. D. Zhang	(a) **108**, 569 (1988)	62.40; S4
Li-Yuan Zhang	(b) **150**, 211 (1988)	74.20 and 75.30; 71.38
Q. R. Zhang	(a) **107**, 63 (1988)	61.60 and 74.70; S10.15
Qirui Zhang	(a) **109**, K129 (1988)	74.70 and 78.70; S10.15
Shu-yi Zhang	(a) **107**, 153 (1988)	63.20 and 68.30; 61.55; S1.2
S. Y. Zhang	(b) **147**, 767 (1988)	78.30; 63.20; S8.15
Xiaobo Zhang	(b) **150**, 623 (1988)	78.20
Yafei Zhang	(a) **109**, K39 (1988)	68.55; 76.30; 78.65; S5
Yan-Zhong Zhang	(a) **105**, 579 (1988)	75.50; S1.1; S1.2
	(a) **106**, 207 (1988)	75.50; S1.1; S1.2
Y. L. Zhang	(a) **110**, 55 (1988)	61.60 and 74.70; S11
Y. N. Zhang	(b) **149**, 691 (1988)	76.30; S10.1
Yupu Zhang	(a) **108**, 337 (1988)	74.70; 75.20; S10.15
Zhang Yun	see under letter "Y"	
Tiesong Zhao	(b) **145**, 679 (1988)	75.30; 71.70; S1.1; S1.4
Wen-qin Zhao	(a) **110**, 515 (1988)	73.40; 78.30; S5.11; S7.12
Y. G. Zhao	(b) **148**, 715 (1988)	61.70 and 78.30; 78.50; S10
Zhao Feng	see under letter "F"	
N. N. Zheleva	(b) **150**, 729 (1988)	78.20; S8.15
A. T. Zheleznyak	(b) **147**, K131 (1988)	73.40
Siming Zheng	(b) **146**, K39 (1988)	75.10
Wen-Chen Zheng	(b) **145**, K99 (1988)	61.70 and 65.70; 71.70; S10.1
	(b) **149**, K105 (1988)	64.30; 71.70; 76.30; S11
Zhen-Ya Li	see under letter "L"	
A. P. Zhernov	(b) **147**, 149 (1988)	71.38; 72.10; S4
V. V. Zhidkov	(a) **106**, K123 (1988)	64.70; S5.11
Zhigang He	see under letter "H"	
E. A. Zhilinskaya	(b) **149**, 427 (1988)	61.60 and 71.55; 76.30; S8.16
V. F. Zhitar	(a) **106**, K181 (1988)	78.40; S8.16
	(a) **109**, K55 (1988)	72.40 and 78.55; S8.16
S. V. Zhokhovets	(a) **106**, 249 (1988)	78.55; 71.55; S7.12
	(a) **110**, 221 (1988)	71.55; 78.55; S7.12
Zhong-nan Lu	see under letter "L"	
Zhongrong Li	see under letter "L"	
J. P. Zhou	(b) **147**, K153 (1988)	74.70; S10.15
Jianshi Zhou	(a) **108**, 337 (1988)	74.70; 75.20; S10.15
Xianyi Zhou	(a) **109**, K129 (1988)	74.70 and 78.70; S10.15
Yi-Yang Zhou	(b) **145**, 227 (1988)	71.70; 76.30; 76.80; 78.30; S11
	(b) **147**, 273 (1988)	71.70 and 76.30; S11
Y.-Y. Zhou	(a) **106**, K1 (1988)	61.70 and 78.70; S12
A. W. Zhu	(a) **108**, 569 (1988)	62.40; S4
Jin-Song Zhu	(a) **108**, 613 (1988)	62.40 and 64.70; S1.2
Yong Zhu	(b) **145**, 679 (1988)	75.30; 71.70; S1.1; S1.4
Y. T. Zhu	(b) **145**, K111 (1988)	73.40; S7.12; S7.15
Y.-Z. Zhu	(a) **106**, K1 (1988)	61.70 and 78.70; S12
Weihua Zhuang	(b) **147**, 421 (1988)	78.55; 73.40; S7.15
Z. C. Zhuang	(b) **145**, 151 (1988)	71.36 and 73.20; 78.30; S9.12
V. P. Zhukov	(b) **149**, 175 (1988)	71.25; S1.61
S. I. Zhukova	(a) **108**, K1 (1988)	61.70 and 62.20; S4
T. B. Zhukova	(a) **105**, K103 (1988)	62.20; S8

ZHUN-ZHI JIN	see under letter "J"	
S. P. ZHVAVYI	(a) **110**, 155 (1988)	61.80 and 73.60; 68.60; S10.1
ZICHENG DENG	see under letter "D"	
K. ŽIČKUS	(b) **150**, K31 (1988)	71.25; S9
A. ZIEGLER	(b) **147**, K83 (1988)	75.10
J. P. ZIELINGER	(a) **108**, 351 (1988)	72.40; 71.55; S8.16
O. ZIEP	(b) **148**, K41 (1988)	72.20
P. ZIESCHE	(b) **145**, K9 (1988)	68.25 and 68.40; 71.10
ZIGING LIN	see under letter "Z"	
A. B. ZIMIN	(b) **150**, 527 (1988)	78.65
L. G. ZIMIN	(b) **148**, K65 (1988)	71.35 and 78.20; S8.12
	(b) **150**, 653 (1988)	78.65; S8.15
	(b) **150**, 761 (1988)	71.35 and 78.65; S8.12
R. ZIMMERMANN	(b) **146**, 357 (1988)	71.25 and 72.20; 72.10
	(b) **146**, 371 (1988)	71.35 and 78.20
	(b) **146**, 545 (1988)	71.35; 78.20
	(b) **150**, 365 (1988)	71.35 and 78.20
E. ZLATEVA	(a) **107**, K99 (1988)	61.80 and 78.60; S9.12
J.-P. ZÖLLNER	(b) **147**, 823 (1988)	73.40; S7.1
	(b) **148**, 421 (1988)	73.40; S5.11; S7.1
	(b) **148**, 611 (1988)	71.25 and 73.40; S8.15
A. V. ZOLOTARYUK	(b) **146**, 487 (1988)	61.70 and 63.90; 66.90; 71.45; S12.1
ZONGRUAN LI	see under letter "L"	
GUANGTIAN ZOU	(b) **146**, K115 (1988)	74.20
XIAN-WU ZOU	(b) **145**, 73 (1988)	61.25; S1; S1.2
	(b) **147**, 45 (1988)	61.25; S1; S1.2
F. ZOUNOVÁ	(a) **109**, K67 (1988)	75.50; S1.5
G. ZUBAUSKAS	(a) **109**, 239 (1988)	68.48; 79.20; S7.11; S7.12
M. ZUBEKOVÁ	(a) **106**, 659 (1988)	73.40 and 85; S7.12
A. ŽUKAUSKAS	(b) **149**, 337 (1988)	63.20 and 78.55; 71.45; S8.12
	(b) **150**, 139 (1988)	71.35; 71.38; 78.55; S8.13
M. L. ZVEREVA	(b) **148**, 197 (1988)	71.30; 72.20; S8.15
C. ZWANZIGER	(b) **145**, 603 (1988)	71.50; 79.20; S5.11; S6
W. ZYCH	(a) **108**, K67 (1988)	76.80; 75.50; S1.1
V. A. ZYULKOV	(b) **150**, 647 (1988)	78.20; S8.15

Substance Classification*)

S1. Transition metals and their alloys and compounds (for oxides, see S10)
- S1.1. Fe
- S1.2. Other 3d metals
- S1.3. Noble metals
- S1.4. Lanthanides
- S1.5. Actinides
- S1.61. Carbides, silicides, germanides
- S1.62. Nitrides, phosphides, arsenides
- S1.63. Sulfides, selenides, tellurides

S2. Alkali metals and their alloys and compounds (for alkali halides, see S9.11)

S3. Alkaline earth metals and their alloys and compounds (for alkaline earth halides, see S9.12)

S4. Other metals and their alloys

S5. Nonmetallic elements
- S5.11. Silicon
- S5.12. Germanium
- S5.2. Solidified Gases

S6. Group IV compounds (for transition metal compounds, see S1.61)

S7. Group V compounds (for transition metal compounds, see S1.62)
- S7.1. III–V compounds
- S7.11. Phosphides
- S7.12. Arsenides
- S7.13. Antimonides
- S7.15. Mixed crystals
- S7.16. Higher compounds

S8. Chalcogenides (for transition metal compounds, see S1.63)
- S8.1. II–VI compounds
- S8.11. Sulfides
- S8.12. Selenides
- S8.13. Tellurides
- S8.15. Mixed crystals
- S8.16. Higher compounds

S9. Halides
- S9.11. Alkali halides
- S9.12. Alkaline earth halides
- S9.15. Mixed crystals
- S9.16. Higher compounds

S10. Oxides
- S10.1. Simple oxides
- S10.15. Mixed crystals

S11. Other higher compounds
- S11.1. Salts of oxy-acids
- S11.2. Spinels, garnets, ferrites

S12. Organic compounds
- S12.1. Quasi-one-dimensional systems

For comments on the use of the classification scheme see p. 10.

---

*) Outside the ICSTI Classification for Physics.

Manuscripts and letters for physica status solidi (b) — basic research and physica status solidi (a) — applied research

should be sent to

Leipziger Straße 3—4, P. B. 1233, DDR-1086 Berlin

or

Kunst und Wissen, Erich Bieber OHG, Postfach 330104, D-1000 Berlin 33.

The object of "physica status solidi (b)" and "physica status solidi (a)" is to publish rapidly important and topical results concerning fundamental research in the field of solid state physics and aspects of application and experimental techniques of fundamental solid state research, respectively. The issues of "physica status solidi (b)" and "physica status solidi (a)" will be published on the 1st and the 16th day of each month, respectively. Either series will contain the following categories:

1. Review articles written by arrangement with the editors. They will be remunerated.
2. Original papers selected from those which have not been submitted for publication elsewhere, after being reviewed. (The latest date for acceptance of contributions is the 8th and the 23rd day of each month for "physica status solidi (b)" and "physica status solidi (a)", respectively, 50 days before the delivery of the journal.)
3. Short notes published in the off-set part of the journal. They are, if necessary after being reviewed, selected by editorial staff. (The latest date for acceptance of manuscripts for this part is the 12th and the 27th day of each month for "physica status solidi (b)" and "physica status solidi (a)", respectively, 16 days before the delivery of the journal.)

Editorial Information: General

Instructions for preparing of figures are given in the second issue of each volume

Manuscripts should be submitted preferentially in English. Besides, papers in German, French, and Russian are admitted. The final decision on the incorporation of the papers in the series (b) or (a) is taken by the editorial staff after consultation of the referee. Papers on liquids are only published in physica status solidi if the problems treated are also relevant for the solid state. Authors are asked to observe the following rules:

1. Manuscripts have to be written as consisely as possible. At most, 20 and 4 pages (28 lines per page; abstracts, tables, references, figures, and captions of figures included, where 3 figures are counted as about one page) are admitted for original papers and short notes, respectively. All manuscripts should be typewritten on one side only. Two copies are required. Badly arranged manuscripts are returned.
2. The author's name, initials, and institution should be given.
3. One abstract of about ten lines in English and one in German, French, or Russian are required for original papers.
4. Because of lack of space conclusions (if any) must not have the character of a summary (i.e. substantially a repetition of the contents of the abstract).
5. Formulae should not be written too small and not with pencil. Separate lines for formulae are desirable. Indistinct writing will lead to a return of the manuscript.
6. Drawings should be enclosed on separate sheets of paper, not exceeding the maximum and minimum sizes of 20×30 and 10×15 cm^2, respectively. Photographs should be submitted, on white paper, only for review articles and original papers. Duplicates are required. Detailed instructions for preparing of figures are given on cover 4 of the second issue of each volume.
7. Captions for the figures should be submitted on separate sheets of paper, and special instructions with regard to the size of illustrations should be given on these.
8. References should be numbered in order of appearance and should be given, on a separate sheet, in the form used in physica status solidi.
9. SI-units should be used.
10. Each paper must contain a proposal for the subject classification according to the classification scheme appearing in the January issues of each annual set.

The authors will be sent galley proofs of review articles and original papers. Proof reading of short notes and pre-printed titles will be done by the editorial staff.

For each review article, original paper, and short note a total of 50 reprints is available free of charge. Requests for additional reprints (against payment) should be given together with the submitted manuscripts. **The deadline for orders of additional reprints is the 15th and the 30th of each month** for "physica status solidi (b)" and "physica status solidi (a)", respectively, two weeks before the delivery of the journal.

Bei Fragen zur Produktsicherheit wenden Sie sich bitte an:
If you have any questions regarding product safety,
please contact:

Walter de Gruyter GmbH
Genthiner Straße 13
10785 Berlin
productsafety@degruyterbrill.com